Bone and Joint Infections

Bone and Joint Infections

From Microbiology to Diagnostics and Treatment

Editor

Werner Zimmerli

WILEY Blackwell

Registered Office
John Wiley & Sons, Ltd, The Atrium, Southern Gate, Chichester, West Sussex, PO19 8SQ, UK

Editorial Offices
9600 Garsington Road, Oxford, OX4 2DQ, UK
The Atrium, Southern Gate, Chichester, West Sussex, PO19 8SQ, UK
111 River Street, Hoboken, NJ 07030-5774, USA

For details of our global editorial offices, for customer services and for information about how to apply for permission to reuse the copyright material in this book please see our website at www.wiley.com/wiley-blackwell.

Library of Congress Cataloging-in-Publication Data

Bone and joint infections (Zimmerli)
Bone and joint infections : from microbiology to diagnostics and treatment / edited by Werner Zimmerli.
p. ; cm.
Includes index.
ISBN 978-1-118-58177-3 (cloth)
I. Zimmerli, Werner, 1948–, editor. II. Title.
[DNLM: 1. Osteomyelitis. 2. Arthritis, Infectious. WE 251]
RC931.O7
616.7′15–dc23

2014031385

A catalogue record for this book is available from the British Library.

Wiley also publishes its books in a variety of electronic formats. Some content that appears in print may not be available in electronic books.

Cover image: Courtesy Werner Zimmerli
Cover design by Soephian Zainal

Set in 10/11.5pt Times New Roman by SPi Publishers Services, Pondicherry, India
Printed and bound in Singapore by Markono Print Media Pte Ltd

1 2015

This book is dedicated to my wife Annelies, and my children, Simone, Joëlle, and Laurent, who sometimes queued behind the kind of patients described in this book.

Contents

Contributors

Yvonne Achermann, MD
Postdoctoral fellow at the Department of Microbial Pathogenesis School of Dentistry
University of Maryland-Baltimore
Baltimore, MD, USA

Mohamed Al-Mayahi, MD
Resident of Orthopaedic Surgery
Division of Orthopaedic and Trauma Surgery
Department of Surgery
Geneva University Hospitals and Faculty of Medicine
Geneva, Switzerland

Jürgen B. Bulitta, PhD
Senior Research Fellow
Drug Delivery, Disposition and Dynamics
Monash Institute of Pharmaceutical Sciences
Monash University (Parkville Campus)
Melbourne, Australia
Department of Pharmaceutical Sciences
School of Pharmacy and Pharmaceutical Sciences
State University of New York at Buffalo
NY, USA

Lorenzo Calabro, MD
Medical Research Fellow
AO Research Institute Davos
AO Foundation
Davos, Switzerland

Nicholas M.P. Clarke, DM, ChM, FRCS, FRCS Ed.
Professor of Orthopaedics
Faculty of Medicine
University of Southampton and University Hospital Southampton NHS Foundation Trust
Southampton, UK

Martin Clauss, MD
Orthopaedic Surgeon, Consultant
Clinic for Orthopaedics and Trauma Surgery
Kantonsspital Baselland
Liestal, Switzerland

Juan D. Colmenero, MD
Associate Professor of Medicine
Infectious Diseases Service
Carlos Haya University Hospital
Malaga, Spain

Mihai A. Constantinescu, MD
Professor in Plastic Surgery
Department of Plastic, Reconstructive, and Hand Surgery
University Hospital of Bern and University of Bern
Bern, Switzerland

Lars Englberger, MD
Lecturer in Cardiovascular Surgery
Department of Cardiovascular Surgery
University Hospital of Bern and University of Bern
Bern, Switzerland

Saul N. Faust, MA, MBBS, FRCPCH, PhD, FHEA
Professor of Paediatric Immunology & Infectious Diseases & Director, NIHR Wellcome Trust Clinical Research Facility
Faculty of Medicine, University of Southampton
University Hospital Southampton NHS Foundation Trust
Southampton, UK

Örjan Friberg, MD
Associate Professor
Department of Cardiothoracic and Vascular Surgery
Örebro University Hospital
Örebro, Sweden

Michael C. Glanzmann, MD
Attending physician at the Upper Extremity Department Schulthess Clinic
Zurich, Switzerland

Bernhard Kessler, MD
Interdisciplinary Unit of Orthopaedic Infections
Basel University Clinic Liestal
Liestal, Switzerland

Markus Knupp, MD
Lecturer in Orthopaedics
Department of Orthopaedic Surgery
Kantonsspital Baselland
Liestal, Switzerland

Todd J. Kowalski, MD
Clinical Assistant Professor
University of Wisconsin School of Medicine and Public Health
Department of Internal Medicine
Gundersen Health System
WI, USA

Cornelia B. Landersdorfer, PhD
Senior Research Fellow
Drug Delivery, Disposition and Dynamics
Monash Institute of Pharmaceutical Sciences
Monash University (Parkville Campus)
Melbourne, Australia
Department of Pharmaceutical Sciences
School of Pharmacy and Pharmaceutical Sciences
State University of New York at Buffalo
NY, USA

Daniel Lew, MD
Professor of Infectiology
Division of Infectious Diseases
Department of Specialties of Internal Medicine
Geneva University Hospitals and Faculty of Medicine
Geneva, Switzerland

Silvin Lito, MD
Assistant of Infectious Diseases
Division of Infectious Diseases
Department of Specialties of Internal Medicine
Geneva University Hospitals and Faculty of Medicine
Geneva, Switzerland

Antoine Lomessy, MD
Resident of Orthopaedic Surgery
Division of Orthopaedic and Trauma Surgery, Department of Surgery
Geneva University Hospitals and Faculty of Medicine
Geneva, Switzerland

Martin McNally, MD, FRCSEd, FRCS(Orth)
Consultant in Limb Reconstruction Surgery, Honorary Senior Lecturer in Orthopaedics
The Bone Infection Unit
Nuffield Orthopaedic Centre
Oxford University Hospitals
Oxford, UK

Beat K. Moor, MD
Shoulder, Elbow and Orthopedic Sports Medicine
Department of Orthopedic and Trauma Surgery
Bern University Hospital and University of Bern
Bern, Switzerland

Pilar Morata, PhD
Professor in Biochemistry
and Molecular Biology
Faculty of Medicine
University of Malaga
Malaga, Spain

T. Fintan Moriarty, PhD
Senior Research Scientist
AO Research Institute Davos,
AO Foundation
Davos, Switzerland

Sophie Nguyen, MD
Infectious Diseases Department
Gustave Dron Hospital
Tourcoing, France

Robin Patel, MD
Professor of Microbiology and Medicine
Division of Clinical Microbiology,
Department of Laboratory Medicine and
Pathology
Mayo Clinic College of Medicine
Rochester, MN, USA

R. Geoff Richards, PhD
Director
AO Research Institute Davos,
AO Foundation
Davos, Switzerland

Seong Yeol Ryu, MD
Chair, Division of Clinical Microbiology,
Department of Laboratory Medicine and
Pathology
Mayo Clinic College of Medicine
Rochester, MN, USA
Associate Professor of Internal Medicine
Division of Infectious Disease,
Department of Medicine
Dongsan Medical Center, Keimyung
University School of Medicine
Daegu, South Korea

Parham Sendi, MD
Lecturer in Infectious Diseases
Department of Infectious Diseases
University Hospital of Bern and
University of Bern
Bern, Switzerland

Eric Senneville, MD, PhD
Professor of Infectious Diseases
Medicine Faculty of Lille
Gustave Dron Hospital
Tourcoing, France

Fritz Sörgel, BSc, PhD
Professor of Pharmacology
IBMP—Institute for Biomedical and
Pharmaceutical Research
Nürnberg-Heroldsberg and
Institute of Pharmacology, Faculty
of Medicine, University of
Duisburg-Essen
Essen, Germany

Priya Sukhtankar, MBChB, MRCPCH, PGDip
Faculty of Medicine
University of Southampton and
NIHR Wellcome Trust Clinical Research
Facility
University Hospital Southampton NHS
Foundation Trust
Southampton, UK

Domizio Suvà, MD
Head of Septic Orthopaedic Unity
Division of Orthopaedic and
Trauma Surgery, Department
of Surgery
Geneva University Hospitals and Faculty
of Medicine
Geneva, Switzerland

Ilker Uçkay, MD
Attending of Infectious Diseases and
Orthopaedic Surgery
Division of Infectious Diseases,
Department of Specialties of Internal
Medicine
Geneva University Hospitals and Faculty
of Medicine
Geneva, Switzerland
Division of Orthopaedic and
Trauma Surgery, Department of Surgery
Geneva University Hospitals and Faculty
of Medicine
Geneva, Switzerland

Pierre Vaudaux, PhD retired
Chief of staphylococcal research
Division of Infectious Diseases,
Department of Specialties of Internal Medicine
Geneva University Hospitals and Faculty of Medicine
Geneva, Switzerland

Pablo Yagupsky, MD
Professor of Pediatrics
Clinical Microbiology Laboratory,
Soroka University Medical Center,
Ben-Gurion University of the Negev
Beer-Sheva, Israel

Werner Zimmerli, MD
Professor of Internal Medicine and Infectious Diseases
Interdisciplinary Unit for Orthopaedic Infections
Kantonsspital Baselland
Liestal, Switzerland

Matthias A. Zumstein, MD
Lecturer in Orthopaedics
Shoulder, Elbow and Orthopedic Sports Medicine, Department of Orthopedic and Trauma Surgery
Bern University Hospital and University of Bern
Bern, Switzerland

Foreword

Correct and rapid diagnosis and treatment of bone and joint infections require the collaboration of many different specialists. In the field of bone and joint infections, there is almost no evidence based on controlled trials. Therefore, learning from the clinical experience of experts is particularly important for the management of such infections. Up to now, a comprehensive internationally available textbook dealing with infections of the locomotor system has been missing. The present book fills this gap.

An editor of a book on bone and joint infections must have a broad knowledge of the pertinent literature. In addition, he or she should know the different competent specialists in the field. When John Wiley & Sons, Publisher, asked Werner Zimmerli to edit this book, they could not have come up with a better choice. He is a competent specialist in the field of infections of the locomotor system and also has contacts with the network of international specialists.

Werner Zimmerli started his professional career in Infectious Diseases in Geneva, in the group of Francis A. Waldvogel, a well-known specialist of osteomyelitis, whose pioneer publications are still widely cited. Under his leadership, Werner Zimmerli started his studies on implant-related infections. In the early 1980s, it was rather an exception for an Infectious Disease specialist to invade a field for which orthopedic surgeons were responsible. Together with Daniel Lew and Pierre Vaudaux, he was able to define host factors that are responsible for the high susceptibility of implants to pyogenic infections. In subsequent studies in Basel, he presented experimental evidence that implants are susceptible not only for exogenous but also for hematogenous infections. Together with Andreas F. Widmer, he could show the special role of rifampin for the treatment of implant-associated infections in vitro and in vivo. In conjunction with my group in Liestal, we started to treat patients with orthopedic implant–associated staphylococcal infections with rifampin combinations. The promising treatment results in observational studies led to the planning of a randomized controlled trial on the role of rifampin in patients with implant retention. In this trial, rifampin showed its superiority in patients suffering from acute orthopedic implant–associated infections treated with debridement and implant retention. After the planned interim analysis, the trial was early stopped, because all treatment failures were observed in the arm without rifampin. This study offers one of the few evidence-based treatment standards in orthopedic infections.

As an orthopedic surgeon specialized in the field of infections of the locomotor system, I met Werner Zimmerli, a partner who is a dedicated clinician. Together with my team in Liestal, I was in permanent contact with him to discuss our cases with bone and joint infections. This contact was even closer when he accepted the position of Head of the Basel University Medical Clinic in Liestal. This allowed us to create an "Interdisciplinary Unit for Orthopedic Infections," the first in Switzerland. Out of this close collaboration resulted a now internationally respected algorithm for the treatment of periprosthetic joint infections. During the last two decades, a large number of Infectious Disease specialists and orthopedic surgeons trained in the field of bone and joint infections in our group. The interdisciplinary core team later included plastic surgeons, microbiologists, and pathologists, a concept of collaboration that is now widely accepted.

This book reflects the concepts of interdisciplinarity. The introductory chapters deal with general fields, which are important for managing bone and joint infections, namely, "Microbiology," "Pharmacokinetics and Pharmacodynamics of Antibiotics in Bone," and "Experimental Preclinical Models." In addition, the book contains general chapters on "Periprosthetic Joint Infections" and "Classification of Osteomyelitis." These chapters offer to the reader detailed knowledge as a basis for competent clinical management of bone and joint infections. The main part of the book deals with the typical clinical entities of infections of the locomotor system. They are written following a common concept and contain all the knowledge needed to better understand the subject treated. Each chapter can be studied independently. Most of the chapters end with an enumeration of key points and some instructive cases illustrating typical situations. This allows the reader to assess whether he or she understood the essentials of management. Several case examples also illustrate common errors that should be avoided. Extensive and updated lists of references help the reader continue his or her studies. The chapters are written by specialists with extensive clinical experience in the field of the infections that have been described. If the treatment is mainly conservative, an Infectious Disease specialist is the author. In the chapters on infections associated with prosthetic joints and internal fixation devices, an orthopedic surgeon joined the Infectious Disease specialist in the writing team.

This book offers clear information on most problems in the management of infections of the locomotor system. Experienced clinicians in the field of Infectious Diseases, Orthopedic Surgery, Trauma Surgery, Rheumatology, and Internal Medicine can use this book as a comprehensive textbook or on a chapter-by-chapter basis. Specialists in the field can benefit from the detailed updated reviews and may find specific help for their own challenging cases.

Peter E. Ochsner
Professor Emeritus in Orthopedic Surgery
of the University of Basel

Acknowledgments

I am grateful to Mindy Okura-Marszycki and Stephanie Dollan, editors from Wiley Blackwell, for inviting me to edit this book and for their continuous help during the realization of this project. My thanks go also to Sandeep Kumar, Project manager at SPi global for his efficiency and professionalism in the production of this book. This textbook would not have been possible without the enthusiastic and competent work of the authors, who are all specialists in different fields of bone and joint infection. My special thanks go to Ruth Wäschle, who was indispensable during the whole editorial process. I would also like to thank all patients whose bone and joint infections were the basis of new diagnostic and therapeutic concepts.

Chapter 1
Introduction

Werner Zimmerli

The prevalence of most bone and joint infections is steadily increasing, mainly due to the rising life expectancy of the population and the increasing use of bone fixation devices and prosthetic joints. For frequent infectious diseases, many diagnostic and therapeutic aspects have been studied in controlled trials [1–3]. In contrast, the management of bone and joint infections is mostly based on expert opinion, since randomized clinical studies comparing different orthopedic techniques, antimicrobial agents, or treatment durations are missing. Thus, diagnostic and therapeutic advice is mainly based on individual clinical expert knowledge and observational studies [4–7].

The optimal diagnostic and therapeutic management of bone and joint infections needs a special know-how in different fields of medicine. Many physicians have only limited clinical experience, since arthritis and osteomyelitis are rare infectious diseases. Therefore, a multidisciplinary approach to these infections is desirable. Guidelines for the management of bone and joint infections are available for only a few topics [8]. In addition, publications on the clinical practice comprising different aspects of these infections are scarce. The aim of this book is to close this gap with texts from a multidisciplinary team of experts in the field. Indeed, specialists in Microbiology, Pharmacology, Preclinical Research, Pediatrics, Pediatric and Adult Orthopedic Surgery, Infectious Diseases, and Cardiovascular Surgery contributed to this book. This broad spectrum of expertise made it possible to cover a wide range of pathophysiological, epidemiological, diagnostic, and therapeutic aspects of bone and joint infection. The principal focus of the book is on clinical practice. It should enable clinicians in managing patients according to the best available evidence.

Beside the routine microbiological tests, novel techniques, such as molecular diagnostic procedures [9] and matrix-assisted laser desorption ionization time-of-flight (MALDI–TOF) mass spectrometry [10], are increasingly used for the diagnosis of infectious diseases including bone and joint infection. The role of the bone/serum ratio in the antimicrobial treatment of bone and joint infections is still a matter of debate. Important methodological

Bone and Joint Infections: From Microbiology to Diagnostics and Treatment, First Edition.
Edited by Werner Zimmerli.

differences have to be considered for adequately judging data on bone penetration. These data are often controversially discussed in the literature, mainly due to the use of various experimental techniques in different studies [11, 12]. Distinct differences in the extent of bone penetration by various classes of antimicrobial agents have been observed. However, proof for the clinical relevance of these differences is still missing. Thus, knowledge about pharmacokinetics and pharmacodynamics of antibiotics in bone should stimulate planning of clinical studies to fill this missing gap. Many current treatment concepts are based on preclinical studies in vitro and in animals [13]. Such data are especially important for the management of implant-associated infections, a field in which controlled clinical trials are lacking.

Septic arthritis encompasses a nonhomogenous group of joint infections. In this book, eight different clinical situations are covered. Many aspects of arthritis in children differ from that in adults. In children, *Kingella kingae*, a microorganism that in adults almost exclusively causes endocarditis, plays a prominent role [14]. In addition, *Streptococcus agalactiae* is still common in neonates. In contrast, *Haemophilus influenzae* type b almost disappeared in young children due to the effective conjugate vaccine. By gathering a careful case history, rare microorganisms such as *Erysipelothrix* sp., *Mycobacterium marinum*, or *Scedosporium* sp. can be suspected and actively looked for. Arthritis of axial joints is rare and difficult to diagnose. Intravenous drug use is the most frequent risk factor for all types of axial arthritis, namely, of the sternoclavicular joint, the symphysis pubis, and the sacroiliac joint. Surgery is rarely needed, if the diagnosis is rapidly made and the patient has no pyogenic complications. Prosthetic joints are increasingly used not only in hip and knee, but also in other joints, mainly shoulder, ankle, and elbow. The perioperative infection rate ranges from about 0.5 to 1.5% after hip or knee arthroplasty up to 10% after elbow or ankle joint replacement. Since many aspects vary between the different joint prostheses, separate chapters deal with periprosthetic joint infection in this book.

Osteomyelitis encompasses a large spectrum of different diseases. Many different classifications are used, depending on different aspects of disease (e.g., pathogenesis, duration, presence of implant) and according to the specialist who is managing the case (e.g., orthopedic surgeon, infectious disease specialist, pediatrician, angiologist). In this book, aspects of age (children, adults), duration of disease (acute, subacute, chronic), presence of implant, anatomic location (long bones, vertebrae, jaws), and presence of diabetes are presented in separate chapters.

My thanks go to the international panel of distinguished authors from different continents, who have been selected for their special expertise in the field of bone and joint infection. All of them respected the deadlines and were mindful of the suggestions for revision. This allowed getting a coherent book text without unnecessary redundancies. In addition, I would like to thank my mentors Francis Waldvogel, Daniel Lew, and John Gallin, as well as many of my former fellows including Andreas Widmer, Reno Frei, Andrej Trampuz, and Parham Sendi who all contributed to the progress that has been achieved in the field of implant-associated bone and joint infections during the last three decades. I am especially grateful to Peter Ochsner from whom I learned about the orthopedic aspects of bone and joint infection and to Peter Graber and Bernhard Kessler who critically revised some of the texts, in order to ensure their clinical utility. Last but not the least, my thanks go to Ruth Wäschle who did important editorial work before submission and to the Editorial Team at John Wiley & Sons Inc. for their efficiency and professionalism in the production of this book.

Together with all authors, I trust that this multidisciplinary book will allow gathering rapid and exhaustive information regarding all types of bone and joint infection. If this book helps in improving patient management, we have reached our goal.

References

[1] Spoorenberg V, Hulscher ME, Akkermans RP, *et al.* Appropriate antibiotic use for patients with urinary tract infections reduces length of hospital stay. Clin Infect Dis 2013; 58(2):164–169.

[2] Ramirez J, Dartois N, Gandjini H, *et al.* Randomized phase 2 trial to evaluate the clinical efficacy of two high-dosage tigecycline regimens versus imipenem-cilastatin for treatment of hospital-acquired pneumonia. Antimicrob Agents Chemother. 2013;57(4):1756–1762.

[3] Dulhunty JM, Roberts JA, Davis JS, *et al.* Continuous infusion of beta-lactam antibiotics in severe sepsis: a multicenter double-blind, randomized controlled trial. Clin Infect Dis. 2013; 56(2):236–244.

[4] Gallagher JC, Huntington JA, Culshaw D, *et al.* Daptomycin therapy for osteomyelitis: a retrospective study. BMC Infect Dis. 2013;12:133.

[5] Peel TN, Cheng AC, Choong PF, Buising KL. Early onset prosthetic hip and knee joint infection: treatment and outcomes in Victoria, Australia. J Hosp Infect 2013;82(4):248–253.

[6] Zimmerli W. Clinical practice. Vertebral osteomyelitis. N Engl J Med 2010;362(11): 1022–1029.

[7] Zimmerli W, Trampuz A, Ochsner PE. Prosthetic-joint infections. N Engl J Med 2004; 351(16):1645–1654.

[8] Osmon DR, Berbari EF, Berendt AR, *et al.* Diagnosis and management of prosthetic joint infection: clinical practice guidelines by the Infectious Diseases Society of America. Clin Infect Dis. 2013;56(1):e1–e25.

[9] Gomez E, Cazanave C, Cunningham SA, *et al.* Prosthetic joint infection diagnosis using broad-range PCR of biofilms dislodged from knee and hip arthroplasty surfaces using sonication. J Clin Microbiol. 2012;50(11):3501–3508.

[10] Patel R. Matrix-assisted laser desorption ionization-time of flight mass spectrometry in clinical microbiology. Clin Infect Dis. 2013;57(4):564–572.

[11] Mouton JW, Theuretzbacher U, Craig WA, *et al.* Tissue concentrations: do we ever learn? J Antimicrob Chemother 2008;61(2):235–237.

[12] Landersdorfer CB, Bulitta JB, Kinzig M, *et al.* Penetration of antibacterials into bone: pharmacokinetic, pharmacodynamic and bioanalytical considerations. Clin Pharmacokinet 2009;48(2):89–124.

[13] Calabro L, Lutton C, Din A, *et al.* Animal models of orthopedic implant-related infection. In: Moriarty TF, Zaat SAJ, Busscher HJ, ediotrs. Biomaterials Associated Infection. New York: Springer; 2013, pp. 273–304.

[14] Yagupsky P. Kingella kingae: from medical rarity to an emerging paediatric pathogen. Lancet Infect Dis. 2004;4(6):358–367.

Chapter 2

Microbiology of Bone and Joint Infections

Seong Yeol Ryu and Robin Patel

Microorganisms in Osteomyelitis

The term osteomyelitis refers to infection involving the bone, a condition that often leads to compromised mobility. It may be caused by a wide variety of bacteria and fungi. Osteomyelitis can occur as a result of contiguous spread from surrounding soft tissues (e.g., progression from deep diabetic ulcers) or joints. Alternatively, it may result from hematogenous seeding or direct inoculation of microorganisms into the bone (e.g., from surgery or trauma). In hematogenous osteomyelitis, infection is typically monobacterial, whereas in contiguous (exogenous) infection, it is often polymicrobial.

Acute Osteomyelitis in Adults

The most common pathogens causing osteomyelitis vary with patient age. *Staphylococcus aureus* is the most frequent organism isolated in adults, with methicillin-resistant *S. aureus* (MRSA) being increasingly detected. Less common pathogens include *Enterococcus* species, *Streptococcus* species, enteric Gram-negative bacilli, nonfermenting Gram-negative bacilli (e.g., *Pseudomonas aeruginosa*), anaerobic bacteria, mycobacteria, and fungi, especially *Candida* species [1]. Fungal and mycobacterial osteomyelitis most commonly occur in the immunocompromised host. Unusual organisms, such as *Bartonella* species and *Coxiella burnetii*, are occasionally detected in adult osteomyelitis [2].

Acute Osteomyelitis in Children

The tibia and femur are the most commonly affected long bones in children. As in adults, *S. aureus* is the most common pathogen associated with acute osteomyelitis, being responsible for 70–90% of cases in children. In contrast to adults, *Kingella kingae* is increasingly reported in children [3]. Other etiological agents include *Streptococcus*

Bone and Joint Infections: From Microbiology to Diagnostics and Treatment, First Edition.
Edited by Werner Zimmerli.

agalactiae (especially in infants), *Streptococcus pyogenes*, *Streptococcus pneumoniae*, enteric Gram-negative bacilli, and anaerobic bacteria. *Haemophilus influenzae* type b (Hib) was historically a common cause of childhood osteomyelitis. However, its prevalence has decreased following the introduction of the Hib conjugate vaccine [3]. Mycobacteria, fungi, *Bartonella* species, and *C. burnetii* are unusual causes of osteomyelitis in children, generally occurring in patients with specific risk factors, such as endemic exposure or immunosuppression [3, 4].

Chronic Osteomyelitis

Chronic osteomyelitis of the long bones is an infrequent but clinically important condition. In adults, it is often the consequence of an open, comminuted fracture with inadequate treatment of the traumatic contamination of the fracture site, or implant-associated infection. The spectrum of causative organisms is poorly described. Sheehy *et al.* [5] reported that *S. aureus* was most commonly isolated (32%) among a wide range of organisms including Gram-negative bacilli, anaerobic bacteria, and coagulase-negative staphylococci. Notably, polymicrobial infections were common (29%). In 28% of the patients, the cultures remained negative.

Diabetic Foot Osteomyelitis

In patients with underlying diabetes or other types of sensory neuropathy, osteomyelitis typically involves the foot. Mendes *et al.* [6] recently examined the microbiology of diabetic foot infection using aspirates, biopsies, and swabs. Aerobes were present in 98% of cases, with Gram-positive bacteria comprising 66% of isolates. *Staphylococcus* was the main genus identified (37%), with MRSA present in 22% of cases. Although coagulase-negative staphylococci were the second most frequently encountered aerobic Gram-positive organisms, data from other investigators suggests that if cultures of bone are exclusively considered, coagulase-negative staphylococci may be less common [7, 8]. Gram-negative aerobes comprised 19% of isolated organisms, with *P. aeruginosa* isolated in 12% of cases. Anaerobes were found in 14% of cases, with *Peptostreptococcus* species accounting for 55% of anaerobic isolates, followed by the *Bacteroides fragilis*, which accounted for 25% of isolates. *Candida* species were infrequently encountered, representing only 1% of isolates.

Vertebral Osteomyelitis

Spondylodiscitis is an uncommon but important infection that represents 3–5% of osteomyelitis cases. Undiagnosed or inadequately treated, it may result in irreversible damage of the spinal cord, including paralysis. Lumbar vertebral bodies are most frequently affected, followed by the thoracic and, infrequently, cervical vertebrae.

The predominant mechanism of spondylodiscitis is hematogenous seeding. The source of infection may be one of many, including genitourinary, cutaneous or subcutaneous sources, respiratory sources, dental sources, intravascular catheter sites, or injection drug use [9]. Although it can occur at any age, vertebral osteomyelitis primarily occurs in adults over the age of 50 years.

S. aureus is the most frequent microorganism encountered in pyogenic vertebral osteomyelitis. Its portal of entry may be the skin, a surgical procedure, or a vascular catheter.

Enteric Gram-negative bacilli, coagulase-negative staphylococci, enterococci, and streptococci may also be implicated. Enteric Gram-negative bacilli (e.g., *Klebsiella* species, *Serratia* species, *Escherichia coli*) are often associated with osteomyelitis complicating urinary tract or intraabdominal infection [10]. Vertebral osteomyelitis due to Gram-negative aerobic bacteria and *Candida* species are common in immunocompromised patients, and also occur as postoperative infections [10]. *Candida* species were quite frequent in injection drug users before the availability of sterile paraphernalia.

Vertebral osteomyelitis can also be acquired during spine surgery. Gram-positive cocci, including *S. aureus* and *Staphylococcus epidermidis*, and *Propionibacterium acnes* are the most common pathogens associated with spinal implant–associated osteomyelitis.

Tuberculous Osteomyelitis

The bone is involved in 1–3% of *Mycobacterium tuberculosis* infections, with approximately half of the cases affecting the spine (i.e., Pott's disease). Tuberculosis is the most frequent etiology of spinal infection globally, and accounts for 9–46% of cases in the developed world. The mean age of patients with tuberculous osteomyelitis ranges from 30 to 40 years. Most cases result from lymphohematogenous spread from a pulmonary source. Risk factors for tuberculous osteomyelitis include diabetes mellitus, chronic renal failure, and corticosteroid therapy [11]. The thoracic segment and the thoracolumbar hinge represent the most frequent localizations of tuberculous osteomyelitis. In contrast to bacterial vertebral osteomyelitis, systemic symptoms are often absent. Extraspinal musculoskeletal tuberculosis has a predilection for large weight-bearing joints, long bones, and the skull [12].

Brucella *Osteomyelitis*

Most cases of brucellosis occur as a result of direct or indirect exposure to animals, the main routes of transmission being eating unpasteurized milk products or drinking unpasteurized milk. Musculoskeletal involvement is common, with the frequency of osteoarticular involvement varying from 10 to 85% and including spondylitis, sacroiliitis, osteomyelitis, arthritis, bursitis, and tenosynovitis. The most frequent osteoarticular presentation in children with brucellosis is monoarticular arthritis, mostly located in the hips or knees, whereas in adults, sacroiliitis is most common [13]. *Brucella* vertebral osteomyelitis most frequently involves the lumbar, followed by the thoracic and cervical spine, and has a high rate of therapeutic failure and functional sequelae [14]. The laboratory should be notified in suspected cases of brucellosis, as lab technicians are at risk for acquiring brucellosis if cultured organisms are not properly handled.

Table 2.1 shows osteomyelitis according to the type, age, and susceptibility factors of the host, and microbial etiology.

Microorganisms in Implant-Associated Bone and Joint Infection

The prototypical foreign body–associated osteoarticular infection is periprosthetic joint infection (PJI). Microorganisms causing foreign body–related infection form biofilms on the surface of the implant, rendering the associated infection challenging to treat and enabling even low-virulence microorganisms, such as coagulase-negative

Table 2.1. Microbiology of osteomyelitis.

Type of osteomyelitis	Age/susceptibility	Etiology
Hematogenous long bones	Children	*Staphylococcus aureus*
		Streptococcus pyogenes
		Streptococcus agalactiae
		Haemophilus influenzae
Contiguous focus	Adults	*S. aureus*
	Diabetes mellitus, sensory neuropathy, vascular insufficiency	*Enterococcus* species
		Coagulase-negative staphylococci
		Streptococcus species
		Aerobic Gram-negative bacilli
		Anaerobes
Vertebral osteomyelitis	Adults	*S. aureus*
		Coagulase-negative staphylococci
		Enterococcus species
		Streptococcus species
	Urinary infection	Aerobic Gram-negative bacilli
	Injection drug user	*Pseudomonas aeruginosa*
		S. aureus
		Candida species
		Serratia marcescens
	Spinal surgery	Coagulase-negative staphylococci
		S. aureus
		Propionibacterium acnes
		Aerobic Gram-negative bacilli

staphylococci or *P. acnes*, to cause infection. Staphylococci (*S. aureus* and coagulase-negative staphylococci, especially *S. epidermidis*) account for more than 50% of the episodes of periprosthetic hip and knee infection (Table 2.2). Other bacteria, including rarely mycobacteria and occasionally fungi, cause the rest of the cases. Approximately 20% of PJI cases are polymicrobial, and in about 7% there is no growth of any microorganism.

PJI after Knee Arthroplasty

The most common organisms associated with prosthetic knee infection are *S. aureus* (27%) followed by *S. epidermidis* (16%), *Proteus mirabilis* and *S. agalactiae* (5% each), *P. aeruginosa*, *Streptococcus mitis*, *Enterococcus faecalis*, *Citrobacter* species, and *Candida albicans* (2% each) [15]. Gram-negative bacilli have been increasingly reported in prosthetic knee infection [16].

Table 2.2. Microbiology of periprosthetic knee and hip infection.

Pathogen	Frequency (%)
Gram-positive cocci	~65
Coagulase-negative staphylococci	
Staphylococcus aureus	
Streptococcus species	
Enterococcus species	
Aerobic gram-negative bacilli	~6
Enterobacteriaceae	
Pseudomonas aeruginosa	
Anaerobic bacteria	~4
Propionibacterium species	
Finegoldia magna	
Polymicrobial	~20
Culture-negative	~7
Fungi	~1

PJI after Hip Arthroplasty

The most frequent organisms causing prosthetic hip infection are *S. aureus* and coagulase-negative staphylococci, followed by mixed flora, streptococci, Gram-negative bacilli, enterococci, and anaerobic bacteria [17].

PJI after Shoulder Arthroplasty

P. acnes and *Staphylococcus* species are the most frequent microorganism associated with shoulder arthroplastics detected from both periprosthetic tissue and sonicated arthroplasty components [18]. The notable presence of *P. acnes* may relate to its prevalence in the skin of the upper body due to the increased density of sebaceous glands in that location (a known habitat of this organism).

PJI after Elbow Arthroplasty

The most frequently isolated pathogen in elbow arthroplasty infection is *S. aureus* (41%), followed by coagulase-negative staphylococci (33%). Polymicrobial infections and culture-negative cases account for 7 and 4% of cases, respectively [19].

PJI after Ankle Arthroplasty

S. aureus is the most common pathogen associated with ankle arthroplasty infection, followed by coagulase-negative staphylococci. Infection is polymicrobial in 15% of cases [20].

Implant-Associated Infection of the Long Bones

Infections subsequent to stabilization of long bones with internal fixation devices are difficult to treat. Implants allow biofilm formation on their surfaces and alter the environment, including local immunity, favoring bacterial invasion. After trauma, damage to soft tissues, decreased vascular supply surrounding fracture sites and delayed healing predispose to infection. Infections linked to internal fixation devices are classified as early (< 2 weeks), delayed (2–10 weeks), and late (> 10 weeks). They usually arise due to exogenous seeding, either during placement of the device or by the trauma itself, or as a result of disturbed wound healing. Early infections are mainly caused by virulent microorganisms such as Gram-negative bacilli or *S. aureus*, whereas delayed and late infections are predominantly caused by organisms of less virulence (e.g., coagulase-negative staphylococci) [21].

Spinal Implant–Associated Infection

Direct inoculation during surgery is the most common route of infection of spinal implants. *S. aureus*, *S. epidermidis*, and *P. acnes* are the most common pathogens [9]. Gram-negative bacteria also play a role and may be associated with systemic illness and multisystem organ failure [22].

Microorganisms in Native Joint Arthritis

Native Joint Infection in Adults

Bacterial arthritis is usually hematogenously acquired. Other routes of infection include direct inoculation into the joint through surgery, trauma, percutaneous puncture, or contiguous spread from adjacent infected soft tissue or bone. Infectious arthritis of single or multiple joints may be caused by a number of microorganisms, the most common in adults being *S. aureus* (37–65% of cases) [23]. *Streptococcus* species are the second most common microorganisms implicated in adults and are often associated with chronic skin infection, trauma, or autoimmune disorders [24]. Gram-negative bacilli are cultured from 5 to 20% of patients with bacterial arthritis, particularly from the very old, injection drug users and immunocompromised hosts. Gonococcal infection was historically a relatively common etiology of septic arthritis in sexually active individuals; in the 1970s, *Neisseria gonorrhoeae* accounted for about two-thirds of septic arthritis and tenosynovitis cases in North America [25]. Its incidence has declined in recent years mainly due to effective control programs; currently, gonococcal arthritis is rare in North America and Europe [24]. Gonococcal arthritis is one of two clinical presentations of disseminated gonococcal infection, the other being the syndrome of tenosynovitis, dermatitis, and polyarthralgia.

In contrast to the acute presentation of bacterial or viral arthritis, patients with joint infection due to mycobacteria and non-*Candida* fungi usually present with subacute or chronic, slowly progressive monoarthritis.

Native Joint Arthritis in Children

Native joint septic arthritis in children typically results from direct inoculation of bacteria into the joint, either following trauma or from iatrogenic causes; however, cases of contiguous extension and hematogenous seeding have been reported. Causative organisms vary with patient age with *S. aureus* being the most common [26], and *S. pyogenes*,

S. pneumoniae, and *H. influenzae* being the other major pathogens. However, with the introduction of the Hib vaccine, the incidence of *H. influenzae* septic arthritis has decreased. *K. kingae* has superseded *H. influenzae* as the most common cause of septic arthritis in this population, especially in the first 2 years of life [27]. *K. kingae* may be found in the oropharynx of infected children. The most common organisms causing septic arthritis in infants are *S. agalactiae*, *S. aureus*, and Gram-negative enteric bacteria. Between the ages of 2 months and 5 years, the predominant agents include *S. pyogenes*, *S. pneumoniae*, *S. aureus*, and *K. kingae*. Articular infection due to anaerobic bacteria is rare. In children older than 5 years, the bacteria most often involved are *S. pyogenes* and *S. aureus* [4].

Diagnostic Approach in Osteomyelitis

Inflammatory Parameters

Although the diagnostic value of serum C-reactive protein (CRP) levels and erythrocyte sedimentation rate (ESR) is limited due to their low specificity, they may be useful to follow the response to therapy.

Microbiological Studies

For adequate treatment, it is important to isolate the infecting agent(s) for identification and antimicrobial susceptibility testing. Specimen collection must be meticulously performed to avoid contamination. As an example, when performing a bone biopsy of diabetic foot osteomyelitis, passing through the ulcer bed with the needle should be avoided to prevent contamination of the specimen. Antimicrobial administration should be withheld (except in the case of a septic patient) until specimens for cultures are collected. Specimen collection is best accomplished by needle aspiration or surgical sampling under imaging guidance. In cases of osteomyelitis, bone biopsy is recommended for microbiological studies [28]. The specimen should be as large as possible and may include intact pieces of bone, shavings, scrapings, and/or excised necrotic material [28]. If the bone is necrotic, a curette may be utilized for specimen collection [28]. Tissues should be submitted for bacterial culture (aerobic and anaerobic), with anaerobic transport containers used for specimens submitted for anaerobic culture [29]. It may be possible to isolate the infecting microorganism from blood if the patient has concomitant bacteremia or fungemia. Swab cultures of sinus tracts poorly correlate with results of bone cultures and are not routinely recommended. Likewise, swab cultures of bone or tissue are not recommended, since larger tissue and/or fluid specimens yield improved results. Polymerase chain reaction (PCR) may be used to detect specific organisms, especially those that are difficult to isolate in culture or require prolonged culture incubation. Colmenero *et al.* [30], for example, used PCR targeting *Brucella* species and *M. tuberculosis* complex to test vertebral and paravertebral tissue samples from subjects with tuberculous or *Brucella* vertebral osteomyelitis; PCR sensitivity was higher (93%) than that of conventional culture (73%).

Histopathological Studies

Although a diagnosis is usually made based on radiographic and clinical features, bone biopsy may be needed when clinical and radiographic features are not diagnostic or to exclude noninfectious causes of bone pathology. Histology is a useful adjunct to culture and can distinguish between granulomatous disease, pyogenic disease, and noninfectious etiologies.

Antigen and Antibody Testing

Serology may be considered for suspected cases of brucellosis and Q fever. Antigen-based tests may be useful for select infectious diseases; an example is the serum cryptococcal antigen test, which is a helpful adjunct to diagnosis of cryptococcosis.

Diagnostic Approach in Foreign Body–Associated Infection

Accurate diagnosis of PJI is important, because its management differs from aseptic failure. Current definitions rely on a number of parameters, including clinical, microbiological, and histopathological features [31]. Traditionally, the diagnosis has been made based on the presence of at least one of the following: (1) acute inflammation on histopathological examination of periprosthetic tissue, (2) presence of a sinus tract communicating with the prosthesis, (3) pus in the joint space, and (4) growth of the same microorganism from two or more cultures of joint aspirates or periprosthetic tissue specimens. A general approach to the diagnosis of PJI is shown in Figure 2.1. Recently, the Musculoskeletal Infection Society (MSIS) and the Infectious Diseases Society of America (IDSA) have released slightly different definitions of PJI [32, 33]. According to the MSIS criteria, definite PJI exists in the presence of (1) a sinus tract communicating with the prosthesis; or (2) positive cultures for the same organism from at least two tissue or fluid samples from the joint; or (3) four of the following: (a) elevated serum ESR and CRP, (b) elevated synovial leukocyte count, (c) elevated synovial fluid neutrophil percentage, (d) purulence in the joint space, (e) growth of a microorganism in one periprosthetic tissue or fluid culture, or (f) greater than five neutrophils per high-power field in five high-power fields on periprosthetic tissue histopathology [31]. PJI may be present, however, even if fewer than four of these criteria are met [31]. The IDSA guidelines consider definitive PJI to be present if there is a sinus tract communicating with the prosthesis, purulence without another known etiology surrounding the prosthesis, or two or more cultures yielding the same organism; the presence of acute inflammation on histopathological examination of periprosthetic tissue is considered highly suggestive of PJI [32]. The authors of these guidelines note that growth of a virulent microorganism, such as *S. aureus*, from a single tissue specimen or synovial fluid may also represent PJI [32].

Histopathological Studies

In patients in whom the diagnosis of PJI has not been established prior to surgery, frozen section examination may be performed at the time of surgery (to look for evidence of acute inflammation). Using a neutrophil count ranging from more than 5 to 10 or more cells per high-power field as a positive criterion, sensitivity for PJI diagnosis ranges from 50 to 93% and specificity from 77 to 100% [34].

Microbiological Studies

Preoperative aspiration of synovial fluid and intraoperative tissue cultures provide accurate specimens for detecting the infecting microorganism. Conversely, cultures of a sinus tract or a superficial wound should be avoided as culture results likely represent microbial colonization, not true infection. At least three and optimally five or six periprosthetic tissue samples should be submitted for culture (aerobic and anaerobic) at surgical debridement to optimize

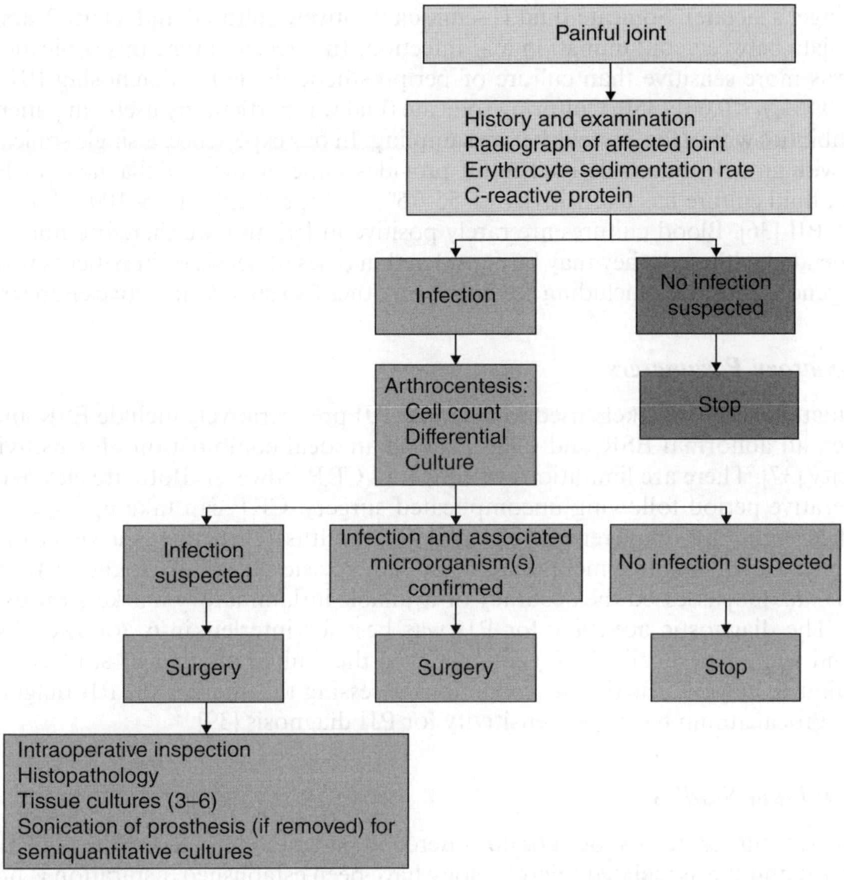

Figure 2.1. Algorithm for the diagnosis of periprosthetic joint infection. (*See insert for color representation of the figure.*)

the likelihood of making a microbiological diagnosis. The sensitivity of cultures of periprosthetic tissue ranges from 65 to 94%, depending on the definition of PJI used [35]. Careful interpretation of tissue cultures is required to avoid considering a pathogen as a contaminant, especially because identical organism types (e.g., coagulase-negative staphylococci) may be isolated as pathogens and contaminants. For this reason, at least two tissue specimens should be culture-positive for the same organism to support a diagnosis of PJI. Cultures for *P. acnes* should be incubated for a week assuming anaerobic thioglycolate broth is used and specimens are collected into and transported to the laboratory using anaerobic tissue and fluid vials [29]; longer incubation times may be needed if only solid media are used. Because of poor sensitivity, Gram staining of the periprosthetic tissue is not recommended. When possible, antibiotic therapy should be withheld for at least 2 weeks prior to collecting operative culture specimens. Because of concerns that tissue cultures are not adequately sensitive to reveal the presence of implant-adhering (biofilm) microorganisms, many experts advise submitting the prosthetic device or modular parts of it for culture [28]. The removed device is vortexed and sonicated in a sterile container with a salt solution (e.g., Ringer's lactate). Sonicate fluid is semiquantitatively cultured and a cutoff applied to differentiate between contamination and infection. In a recent study, this sonication technique was more sensitive than culture of periprosthetic tissue for diagnosing PJI (78.5% versus 60.8%, $p < 0.001$ [36]). Culturing sonicate fluid was particularly useful in patients who had antibiotics within the 2 weeks before sampling. In our experience, a single sonicate fluid culture with growth above a defined cutoff provides a microbiological diagnosis of PJI [36]. Synovial fluid culture has a sensitivity of 56–75% and specificity of 95–100% for the diagnosis of PJI [36]. Blood cultures are rarely positive in PJI and are therefore not routinely recommended, although they may be considered in cases of sepsis or in patients with likely hematogenous infection, including seeding from other foci such as infective endocarditis.

Inflammatory Parameters

Inflammatory serum markers used to diagnose PJI preoperatively include ESR and CRP. Together, an abnormal ESR and CRP provide an ideal combination of sensitivity and specificity [37]. There are limitations to ESR and CRP, however. Both are elevated in the postoperative period following uncomplicated surgery. CRP can take up to 3 weeks to normalize, and ESR may remain elevated for months (even up to a year) following uncomplicated surgery. In a meta-analysis of 3909 revision total hip or knee arthroplasty, Berbari *et al.* [38] assessed the accuracy of available inflammatory markers for diagnosis of PJI. The diagnostic accuracy for PJI was best for interleukin-6, followed by CRP, ESR, and white blood cell (WBC) count. Given the limited number of studies assessing interleukin-6, however, further investigations assessing this marker for PJI diagnosis are needed. Procalcitonin has a low sensitivity for PJI diagnosis [39].

Synovial Fluid Studies

Diagnostic arthrocentesis should be considered in patients suspected as having PJI unless a diagnosis and the associated microbiology have been established, aspiration is not technically feasible, or imminent surgery is planned. Aspirated fluid should be submitted for total and differential cell count and culture. Synovial fluid cultures are ideally performed in blood culture bottles. Previous treatment with antimicrobial agents lowers the sensitivity of synovial fluid culture to detect PJI. Interpretive criteria for synovial fluid cell count and differential in PJI differ from those of native joint septic arthritis and vary with the time following joint implantation and the anatomic location of the implant. A synovial fluid leukocyte count of more than 1700/μl or a differential count of more than 65% neutrophils is consistent with infection of knee arthroplasties that have been in place for at least 6 months in subjects without inflammatory joint diseases [40]. In the first 6 weeks following knee arthroplasty, however, a synovial fluid leukocyte count more than 27,800/μl or a differential of more than 89% neutrophils is predictive of prosthetic knee infection [41]. A synovial fluid leukocyte count of more than 4200/μl or a differential count of more than 80% neutrophils is consistent with prosthetic hip infection [42].

Molecular Studies

Cultures of periprosthetic tissue and synovial fluid represent the standard method for diagnosis of PJI. However, in approximately 7% of patients, no growth can be detected, representing so-called culture-negative PJI [43]. Therefore, molecular methods have been developed in an attempt to improve the yield of microbiological diagnoses of PJI, particularly in the setting of antecedent antimicrobial exposure and culture negativity. Most reports on the use of molecular tools to detect bacteria in PJI have been based on amplification of a broad-range bacterial target, such as the 16S ribosomal RNA (rRNA) gene, with subsequent identification of the organism detected by sequencing of amplified DNA. In one study, broad-range PCR of synovial fluid had higher sensitivity (71%) than synovial fluid culture (44%) [44]. Gomez *et al.* [45] compared broad-range PCR of material dislodged from replaced orthopedic implants using vortexing and sonication (sonicate fluid) with synovial fluid culture, tissue culture and culture of sonicate fluid. The combination of sonicate fluid culture and PCR (78.5%) had a higher sensitivity than synovial fluid (64.7%) or tissue culture (70.4%). Data from studies evaluating broad-range PCR of tissue for PJI diagnosis may be conflicting because of technical differences and differences in criteria used for diagnosis of PJI. De Man *et al.* [46] compared 16S rRNA PCR to culture using synovial fluid and periprosthetic tissue; PCR and culture sensitivity were 50 and 58%, respectively. Vandercam *et al.* [47] analyzed tissue from 69 subjects (34 with PJI); 16S rRNA PCR sensitivity was higher (91%) than that of culture (65%). Broad-range PCR does not identify detected organisms (unless amplified DNA is sequenced) or signal the presence of a polymicrobial infection, and false-positive results may occur from background DNA in clinical specimens or reagents [48]. For these reasons, targeted PCR has been explored for PJI diagnosis. Portillo *et al.* [49] compared real-time multiplex PCR of sonicate fluid with sonicate fluid and periprosthetic tissue culture. Multiplex PCR of sonicate fluid had a higher sensitivity (96%) than tissue (71%) or sonicate fluid (67%) culture for diagnosing PJI. Cazanave *et al.* [50] used a genus-/group-specific rapid PCR assay panel targeting PJI bacteria to test sonicate fluids from subjects with infected and uninfected hip and knee arthroplasties undergoing resection or revision arthroplasty. Sensitivities of tissue culture, sonicate fluid culture, and PCR were 70.1, 72.9, and 77.1%, respectively, with sonicate fluid PCR being more sensitive than tissue culture ($P = 0.04$) [50]. A limitation of PCR is that it does not give information about antimicrobial susceptibility (or it may provide limited data, such as the presence of *mecA*); therefore, it should not supplant culture, but rather it should be used as an adjunct to culture.

Diagnostic Approach in Native Joint Infection

Septic arthritis is considered an emergency because of the potential for rapid joint destruction with irreversible loss of function. The definitive diagnosis of septic arthritis is made by direct demonstration of a microorganism in or by microbial growth from synovial fluid. A general approach to diagnosis is shown in Figure 2.2.

Inflammatory Parameters

Blood tests typically show increased levels of white cell count, ESR, and CRP. However, the lack of elevation of acute phase reactants does not eliminate the possibility of septic arthritis [51]. ESR, CRP, and WBC are useful for monitoring response to therapy. In a recent study, Maharajan *et al.* [52] analyzed white cell counts, procalcitonin, CRP, ESR, and bacterial culture in 82 subjects with acute osteomyelitis or septic arthritis. In contrast to PJI, a serum procalcitonin, at a cutoff of 0.4 ng/ml, is a sensitive and specific marker

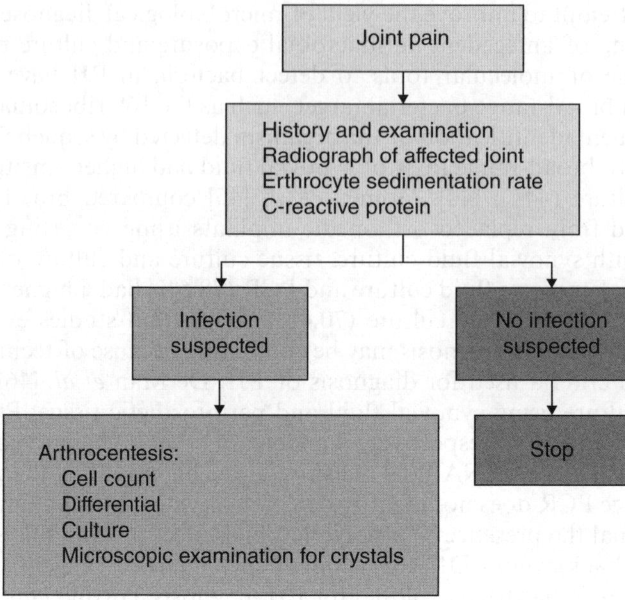

Figure 2.2. Algorithm for the diagnosis of native joint arthritis. (*See insert for color representation of the figure.*)

for the diagnosis of acute osteomyelitis and septic arthritis. Thus, serum procalcitonin may be used as a diagnostic marker for initiation of treatment in the management of acute osteomyelitis and septic arthritis.

Synovial Fluid Studies

Aspiration of synovial fluid from potentially infected joints is mandatory for establishing the correct diagnosis. Joint fluid analysis is useful in assessing the etiology of effusions, although there may be overlap in the clinical and laboratory findings in patients with infected joints and those with crystal arthropathy (and these conditions may coexist). Arthrocentesis of an affected joint usually reveals purulent, low-viscosity synovial fluid with an elevated neutrophil count. In septic arthritis, the synovial fluid usually has a synovial fluid leukocyte count greater than 50,000/mm^3. Low joint fluid glucose levels may be found in septic arthritis, but this is a nonspecific finding that may be present in other inflammatory processes. Synovial fluid should be cultured (aerobically and anaerobically) and crystal analysis performed. In nongonococcal infection, synovial fluid culture will yield bacterial growth in up to 80–90% of cases. Synovial fluid culture sensitivity declines after antimicrobial therapy has been initiated. Gram staining of the fluid may be helpful but is diagnostic in only 50% of cases [53].

Blood Cultures

Blood cultures must be collected before beginning antimicrobial agents to optimize the possibility of isolating the pathogen(s); an exception is the septic patient, in whom delayed antibiotic therapy is associated with patient mortality. In such cases, an attempt to obtain blood cultures may be made, but it should not delay initiation of antibiotic therapy. Blood cultures are positive in 50–70% of patients with nongonococcal arthritis [23].

Molecular Studies

In general, molecular studies are not needed for the etiological diagnosis of native arthritis. However, PCR may be useful for the diagnosis of difficult-to-grow organisms, such as *K. kingae*, a common microorganism in pediatric bone and joint infections [54].

Novel Diagnostic Procedures

Matrix-assisted laser desorption ionization time-of-flight analysis mass spectrometry (MALDI–TOF MS) is a relatively new technology that provides fast, accurate, and relatively inexpensive identification of bacteria and fungi growing in culture [55]. Starting from a single bacterial or fungal colony on a culture plate, the colony is moved onto a MALDI–TOF MS plate and overlain with 1–2 μl of matrix (e.g., α-cyano-4-hydroxycinnamic acid dissolved in 50% acetonitrile and 2.5% trifluoroacetic acid). The plate is dried and placed into the chamber of a mass spectrometer, in which the dried spot is hit by a laser. The matrix protects the microbial proteins, and aids in desorbing and ionizing microbial proteins (primarily highly abundant ribosomal proteins). The ionized microbial proteins are accelerated into a flight tube where they are separated by their mass-to-charge ratios. The smallest analytes travel the fastest and hit the detector first, followed by progressively larger analytes. A mass spectrum is generated representing the numbers of ions hitting the detector at the end of the flight tube at a specific time. Computer software is used to compare the generated mass spectrum against a database of reference spectra. The analytical turnaround time is ≤3 min per isolate. Commercial MALDI–TOF MS systems are available from Bruker Daltonics (Billerica, MA), that is, the Bruker Biotyper, and bioMérieux (Durham, NC), that is, VITEK® MS. MALDI–TOF MS has been shown to be useful for identifying PJI-causing staphylococci [56].

PCR–electrospray ionization mass spectrometry may be used to measure the mass-to-charge ratio of PCR amplicons generated from several loci on microbial genomes, including conserved and species-specific regions (as well as antimicrobial resistance genes), as a means to determine the base composition of PCR-amplified DNA. That base composition is compared to a database of microbial base compositions to determine the source of the amplified DNA [57]. The technique has recently been applied to synovial fluid [58, 59] and sonicate fluid as a diagnostic for PJI [60].

Acknowledgment

The authors thank Dr. Trisha N. Peel for her thoughtful review of this chapter.

References

[1] Calhoun JH, Manring MM. Adult osteomyelitis. Infect Dis Clin North Am 2005;19(4): 765–786.

[2] Merhej V, Tattevin P, Revest M, *et al.* Q fever osteomyelitis: a case report and literature review. Comp Immunol Microbiol Infect Dis 2012;35(2):169–172.

[3] Dartnell J, Ramachandran M, Katchburian M. Haematogenous acute and subacute paediatric osteomyelitis: a systematic review of the literature. J Bone Joint Surg 2012;94(5):584–595.
[4] Gutierrez K. Bone and joint infections in children. Pediatr Clin North Am 2005;52(3): 779–794.
[5] Sheehy SH, Atkins BA, Bejon P, *et al.* The microbiology of chronic osteomyelitis: prevalence of resistance to common empirical anti-microbial regimens. J Infect 2010;60(5):338–343.
[6] Mendes JJ, Marques-Costa A, Vilela C, *et al.* Clinical and bacteriological survey of diabetic foot infections in Lisbon. Diabetes Res Clin Pract 2012;95(1):153–161.
[7] Senneville E, Morant H, Descamps D, *et al.* Needle puncture and transcutaneous bone biopsy cultures are inconsistent in patients with diabetes and suspected osteomyelitis of the foot. Clin Infect Dis 2009;48(7):888–893.
[8] Zuluaga AF, Galvis W, Jaimes F, Vesga O. Lack of microbiological concordance between bone and non-bone specimens in chronic osteomyelitis: an observational study. BMC Infect Dis 2002;2;8.
[9] Fantoni M, Trecarichi EM, Rossi B, *et al.* Epidemiological and clinical features of pyogenic spondylodiscitis. Eur Rev Med Pharmacol Sci 2012;16(Suppl 2):2–7.
[10] Mete B, Kurt C, Yilmaz MH, *et al.* Vertebral osteomyelitis: eight years' experience of 100 cases. Rheumatol Int 2012;32(11):3591–3597.
[11] Alavi SM, Sharifi M. Tuberculous spondylitis: risk factors and clinical/paraclinical aspects in the south west of Iran. J Infect Public Health 2010;3(4):196–200.
[12] Shikhare SN, Singh DR, Shimpi TR, Peh WC. Tuberculous osteomyelitis and spondylodiscitis. Semin Musculoskelet Radiol 2011;15(5):446–458.
[13] Arkun R, Mete BD. Musculoskeletal brucellosis. Semin Musculoskelet Radiol 2011;15(5): 470–479.
[14] Colmenero JD, Ruiz-Mesa JD, Plata A, *et al.* Clinical findings, therapeutic approach, and outcome of brucellar vertebral osteomyelitis. Clin Infect Dis 2008;46(3):426–433.
[15] Mortazavi SM, Schwartzenberger J, Austin MS, *et al.* Revision total knee arthroplasty infection: incidence and predictors. Clin Ortho Rel Res 2010;468(8):2052–2059.
[16] Sendi P, Zimmerli W. Challenges in periprosthetic knee-joint infection. Int J Artif Organs 2011;34(9):947–956.
[17] Hudetz D, Rod E, Radic A, Ivkovic A. Diagnosis and treatment of peri-prosthetic infections in total hip replacement. Med Glas (Zenica) 2012;9(1):152–159.
[18] Piper KE, Jacobson MJ, Cofield RH, *et al.* Microbiologic diagnosis of prosthetic shoulder infection by use of implant sonication. J Clin Microbiol 2009;47(6):1878–1884.
[19] Achermann Y, Vogt M, Spormann C, *et al.* Characteristics and outcome of 27 elbow periprosthetic joint infections: results from a 14-year cohort study of 358 elbow prostheses. Clin Microbiol Infect 2011;17(3):432–438.
[20] Kessler B, Sendi P, Graber P, *et al.* Risk factors for periprosthetic ankle joint infection: a case-control study. J Bone Joint Surg 2012;94(20):1871–1876.
[21] Trampuz A, Zimmerli W. Diagnosis and treatment of implant-associated septic arthritis and osteomyelitis. Curr Infect Dis Rep 2008;10(5):394–403.
[22] Nasto LA, Colangelo D, Rossi B, *et al.* Post-operative spondylodiscitis. Eur Rev Med Pharmacol Sci 2012;16:50–57.
[23] Goldenberg DL. Septic arthritis. Lancet 1998;351(9097):197–202.
[24] Kaandorp CJ, Dinant HJ, van de Laar MA, *et al.* Incidence and sources of native and prosthetic joint infection: a community based prospective survey. Ann Rheum Dis 1997;56(8): 470–475.
[25] Sharp JT, Lidsky MD, Duffy J, Duncan MW. Infectious arthritis. Arch Intern Med 1979;139(10):1125–1130.
[26] Chen WL, Chang WN, Chen YS, *et al.* Acute community-acquired osteoarticular infections in children: high incidence of concomitant bone and joint involvement. J Microbiol Immunol Infection 2010;43(4):332–338.

[27] Yagupsky P. *Kingella kingae*: an emerging pediatric pathogen. Adv Exp Med Biol 2006;582: 179–190.

[28] Baron EJ, Miller JM, Weinstein MP, *et al.* A guide to utilization of the microbiology laboratory for diagnosis of infectious diseases: 2013 recommendations by the Infectious Diseases Society of America (IDSA) and the American Society for Microbiology (ASM). Clin Infect Dis 2013;57:e22–e121.

[29] Shannon SK, Mandrekar J, Gustafson DR, *et al.* Anaerobic thioglycolate broth culture for recovery of *Propionibacterium acnes* from shoulder tissue and fluid specimens. J Clin Microbiol 2013;51(2):731–732.

[30] Colmenero JD, Morata P, Ruiz-Mesa JD, *et al.* Multiplex real-time polymerase chain reaction: a practical approach for rapid diagnosis of tuberculous and brucellar vertebral osteomyelitis. Spine 2010;35(24):E1392–E1396.

[31] Parvizi J, Zmistowski B, Berbari EF, *et al.* New definition for periprosthetic joint infection: from the Workgroup of the Musculoskeletal Infection Society. Clin Ortho Rel Res 2011;469(11):2992–2994.

[32] Osmon DR, Berbari EF, Berendt AR, *et al.* Diagnosis and management of prosthetic joint infection. Clin Infect Dis 2013;56(1):e1–e25.

[33] Parvizi J. New definition for periprosthetic joint infection. Am J Orthop 2011;40(12): 614–615.

[34] Ko PS, Ip D, Chow KP, *et al.* The role of intraoperative frozen section in decision making in revision hip and knee arthroplasties in a local community hospital. J Arthroplasty 2005;20(2):189–195.

[35] Trampuz A, Osmon DR, Hanssen AD, *et al.* Molecular and antibiofilm approaches to prosthetic joint infection. Clin Ortho Rel Res 2003;414:69–88.

[36] Trampuz A, Piper KE, Jacobson MJ, *et al.* Sonication of removed hip and knee prostheses for diagnosis of infection. N Engl J Med 2007;357:654–663.

[37] Piper KE, Fernandez-Sampedro M, Steckelberg KE, *et al.* C-reactive protein, erythrocyte sedimentation rate and orthopedic implant infection. PLoS One 2010;5(2):e9358.

[38] Berbari E, Mabry T, Tsaras G, *et al.* Inflammatory blood laboratory levels as markers of prosthetic joint infection: a systematic review and meta-analysis. J Bone Joint Surg 2010;92(11):2102–2109.

[39] Bottner F, Wegner A, Winkelmann W, *et al.* Interleukin-6, procalcitonin and TNF-alpha: markers of peri-prosthetic infection following total joint replacement. J Bone Joint Surg Br 2007;89(1):94–99.

[40] Trampuz A, Hanssen AD, Osmon DR, *et al.* Synovial fluid leukocyte count and differential for the diagnosis of prosthetic knee infection. Am J Med 2004;117(8):556–562.

[41] Bedair H, Ting N, Jacovides C, *et al.* The Mark Coventry Award: diagnosis of early postoperative TKA infection using synovial fluid analysis. Clin Ortho Rel Res 2011;469(1):34–40.

[42] Schinsky MF, Della Valle CJ, Sporer SM, Paprosky WG. Perioperative testing for joint infection in patients undergoing revision total hip arthroplasty. J Bone Joint Surg 2008;90(9): 1869–1875.

[43] Tsukayama DT, Goldberg VM, Kyle R. Diagnosis and management of infection after total knee arthroplasty. J Bone Joint Surg 2003;85-A:S75–S80.

[44] Gallo J, Kolar M, Dendis M, *et al.* Culture and PCR analysis of joint fluid in the diagnosis of prosthetic joint infection. New Microbiol 2008;31(1):97–104.

[45] Gomez E, Cazanave C, Cunningham SA, *et al.* Prosthetic joint infection diagnosis using broad-range PCR of biofilms dislodged from knee and hip arthroplasty surfaces using sonication. J Clin Microbiol 2012;50(11):3501–3508.

[46] De Man FH, Graber P, Luem M, *et al.* Broad-range PCR in selected episodes of prosthetic joint infection. Infection 2009;37(3):292–294.

[47] Vandercam B, Jeumont S, Cornu O, *et al.* Amplification-based DNA analysis in the diagnosis of prosthetic joint infection. J Mol Diagn 2008;10(6):537–543.

[48] Achermann Y, Vogt M, Leunig M, *et al.* Improved diagnosis of periprosthetic joint infection by multiplex PCR of sonication fluid from removed implants. J Clin Microbiol 2010;48(4): 1208–1214.
[49] Portillo ME, Salvado M, Sorli L, *et al.* Multiplex PCR of sonication fluid accurately differentiates between prosthetic joint infection and aseptic failure. J Infect 2012;65(6):541–548.
[50] Cazanave C, Greenwood-Quaintance KE, Hanssen AD, *et al.* Rapid molecular microbiologic diagnosis of prosthetic joint infection. J Clin Microbiol 2013;51(7):2280–2287.
[51] Le Dantec L, Maury F, Flipo RM, *et al.* Peripheral pyogenic arthritis. A study of one hundred seventy-nine cases. Rev Rhum Engl Ed 1996;63(2):103–110.
[52] Maharajan K, Patro DK, Menon J, *et al.* Serum procalcitonin is a sensitive and specific marker in the diagnosis of septic arthritis and acute osteomyelitis. J Orthop Surg Res 2013;8(1):19.
[53] Swan A, Amer H, Dieppe P. The value of synovial fluid assays in the diagnosis of joint disease: a literature survey. Ann Rheum Dis 2002;61(6):493–498.
[54] Cherkaoui A, Ceroni D, Emonet S, *et al.* Molecular diagnosis of *Kingella kingae* osteoarticular infections by specific real-time PCR assay. J Med Microbiol 2009;58(Pt 1):65–68.
[55] Patel R. Matrix-assisted laser desorption ionization-time of flight mass spectrometry in clinical microbiology. Clin Infect Dis 2013;57:564–572.
[56] Harris LG, El-Bouri K, Johnston S, *et al.* Rapid identification of staphylococci from prosthetic joint infections using MALDI-TOF mass-spectrometry. Int J Artif Organs 2010;33(9): 568–574.
[57] Wolk DM, Kaleta EJ, Wysocki VH. PCR-electrospray ionization mass spectrometry: the potential to change infectious disease diagnostics in clinical and public health laboratories. J Mol Diagn 2012;14(4):295–304.
[58] Jacovides CL, Kreft R, Adeli B, *et al.* Successful identification of pathogens by polymerase chain reaction (PCR)-based electron spray ionization time-of-flight mass spectrometry (ESI-TOF-MS) in culture-negative periprosthetic joint infection. J Bone Joint Surg 2012;94(24): 2247–2254.
[59] Melendez DP, Uhl JR, Greenwood-Quaintance KE, *et al.* Detection of prosthetic joint infection by use of PCR-electrospray ionization mass spectrometry applied to synovial fluid. J Clin Microbiol 2014;52(6):2202–2205.
[60] Greenwood-Quaintance KE, Uhl JR, Hanssen AD, *et al.* Diagnosis of Prosthetic Joint Infection (PJI) Using Polymerase Chain Reaction-Electrospray Ionization Mass Spectrometry (PCR-ESI/MS). J Clin Microbiol 2014;52(2):642–649.

Chapter 3

Pharmacokinetics and Pharmacodynamics of Antibiotics in Bone

Cornelia B. Landersdorfer, Jürgen B. Bulitta, and Fritz Sörgel

Chronic osteomyelitis requires prolonged antibiotic treatment, has a high recurrence rate, and can cause irreversible damage. Also, the number of orthopedic device–related infections continues to increase [1]. Therefore, adequate antibiotic treatment and surgical prophylaxis are critical. Therapeutic success is primarily determined by the antimicrobial activity against the infecting pathogen and the rate and extent of antibiotic penetration into bone. Adequate bone penetration has to be ensured as antibiotics need to reach effective concentrations at the infection site to kill bacterial pathogens. Therefore studying the time course and extent of bone penetration before launching a clinical effectiveness trial is important. The aim of this chapter is to review the pharmacokinetics (PK) and pharmacodynamics (PD) of antibiotics in bone and present methods that support optimized evidence-based selection of antibiotic dosage regimens.

Pharmacokinetics

Time course and magnitude of drug concentrations in the body and particularly at the site of action determine the drug effects. Therefore it is important to study PK, which describes the relationship between the dose of a drug and the resulting time course of drug concentrations at various spaces in the body [2, 3]. PK processes include drug absorption from the site of administration into the systemic circulation (except if administered directly into the bloodstream), distribution from the systemic circulation into tissues, and elimination via metabolism, renal excretion, or both. Most frequently PK is characterized based on drug concentrations measured in plasma or serum. However, in treating bone infections, adequate antibiotic concentrations need to be achieved at the site of infection in bone. Numerous clinical studies have been conducted to quantify antibiotic concentrations in bone.

Bone and Joint Infections: From Microbiology to Diagnostics and Treatment, First Edition.
Edited by Werner Zimmerli.

Bone is a heterogeneous tissue, where the organic bone matrix represents 30–35% of total bone mass and includes collagen fibrils (~90%), glycoproteins, proteoglycans, and extracellular fluid. Blood vessels in bone are located in Haversian and Volkmann's canals that transverse the bone matrix. Bone cells represent only 1–2% of total bone mass and in their most mature form as osteocytes are trapped inside the bone matrix. The inorganic matrix (65–70%) consists of calcium phosphate crystals (hydroxyapatite) deposited inside the organic matrix. Due to this heterogeneous composition, most likely neither bacteria nor antibiotics distribute evenly throughout the bone tissue.

The site of the pathogens in bone is not well-known. Based on their size (e.g., ~1 μm for *Staphylococcus aureus*), bacteria are expected to distribute through the Haversian and Volkmann canals (~70-μm diameter) in bone, but not into the hydroxyapatite crystals. *S. aureus* can enter into and survive in osteoblasts, which may explain relapses. In addition, it adheres to components of the bone matrix such as collagen [4].

Techniques to separate the different components of bone and measure concentrations in each are lacking. Therefore the vast majority of published studies are based on homogenized bone samples, and the total drug concentrations in bone homogenate are reported. For the interpretation of bone penetration results, it is important to note that only free drug is microbiologically active. However, total drug concentrations in bone homogenate, provided they are reliably determined and analyzed by population modeling and Monte Carlo simulations, may be more predictive of therapeutic success than serum concentrations.

Bone Sample Preparation and Analysis

In contrast to plasma or serum, there is no specific guidance available for drug analysis in bone or other tissues. However, validated and reproducible sample preparation and drug determination procedures are undoubtedly critical. It is important to consider the techniques used in published studies, when interpreting the results of these trials.

After bone resection, adhering blood and soft tissue is often removed from the sample. Excess blood due to intraoperative soaking can result in biased results, for example, artificially high bone concentrations for a drug with low bone penetration but high blood concentrations. Samples are usually separated into cancellous bone (the inner part of the long bones) and cortical bone. Cancellous bone has a higher degree of vascularization, a higher percentage of extravascular fluid, and a lower percentage of inorganic matrix than cortical bone, which can cause differences in antibiotic penetration.

For efficient extraction of the antibiotic, bone samples need to be homogenized. When bone samples are pulverized under liquid nitrogen in a cryogenic mill, this provides a very fine powder, is highly reproducible, and is applicable to thermally unstable drugs (e.g., β-lactam antibiotics) that are prone to degradation during grinding without freezing. Therefore, this method is preferable to slicing, grinding by mortar and pestle, or using mixers without cooling, as frequently applied in earlier studies before more recent technology was developed. During drug extraction from the homogenized sample, sufficient recovery and stability of the drug need to be ensured.

Calibration standards and quality control samples are necessary for accurate drug determination and should be prepared in drug-free bone powder instead of plasma, serum, or buffer. An internal calibration standard should be added to each sample to improve the analytical accuracy and precision. Older studies frequently determined drug concentrations by bioassay. Newer studies have mainly employed high-performance liquid chromatography (HPLC) and recently also liquid chromatography–tandem mass spectrometry (LC–MS/MS), offering improved sensitivity and specificity. HPLC was shown to be generally superior to bioassays when analyzing bone samples [5]. Bone penetration studies should report details on the chosen methods for sample preparation and analysis, and the recovery, bias, and precision.

Concentrations in bone are typically reported as mg/kg of total bone mass. Some studies report concentrations in relation to bone volume, organic bone mass, or interstitial fluid or correct for blood content. Potential differences in reporting need to be taken into account when comparing results between studies.

Pharmacokinetic Sampling and Data Analysis

Usually, only one bone sample can be taken per patient, and a blood sample is taken at the same time. Most studies report bone penetration as the concentration ratio between bone and serum or plasma at one time point. However, due to different kinetics of drug concentrations in plasma and bone, the concentration ratios change over time until eventually an equilibrium has been reached during the terminal phase. This phenomenon (system hysteresis) hampers the interpretation of results and comparison between drugs and studies when samples are taken at different times post dosing.

A better measure for the extent of bone penetration is to calculate the area under the concentration–time curve (AUC) in bone and compare it to the AUC in plasma or serum. This takes into account the full time course of the concentration profiles in bone, plasma, and serum. Instead of collecting the bone and blood samples at the same time point after the dose for all patients, samples should be spread out over a time period to support PK modeling. Based on such study designs, investigators have averaged the concentrations at each time point and derived the average AUCs in bone and plasma (naive averaging) [6–8]. Alternatively, one PK function was fit to the concentration–time data from all patients (naive pooling) and the AUC integrated [9, 10]. While these approaches remove the issue of time-dependent concentration ratios, they only consider the average concentration–time profile and ignore the true biological variability between patients.

Population PK analysis is the most powerful approach for the analysis of sparse data (e.g., one bone sample per patient) and accounts for the average rate and extent of bone penetration and interpatient variability [11, 12]. By fitting each patient's data in the perspective of the concentrations from the other patients, the most likely concentration–time course in bone and serum and the AUC can be predicted for each patient. Estimating the rate of bone penetration enables recommendations on the administration time of antibiotic prophylaxis before surgery. An existing population PK model can also be used to identify the optimal timing of bone and plasma samples in future bone penetration studies.

Bone penetration is usually studied in joint replacement patients with uninfected bone as such patients are more easily recruited than osteomyelitis patients. The condition of the bone samples is likely more homogeneous among joint replacement patients than patients with various stages and locations of bone infections; therefore, results of different studies can be more readily compared. However, antibiotic concentrations might differ between infected and uninfected bone. Reactive hyperemia could increase the blood flow into bone, whereas pus or sequesters might limit the distribution of antibiotics into bone. To date, few studies have been performed in patients with bone infections, which does not enable a systematic comparison of penetration between infected and uninfected bone. Presence of ischemic, calcified, or arthritic tissues, bone cysts, or fat in the cancellous bone may affect antibiotic distribution. Different types of bone (e.g., hip, knee, sternum) and influences on blood circulation, for example, tourniquet application or internal mammary artery harvesting, may also affect antibiotic bone concentrations.

Penetration of Antibiotics into Bone

Figure 3.1 presents an overview of the extent of bone penetration by antibiotic group. Each symbol represents the median bone-to-serum (or plasma) concentration ratio from one clinical study, and the lines indicate the median per antibiotic group. In total, 126 studies (until July 2013) were included. Most concentration ratios were reported directly in the published studies; sometimes they were calculated from the reported concentrations or read from plots. A comprehensive reference list can be found in Ref. [13]. Tables 3.1 and 3.2 list the range of average concentration ratios for various antibiotics. Concentration ratios are average ± standard deviation and based on total concentrations in bone homogenate from at least five samples, unless indicated otherwise. AUC ratios are reported in the text where available. Studies published since the previous review [13] are discussed here in more detail.

Systematic differences can be observed between antibiotic groups, which may be due to different physicochemical and binding characteristics. Median bone-to-serum concentration ratios were 0.48 for quinolones and 0.40 for linezolid. Macrolides span a

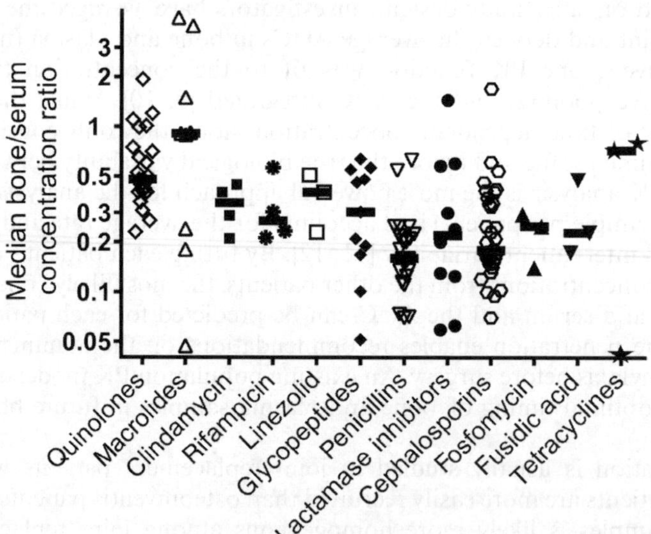

Figure 3.1. Bone penetration for different antibiotic groups [13]. Each symbol represents the median or average bone-to-serum or bone-to-plasma concentration ratio from one clinical trial. The lines represent the group medians.

Table 3.1. Bone penetration of quinolones and macrolides.

Antibiotic and bone condition	Range of time since last dose	Range of average bone/serum concentration ratios	Bone or surgery type	Bio-analytical method
Ciprofloxacin				
uninfected	0.5–13 h	0.27–1.2	Hip, knee, skull,	HPLC [8, 14, 15]
osteomyelitis	2–4.5 h	0.42	debridement surgery	HPLC [14]
Levofloxacin				
uninfected	0.7–2 h	0.36–1.0	Hip, other	HPLC [16–18]
Ofloxacin				
uninfected	0.5–12 h	0.09–1.04	Hip, nasal bone, mastoid process	HPLC [19–21]
Moxifloxacin				
uninfected	1.5–5 h	0.33–1.05	Hip, knee, sternum, manubrium	HPLC [11, 18, 22, 23]
Azithromycin				
uninfected	0.5–6.5 days	2.5–6.3	Alveolar bone	Bioassay [24, 25]
Telithromycin				
uninfected	3.3–24 h	1.5–2.6	Ethmoid bone	HPLC [6]

HPLC, high-performance liquid chromatography.

wide range of concentration ratios. Despite large differences in chemical structure, clindamycin, rifampicin, glycopeptides, fosfomycin, and fusidic acid had comparable median concentration ratios of 0.23–0.35. Penicillins, cephalosporins, and β-lactamase inhibitors showed median concentration ratios of 0.16, 0.18, and 0.22. Figure 3.1 also demonstrates a large variability between antibiotics within each group, which may in part be caused by different bioanalytical methodologies and sampling times (as described earlier).

Fluroquinolones

Fluoroquinolones are frequently used in bone infections and show one of the highest median extents of bone penetration of all antibiotic groups with bone-to-serum concentration ratios mostly between 0.3 and 1.2 (Figure 3.1). The high penetration may be partly due to binding of quinolones to the calcium in bone. As only free antibiotic is considered microbiologically active, the quinolone concentrations available for antimicrobial action are likely lower than the total bone concentrations. The concentration ratios of most quinolones tend to increase with time since the last dose, indicating slow redistribution from bone back into the bloodstream. Quinolones generally penetrate well into cells. This could be advantageous for treatment of *S. aureus* osteomyelitis, since *S. aureus* was shown to penetrate into and survive in osteoblasts in vitro [4, 34].

Multiple studies in different patient groups have examined the bone penetration of ciprofloxacin (Table 3.1). Massias *et al.* [8] took cortical bone of the mastoid process plus serum samples at five different time points from 21 patients suffering from chronic otitis.

Table 3.2. Bone penetration of beta-lactams.

Antibiotic and bone condition	Range of time since last dose	Range of average bone/serum concentration ratios	Bone or surgery type	Bio-analytical method
Amoxicillin				
Uninfected	0.5–6 h	0.03–0.31	Hip, jaw	bioassay [9, 26, 27], LC-MS/MS [12]
Clavulanic acid				
Uninfected	0.5–6 h	0.01–0.14	Hip	bioassay [9, 26], LC-MS/MS [12]
Ampicillin				
Uninfected	0.25–1 h	0.11–0.20	Hip, knee, vertebrae	Bioassay [10, 28]
Uninfected (with blood washing)	1–4 h	0.44–0.71	orthopedic surgery	Bioassay [29]
Sulbactam				
Uninfected	0.25–1 h	0.17–0.58	Hip, knee, vertebrae	Gas chromatography [10, 28]
Uninfected (with blood washing)	1–4 h	0.58–0.71	orthopedic surgery	Gas chromatography [29]
Cefotiam				
Uninfected (with blood washing)	1–4 h	0.27–0.44	Orthopedic surgery	HPLC [29]
Cefepime				
Uninfected	1–2 h	0.46–0.76[a]	Hip	HPLC [30]
Ceftazidime				
Uninfected	2 h	0.54	Cardiac surgery	Bioassay [31]
Ischemic bone	1–2 h	0.04–0.08	foot	HPLC [32, 33]

HPLC, high-performance liquid chromatography; LC–MS/MS, liquid chromatography–mass spectrometry.
[a]Assuming a bone density of 1 kg/l.

The AUCs in bone and serum were calculated by naive averaging and the trapezoidal rule and the bone-to-serum AUC ratio was 0.63. Average bone-to-serum concentration ratios increased from 0.27 to 1.2 between 1 and 12 h after the dose, suggesting slow redistribution from bone to blood. Fong *et al.* [14] compared ciprofloxacin concentrations in cortical bone from patients without ($n = 18$, hip or knee replacement or osteotomy) and with ($n = 10$) osteomyelitis. Concentrations in infected bone were 30–100% higher than in uninfected bone. As serum concentrations were also higher in osteomyelitis patients, the average bone-to-serum concentration ratios were approximately 0.4 in both patient groups. Studies in various bone types indicate a penetration of 0.40 or higher for ciprofloxacin.

Bone penetration of levofloxacin was evaluated by three relatively recent studies (Table 3.1). In patients undergoing bone surgery ($n = 9$) or decubitus ulcer debridement ($n = 12$) the bone-to-serum concentration ratios were 0.36 for cortical ($n = 6$) and 0.85 ± 0.40 for cancellous ($n = 14$) bone [16]. In 12 hip replacement patients, ratios of

1.0 ± 0.4 for cortical and 0.5 ± 0.1 for cancellous bone were reported [17]. Concentration ratios of 0.42 ± 0.04 in cortical and 0.54 ± 0.05 in cancellous bone were found in eight hip replacement patients [18]. The differences in penetration to cortical versus cancellous bone might be partly due to relatively small sample sizes.

Moxifloxacin was studied in four trials, which showed consistently high penetration considering the range of different bone types (Table 3.1) and methods for sample homogenization (hand mincing, sonication, cryogenic mill). In all moxifloxacin studies, the penetration into cancellous and cortical bone was similar. Utilizing a cryogenic mill and population PK analysis, the bone-to-serum AUC ratios in 24 hip replacement patients were 0.80 (10th–90th percentile for between-patient variability: 0.51–1.26) for cortical and 0.78 (0.42–1.44) for cancellous bone [11].

Macrolides and Telithromycin

Macrolides demonstrate the largest range of penetration of all antibiotic groups (Figure 3.1). All studies utilized bioassays, and most were performed decades ago when contemporary bioanalytical technology was not yet available. The analytical recovery from bone samples was often low, for example, for erythromycin. In two more recent studies [24, 25], patients received 500 mg azithromycin once daily for 3 days before periodontal surgery. In both studies, the average concentration ratios increased slightly from 12 h to 2.5 days, when they reached greater than 6.0, and then slowly decreased to approximately 2.5 at 6.5 days. Azithromycin bone concentrations decreased from 1.61 ± 0.22 mg/kg at 12 h to 0.44 ± 0.05 mg/kg at 6.5 days [25]. The rate of azithromycin penetration into bone is not well-known as the first samples were taken at 12 h. Azithromycin is known to accumulate in cells, for example, macrophages. However, bone cells represent only 1–2% of total bone weight. Depending on the bone type, differences in the content of red bone marrow as part of a cancellous bone sample might potentially lead to variations in macrolide concentrations due to their accumulation in leukocytes.

Telithromycin penetration into ethmoid bone was studied in 29 patients [6]. Using naive averaging, the average AUC in bone was 6730 mg·h/l and the AUC in plasma was 4230 mg·h/l, indicating a bone-to-serum AUC ratio of 1.6. This suggests one of the highest extents of penetration of all studied antibiotics. The concentration ratio increased between 3 and 24 h, indicating slow equilibrium that may favor administration at least approximately 12 h ahead of surgery.

Clindamycin

Clindamycin is often referred to as possessing exceptionally high bone penetration. The median bone-to-serum concentration ratio of 0.35 from four clindamycin studies, however, appears to be lower than for quinolones (median 0.50) and linezolid (median 0.40) (Figure 3.1). As most clindamycin studies were performed in the 1970s, that is, before the introduction of fluoroquinolones, linezolid, and azithromycin, the bone penetration of clindamycin was higher than that of other available antibiotics at that time. A more recent study reports bone concentrations between 3.4 and ~0.2 mg/l in 13 maxillofacial surgery patients at 0.5–8 h after a 600 mg dose. Concentration ratios were not reported; however, based on plots the bone concentrations were less than 50% of plasma concentrations at all times [35]. All available clindamycin studies used bioassays, which have the potential to be confounded by active metabolites of clindamycin. Results from multiple studies suggest an extent of bone penetration of clindamycin of 0.21–0.45, similar to or slightly higher than cephalosporins.

Rifampicin

A wide range of bone-to-serum concentration ratios (0.08–0.56 at 2–14 h after the dose) was found for rifampicin in four studies in uninfected bone from the 1970s/1980s. One of the trials also investigated infected bone, and concentration ratios were similar to those for uninfected bone (0.57 versus 0.46). All studies utilized bioassays and had high inter-patient variability [13].

Tetracyclines and Tigecycline

Few studies are available for tetracyclines. Results vary despite the high binding affinity of tetracyclines to calcium. For tigecycline, initially a bone-to-serum AUC ratio of 0.41 (by naïve averaging) or 0.28 (calculated from median concentrations) over 24 h was found in 25 uninfected surgical patients [7]. Concentration ratios increased from 4 to 24 h. Re-analysis of the samples by a new LC–MS/MS assay, including a stabilizing agent, resulted in bone concentrations that were on an average 9.5-fold higher as compared to the previous method [36]. These results are also consistent with radiolabeling studies in animals. This study highlights the importance of bioanalytical methods and of taking into account the full concentration–time course.

Cephalosporins

Numerous studies were performed with cephalosporins, most frequently cefuroxime. Its overall average bone-to-serum concentration ratio was 0.32 (range 0.09–0.55, 10 min to 6.5 h post dose) in five studies that reported concentrations in serum and uninfected bone and in which the majority of samples were above the detection limit [13]. One study found very low penetration into sternum due to a high detection limit [37]. Median concentration ratios at 1 h after the dose were approximately 0.18 in 14 trauma surgery patients and approximately 0.06 in 7 osteomyelitis patients [38].

Ceftriaxone and cefamandole were evaluated in the same study in hip replacement patients [39]. At 10–30 min after the dose, average (95% confidence interval) bone-to-serum concentration ratios were 0.156 (0.123–0.190) for ceftriaxone and 0.184 (0.156–0.212) for cefamandole. The bone-to-plasma concentration ratios based on total drug were similar despite a sixfold difference in the non-protein-bound fractions in plasma (0.05 for ceftriaxone, 0.30 for cefamandole), although only unbound drug is believed to distribute between plasma and tissues. This issue is discussed in more detail in our previous review [13].

At 8 h after the dose, ceftriaxone bone-to-serum concentration ratios were 0.142 (0.073–0.210), very similar to those at 10–30 min, suggesting a fast equilibrium between serum and bone [39]. Recently, in 11 patients undergoing debridement for septic non-union of the tibia, ceftriaxone concentrations were measured by HPLC, and average AUCs over 24 h were calculated by the trapezoidal rule [40]. Average bone-to-plasma AUC ratios were 0.093 in cortical and 0.241 in cancellous bone. This 2.6-fold difference between cortical and cancellous bone is considerably larger than in most other studies.

Average bone-to-serum concentration ratios for cefamandole in hip replacement patients were 0.227–0.249 at 10–30 min after the dose [41]. Another trial found cefamandole bone-to-serum concentration ratios increasing from 0.8 at 1 h after the dose to 2.3 at 4 h [29]. Considering the lower values at early time points [39, 41] and the short elimination half-life (0.8 h), this could indicate slow redistribution from bone to blood. However, modeling the full concentration–time course and additional data would be required to support a sound time–course analysis. Alternatively, the high ratios [29] could be partly due to low serum concentrations as intraoperative blood saving including washing of the drained blood was applied in this study. The overall range of concentration ratios reported for cefamandole was 0.12–2.3 at 10 min to 4 h [13].

Two recent studies investigated cefazolin. The median bone-to-serum concentration ratios in eight infected patients were 0.25 (range 0.06–0.41) during continuous cefazolin infusion, with concentrations determined by bioassay [42]. Yamada *et al.* [43] utilized HPLC and found cancellous bone concentrations of 22.4 ± 14.8 mg/kg at 63 ± 25 min and serum concentrations of 170.3 ± 51.3 mg/l at 49 ± 13 min in 42 patients. Bone concentrations in knee replacement (16.0 mg/kg) tended to be lower than those in hip replacement (32.3 mg/kg) during a similar sampling time period. In an earlier study, the average bone-to-serum concentration ratio was 0.18 in 20 hip replacement patients at 0.9 h after the dose [44]. Additional results for several β-lactams are presented in Table 3.2.

Overall cephalosporins achieved concentration ratios of 0.1-0.5. Penetration was higher into cancellous bone than into cortical bone in all studies that analyzed both, potentially due to the higher proportion of extracellular fluid in cancellous bone [13]. β-Lactams, including cephalosporins, are assumed to distribute mainly into extracellular fluid and were found to exhibit limited binding to the inorganic bone matrix.

Penicillins and β-Lactamase Inhibitors

The most frequently studied β-lactams/β-lactamase inhibitors are amoxicillin/clavulanic acid, ampicillin/sulbactam, and piperacillin/tazobactam.

Two studies by different groups from 1994 and 2001 evaluated piperacillin/tazobactam penetration into uninfected hip bone in 12 patients each, used the same sample preparation methods and analysis by HPLC, and found consistent results. Bone-to-plasma concentration ratios were 0.2–0.3 for piperacillin and tazobactam in cortical and cancellous bone at 1–1.5 h after the dose [45, 46]. More recently, penetration of piperacillin/tazobactam into uninfected jaw ($n = 7$) and hip ($n = 2$) bone was studied [47]. Sample preparation was similar to the previous studies and concentrations were analyzed by LC–MS/MS. At an average of 3 h (range 1–7 h) after the start of the infusion, bone-to-plasma concentration ratios were 0.15 for piperacillin and 0.13 for tazobactam. These results were slightly lower and more variable than those from the previous studies, potentially due to different bone types and the wider range of sampling times.

A wide range of average amoxicillin bone-to-serum concentration ratios was reported in studies utilizing bioassays (Table 3.2). In a study in 20 hip replacement patients analyzed by LC–MS/MS and population PK analysis, the bone-to-serum AUC ratios were 0.20 (10th–90th percentile for between-patient variability 0.16–0.25) for cortical and 0.18 (0.11–0.29) for cancellous bone [12]. In the same study, the bone-to-serum AUC ratios of clavulanic acid were 0.15 (0.11–0.21) for cortical and 0.10 (0.051–0.21) for cancellous bone. Penetration of clavulanic acid tended to be slightly lower in most studies, although it is the smaller molecule and might be expected to distribute more freely. However, lipophilicity information is not available, which may also play an important role.

Linezolid

Linezolid is comparatively stable, as opposed to many β-lactams, and the available studies were performed utilizing HPLC. Average (95% confidence interval) bone-to-serum concentration ratios were 0.51 (0.43–0.75) in 12 hip replacement patients at 30–50 min after the start of the infusion [41]. Relatively high penetration (0.40 ± 0.24) was also found in 12 elderly patients during knee replacement at 1.5 h [48]. At the same dose as the two joint replacement studies [41, 48], and 0.5–1.5 h after the dose, lower linezolid concentrations were found in 11 patients with implant-associated infections. The average bone-to-plasma concentration ratio was approximately 0.23 [49]. Inflammation-related decreased blood supply or differences in sample preparation were considered as potential reasons for the differences between uninfected and infected bone [49].

Recently, linezolid in bone was also analyzed by microdialysis, which allows determination of unbound drug concentrations in interstitial fluid, serial sampling, and calculation of individual AUCs in bone. For insertion of the microdialysis catheter, a hole is drilled into the bone, which results in a dead space that fills with blood clots and extracellular fluid exudations [50]. Measured concentrations are therefore assumed to represent concentrations in the dead space and the interstitial fluid of the adjacent bone tissue [50]. The ratio of unbound AUCs (*f*AUC) in vital cancellous bone/plasma over 12 h was 1.09 ± 0.11 in three diabetic patients with severe foot infections, suggesting similar exposure to microbiologically active linezolid for pathogens in interstitial bone fluid and in the bloodstream [51]. The higher AUC ratio from microdialysis, as compared to concentration ratios based on bone homogenate, is in keeping with a low propensity of linezolid to form chelate complexes with the inorganic bone matrix.

Daptomycin

Bone penetration of daptomycin was evaluated in diabetic foot infections [52]. Serial microdialysis samples at the steady state were collected from 0 to 8 h after the dose in five patients and from 8 to 16 h in another four patients. The average ratio of the *f*AUC (0–16 h) in interstitial fluid of metatarsal bone/plasma was 1.08. This ratio of unbound bone-to-unbound plasma concentrations suggests high penetration of daptomycin into interstitial fluid, that is, the most likely site of infection, and was achieved for a drug with high plasma protein binding (~90%) and a high molecular weight. The authors stated that the effect of drilling a hole into the bone for insertion of the microdialysis catheter on bone concentrations is not yet clarified [52]. Thus, a comparison with total bone-to-plasma concentrations would be interesting to compare the results of different methods.

Fosfomycin

Bone homogenate-to-serum concentration ratios of 0.13–0.45 were reported in three trials from 1980 to 1983 utilizing bioassays (Figure 3.1). Fosfomycin binds to hydroxyapatite in bone, suggesting that not all fosfomycin in bone homogenate is microbiologically active. However, a recent microdialysis study found *f*AUC ratios of 0.43 ± 0.04 in nine osteomyelitis patients with diabetic foot infection, which is higher than or similar to the reported bone homogenate-to-plasma concentration ratios [53]. Considering the very limited or no binding of fosfomycin to plasma proteins, this would indicate that the average concentration bound to various components of bone tissue is lower than or similar to the interstitial fluid concentrations. However, comparison among studies is hampered by differences in study designs and methodologies.

Glycopeptides

A wide range of average concentration ratios, mostly between 0.1 and 0.6, has been reported for glycopeptides in hip, knee, or sternal bone (Figure 3.1). A recent study investigated both glycopeptides in septic pseudoarthritis of the tibia [54]. Validated HPLC assays were used. However, samples were ground without cooling, and the PK methods to calculate individual AUCs were reported in limited detail. Average bone-to-plasma AUC ratios for vancomycin were 0.21 for cortical and 1.04 for cancellous bone. Average bone-to-plasma AUC ratios for teicoplanin were 0.12 for cortical ($n=17$ patients) and 0.56 for cancellous bone ($n=15$). Such large differences between cortical and cancellous bone were also seen in a ceftriaxone study by the same authors (described earlier), who suggest that this may be due to infected bone or different methods of drug extraction or analysis. Glycopeptide bone concentrations increased with inflammatory marker concentrations, potentially due to increased tissue vascularization. The results from this study fall into the range of previous reports; however, drawing overall conclusions on glycopeptide bone penetration remains difficult.

Pharmacodynamics and Monte Carlo Simulations

PD describes the relationship between drug concentrations in plasma or at the target (i.e., infection) site and the time course of drug effect(s). For β-Lactams the time during which the unbound antibiotic concentration remains above the minimum inhibitory concentration ($fT_{>MIC}$) of the pathogen has been shown to be predictive of the extent of antibiotic effect. For other antibiotics, such as quinolones, the *f*AUC/MIC best correlates with effect.

For bone penetration studies that used the same sampling time for all samples, it is not feasible to perform a PD analysis because comparing the bone concentration at a specific time to the MIC provides limited information. Irrespective of the type of data analysis used, adequate reporting of the methods and assumptions is important.

Naive pooling or averaging approaches (as described earlier) can calculate the average AUC/MIC and time above MIC. This indicates whether an "average" patient would attain the PK/PD target. However, these naive methods have the disadvantage that they do not consider the true variability between patients, which tends to be large for bone penetration.

Population modeling accounts for both the average penetration and its variability between subjects (Figure 3.2). Once a population PK model for plasma and bone has been developed, it can be employed in Monte Carlo simulations to predict the expected concentration time profiles for other than the studied dosage regimens. This includes predicting the variability in concentration time profiles between patients. Thereby, the probability of achieving a PK/PD target can be predicted and recommendations be made on how to dose an antibiotic to maximize the probability of successful therapeutic outcome.

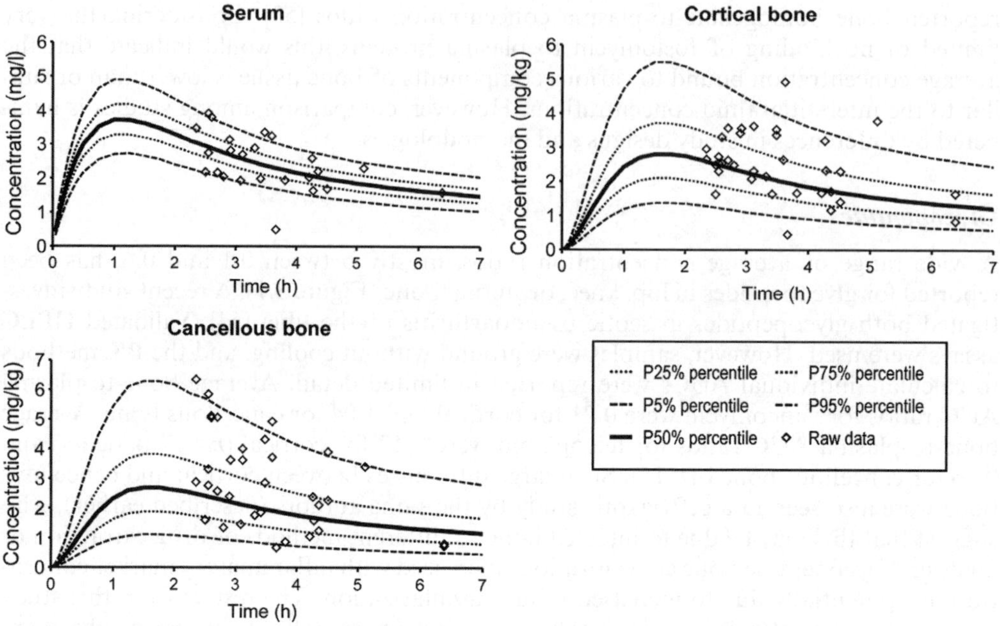

Figure 3.2. Visual predictive checks for the moxifloxacin population model presenting both the median concentration–time profiles and their between-subject variability in serum, cortical bone, and cancellous bone. The diamonds represent the individual observed data from 24 hip replacement patients. The solid lines represent the model-predicted median profiles, and the dashed lines are the model-predicted 5, 25, 75, and 95% percentiles. (From Ref. [11]). © American Society for Microbiology).

The PK/PD target values for plasma and bone concentrations to successfully treat bone infections are most often unknown, and the target values for other types of infections can likely not be used. For moxifloxacin, no published clinical studies in osteomyelitis were available.

Moxifloxacin: To address the lack of known PK/PD target values for bone, a reverse engineering approach [55] was applied for moxifloxacin to identify the most likely PK/PD target required to achieve clinical and microbiological cure of osteomyelitis [11]. This approach combined effectiveness data from clinical studies with ciprofloxacin in osteomyelitis, the expected plasma AUCs from these studies, the AUCbone-to-AUCplasma ratio for ciprofloxacin [8], and bacterial susceptibility data from the time of the clinical studies. Reverse engineering suggested a *f*AUC/MIC of 40 in serum and an AUC/MIC of 33 in bone as the most likely PK/PD targets for successful clinical and microbiological outcome. No assumptions are made regarding the numerical value of the free fraction of moxifloxacin in bone. It is assumed that binding and distribution within the bone tissue is similar for moxifloxacin and ciprofloxacin, two quinolones with the same essential chemical structure that are expected to be responsible for binding characteristics. The population PK model for moxifloxacin in serum and bone was utilized to predict likely probabilities of target attainment.

A ≥90% probability of successful clinical and microbiological outcome was predicted for 400 mg moxifloxacin once daily up to an MIC of 0.375 mg/l (mg/kg) in serum and

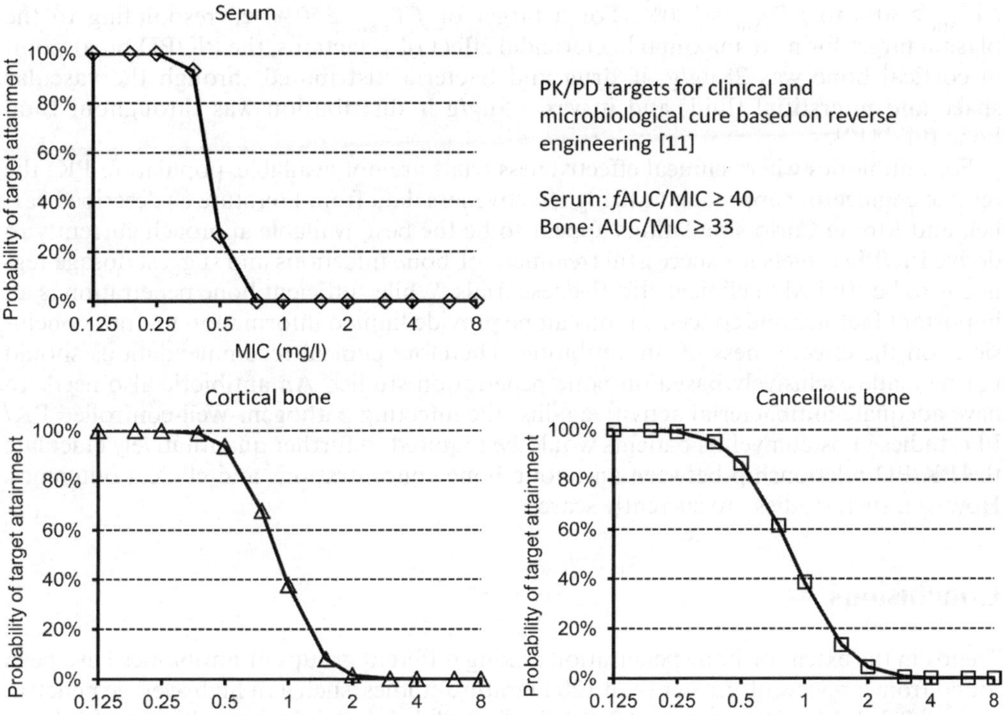

Figure 3.3. Probabilities of target attainment to achieve successful clinical and microbiological outcome. The PK/PD targets for serum, cortical bone, and cancellous bone are based on a reverse engineering approach [11].

cancellous bone and 0.5 mg/l in cortical bone (Figure 3.3). Compared to, for example, an MIC_{90} of 0.125 mg/l for *S. aureus*, these are favorable results and suggest clinical trials are warranted. The antibiotic susceptibility of the local hospital should be considered when published probabilities of target attainment are used to decide about antibiotic therapy in patients. The population PK and Monte Carlo simulation approach described for bone is applicable to other matrices, for example, synovial fluid.

As an additional complexity, no methods are currently available to measure concentrations in different compartments of bone. Also, assumptions for binding (e.g., to calcium) and distribution in bone have to be made until techniques that reliably determine free antibiotic concentrations in bone become available. While microdialysis methods have been applied for bone, it is unknown whether drilling a hole into bone affects the measured unbound concentrations in bone [52].

Amoxicillin/clavulanic acid: To address this situation, various scenarios for distribution of amoxicillin and bacteria into interstitial fluid, total bone fluid, bone cells, and organic and inorganic matrix were considered and volumes of the bone compartments were taken from literature. Thereby the likely PK/PD breakpoints, that is, the MICs up to which a 90% probability for successful treatment is expected, could be predicted for various scenarios [12]. As clinical effectiveness studies were not available for amoxicillin/clavulanic acid, breakpoints were calculated for potential targets ranging from

$fT_{>MIC} \geq 30\%$ to $fT_{>MIC} = 100\%$. For a target of $fT_{>MIC} \geq 50\%$, corresponding to the plasma target for near-maximal bactericidal effect of β-lactams, the PK/PD breakpoint in cortical bone was 28 μg/g, if drug and bacteria distributed through the vascular space and interstitial fluid, and it was 7.5 μg/g if distribution was throughout total bone fluid [12].

For antibiotics where clinical effectiveness trials are not available, population PK, the reverse engineering approach utilizing effectiveness data from literature as described earlier, and Monte Carlo simulations appear to be the best available approach currently to derive PK/PD targets for successful treatment of bone infections and suggest dosage regimens to be studied in clinical effectiveness trials. While sufficient bone penetration is an important factor, bone concentrations alone provide limited information to draw conclusions on the effectiveness of an antibiotic. Therefore clinical recommendations should not be made exclusively based on bone penetration studies. An antibiotic also needs to have adequate antibacterial activity against the infecting pathogen. Well-controlled PK/PD studies in osteomyelitis patients would be required to further quantitatively elucidate the PK/PD relationship between antibiotic bone concentrations and clinical outcomes. However, such studies are currently scarce.

Conclusions

Trends in the extent of bone penetration among different groups of antibiotics have been found from a review of greater than 120 literature studies, such as a high average penetration of 0.3–1.2 for quinolones, 0.3–0.4 for linezolid, 0.1–0.3 for penicillins, and 0.1–0.5 for cephalosporins. These differences are most likely due to different physicochemical characteristics of the antibiotic groups. High variability between studies for a particular antibiotic group is likely partly due to a lack of standardization of bioanalytical methods and study design. The variability between patients within a study needs to be taken into account, and this can be achieved by population PK modeling. Developing approaches that provide insights into distribution and binding of antibiotics in bone is warranted. In 20 of 25 antibiotics, the measured concentrations were slightly higher in cancellous bone than those in cortical bone. More data are needed to characterize the effect of infected versus uninfected bone, the presence of an implant, and the type of bone (e.g., hip, knee, sternum) on the PK. Future clinical trials should focus on validated bioanalytical methods, as well as study designs, and apply PK/PD analyses that take into account the time course of bone concentrations to contribute to evidence-based care for patients with bone infections.

Key Points

- There are differences in the extent of bone penetration among various antibiotic groups, such as high median bone-to-serum concentration ratios of 0.3–1.2 for quinolones and lower ratios of 0.1–0.5 for β-lactams. These trends are likely related to different physicochemical and pharmacokinetic characteristics.
- The variability within antibiotic groups and between different studies for the same agent is high. Therefore utilizing standardized, validated methods for sample preparation and bioassay as well as calculating the bone-to-serum AUC ratios instead of concentration ratios would be advantageous.

- Based on the currently available data, population PK modeling and Monte Carlo simulations appear to be the most promising approach to elucidate the extent and time course of bone penetration and its relationship with likely clinical outcomes. Well-controlled PK/PD studies in osteomyelitis patients are required to directly identify PK/PD targets.

References

[1] Sendi P, Zimmerli W. Antimicrobial treatment concepts for orthopaedic device-related infection. Clin Microbiol Infect 2012;18(12):1176–1184.

[2] Holford N. Clinical pharmacokinetics and pharmacodynamics: the quantitative basis for therapeutics. In: Melmon K, Morrelli H, Nierenberg D, Hoffman B, editors. Melmon and Morrelli's Clinical Pharmacology: Basic Principles in Therapeutics, 3rd edn. New York: McGraw-Hill; 1992.

[3] Schoenwald R. Basic principles. In: Schoenwald R, editors. Pharmacokinetics in Drug Discovery and Development. New York: CRC Press; 2002.

[4] Jevon M, Guo C, Ma B, *et al.* Mechanisms of internalization of *Staphylococcus aureus* by cultured human osteoblasts. Infect Immun 1999;67(5):2677–2681.

[5] Bottcher S, von Baum H, Hoppe-Tichy T, *et al.* An HPLC assay and a microbiological assay to determine levofloxacin in soft tissue, bone, bile and serum. J Pharm Biomed Anal 2001; 25(2):197–203.

[6] Kuehnel TS, Schurr C, Lotter K, Kees F. Penetration of telithromycin into the nasal mucosa and ethmoid bone of patients undergoing rhinosurgery for chronic sinusitis. J Antimicrob Chemother 2005;55(4):591–594.

[7] Rodvold KA, Gotfried MH, Cwik M, *et al.* Serum, tissue and body fluid concentrations of tigecycline after a single 100 mg dose. J Antimicrob Chemother 2006;58(6):1221–1229.

[8] Massias L, Buffe P, Cohen B, *et al.* Study of the distribution of oral ciprofloxacin into the mucosa of the middle ear and the cortical bone of the mastoid process. Chemotherapy 1994;40 (Suppl 1):3–7.

[9] Weismeier K, Adam D, Heilmann HD, Koeppe P. Penetration of amoxycillin/clavulanate into human bone. J Antimicrob Chemother 1989;24(Suppl B):93–100.

[10] Wildfeuer A, Mallwitz J, Gotthardt H, *et al.* Pharmacokinetics of ampicillin, sulbactam and cefotiam in patients undergoing orthopedic surgery. Infection 1997;25(4):258–262.

[11] Landersdorfer CB, Kinzig M, Hennig FF, *et al.* Penetration of moxifloxacin into bone evaluated by Monte Carlo simulation. Antimicrob Agents Chemother 2009;53(5):2074–2081.

[12] Landersdorfer CB, Kinzig M, Bulitta JB, *et al.* Bone penetration of amoxicillin and clavulanic acid evaluated by population pharmacokinetics and Monte Carlo simulation. Antimicrob Agents Chemother 2009;53(6):2569–2578.

[13] Landersdorfer CB, Bulitta JB, Kinzig M, *et al.* Penetration of antibacterials into bone: pharmacokinetic, pharmacodynamic and bioanalytical considerations. Clin Pharmacokinet 2009;48(2):89–124.

[14] Fong IW, Ledbetter WH, Vandenbroucke AC, *et al.* Ciprofloxacin concentrations in bone and muscle after oral dosing. Antimicrob Agents Chemother 1986;29(3):405–408.

[15] Leone M, Sampol-Manos E, Santelli D, *et al.* Brain tissue penetration of ciprofloxacin following a single intravenous dose. J Antimicrob Chemother 2002;50(4):607–609.

[16] von Baum H, Bottcher S, Abel R, *et al.* Tissue and serum concentrations of levofloxacin in orthopaedic patients. Int J Antimicrob Agents 2001;18(4):335–340.

[17] Rimmele T, Boselli E, Breilh D, *et al.* Diffusion of levofloxacin into bone and synovial tissues. J Antimicrob Chemother 2004;53(3):533–535.

[18] Metallidis S, Topsis D, Nikolaidis J, *et al.* Penetration of moxifloxacin and levofloxacin into cancellous and cortical bone in patients undergoing total hip arthroplasty. J Chemother 2007;19(6):682–687.

[19] Tolsdorff P. Penetration of ofloxacin into nasal tissues. Infection 1993;21(1):66–70.
[20] Meissner A, Borner K, Koeppe P. Concentrations of ofloxacin in human bone and in cartilage. J Antimicrob Chemother 1990;26(Suppl D):69–74.
[21] Tolsdorff P. Tissue and serum concentrations of ofloxacin in the ear region following a single daily oral dose of 400 mg. Infection 1993;21(1):63–65.
[22] Malincarne L, Ghebregzabher M, Moretti MV, *et al.* Penetration of moxifloxacin into bone in patients undergoing total knee arthroplasty. J Antimicrob Chemother 2006;57(5): 950–954.
[23] Metallidis S, Charokopos N, Nikolaidis J, *et al.* Penetration of moxifloxacin into sternal bone of patients undergoing routine cardiopulmonary bypass surgery. Int J Antimicrob Agents 2006;28(5):428–432.
[24] Malizia T, Tejada MR, Ghelardi E, *et al.* Periodontal tissue disposition of azithromycin. J Periodontol 1997;68(12):1206–1209.
[25] Malizia T, Batoni G, Ghelardi E, *et al.* Interaction between piroxicam and azithromycin during distribution to human periodontal tissues. J Periodontol 2001;72(9):1151–1156.
[26] Grimer RJ, Karpinski MR, Andrews JM, Wise R. Penetration of amoxycillin and clavulanic acid into bone. Chemotherapy 1986;32(3):185–191.
[27] Akimoto Y, Kaneko K, Tamura T. Amoxicillin concentrations in serum, jaw cyst, and jawbone following a single oral administration. J Oral Maxillofac Surg 1982;40(5):287–293.
[28] Warnke JP, Wildfeuer A, Eibel G, *et al.* Pharmacokinetics of ampicillin/sulbactam in patients undergoing spinal microneurosurgical procedures. Int J Clin Pharmacol Ther 1998;36(5): 253–257.
[29] Dehne MG, Muhling J, Sablotzki A, *et al.* Pharmacokinetics of antibiotic prophylaxis in major orthopedic surgery and blood-saving techniques. Orthopedics 2001;24(7):665–669.
[30] Breilh D, Boselli E, Bel JC, *et al.* Diffusion of cefepime into cancellous and cortical bone tissue. J Chemother 2003;15(2):134–138.
[31] Adam D, Reichart B, Williams KJ. Penetration of ceftazidime into human tissue in patients undergoing cardiac surgery. J Antimicrob Chemother 1983;12(Suppl A):269–273.
[32] Raymakers JT, Houben AJ, van der Heyden JJ, *et al.* The effect of diabetes and severe ischaemia on the penetration of ceftazidime into tissues of the limb. Diabet Med 2001;18(3):229–234.
[33] Raymakers JT, Schaper NC, van der Heyden JJ, *et al.* Penetration of ceftazidime into bone from severely ischaemic limbs. J Antimicrob Chemother 1998;42(4):543–545.
[34] Hudson MC, Ramp WK, Nicholson NC, *et al.* Internalization of *Staphylococcus aureus* by cultured osteoblasts. Microb Pathog 1995;19(6):409–419.
[35] Mueller SC, Henkel KO, Neumann J, *et al.* Perioperative antibiotic prophylaxis in maxillofacial surgery: penetration of clindamycin into various tissues. J Craniomaxillofac Surg 1999;27(3):172–176.
[36] Ji AJ, Saunders JP, Amorusi P, *et al.* A sensitive human bone assay for quantitation of tigecycline using LC/MS/MS. J Pharm Biomed Anal 2008;48(3):866–875.
[37] Vuorisalo S, Pokela R, Satta J, Syrjala H. Internal mammary artery harvesting and antibiotic concentrations in sternal bone during coronary artery bypass. Int J Angiol 2000;9(2):78–81.
[38] Ketterl R, Wittwer W. Possibilities for the use of the basic cephalosporin cefuroxime in bone surgery. Tissue levels, effectiveness and tolerance. Infection 1993;21 (Suppl 1):S21-S27.
[39] Lovering AM, Walsh TR, Bannister GC, MacGowan AP. The penetration of ceftriaxone and cefamandole into bone, fat and haematoma and relevance of serum protein binding to their penetration into bone. J Antimicrob Chemother 2001;47(4):483–486.
[40] Garazzino S, Aprato A, Baietto L, *et al.* Ceftriaxone bone penetration in patients with septic non-union of the tibia. Int J Infect Dis 2011;15(6):e415-e421.
[41] Lovering AM, Zhang J, Bannister GC, *et al.* Penetration of linezolid into bone, fat, muscle and haematoma of patients undergoing routine hip replacement. J Antimicrob Chemother 2002;50(1):73–77.

[42] Zeller V, Durand F, Kitzis MD, *et al.* Continuous cefazolin infusion to treat bone and joint infections: clinical efficacy, feasibility, safety, and serum and bone concentrations. Antimicrob Agents Chemother 2009;53(3):883–887.

[43] Yamada K, Matsumoto K, Tokimura F, *et al.* Are bone and serum cefazolin concentrations adequate for antimicrobial prophylaxis? Clin Orthop Relat Res 2011;469(12):3486–3494.

[44] Polk R, Hume A, Kline BJ, Cardea J. Penetration of moxalactam and cefazolin into bone following simultaneous bolus or infusion. Clin Orthop Relat Res 1983;(177):216–221.

[45] Incavo SJ, Ronchetti PJ, Choi JH, *et al.* Penetration of piperacillin-tazobactam into cancellous and cortical bone tissues. Antimicrob Agents Chemother 1994;38(4):905–907.

[46] Boselli E, Breilh D, Biot L, *et al.* Penetration of piperacillin/tazobactam into cancellous and cortical bone tissue. Curr Ther Res 2001;62(7):538–545.

[47] Al-Nawas B, Kinzig-Schippers M, Soergel F, Shah PM. Concentrations of piperacillin-tazobactam in human jaw and hip bone. J Craniomaxillofac Surg 2008;36(8):468–472.

[48] Rana B, Butcher I, Grigoris P, *et al.* Linezolid penetration into osteo-articular tissues. J Antimicrob Chemother 2002;50(5):747–750.

[49] Kutscha-Lissberg F, Hebler U, Muhr G, Koller M. Linezolid penetration into bone and joint tissues infected with methicillin-resistant staphylococci. Antimicrob Agents Chemother 2003;47(12):3964–3966.

[50] Bogehoj M, Emmeluth C, Overgaard S. Blood flow and microdialysis in the human femoral head. Acta Orthop 2007;78(1):56–62.

[51] Traunmuller F, Schintler MV, Spendel S, *et al.* Linezolid concentrations in infected soft tissue and bone following repetitive doses in diabetic patients with bacterial foot infections. Int J Antimicrob Agents 2010;36(1):84–86.

[52] Traunmuller F, Schintler MV, Metzler J, *et al.* Soft tissue and bone penetration abilities of daptomycin in diabetic patients with bacterial foot infections. J Antimicrob Chemother 2010;65(6):1252–1257.

[53] Schintler MV, Traunmuller F, Metzler J, *et al.* High fosfomycin concentrations in bone and peripheral soft tissue in diabetic patients presenting with bacterial foot infection. J Antimicrob Chemother 2009;64(3):574–578.

[54] Garazzino S, Aprato A, Baietto L, *et al.* Glycopeptide bone penetration in patients with septic pseudoarthrosis of the tibia. Clin Pharmacokinet 2008;47(12):793–805.

[55] Blumer JL, Reed MD, Kaplan EL, Drusano GL. Explaining the poor bacteriologic eradication rate of single-dose ceftriaxone in group a streptococcal tonsillopharyngitis: a reverse engineering solution using pharmacodynamic modeling. Pediatrics 2005;116(4):927–932.

Chapter 4

Preclinical Models of Infection in Bone and Joint Surgery

Lorenzo Calabro, R. Geoff Richards, and T. Fintan Moriarty

Introduction

Increasing placement of medical devices, coupled with rising antibiotic resistance amongst bacteria within the hospital and community environments, will ensure that bone and joint infection continues to pose a major challenge to clinicians across numerous medical specialties in the decades to come. Current treatment algorithms have benefited from extensive preclinical studies providing in vivo evidence with regard to antimicrobial selection and dosing. In the near future, further preclinical studies are expected to contribute vital efficacy data for new technologies such as antimicrobial loaded coatings, vaccines, as well as rapid, sensitive, and specific diagnostics. A robust preclinical assessment of any antimicrobial strategy, and safe and expedient implementation of any such technology, relies on a well-designed and clinically relevant in vivo simulation using animal models.

Many animal models of musculoskeletal infection have been described in the literature. However, there remains a dearth of fully standardized or universally accepted reference models. The design variables involved in creating an animal model for bone and joint infection are multiple and inevitably require some compromise. In the orthopedic and in particular in the trauma field, standardization and refinement of fixation methods would clearly improve existing models. Stable, repeatable fixation systems, which mirror clinical practice and allow reliable healing of fractures without complication, should be the starting point for clinically relevant research and development into anti-infective strategies.

Recent advances in the area of custom-made fixation systems for small laboratory animals and the burgeoning availability of tools for longitudinal in vivo monitoring of infection via the use of bioluminescent bacteria are exciting developments in the field. These refinements and others should enable the creation of robust, controlled, and consistent models, which allow strong scientific conclusions with a minimum of harm to animals.

Bone and Joint Infections: From Microbiology to Diagnostics and Treatment, First Edition.
Edited by Werner Zimmerli.

Influence of Species in Preclinical Models of Bone and Joint Infection

A large number of different animal species have been used in preclinical studies as surrogates for bone and joint infection in humans [1, 2]. The central premise in performing a preclinical in vivo trial is that the pathophysiological and therapeutic response in a chosen animal model is sufficiently similar to that in humans to allow valid extrapolation of findings.

The ideal animal model for bone and joint infection research would (i) have molecular, cellular, structural, and mechanical features akin to human bone, (ii) have a size and temperament allowing low-cost maintenance and handling, (iii) have a well-documented genetic and immunological profile, and (iv) be sufficiently robust to endure medical and surgical interventions that reflect current clinical practice. In reality, many of these features are highly variable between species, and effective study design requires awareness of these differences. The clinical reality of many bone and joint infection cases involves chronic infection, implant loosening, long-term antibiotic administration, and multiple surgeries. It is clear that replicating such scenarios in preclinical animal models would inevitably result in a high burden for the animals involved, a burden that may not be justifiable in many preclinical studies. The extent to which a particular model is required to reflect clinical conditions should therefore be based on a considered approach dependent upon the research goals at hand along with consideration of the burden upon the animal.

Historically, the success of animal osteomyelitis models has been determined by the degree to which radiographic, histological, and microbiological outcomes mirror those found in human disease. Host response to infection is dependent on the innate immunity of the host organism, the mobilization of adaptive immune defenses, and bone modeling and remodeling mechanisms, which are all species-dependent. While phenotypic differences have been described and are summarized later, it is important to recognize that their independent impact on the pathophysiology of osteomyelitis is not well-understood. Some aspects of model design, such as animal choice, type and amount of inoculum, and study duration, are invariably based on empirical evidence or for consistency with historical controls, and results are not always predictable. This perhaps accounts for some of the vast and disparate array of models in the literature.

Bone composition and micro- and macrostructure are important determinants of its mechanical properties and vary between species, within species, and between anatomical loci of individual animals [3, 4]. A study by Aerssens *et al.* [3] looked at the bone composition of cows, sheep, chickens, and rats. The mineral content of femoral cortical bone was significantly greater in all four species than in humans, with rats having the highest content. The proportion of collagen content showed the inverse relationship with rats having the lowest content. Micro-architecture of bone is related to remodeling characteristics and also differs between species [5]. Human cortical bone, from the fetal stage onward, shows a high degree of remodeling with a secondary osteonal structure [6]. Nonhuman primates and dogs share a similar microstructure [7], while sheep, pigs, goats, and cows begin life with a plexiform bone structure, and develop only secondary osteons in certain locations later in life [8]. Rodents and rabbits have a primary, lamellar bone structure, and secondary osteons are rare [5, 7].

Another important requirement for an animal model of bone and joint infection is the capability of the test microorganism to cause an infection in the species under study. Human pathogens do not necessarily cause predictable disease in a particular animal or any disease at all [9]. In the case of bone and joint infections, *Staphylococcus aureus* is the pathogen cultured most frequently from clinical osteomyelitis cases [10, 11] and used most commonly in animal osteomyelitis models. *S. aureus* is also an important cause of infection in other mammals [12]. Indeed, its role in humans as colonizer and pathogen seems to be mirrored in the animal kingdom, at least for household pets and horses and in food-producing animals [13]. Comparison of *S. aureus* strains from human and animal clinical isolates revealed that veterinary *S. aureus* infections are caused, to a large extent by genetically and phenotypically distinct strains [14]. The potential implication is that pathogenic strains have evolved separately with different immunological selection pressures exerted by different immune defenses in different species. Furthermore, staphylococcal species such as *Staphylococcus pseudintermedius* appear to be more commonly found as animal pathogens, particularly dogs, cats, horses, and goats [13, 15–17]. *S. pseudintermedius* is a catalase-positive staphylococcus and has been found to result in a large number of infection cases in animals, for example, in tibial plateau-leveling osteotomy in dogs, where it was found to account for over 50% of infections [18].

Interspecies differences in immune response to bacterial invasion are also important to acknowledge in infection trials. With specific reference to the causative microorganisms involved, most animal species used in preclinical research are specific pathogen-free, and may have limited exposure to bacterial infection, except for, perhaps, minor skin abrasions when skin commensals may invade the broken skin. Therefore, most animals enrolled in a bone and joint infection study will have a very limited repertoire of circulating antibodies to the causative pathogen. This is in contrast to the human situation, where it is considered most people will encounter *S. aureus* and many coagulase-negative staphylococci (CNS) throughout their lifetime, and consequently have a significant adaptive immune memory for many staphylococcal antigens. The impact this has on the progression of infection and the importance of this in the clinical situation remain to be determined, and has not been greatly considered in preclinical research.

Murine models clearly represent an attractive option for preclinical investigation into the host response to infection. Their big advantage is the variety of genetically defined mouse strains, both wild type and mutant, the availability of humanized strains, and the great array of molecular biology tools. For example, T helper (TH)–type responses are crucial in the immune response to infection and may be closely controlled in murine studies by careful selection of mouse strains. It has been shown, for example, that C57BL/6 mice exhibit a bias toward the TH1-type immune response and are susceptible to chronic implant-associated *S. aureus* infections (19). In the same experiment, the infection rate amongst the test animals decreased from 100% in C57BL/6 mice to approximately 50 and 25% after 21 and 49 days, respectively, when BALB/c mice with a bias to a TH2 response were used [19, 20] Similar studies would not be possible in other animal species due to the lack of defined genotypes.

Species-specific activity of bacterial toxins has recently emerged as a potential confounding factor in preclinical trials of at least some virulence factors [21]. The *S. aureus* exotoxin Panton–Valentine leukocidin (PVL) is found in a majority of methicillin-resistant *S. aureus* (MRSA) strains that cause community-associated (CA)-MRSA infections, such as necrotizing pneumonia and skin and soft tissue infections [22, 23]. In numerous mouse studies, and in in vitro studies using murine cells, PVL was not found to significantly activate or kill murine neutrophils [24, 25]. It was then discovered in an in vitro study that PVL did activate and kill human and rabbit neutrophils but was inactive against mouse or monkey neutrophils [21]. This specificity for rabbit neutrophils is somewhat corroborated by preclinical studies in vivo, whereby rabbits were in fact found to display a differential response to infection with PVL-positive and corresponding PVL-negative mutants [26–28], which was undetected in the murine models. The reason for the species-specific sensitivity to this toxin is unknown, but varying receptors or signal transduction pathways between species are likely to be responsible. This highlights the importance of species selection, since some animals do not necessarily correctly replicate all facets of *S. aureus* diseases in humans.

With regard to bone and joint infection, and the implant systems used, larger animals such as nonhuman primates, sheep, goats, and dogs are better able to tolerate surgical interventions and their bony geometry can accommodate human-scale prostheses [29]. This is important as mechano-biological variables including fixation stability are known to influence both bone healing and infection susceptibility [30, 31]. Relatively recent technological advancements in small-scale fixation devices have challenged this conception [32, 33]. For example, an internal fixator analogous to a locked plate has been developed for use in mouse femoral osteotomy models [33]. Variations on this implant allow the investigator to choose between stable or flexible fixation, and hence choose between a model with primarily intramembranous or endochondral fracture healing [34]. Garcia *et al.* [32] similarly developed an interlocking nail for mouse femoral models with primary or secondary fracture healing. These devices enable small animal fracture models to better emulate clinical conditions.

Overview of Animal Models

Experimental osteomyelitis models have been created in animals for diverse purposes, by a range of means and with varied results. Common goals include profiling infection parameters such as bacterial virulence factors or the performance of novel diagnostic tools, interventions, or biomaterials. Infection is typically created by bacterial inoculation coupled with a local perturbation in bone physiology. This can be done by an implanted foreign body, experimentally induced ischemia, or administration of a sclerosing agent. Study design in bone and joint infection research aims to reflect the clinical situation and fit within clinical classifications (see Chapter 13). Clinical osteomyelitis can be classified according to Waldvogel *et al.* [35] who distinguished between infection by the hematogenous route or infection locally acquired by exogenous invasion of bacteria. Alternatively, Cierny *et al.* [36] developed a classification based on key indices of surgical decision making. The major disease burden of adult osteomyelitis is related to orthopedic surgery, where risk profiles for patients undergoing elective surgery such as knee arthroplasty are distinct from those with traumatic injuries [37, 38]. Research studies investigating osteomyelitis should also reflect these clinical descriptions. For this reason, we have broadly categorized the available preclinical models based upon their means of bacterial inoculation (hematogenous versus exogenous), and whether controlled trauma beyond the minimum required for surgery was applied (Table 4.1) [39–43].

Direct Inoculation with Minimal Trauma

Most preclinical models of bone and joint infection involve surgical incision followed by placement of a foreign body into the bone. The models included in this section do not create any additional injury such as bone fracture or soft tissue damage secondary to

Table 4.1. Select examples of animal models used for bone and joint infection research classified according to clinical situation.

Clinical situation	Study goal	Implant and species	Trauma	Clinical relevance
Prevention of implant-related osteomyelitis (Lucke *et al.*) [41]	Evaluation of a gentamicin-loaded coating to prevent implant-related osteomyelitis	Kirschner wire in rat tibia	Trauma limited to insertion of implant	This model reflects perioperative contamination of bone, prior to insertion of an implant. The lack of fracture or other trauma simplifies the model and minimizes the risk of complications. The pathology is, nevertheless, reflective of osteomyelitis.
Treatment of implant-related osteomyelitis (Isiklar *et al.*) [39]	Comparative study of different systemic antibiotic regimens for treatment of osteomyelitis	Cancellous screw in rabbit femur	Trauma limited to insertion of implant	This model allows an infection to develop, prior to intervention. Such models allow evaluation of different treatment strategies, which should only be tested against established infections.
Posttraumatic osteomyelitis with a fracture (Schaer *et al.*) [42]	Evaluation of an antimicrobial coating to prevent osteomyelitis in a fracture model	Internal fixation plate in sheep tibia	Unilateral tibial mid-diaphyseal osteotomy	This large animal model uses human implants to fix an osteotomy. The model therefore displays relevant biomechanics, and the septic complications that are observed at the osteotomy site (control group) are reflective of clinical reality.
Posttraumatic osteomyelitis with soft tissue damage (Kälicke *et al.*) [40]	Comparison of the infection rate in animals with and without soft tissue damage	Rat, no implant	Standardized, closed, soft tissue trauma to tibialis anterior muscle	This model does not involve an implant or bone damage, yet shows soft tissue condition will independently influence infection risk.
Hematogenous osteomyelitis (Whalen *et al.*) [43]	Comparison of the incidence of osteomyelitis with and without localized bone damage	Rabbit, no implant	Localized tibial growth plate injury	This is one of comparatively few hematogenous osteomyelitis models in the literature. The model successfully achieves osteomyelitis in test groups, with low infection rates in control animals.

trauma. These models are particularly suitable for research questions regarding prosthetic joint infection.

Early attempts at developing in vivo models of osteomyelitis had found that intravenous (IV) inoculation of bacteria alone to young healthy animals caused inconsistent results [44, 45] and that direct inoculation of *S. aureus* alone into bone failed to create pathology mimicking chronic osteomyelitis [44]. In order to get a consistent and progressive osteomyelitis, a sclerosing agent such as sodium morrhuate (SM) is administered to initiate tissue damage to the bone [44, 45]. Sclerosing agents cause local vascular thrombosis, leading to necrotic tissue formation, and potentiate an infection in bone by limiting the ability of the immune system to respond to the bacteria. Scheman *et al.* [44] first developed a model of osteomyelitis resembling human disease in rabbit tibiae using SM as a sclerosing agent, although radiographic signs due to infection were difficult to distinguish from radiographic signs due to the sclerosing agent The principle of combining inoculation with local ischemia was developed further, using lower doses of SM, which sustained chronic osteomyelitis over longer periods [46] and has subsequently been replicated in rats and goats [47, 48]. The use of sclerosing agents is now considered somewhat controversial due to the unknown effects of the sclerosing agent, which may confound results.

The addition of SM is no longer required in animal models, provided that appropriate pathophysiological experimental conditions are used. Smeltzer *et al.* [49] showed that a chronic osteomyelitis could be generated in an otherwise healthy rabbit, by using a devascularized segment of rabbit radius. The devascularized bone segment was inoculated with *S. aureus*, and it was found that the bone served as a focus for infection propagation, which did not occur to the same extent when the inoculum was added to the empty defect [49]. Equivalent early work by Andriole *et al.* [50] had demonstrated that a foreign body implanted in bone played a similar role, whereby in the presence of a piece of steel within the rabbit tibia, an infection could develop with greater frequency than in the absence of the foreign body. Petty *et al.* [51] confirmed the increased likelihood of infection in the presence of a foreign body and went a step further by evaluating susceptibility with different biomaterials such as bone cement, titanium, and stainless steel. It was found that polymethylmethacrylate (PMMA), the material used as bone cement, was the most likely material to get an infection when placed in the animal prior to polymerization. The increased tissue damage, caused by the PMMA polymerizing in vivo, was probably responsible for this increased susceptibility to infection. It is at least partially due to heat production during polymerization. In a similar study, Cordero *et al.* [52], using a rabbit femoral model, determined that not only the composition but the surface texture of an implant influenced infection. It was found that porous surfaces of cobalt chrome or titanium implants were significantly more likely to get infected than smooth polished surfaces. The porous surface may provide additional bacterial attachment opportunities and microcolony formation within the pores, which are less accessible to host defense mechanisms. More recent examples using similar models of infection have elucidated the role of implant architecture and design along with more subtle effects including surface topography and chemistry [53–55].

Systemic antibiotics are a cornerstone in both prevention and treatment of osteomyelitis. Animal models have been used predominantly to confirm their efficacy, tailor regimens, and characterize the pharmacological parameters involved [56]. For example, the efficacy of rifampicin, a crucial antibiotic in the medical treatment of implant-related bone infection, was described in a subcutaneous tissue cage model using guinea pigs [57]. Because of rapid emergence of resistance, when rifampicin is used as single agent, a combination antibiotic regimen for MRSA was evaluated in this model. The data showed that daptomycin and levofloxacin are particularly effective combination partners that are able to prevent the emergence of rifampin resistance. It should be noted that subcutaneous tissue cages are commonly used as a foreign body infection model. However, this animal model does not consider the special case of bone infection. Nevertheless, it provides preclinical data that facilitate extrapolation to implant-related bone infections.

Schwank *et al.* [58] performed an interesting approach to antibiotic therapy in preclinical testing. The novel approach involved growing bacterial biofilms on small glass beads in vitro and exposing them to antibiotic concentrations based on normal human pharmacokinetics. The authors were able to identify antibiotic combinations that could be shown to result in eradication of biofilm in vitro, and after replicating these scenarios in the guinea pig, there was a correlation between the regimens found to work in vitro with clinical outcome in biofilm infections in vivo.

Numerous animal studies have also investigated the efficacy of local antibiotic delivery vehicles. Over the last two decades, there has been increased focus on local antibiotic delivery. Infection prophylaxis with antibiotic-loaded bone cement, combined with systemic antibiotic administration, is supported by large-scale registry data [59]. The trend toward cementless arthroplasty has driven the search for alternative methods of drug delivery. Technology explored using in vivo models for prophylactic local delivery of antiseptic or antibiotic agents includes collagen sheets [60], calcium phosphate pellets [61], polysaccharide (chitosan) beads, [62] cross-linked high amylose starch implants [63], alginate beads, biodegradable polymer beads and coatings [64–66], cytokine nanocoating [67], and covalently bonded antibiotics [68]. In this situation, the in vivo models used are designed to analyze outcomes such as drug release profiles, biocompatibility, effects on fracture healing, infection susceptibility, and drug resistance.

A rat model designed by Lucke *et al.* [41] to evaluate a gentamicin-impregnated poly-D,L-lactic acid (PDLLA)–coated nail, which has since been approved for clinical use, provides a good example. Kirschner wires were coated with a PDLLA polymer containing gentamicin and implanted in the tibial medullary canal along with 1×10^3 colony forming units (CFU) of *S. aureus.* There was a significant reduction in clinical symptoms in the treatment group compared with controls. In follow-up experiments, using the same rat model, the antibiotic burst release profile and osseous drug concentrations at progressive time points were described. This study was conducted in parallel with a prospective clinical trial of the nail in open tibial fractures [69]. This particular antibiotic-containing medical device is only one of many currently being developed, but this series of studies demonstrates the integral role of the animal model in proving the efficacy of the coating prior to successful introduction into the clinic.

Successfully treating or managing an established active bone or joint infection is a distinctly more challenging undertaking than preventing infection. Biofilm formation, poor osseous perfusion of antibiotics, and abscess formation necessitate a multimodal medical and surgical approach. In vivo models for treatment of infection are correspondingly more complex. At least two surgical procedures are required – one to inoculate, and the second to treat, with an intervening period to allow infection to develop. It is clear that interventions, which are effective in prophylaxis, will not necessarily be effective in treatment. For example, using a canine bone infection model, it was shown that PMMA loaded with gentamicin was successful in preventing the development of an infection. However, it was unable to successfully treat an active infection [70]. The clinical conditions surrounding the treatment of an active infection, including biofilm formation, intracellular bacteria, and tissue necrosis, represent a significantly more challenging target for any antibiotic-loaded biomaterial.

The therapeutic value of degradable local delivery options for established infection has been evaluated in in vivo models with mixed results. Antibiotic-loaded calcium hydroxyapatite implants were used to treat established *S. aureus* osteomyelitis in rabbit tibiae in two separate trials [71, 72]. Even when combined with debridement, neither study demonstrated reliable eradication of infection, although it must be noted that the hydroxyapatite implant performed as well as antibiotic-loaded bone cement when the two were compared [72].

Other authors have sought to refine treatment models by selecting specific strains of bacteria known to form biofilms. Isiklar *et al.* [39] inoculated a biofilm-forming strain of *Staphylococcus epidermidis* into rabbit femora along with a stainless-steel screw to evaluate the efficacy of different antibiotic regimes after a 2-week period to allow the infection to progress. They found that vancomycin alone was ineffective in treating the infection but when combined with rifampicin it appeared to eradicate infection in 90% of subjects.

Animal Osteomyelitis Models Incorporating Trauma

Deep infection following open fractures causes significant clinical morbidity and creates unique management challenges [37]. Fractures or soft tissue trauma add an additional level of complexity to infection models and are incorporated in only a minority of studies, primarily due to the burden upon the experimental animal. Clinical trauma studies are confounded by the heterogeneity of injury mechanisms and fracture patterns, and indeed the first methodological challenge in an animal model is the need for creation of a repeatable injury. Some researchers performed an osteotomy to simulate fracture [42, 73], while others have used instruments such as blunt impact guillotines [74] or three-point bending apparatus to mimic traumatic forces [75].

The potential ethical issues and complications inherent in these models are exemplified in an early study, where efforts were made to reliably induce either *Escherichia coli* or *S. aureus* osteomyelitis in a guinea pig femoral osteotomy [76]. Subject animals were divided into small groups depending on which pathogen was used and whether the fracture was fixed. The authors reported an attrition rate in some groups of greater than 50% and microbiological outcomes, which were potentially confounded by a high rate of fracture nonunion and contaminant bacteria.

Other early models were more successful in larger animals. Rittman and Perren [73] established an early experimental model in sheep to evaluate the impact of fixation stability on healing in an infected fracture. Their priority, reflecting clinical focus at the time, was on histological evidence of primary bone healing, and the model demonstrated that a large implant offering greater stability was more likely to result in fracture union in this setting. Hill *et al.* [77] also used sheep in a model, which confirmed that primary intramedullary nailing in severely contaminated open long bone fractures was unsafe, regardless of debridement. Curtis *et al.* [78] created a goat model to investigate a similar clinical question, with results indicating that external fixation was safer than intramedullary nail fixation in contaminated tibial fractures, although supportive clinical data is required to confirm these findings.

Small animal trauma and osteomyelitis models have also been described. Worlock *et al.* [31] first demonstrated that they could reliably create chronic osteomyelitis in a rabbit tibia osteotomy. They used this model to show the impact of fixation stability on infection susceptibility. The limitation of their experimental design was the unstable construct using a Kirschner wire in the tibia without interlocking bolt. Thus, it was a rather extreme case scenario, not replicating the clinical situation. Lindsey *et al.* [79] reported on a rat femur model, where they were able to create a reproducible closed fracture with a blunt impact guillotine, fix it with a Kirschner wire, and inoculate *S. aureus* to create local infection without overwhelming sepsis. In a study using a rat bone fracture model, it was shown that the timing of antibiotic administration and the timing of surgery affect the rate of infection in wounds contaminated with *S. aureus*. By varying the timing of both interventions, it was shown that early antibiotic therapy was the single most important factor for the risk of infection [80]. A delay in surgery did result in an increase in infection rate, though there was no significant increase between a delay of 6 and 24 h.

Rat femur models with infected critical size cortical bone defects have also been created, allowing the evaluation of osteo-inductive agents and/or osteo-conductive scaffold materials in the context of osteomyelitis. However, stabilizing these defects creates a major challenge [81–83]. Large and perhaps unconventional internal fixation constructs are needed, and in some cases the authors acknowledge that high rates of fixation failure may confound results [82, 84].

Darouiche *et al.* [85] created a model, which is noteworthy as one of the first in vivo models evaluating prophylactic implant coatings in the presence of a fracture. A saw osteotomy was made in rabbit tibiae, which were subsequently fixed with a chlorhexidine- and chloroxylenol-coated, intramedullary Kirschner wire and inoculated with *S aureus*. After 6 weeks, the coated nail group was significantly less likely to have microbiological evidence of implant-related infection than the uncoated control group.

In an example of a study incorporating a soft tissue injury, Kälicke *et al.* [40] demonstrated that the infection rate was significantly increased after a standardized closed soft tissue injury, demonstrating that the soft tissue damage and its pathophysiological consequences result in decreased infection resistance.

Hematogenous Models

Hematogenous osteomyelitis is a particular issue in pediatric medicine where septic arthritis and infection in the adjacent metaphyses of long bones is relatively common [86]. It is also a common cause for late infection in previously well-functioning prosthetic joints, with a recent study showing that the risk of prosthetic joint infection following a *S. aureus* bacteremia was 39% [87]. Acknowledging this distinct etiology, a number of authors have attempted to create models of hematogenous osteomyelitis. *S. aureus* is most commonly used, and a focal bone lesion is typically created either concurrently or prior to inoculation. An early model, created by Deysine *et al.* [88], involved injecting a combination of barium and 5×10^5 CFU of *S. aureus* simultaneously into the tibial nutrient artery of dogs. The authors reliably created acute osteomyelitis, but there was an unacceptably high mortality rate from sepsis. Other similar models in chickens [89] and rabbits [90] encountered the same problem, describing narrow safety margins in inoculum dose when administered systemically. Hienz *et al.* [91], however, were able to create a model using rats with no mortality during the 14-day study period. They injected SM locally into the mandible and tibia of rats and inoculated varying doses of *S. aureus* intravenously into the tail vein to determine the dose required to infect 50% (ID_{50}) and 100% (ID_{100}) of animals. Animals inoculated with bacteria, but spared the focal SM injection, did not develop osteomyelitis, again demonstrating the importance of local perturbation of bone physiology. An alternative approach was taken by Whalen *et al.* [43] who designed a model to mimic pediatric hematogenous osteomyelitis. Using skeletally immature rabbits, they created a partial growth plate fracture at the proximal tibial metaphysis and administered between 7×10^7 and 1×10^8 CFU of *S. aureus* intravenously via an ear vein, reliably causing focal acute osteomyelitis without metastatic infection. Skeletally immature (20–24 weeks) rabbits were used in this study, in order to simulate pediatric hematogenous osteomyelitis. In general, for most preclinical studies, skeletally mature animals are preferred in order to minimize variation between animals and to better replicate clinical cases in skeletally mature adult humans. The hematogenous osteomyelitis model described earlier has since been adapted to mice in order to study T cell differentiation [92], again highlighting the opportunities available for murine studies due to the availability of molecular biology tools that are not always available in larger animal species.

Future Directions

Animal models of osteomyelitis are developed with the primary aim of improving outcomes in clinical medicine. In theory, they allow in vivo evaluation of potential therapies, prophylactics, and diagnostics without the costs, safety, and ethical issues associated with human clinical trials.

Prophylactic strategies in orthopedic surgery, such as systemic and local antibiotic therapy, have evolved over the last four decades. Clinical use of IV antibiotics and antibiotic-loaded bone cement in arthroplasty grew sporadically throughout the 1970s and was followed only later with detailed characterization in controlled animal models [56]. Mainstay surgical techniques in the treatment of established infection such as debridement, stabilization, and lavage have also been characterized after the fact, rather than developed with the help of animal models [93]. The limitations involved with these traditional strategies, and an ever-expanding understanding of bacterial virulence factors, have, however, fueled significant efforts to develop new technology to combat osteomyelitis, which inevitably increases the demand for reliable and valid in vivo models [94].

Promising new approaches to prevent infection include (i) modification of implant surface to deter bacterial adhesion [68, 95], (ii) coating implants with degradable polymers that elute high concentrations of antibiotic into the local milieu (without causing toxic systemic concentrations) [41], (iii) new drugs targeted at either the genes or effector molecules of adhesion [96, 97], quorum sensing [98] or RNA processing [99, 100], and (iv) vaccine development against biofilm-forming bacteria [101, 102].

Modern technologies also have the potential to improve osteomyelitis animal models in the next decade. For example, in vivo longitudinal quantitative monitoring of infection via the use of bioluminescent bacteria offers one such opportunity. Previously, thorough quantitative microbiological or polymerase chain reaction (PCR) evaluation of osteomyelitis required the subject animal to be euthanized. Where data at different time points is required, this necessitates cross-sectional evaluation of groups, and increases the animal numbers required. Li *et al.* [103] characterized the longitudinal progression of tibial osteomyelitis in a murine model, using bioluminescent *S. aureus* and the Xenogen IVIS camera system (Xenogen Co. Alameda, CA). The results were compared to radiography along with high-resolution microcomputed tomography (micro-CT), histology, and PCR after sacrifice, allowing the authors to chart temporal variation in bacterial metabolic activity as well as number. Deep bioluminescent imaging is only available in small animal models, such as mice, due to the lack of penetration of the signal through the significantly deeper soft and hard tissues in large animals. This may be seen as a disadvantage, but as mentioned earlier, the availability of custom-designed, stable fixation options and genetically engineered variants for mice and rats has increased their utility in orthopedic trauma research. Bioluminescent imaging is also suitable for surface measurements from larger animals in real time. For example, Stinner *et al.* [104] used bioluminescent imaging to monitor the surface of large open wounds in a goat model treated either with negative pressure wound therapy (NPWT) or with the bead pouch technique. Again, benefitting from real-time quantification of bacterial burden upon the open wound, it was demonstrated that NPWT may reduce the effectiveness of antibiotic-laden PMMA, by removing the antibiotic from the wound directly. In a similar study, it was also shown that NPWT can reduce bacterial numbers; however, the effect is somewhat species-specific as *Pseudomonas aeruginosa* was more effectively removed than *S. aureus*, when compared with routine dressings [105]. The risk of superinfection with multiple microorganisms including *Candida* species increases if NPWT is applied for several days in patients with an open wound in the presence of implants. Regardless of the controversies of NPWT, it is clear that in vivo bioluminescent imaging of bacteria has massive potential as a tool for real-time monitoring of infection and a potential to significantly reduce the number of animals required in preclinical studies.

Conclusion

Animal models in modern biomedical research are indispensable in the development of novel interventional and diagnostic technologies. The success of any future anti-infective technology will depend upon proper evaluation in appropriate animal models. Robust assessment of the performance of any clinical device may require testing in comparatively low-burden animal models, although higher-burden models, including, for example, fracture creation and localized tissue damage, will be required in certain cases. The development of refined small animal models will enable screening of candidate technologies that are gated at an early stage to reduce the need for more burdensome investigations of any but the most promising candidates. Real-time, in vivo estimation of bacterial burden is also likely to be a key area for a reduction in the number of animals required in the future.

Key Points

- Different animal species may vary in bone structure, susceptibility to infection, adaptive immune response to bacteria, and specificity of bacterial toxins.
- The implant systems available for laboratory animals are improving, with biomechanically defined fracture fixation now available in rodent models.
- Investigators should weigh the importance of clinical relevance versus burden upon the animal when deciding upon the particular model chosen.
- Testing novel prophylactic measures requires different models in comparison with testing of novel treatments for bone and joint infection.
- In vivo monitoring of infection via the use of bioluminescent bacteria is available for small laboratory animals and for open wounds in larger animals. This technology has contributed to a significant reduction in animal numbers required for preclinical research.

References

[1] An YH, Friedman RJ. Animal models of orthopedic implant infection. J Invest Surg Off J Acad Surgical Res 1998;11(2):139–146.

[2] Calabro L, Lutton C, Din A, *et al.* Animal Models of Orthopedic Implant-Related Infection. Biomaterials Associated Infection. New York: Springer; 2013, pp. 273–304.

[3] Aerssens J, Boonen S, Lowet G, *et al.* Interspecies differences in bone composition, density, and quality: potential implications for in vivo bone research. Endocrinology 1998; 139(2):663–670.

[4] Pearce AI, Richards RG, Milz S, *et al.* Animal models for implant biomaterial research in bone: a review. Eur Cell Mater 2007;13:1–10.

[5] Jowsey J. Studies of Haversian systems in man and some animals. J Anat 1966;100(Pt 4): 857–864.

[6] Currey JD. Bones: Structure and Mechanics. Princeton: Princeton University Press; 2006.

[7] Wang X, Mabrey JD, Agrawal CM. An interspecies comparison of bone fracture properties. Biomed Mater Eng 1998;8(1):1–9.

[8] Reinwald S, Burr D. Review of nonprimate, large animal models for osteoporosis research. J Bone Miner Res 2008;23(9):1353–1368.

[9] Legrand N, Ploss A, Balling R, *et al.* Humanized mice for modeling human infectious disease: challenges, progress, and outlook. Cell Host Microbe 2009;6(1):5–9.

[10] Patzakis MJ, Wilkins J, Kumar J, *et al.* Comparison of the results of bacterial cultures from multiple sites in chronic osteomyelitis of long bones. A prospective study. J Bone Joint Surg (American) 1994;76(5):664–666.

[11] Trampuz A, Zimmerli W. Diagnosis and treatment of infections associated with fracture-fixation devices. Injury 2006;37(Suppl 2):S59–S66.

[12] Walker RD, Richardson DC, Bryant MJ, *et al.* Anaerobic bacteria associated with osteomyelitis in domestic animals. J Am Vet Med Assoc 1983;182(8):814–816.

[13] Weese JS, van DuijkerenDE. Methicillin-resistant *Staphylococcus aureus* and Staphylococcus pseudintermedius in veterinary medicine. Vet Microbiol 2010;140(3–4):418–429.

[14] Sung JM, Lloyd DH, Lindsay JA. *Staphylococcus aureus* host specificity: comparative genomics of human versus animal isolates by multi-strain microarray. Microbiology 2008;154 (Pt 7):1949–1959.

[15] Cox HU, Hoskins JD, Newman SS, *et al.* Distribution of staphylococcal species on clinically healthy cats. Am J Vet Res 1985;46(9):1824–1828.

[16] Cox HU, Hoskins JD, Newman SS, *et al.* Temporal study of staphylococcal species on healthy dogs. Am J Vet Res 1988;49(6):747–751.

[17] Vengust M, Anderson ME, Rousseau J, *et al.* Methicillin-resistant staphylococcal colonization in clinically normal dogs and horses in the community. Lett Appl Microbiol 2006;43(6):602–606.

[18] Savicky R, Beale B, Murtaugh R, *et al.* Outcome following removal of TPLO implants with surgical site infection. Vet Comp Orthop Traumatol 2013;26(4):260–265.

[19] Prabhakara R, Harro JM, Leid JG, *et al.* Murine immune response to a chronic *Staphylococcus aureus* biofilm infection. Infect Immun 2011;79(4):1789–1796.

[20] Prabhakara R, Harro JM, Leid JG, *et al.* Suppression of the inflammatory immune response prevents the development of chronic biofilm infection due to methicillin-resistant *Staphylococcus aureus*. Infect Immun 2011;79(12):5010–5018.

[21] Loffler B, Hussain M, Grundmeier M, *et al.* *Staphylococcus aureus* panton-valentine leukocidin is a very potent cytotoxic factor for human neutrophils. PLoS Pathogen 2010;6(1): e1000715.

[22] Gillet Y, Issartel B, Vanhems P, *et al.* Association between *Staphylococcus aureus* strains carrying gene for Panton-Valentine leukocidin and highly lethal necrotising pneumonia in young immunocompetent patients. Lancet 2002;359(9308):753–759.

[23] Lina G, Piemont Y, Godail-Gamot F, *et al.* Involvement of Panton-Valentine leukocidin-producing *Staphylococcus aureus* in primary skin infections and pneumonia. Clin Infect Dis 1999;29(5):1128–1132.

[24] Bubeck WJ, Bae T, Otto M, *et al.* Poring over pores: alpha-hemolysin and Panton-Valentine leukocidin in *Staphylococcus aureus* pneumonia. Nat Med 2007;13(12):1405–1406.

[25] Voyich JM, Otto M, Mathema B, *et al.* Is Panton-Valentine leukocidin the major virulence determinant in community-associated methicillin-resistant *Staphylococcus aureus* disease? J Infect Dis 2006;194(12):1761–1770.
[26] Cremieux AC, Dumitrescu O, Lina G, *et al.* Panton-valentine leukocidin enhances the severity of community-associated methicillin-resistant *Staphylococcus aureus* rabbit osteomyelitis. PLoS One 2009;4(9):e7204.
[27] Diep BA, Palazzolo-Ballance AM, Tattevin P, *et al.* Contribution of Panton-Valentine leukocidin in community-associated methicillin-resistant *Staphylococcus aureus* pathogenesis. PLoS One 2008;3(9):e3198.
[28] Lipinska U, Hermans K, Meulemans L, *et al.* Panton-Valentine leukocidin does play a role in the early stage of *Staphylococcus aureus* skin infections: a rabbit model. PLoS One 2011;6(8):e22864.
[29] Auer JA, Goodship A, Arnoczky S, *et al.* Refining animal models in fracture research: seeking consensus in optimising both animal welfare and scientific validity for appropriate biomedical use. BMC Musculoskelet Disord 2007;8:72.
[30] Claes L, Eckert-Hubner K, Augat P. The effect of mechanical stability on local vascularization and tissue differentiation in callus healing. J Orthop Res 2002;20:1099–1105.
[31] Worlock P, Slack R, Harvey L, *et al.* The prevention of infection in open fractures. An experimental study of the effect of antibiotic therapy. J Bone Joint Surg (Am) 1988;70(9): 1341–1347.
[32] Garcia P, Holstein J, Histing T, *et al.* A new technique for internal fixation of femoral fractures in mice: impact of stability on fracture healing. J Biomech 2008;41:1689–1696.
[33] Matthys R, Perren S. Internal fixator for use in the mouse. Injury 2009;40:S103–S109.
[34] Histing T, Garcia P, Matthys R, *et al.* An internal locking plate to study intramembranous bone healing in a mouse femur fracture model. J Orthop Res 2010;28(3):397–402.
[35] Waldvogel FA, Medoff G, Swartz MN. Osteomyelitis: a review of clinical features, therapeutic considerations and unusual aspects. N Engl J Med 1970;282(4):198–206.
[36] Cierny G, III, Mader JT, Penninck JJ. A clinical staging system for adult osteomyelitis. Clin Orthop Relat Res 2003;(414):7–24.
[37] Patzakis MJ, Wilkins J. Factors influencing infection rate in open fracture wounds. Clin Orthop Relat Res 1989;(243):36–40.
[38] Zimmerli W, Trampuz A, Ochsner PE. Prosthetic-joint infections. N Engl J Med 2004;351(16):1645–1654.
[39] Isiklar ZU, Darouiche RO, Landon GC, *et al.* Efficacy of antibiotics alone for orthopaedic device related infections. Clin Orthop Relat Res 1996;(332):184–189.
[40] Kälicke T, Schlegel U, Printzen G, *et al.* Influence of a standardized closed soft tissue trauma on resistance to local infection. An experimental study in rats. J Orthopaed Res 2003;21(2):373–378.
[41] Lucke M, Schmidmaier G, Sadoni S, *et al.* Gentamicin coating of metallic implants reduces implant-related osteomyelitis in rats. Bone 2003;32(5):521–531.
[42] Schaer TP, Stewart S, Hsu BB, *et al.* Hydrophobic polycationic coatings that inhibit biofilms and support bone healing during infection. Biomaterial 2012;33(5):1245–1254.
[43] Whalen JL, Fitzgerald RH, Jr., Morrissy RT. A histological study of acute hematogenous osteomyelitis following physeal injuries in rabbits. J Bone Joint Surg (American) 1988;70(9):1383–1392.
[44] Scheman L, Janota M, Lewin P. The production of experimental osteomyelitis. JAMA 1941;117(18):1525–1529.
[45] Rodet A. THE CLASSIC: an experimental study on infectious osteomyelitis. 1884. Clin Orthop Relat Res 2005;439:11–12.
[46] Norden CW. Experimental osteomyelitis. I. A description of the model. J Infect Dis 1970;122(5):410–418.

[47] Rissing JP. Animal models of osteomyelitis. Knowledge, hypothesis, and speculation. Infect Dis Clin North Am 1990;4(3):377–390.
[48] Salgado CJ, Jamali AA, Mardini S, *et al.* A model for chronic osteomyelitis using Staphylococcus aureus in goats. Clin Orthop Relat Res 2005;(436):246–250.
[49] Smeltzer MS, Thomas JR, Hickmon SG, *et al.* Characterization of a rabbit model of staphylococcal osteomyelitis. J Orthopaed Res 1997;15(3):414–421.
[50] Andriole VT, Nagel DA, Southwick WO. A paradigm for human chronic osteomyelitis. J Bone Joint Surg (American) 1973;55(7):1511–1515.
[51] Petty W, Spanier S, Shuster JJ, *et al.* The influence of skeletal implants on incidence of infection. Experiments in a canine model. J Bone Joint Surg (American) 1985;67(8):1236–1244.
[52] Cordero J, Munuera L, Folgueira MD. Influence of metal implants on infection. An experimental study in rabbits. J Bone Joint Surg (Br) 1994;76(5):717–720.
[53] Arens S, Schlegel U, Printzen G, *et al.* Influence of materials for fixation implants on local infection. An experimental study of steel versus titanium DCP in rabbits. J Bone Joint Surg Br 1996;78(4):647–651.
[54] Arens S, Eijer H, Schlegel U, *et al.* Influence of the design for fixation implants on local infection: experimental study of dynamic compression plates versus point contact fixators in rabbits. J Orthop Trauma 1999;13(7):470–476.
[55] Moriarty TF, Debefve L, Boure L, *et al.* Influence of material and microtopography on the development of local infection in vivo: experimental investigation in rabbits. Int J Artif Organs 2009;32(9):663–670.
[56] Norden CW. Experimental osteomyelitis. IV. Therapeutic trials with rifampin alone and in combination with gentamicin, sisomicin, and cephalothin. J Infect Dis 1975;132(5):493–499.
[57] John AK, Baldoni D, Haschke M, *et al.* Efficacy of daptomycin in implant-associated infection due to methicillin-resistant *Staphylococcus aureus*: importance of combination with rifampin. Antimicrob Agents Chemother 2009;53(7):2719–2724.
[58] Schwank S, Rajacic Z, Zimmerli W, *et al.* Impact of bacterial biofilm formation on in vitro and in vivo activities of antibiotics. Antimicrob Agents Chemother 1998;42(4):895–898.
[59] Espehaug B, Engesaeter LB, Vollset SE, *et al.* Antibiotic prophylaxis in total hip arthroplasty. Review of 10,905 primary cemented total hip replacements reported to the Norwegian arthroplasty register, 1987 to 1995. J Bone Joint Surg (Br) 1997;79(4):590–595.
[60] Riegels-Nielsen P, Espersen F, Holmich LR, *et al.* Collagen with gentamicin for prophylaxis of postoperative infection. *Staphylococcus aureus* osteomyelitis studied in rabbits. Acta Orthop Scand 1995;66(1):69–72.
[61] Beardmore AA, Brooks DE, Wenke JC, *et al.* Effectiveness of local antibiotic delivery with an osteoinductive and osteoconductive bone-graft substitute. J Bone Joint Surg (Am) 2005;87(1):107–112.
[62] Aimin C, Chunlin H, Juliang B, *et al.* Antibiotic loaded chitosan bar. An in vitro, in vivo study of a possible treatment for osteomyelitis. Clin Orthop Relat Res 1999;(366):239–247.
[63] Huneault LM, Lussier B, Dubreuil P, *et al.* Prevention and treatment of experimental osteomyelitis in dogs with ciprofloxacin-loaded crosslinked high amylose starch implants. J Orthop Res 2004;22(6):1351–1357.
[64] Ambrose CG, Clyburn TA, Louden K, *et al.* Effective treatment of osteomyelitis with biodegradable microspheres in a rabbit model. Clin Orthop Relat Res 2004;(421):293–299.
[65] Diefenbeck M, Mückley T, Hofmann GO. Prophylaxis and treatment of implant-related infections by local application of antibiotics. Injury 2006;37(Suppl 2):S95–S104.
[66] Ueng SW, Yuan LJ, Lee N, *et al.* In vivo study of biodegradable alginate antibiotic beads in rabbits. J Orthop Res 2004;22(3):592–599.
[67] Li BC, Zhang JJ, Xu C, *et al.* Treatment of rabbit femoral defect by firearm with BMP-4 gene combined with TGF-beta1. J Trauma 2009;66(2):450–456.
[68] Antoci V, Jr., Adams CS, Parvizi J, *et al.* The inhibition of Staphylococcus epidermidis biofilm formation by vancomycin-modified titanium alloy and implications for the treatment of periprosthetic infection. Biomaterials 2008;29(35):4684–4690.

[69] Fuchs T, Stange R, Schmidmaier G, *et al.* The use of gentamicin-coated nails in the tibia: preliminary results of a prospective study. Arch Orthop Trauma Surg 2011;131(10): 1419–1425.
[70] Fitzgerald RH, Jr. Experimental osteomyelitis: description of a canine model and the role of depot administration of antibiotics in the prevention and treatment of sepsis. J Bone Joint Surg (Am) 1983;65(3):371–380.
[71] Cornell CN, Tyndall D, Waller S, *et al.* Treatment of experimental osteomyelitis with antibiotic-impregnated bone graft substitute. J Orthop Res 1993;11(5):619–626.
[72] Shirtliff ME, Calhoun JH, Mader JT. Experimental osteomyelitis treatment with antibiotic-impregnated hydroxyapatite. Clin Orthop Relat Res 2002;(401):239–247.
[73] Rittman WW, Perren S. Cortical Bone Healing after Internal Fixation and Fracture Springer-Verlag. Berlin/Heidelberg/New York, 1974.
[74] Bonnarens F, Einhorn TA. Production of a standard closed fracture in laboratory animal bone. J Orthopaed Res 1984;2(1):97–101.
[75] Ashhurst DE, Hogg J, Perren SM. A method for making reproducible experimental fractures of the rabbit tibia. Injury 1982;14(3):236–242.
[76] Passl R, Muller C, Zielinski CC, *et al.* A model of experimental post-traumatic osteomyelitis in guinea pigs. J Trauma 1984;24(4):323–326.
[77] Hill PF, Clasper JC, Parker SJ, *et al.* Early intramedullary nailing in an animal model of a heavily contaminated fracture of the tibia. J Orthopaed Res 2002;20(4):648–653.
[78] Curtis MJ, Brown PR, Dick JD, *et al.* Contaminated fractures of the tibia: a comparison of treatment modalities in an animal model. J Orthopaed Res 1995;13(2):286–295.
[79] Lindsey BA, Clovis NB, Smith ES, *et al.* An animal model for open femur fracture and osteomyelitis: Part I. J Orthopaed Res 2010;28(1):38–42.
[80] Penn-Barwell JG, Murray CK, Wenke JC. Early antibiotics and debridement independently reduce infection in an open fracture model. J Bone Joint Surg (British) 2012;94(1): 107–112.
[81] Chen X, Kidder LS, Lew WD. Osteogenic protein-1 induced bone formation in an infected segmental defect in the rat femur. J Orthop Res 2002;20(1):142–150.
[82] Southwood LL, Frisbie DD, Kawcak CE, *et al.* Evaluation of Ad-BMP-2 for enhancing fracture healing in an infected defect fracture rabbit model. J Orthop Res 2004;22(1):66–72.
[83] Stewart PS, Costerton JW. Antibiotic resistance of bacteria in biofilms Lancet 2001;358(9276):135–138.
[84] Chen X, Tsukayama DT, Kidder LS, *et al.* Characterization of a chronic infection in an internally-stabilized segmental defect in the rat femur. J Orthop Res 2005;23(4):816–823.
[85] Darouiche RO, Farmer J, Chaput C, *et al.* Anti-infective efficacy of antiseptic-coated intramedullary nails. J Bone Joint Surg (Am) 1998;80(9):1336–1340.
[86] Lew DP, Waldvogel FA. Osteomyelitis. Lancet 2004;364(9431):369–379.
[87] Sendi P, Banderet F, Graber P, *et al.* Periprosthetic joint infection following *Staphylococcus aureus* bacteremia. J Infect 2011;63(1):17–22.
[88] Deysine M, Rosario E, Isenberg HD. Acute hematogenous osteomyelitis: an experimental model. Surgery 1976;79(1):97–99.
[89] Emslie KR, Nade S. Pathogenesis and treatment of acute hematogenous osteomyelitis: evaluation of current views with reference to an animal model. Rev Infect Dis 1986;8(6): 841–849.
[90] Poultsides LA, Papatheodorou LK, Karachalios TS, *et al.* Novel model for studying hematogenous infection in an experimental setting of implant-related infection by a community-acquired methicillin-resistant *S. aureus* strain. J Orthop Res 2008;26(10):1355–1362.
[91] Hienz SA, Sakamoto H, Flock JI, *et al.* Development and characterization of a new model of hematogenous osteomyelitis in the rat. J Infect Dis 1995;171(5):1230–1236.
[92] Chadha HS, Fitzgerald RH, Jr., Wiater P, *et al.* Experimental acute hematogenous osteomyelitis in mice. I. Histopathological and immunological findings. J Orthop Res 1999; 17(3):376–381.

[93] Rodeheaver GT, Rukstalis D, Bono M, *et al.* A new model of bone infection used to evaluate the efficacy of antibiotic-impregnated polymethylmethacrylate cement. Clin Orthop Relat Res 1983;(178):303–311.
[94] Montanaro L, Campoccia D, Arciola CR. Advancements in molecular epidemiology of implant infections and future perspectives. Biomaterials 2007;28(34):5155–5168.
[95] Variola F, Vetrone F, Richert L, *et al.* Improving biocompatibility of implantable metals by nanoscale modification of surfaces: an overview of strategies, fabrication methods, and challenges. Small 2009;5(9):996–1006.
[96] Hatamoto M, Ohashi A, Imachi H. Peptide nucleic acids (PNAs) antisense effect to bacterial growth and their application potentiality in biotechnology. Appl Microbiol Biotechnol 2010;86(2):397–402.
[97] Xue-Wen H, Jie P, Xian-Yuan A, *et al.* Inhibition of bacterial translation and growth by peptide nucleic acids targeted to domain II of 23S rRNA. J Pept Sci 2007;13(4):220–226.
[98] Teiber JF, Horke S, Haines DC, *et al.* Dominant role of paraoxonases in inactivation of the Pseudomonas aeruginosa quorum-sensing signal N-(3-oxododecanoyl)-L-homoserine lactone. Infect Immun 2008;76(6):2512–2519.
[99] Eidem TM, Roux CM, Dunman PM. RNA decay: a novel therapeutic target in bacteria. Wiley Interdiscip Rev RNA 2012;3(3):443–454.
[100] Olson PD, Kuechenmeister LJ, Anderson KL, *et al.* Small molecule inhibitors of *Staphylococcus aureus* RnpA alter cellular mRNA turnover, exhibit antimicrobial activity, and attenuate pathogenesis. PLoS Pathogen 2011;7(2):e1001287.
[101] Kuklin NA, Clark DJ, SecoreS, *et al.* A novel *Staphylococcus aureus* vaccine: iron surface determinant B induces rapid antibody responses in rhesus macaques and specific increased survival in a murine S. aureus sepsis model. Infect Immun 2006;74(4):2215–2223.
[102] McCarthy AJ, Lindsay JA. Genetic variation in *Staphylococcus aureus* surface and immune evasion genes is lineage associated: implications for vaccine design and host-pathogen interactions. BMC Microbiol 2010;10:173.
[103] Li D, Gromov K, Soballe K, *et al.* Quantitative mouse model of implant-associated osteomyelitis and the kinetics of microbial growth, osteolysis, and humoral immunity. J Orthop Res 2008;26(1):96–105.
[104] Stinner DJ, Hsu JR, Wenke JC. Negative pressure wound therapy reduces the effectiveness of traditional local antibiotic depot in a large complex musculoskeletal wound animal model. J Orthop Trauma 2012;26(9):512–518.
[105] Lalliss SJ, Stinner DJ, Waterman SM, *et al.* Negative pressure wound therapy reduces Pseudomonas wound contamination more than *Staphylococcus aureus*. J Orthop Trauma 2010;24(9):598–602.

Chapter 5
Native Joint Arthritis in Children

Pablo Yagupsky

Introduction

Bacterial and fungal joint infections in children are medical emergencies. If diagnosis and treatment are delayed or inadequate, severe morbidity, irreversible joint damage, and even fatalities may result. Septic arthritis is more common in childhood than in any other age period, and more than half the cases are diagnosed in individuals younger than 20 years of age.

Epidemiology

The estimated annual incidence of pediatric joint infections in the Western world ranges between 2 and 10 cases per 100,000 and is much higher in the developing world and indigent populations [1, 2]. A male-to-female ratio greater than 1 has been consistently reported in large patients' series [3].

Because over 95% of cases of pediatric septic arthritis are of hematogenous origin, the age distribution of patients with joint infection is markedly skewed, reflecting the increased attack rate of bacteremia in early childhood (Figure 5.1) [3]. The delayed maturation of the T cell–independent arm of the immune system in humans results in impaired production of antibodies to bacterial polysaccharides below 2–4 years of age [4]. Thus, this age group has an increased susceptibility to encapsulated organisms such as *Haemophilus influenzae* type b or pneumococci [5, 6]. On the other hand, the incidence of infectious arthritis in infants younger than 6 months is low, indicating vertically acquired immunity and relative lack of social contacts, resulting in reduced exposure to potential pathogens in early life.

Bone and Joint Infections: From Microbiology to Diagnostics and Treatment, First Edition.
Edited by Werner Zimmerli.

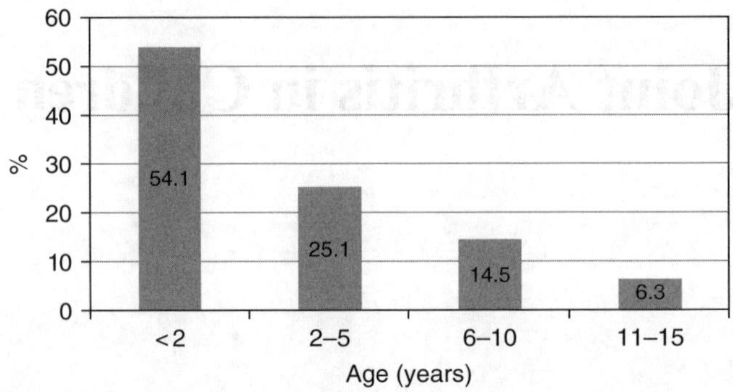

Figure 5.1. Age distribution of children with septic arthritis [5]. (*See insert for color representation of the figure.*)

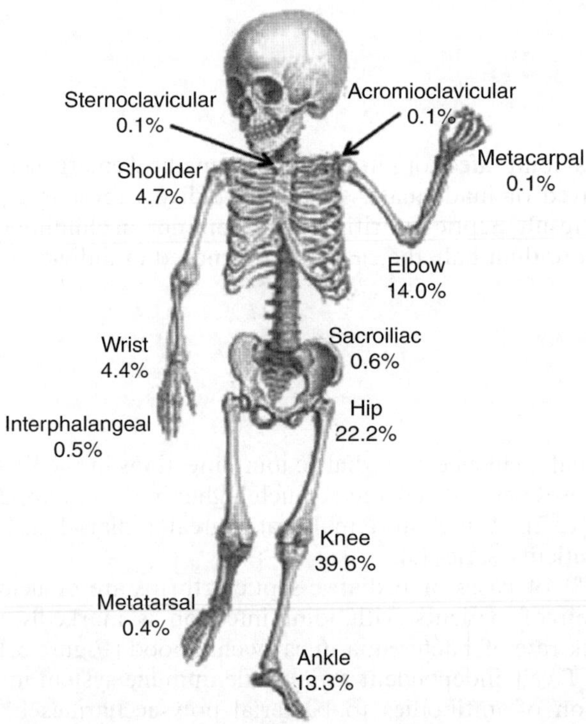

Figure 5.2. Anatomical distribution of 781 septic joints diagnosed in 725 children [5]. (*See insert for color representation of the figure.*)

Microbiology

Specific Microorganisms and Predisposition

Because joint infections in children usually result from hematogenous seeding, the etiology of septic arthritis frequently overlaps with that of pediatric bacteremia (Figure 5.2). However, some microorganisms such as *Staphylococcus aureus* or *Kingella kingae* are remarkably overrepresented in childhood arthritis, indicating joint tissue tropism.

As a rule, pediatric septic arthritis is caused by the intra-articular invasion of a single bacterial (or fungal) species. Isolation of multiple organisms should raise the suspicion of culture contamination, immunodeficiency, intravenous (IV) drug use, or penetrating trauma with direct inoculation of microorganisms into the joint space. The patient's age and presence of associated extra-articular symptoms and signs may provide clues to the likely bacterial etiology, as shown in Tables 5.1 and 5.2.

S. aureus is the most common cause of joint infections in neonates, as well as in children older than 4 years. This organism is characterized by a wide array of virulence factors, skeletal system invasiveness, and genetic determinants of antibiotic resistance. In recent years, methicillin-resistant strains of *S. aureus* (MRSA) are being increasingly detected in many regions, whereas the rate of infection caused by methicillin-susceptible *S. aureus* remains stable. Community-associated MRSA (CA-MRSA) affect patients lacking traditional risk factors for nosocomial MRSA infections and are usually susceptible to antibiotics other than β-lactams [7–9]. Infections with CA-MRSA involve skin, soft tissues, the lung, and the skeletal system and are characterized by remarkable tissue destruction.

Streptococcus pyogenes (group A *Streptococcus*) is isolated in 10–20% of preschool and early school children with septic arthritis, and is especially common in patients with concomitant skin infections or chickenpox [10].

Table 5.1. Etiology of pediatric hematogenous septic arthritis by age group.

Age	Organism
<2 months	*Staphylococcus aureus*
	Streptococcus agalactiae
	Enterobacteriaceae
	Candida species[a]
	Coagulase-negative staphylococci[a]
	Neisseria gonorrhoeae
2 months to ≤2 years	*Kingella kingae*
	Haemophilus influenzae[b]
	Staphylococcus aureus
	Streptococcus pneumoniae[b]
	Streptococcus pyogenes
2–4 years	*Staphylococcus aureus*
	Streptococcus pyogenes
	Kingella kingae
5–15 years	*Staphylococcus aureus*
	Streptococcus pyogenes
>15 years	*Staphylococcus aureus*
	Neisseria gonorrhoeae[c]

[a]In premature babies with indwelling vascular catheters.
[b]Among unvaccinated and incompletely vaccinated children.
[c]In sexually active adolescents.

Table 5.2. Etiology of pediatric septic arthritis and associated clinical conditions.

Associated condition	Possible etiology
Skin	
Cellulitis, erysipelas	*Staphylococcus aureus*, *Streptococcus pyogenes*
Varicella lesions	*Streptococcus pyogenes*, *Kingella kingae*
Erythematous rash	*Streptococcus pyogenes*
Erythema migrans	*Borrelia burgdorferi*
Petechial rash	*Neisseria meningitidis*
Mucosae	
Gingivostomatitis	*Kingella kingae*
Urethritis/genital infection	*Neisseria gonorrhoeae, Ureaplasma* spp., *Mycoplasma hominis*
Respiratory system	
Pneumonia	*Staphylococcus aureus*, *Streptococcus pneumoniae*, *Haemophilus influenzae* type b
Necrotizing pneumonia	*Staphylococcus aureus* (especially community-acquired MRSA)
Cardiovascular	
Endocarditis	*Staphylococcus aureus*, *Kingella kingae*
Central nervous system	
Meningitis	*Streptococcus agalactiae*, *Streptococcus pneumoniae*, *Haemophilus influenzae* type b, *Neisseria meningitidis*, *Borrelia burgdorferi*
Hepatosplenomegaly	*Brucella* spp.
Multifocal involvement	*Staphylococcus aureus*, *Haemophilus influenzae* type b, *Neisseria gonorrhoeae*, *Brucella* spp.
Hemoglobinopathies	*Salmonella enterica*, *Streptococcus pneumoniae,* other *Enterobacteriaceae*
Immune system	
HIV infection	*Staphylococcus aureus*, *Streptococcus pneumoniae*, *Mycobacteria*, *Nocardia* spp., fungi
X-linked agammaglobulinemia	Encapsulated bacteria, *Mycoplasma* and *Ureaplasma* spp.
Common variable immunodeficiency	*Mycoplasma* and *Ureaplasma* spp.
Chronic granulomatous disease	*Staphylococcus aureus*, *Serratia marcescens*, *Pseudomonas aeruginosa*, nontuberculous mycobacteria, *Aspergillus* spp.
Addictions	
Intravenous drug users	*Staphylococcus aureus*, *Pseudomonas aeruginosa*

Streptococcus agalactiae (group B *Streptococcus*) is diagnosed almost exclusively in the neonatal period [11]. Remarkably, *S. agalactiae* arthritis commonly affects the shoulder. In children born in breech presentation, it typically involves the hip joint, suggesting that trauma and local hyperemia in the course of bacteremia facilitate seeding of the organism into the articular space.

Streptococcus pneumoniae arthritis is most common between the ages of 6 months and 2 years [12, 13]. In countries, where the conjugate pneumococcal vaccine has been introduced, the incidence of invasive *S. pneumoniae* diseases, including those affecting the skeletal system, has substantially decreased.

H. influenzae type b was the most common cause of septic arthritis in children younger than 2 years prior to the advent of the conjugate vaccine, accounting for almost one-half of the cases. Nowadays, the disease has become rare in countries where immunization coverage is high [14, 15]. Children with *H. influenzae* type b arthritis frequently present with other foci of infection such as meningitis (in 30% of patients), osteomyelitis (in 22%), cellulitis (in 30%), pneumonia (in 4%), and otitis media (in 35%) [14, 15]. Arthritis due to non–type b, nonencapsulated *H. influenzae* and *Haemophilus* species other than *influenzae* almost exclusively affects immunocompromised children [16].

In recent years, increasing use of blood culture vials for seeding skeletal system exudates and nucleic acid amplification assays has resulted in the recognition of *K. kingae*, a Gram-negative member of the normal pharyngeal flora, as the most common etiology of septic arthritis below 3–4 years of age [17, 18]. Antecedent or concomitant stomatitis and/or signs of an upper respiratory tract infection are frequent, suggesting that invasion of the bloodstream and dissemination of the organism to the joints is facilitated by breaching of the mucosal layer by an intercurrent viral disease.

Although the incidence of arthritis in invasive meningococcal disease is as high as 14%, true invasion of the joint by *Neisseria meningitidis* is uncommon [19]. In most cases, signs of joint inflammation develop several days after initiation of antibiotic therapy and the synovial fluid is usually sterile, suggesting an immune complex–mediated phenomenon [19]. Recurrent disease, a prolonged course, isolation of uncommon *N. meningitidis* serogroups, and family clustering of cases should raise the possibility of complement or properdin deficiencies [20].

Neisseria gonorrhoeae becomes common in sexually active adolescents, and its isolation in children beyond the neonatal period is a definitive proof of sexual abuse. Rarely gonococci may infect the joint in the course of a disseminated disease in neonates born to infected mothers [21].

Invasion of the joint space by *Salmonella enterica* has been reported in children suffering from sickle cell anemia and other hemoglobinopathies and in those living in poverty in developing world countries. Other *Enterobacteriaceae*, and especially *Escherichia coli* and *Klebsiella pneumoniae*, are associated with suppurative arthritis in the neonatal period and in immunocompromised patients [22].

Pseudomonas aeruginosa is a rare cause of septic arthritis in the general pediatric population. However, it may cause joint infection in neonates, in patients with vascular catheter-related infection, immunodeficient children, and IV drug-using adolescents [23].

As the result of effective public health measures, human brucellosis has been eradicated from most Western world countries. In children who are arthritis residents of endemic countries (Latin America, the Middle East, the Mediterranean Basin, Eastern Europe, Asia, and Africa), and travelers returning from these regions, the possibility of brucellar arthritis should be considered (see Chapter 16). The disease is characterized by pain, limited mobility, and swelling, whereas local redness or warmth are rarely found. Brucellosis usually affects the weight-bearing articulations, specially the hip (in half of the cases). Involvement of multiple joints is seen in one quarter of patients [24].

Lyme disease should be included in the differential diagnosis of children exposed to ticks in endemic areas who present with arthritis involving large joints (with the noticeable exception of the hip). Migratory arthralgia is present in 18% of children with *Borrelia burgdorferi* infections and frank arthritis in 10% [25]. The clinical presentation of Lyme arthritis is typically milder than that induced by pyogenic bacteria. Despite the presence of impressive joint inflammation and large effusions, children do not look ill, motion is possible, fever is absent in half of the cases, and the leukocyte counts are within normal limits [25].

Septic arthritis caused by *Mycoplasma* spp. and *Ureaplasma* species is almost exclusively detected in patients with X-linked agammaglobulinemia, common variable immunodeficiency, or after organ transplantation [26].

Hematogenous septic arthritis caused by anaerobic organisms is exceptionally seen in children and is usually caused by a single bacterial species, generally a Gram-negative bacillus. Whenever a penetrating wound or bite is the mechanism of infection, multiple organisms, including both aerobes and anaerobes, may be isolated in the joint fluid culture [27].

Arthritis is the most common manifestation of tuberculosis of the skeletal system after Pott's disease [28]. Usually, *Mycobacterium tuberculosis* bacilli are seeded in synovial tissues by the hematogenous route during a primary infection. More rarely, the disease spreads from a contiguous focus such as invasion of the atlantoaxial joint from an apical pulmonary infection. Tubercular arthritis is monoarticular in 90% of cases and, although it can affect virtually any joint, usually involves the hip or knee [28]. Constitutional symptoms, such as fever and weight loss, occur in only a minority of children. Granulomatous changes and cartilage erosion result in chronic effusion and progressive joint destruction. Signs of acute inflammation are frequently absent, whereas local deformity and restricted motion range are typically observed.

Rat-bite fever is a rare zoonosis caused by two members of rodents' oral flora, by *Streptobacillus moniliformis*, mostly in Western countries and Australia, and by *Spirillum minus* in Asia [29]. The site of inoculation of the disease usually heals before a septicemic disease, frequently characterized by fever and rash, develops. Arthritis involving multiple joints is commonly seen in rat-bite disease caused by *S. moniliformis* but is rare in spirillar infection. Culture of the synovial fluid exudate is frequently negative, suggesting a reactive mechanism [29].

Candida species and coagulase-negative staphylococci are pathogens of low virulence that can cause infectious arthritis in premature babies and neonates in the intensive care setting and in young infants with indwelling vascular catheters [30].

Culture-Negative Septic Arthritis

On average, in 33% of children with presumptive joint infections, blood and synovial fluid cultures reveal no growth [5], with percentages ranging between 16% [31] and 60% [32]. This wide variation reflects not only differences in the sensitivity of the microbiological methods, but also the wide array of inclusion criteria employed in the different studies or previous administration of antibiotic therapy [33]. The epidemiological profile and clinical presentation of children with negative cultures is similar to that of those with culture-confirmed arthritis. However, in patients with culture-negative disease, a trend toward younger age, lower body temperature, leukocyte counts, and C-reactive protein (CRP) values on admission, a milder clinical course, shorter hospital stay, and better prognosis have been consistently reported [34–36]. It is to be expected that improved microbiological culture methods and widespread use of nucleic acid amplification assays will reduce or even eliminate these ill-defined cases altogether.

Pathogenesis

The synovial membrane is highly vascular and lacks a limiting basement membrane, enabling easy bacterial access to the joint space in the course of a bacteremic episode. Once organisms have penetrated into the joint, the low fluid shear conditions facilitate microbial adherence [37]. Uncommonly, pediatric septic arthritis may also result from direct inoculation of organisms in the joint by human or animal bites, joint taps especially with injection of corticosteroids, or surgical procedures.

Invasion of the joint space in neonates occurs in the majority of cases as a result of dissemination of infection from a contiguous metaphyseal focus of osteomyelitis [38]. In young children, the cartilaginous epiphyses receive their blood supply from a metaphyseal capillary network that obliterates between 6 and 9 months of age. Therefore, infection of a metaphyseal site can easily spread across the growth plate to the epiphysis and joint space. Because in older children the epiphyses and metaphyses have separate blood supply and only the metaphyses of the hip, shoulder, and ankle bones remain intracapsular, spread of infection from bone to joints becomes less common [38]. Occasionally, the neonatal joint may be directly invaded during a bacteremic episode and, in the hospital setting, by direct inoculation of skin organisms during a femoral venipuncture [39]. The source of the preceding bacteremia may be the result of nosocomial transmission of virulent *S. aureus*, the newborn's normal skin flora, or acquisition of maternal organisms when delivered through a birth canal colonized by *Enterobacteriaceae*, *S. agalactiae*, or *N. gonorrhoeae* [5, 11, 21].

Bacteria implicated in septic arthritis usually display a variety of surface-exposed receptors that recognize adhesive matrix molecules, such as collagen, fibronectin, and elastin, facilitating invasion by firmly anchoring the organism to the synovial layer [37]. Local trauma may unveil these tissue components, promoting bacterial adherence and increasing the risk of suppurative arthritis. Inactivation of the genes encoding bacterial adhesins significantly reduces the capability of the organisms to establish a joint infection [37].

In some cases, the immune system may contain incipient joint infections caused by low-grade pathogens such as *K. kingae*, resulting in a transient arthralgia suggestive of an abortive course [17]. More commonly, microorganisms multiply in the joint space, and the infection advances to full-blown arthritis.

Both bacterial factors and the host's immune response contribute to the progressive destruction of joint tissues [37]. Bacteria such as *S. aureus* release potent cytotoxins, are internalized by osteoblasts causing apoptosis, or evade the immune response by surviving and multiplying in the intracellular milieu. Presence of bacteria in the joints induces a strong inflammatory response consisting of proliferation of the synovial cells, leukocyte migration, and formation of granulation tissue and abscesses. Synoviocytes and infiltrating leukocytes release proteases and secrete cytokines such as interleukin-1-β, interleukin-6, and tumor necrosis factor-α [37]. These cytokines activate an inflammatory cascade releasing acute-phase reactants from the liver, such as CRP, that adhere to invading bacteria and facilitate opsonization and complement activation. On the other hand, cytokines increase the release of host matrix metalloproteinases, such as stromelysin, and other collagen-degrading enzymes. The inflammatory process triggers fluid accumulation, increasing intra-articular pressure and inducing tissue ischemia and necrosis [37]. The resulting cartilage destruction causes narrowing of the joint space and further erosive damage, leading to disabling orthopedic sequelae.

Clinical Presentation

Typically, septic arthritis exhibits a more acute presentation than osteomyelitis, and most children with joint infections are brought to medical attention within 2–5 days from the onset of symptoms. Hematogenous pediatric septic arthritis affects a single joint in 95% of cases. Involvement of multiple articulations suggests a viral, reactive, or inflammatory arthropathy or an immunocompromising condition. Polyarticular septic arthritis, however, has been noted in neonates, in half of the cases caused by gonococci, in 7% of those caused by *S. aureus* or *H. influenzae* type b, and in infections by *Candida* species [5].

Septic arthritis usually affects the large weight-bearing joints of the lower extremities (Figure 5.3). Small joints of the hand and feet are overrepresented in *K. kingae* infections,

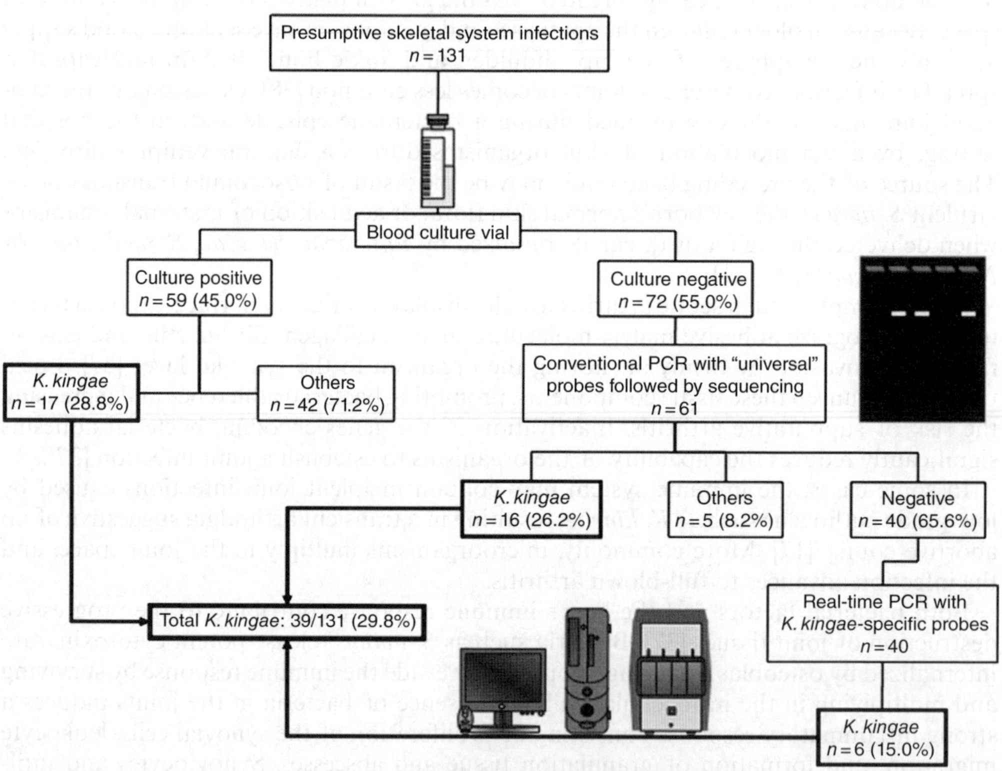

Figure 5.3. Combined use of the blood culture vial technique and nucleic acid amplification assays for diagnosing *Kingella kingae* arthritis [18]. (*See insert for color representation of the figure.*)

the sacroiliac joints are typically affected in brucellosis, and the sternoclavicular joint may be invaded by *K. kingae* in young children and by *P. aeruginosa* in intravenous drug users (IVDU) [23], and as a rare complication of subclavian vein catheterization [40] (see Chapter 7).

Most children with septic arthritis present with acute onset of fever and local inflammatory changes, such as swelling or localized erythema of the overlying skin. Irritability, pain, abnormal (antalgic) posture, restricted range of motion or refusal to move the affected extremity or bear weight, and limping are frequent complaints. The pain of untreated septic arthritis is continuous and progressive, in contrast to inflammatory arthropathies such as juvenile idiopathic arthritis, where symptoms worsen upon rising in the morning.

Infected joints are splinted by muscle contraction to limit motion and reduce pressure and the resulting pain. When the hip joint is involved, the extremity is held in flexion, external rotation, and abduction, the infected knee or ankle in slight flexion, and the shoulder in adduction and internal rotation. While examining the child, it should be kept in mind that arthritis of the hip is frequently difficult to localize and patients may present with pain referred to the knee or anterior thigh [5]. Patients with sacroiliitis exhibit a positive FABERE (flexion, abduction, external rotation, extension) test. Painful palpation of the joint may also be elicited by direct compression of the iliac wing or by digital dorsal compression in rectal examination. Newborns and young patients infected with low-grade virulence pathogens such as *K. kingae* or *Brucella* spp. may be afebrile at the time of diagnosis, requiring an increased awareness of the possibility of a joint infection [17, 24]. In neonates, and especially in premature babies, the clinical picture may be dominated by nonspecific signs such as poor feeding, vomiting, abdominal distention, tachycardia, tachypnea, hypothermia, irritability or apathy, hypotension, poor perfusion, and acidosis [30]. Meticulous physical examination may disclose limited use of an extremity or pseudoparalysis, and subtle signs of local inflammation over the affected joint, such as discomfort when handled or having the diaper changed, or swelling of the buttock, genitalia, thigh, or the entire extremity. In addition to obtaining synovial fluid specimens for culture, a complete sepsis workout, including obtaining blood and urine cultures and performance of a lumbar puncture, are indicated before administering empiric broad-spectrum antimicrobial therapy (Figure 5.4).

Laboratory Investigation

The key to the diagnosis of bacterial arthritis in children is a high index of clinical suspicion. The diagnosis should be confirmed without delay by aspiration of the joint, performed with a large-bore needle (20 gauge or larger). A comprehensive microbiological, biochemical, and cytological study of the synovial fluid should be ordered [41, 42]. However, the only definitive proofs of an infectious etiology of the joint inflammation are demonstration of bacteria in the Gram's stain, growth of an irrefutable pathogen in culture, or detection of pathogen-specific DNA sequences by a nucleic acid amplification test.

If synovial fluid cannot be obtained by close needle aspiration, the procedure should be attempted again with imaging guidance, especially for sites that are not easily accessible such as the hips, shoulders, or sacroiliac joints [42]. Aspiration of an amount of fluid insufficient for an extensive laboratory workup is common in young children with arthritis or when a small joint is drained. In this case, performance of a Gram's stain,

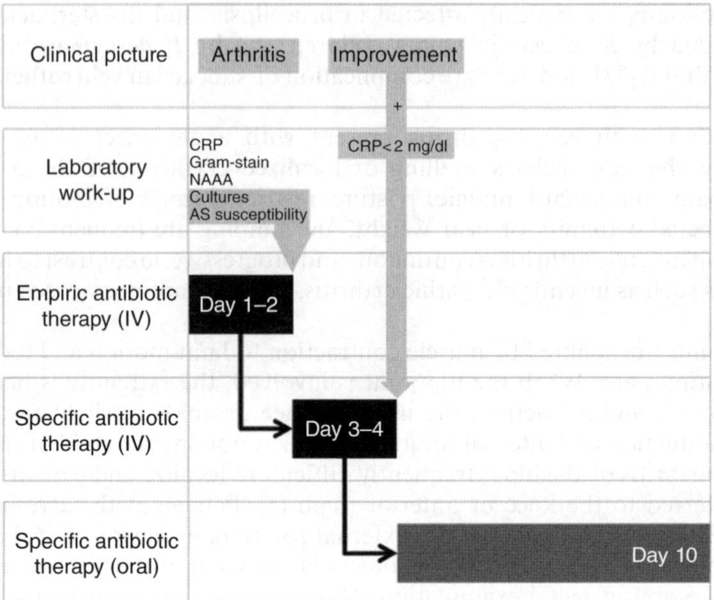

Figure 5.4. Shortened protocol for guiding antibiotic therapy in uncomplicated pediatric septic arthritis [52, 55, 65]. AB, antibiotic; CRP, C-reactive protein; NAAA, nucleic acid amplification assays. (*See insert for color representation of the figure.*)

inoculation of a blood culture vial, and setting of a polymerase chain reaction (PCR) assay employing "universal" bacterial primers are probably the best diagnostic options. Whenever bacteriological laboratory services are not readily available, the aspirated specimen should be inoculated bedside into a pediatric blood culture vial.

Any cloudy joint effusion should be considered infectious until proven otherwise. Although acute rheumatic fever, Reiter's disease, and juvenile idiopathic arthritis can cause a markedly inflammatory synovial fluid, the highest leukocyte counts are seen in patients with septic arthritis, usually in the 50,000-200,000 cells per mm^3 range, of which more than 90% are polymorphonuclear leukocytes. Leukocyte counts higher than 50,000/mm^3 are generally proposed as a cutoff to differentiate septic arthritis from noninfectious joint exudates. Yet lower counts may be seen in infections caused by Gram-negative organisms such as *N. gonorrhoeae*, *K. kingae*, and *Brucella* species early in the course of bacterial arthritis of any etiology and in neutropenic patients [43, 44]. Conversely, leukocyte counts greater than 50,000/mm^3 of synovial fluid may be observed in children with juvenile idiopathic arthritis, serum sickness, or reactive arthritis.

A low synovial glucose concentration (<1.7–2.2 mmol/l) is suggestive of infection, but the sensitivity of this criterion is only 50%, and a low glucose level can also occur in patients with juvenile idiopathic arthritis. Measurements of the synovial fluid protein or lactate contents are neither sensitive nor specific for bacterial arthritis [45].

The aspirate should be transported to the microbiology laboratory without delay in the original syringe or in a sterile tube. Use of swabs, although inexpensive and easy to use, should be discouraged. They are more likely to be contaminated; certain fibers, such as cotton, may inhibit bacterial growth, and organisms may remain adherent to swabs resulting in a false-negative Gram's stain examination and reducing the culture's sensitivity [43].

A Gram's stain should be prepared from a centrifuged synovial specimen and carefully examined. The test is positive in 75% of patients with staphylococcal arthritis, but in less than half of those infected by Gram-negative organisms [41], probably because of a lower bacterial load and the difficulties in recognizing the presence of bacteria against the pink-stained fibrin and cell background.

The fluid should be seeded onto appropriate media (including a chocolate agar plate), and incubated in a CO_2-enriched atmosphere to enable growth of capnophilic bacteria, such as pneumococci or neisseriae. Inoculation of a pediatric blood culture vial, and preferably of one containing antibiotic-binding resins such as the BACTEC 9240 Peds Plus bottle [46] or the BacT/Alert vial [47], is also recommended because this method significantly improves the recovery of fastidious organisms such as *K. kingae* and is recommended for patients receiving antimicrobial therapy. Anaerobic cultures are not routinely indicated in children unless there are associated risk factors such as penetrating wounds or bites.

Nowadays, nucleic acid amplification assays and DNA sequencing are making a profound impact in the management of septic arthritis. This novel technology improves detection of difficult-to-culture organisms, allows diagnosis in patients already treated with antibiotics, reduces time to detection, and enables precise identification of unusual pathogens [48]. Most clinical microbiology laboratories employ amplification and sequencing of the *16S rDNA* gene. As an alternative to this approach, species-specific primers that detect the most plausible pathogens (such as *K. kingae* in young children or *N. gonorrhoeae* in a sexually active adolescent) may be used. This strategy shows improved sensitivity compared to the use of universal primers, but requires epidemiological and clinical expertise to choose the most adequate primers.

Blood cultures should be obtained in all children with suspected suppurative arthritis, not only because of convenient accessibility of the specimen compared to obtaining a joint aspirate, but also because the etiologic agent may be recovered from the bloodstream in up to 50% of cases, even when cultures of synovial exudates are sterile [41]. When a gonococcal infection is suspected in an adolescent, cultures from the cervix, urethra, rectum, and oropharynx, as well as sensitive nucleic acid amplification tests of the urine, should also be performed [21].

Leukocyte counts, erythrocyte sedimentation rate (ESR), CRP, and procalcitonin levels are usually elevated in pediatric patients with septic joints. However, infections caused by low-grade pathogens such as *K. kingae* or *Brucella* species are frequently accompanied by normal or only moderately elevated laboratory markers [17, 49]. The sensitivity of CRP and ESR for diagnosing septic arthritis reaches greater than 90% but only at low thresholds (≥ 20 mg/l and ≥ 20 mm in the first hour, respectively) [50, 51]. CRP increases and decreases more rapidly than ESR and, therefore, sequential measurement of CRP levels is preferred to assess the response to antimicrobial treatment and switch from parenteral to oral antibiotics. Normalization of the CRP levels to less than 20 mg/l within 10 days characterize a favorable clinical course, whereas persistently elevated levels may represent therapeutic failure and help early recognition of complications [52]. The ESR, on the contrary, may continue to increase for a few days even in adequately treated patients, and requires approximately 1 month to normalize [51].

Serum procalcitonin values of 0.2–0.3 ng/ml have 90% sensitivity and 87% specificity for distinguishing between septic versus nonseptic arthritis [53]. Higher cutoff values (0.5 ng/ml) reduce sensitivity to 46% with no substantial specificity gain (91%). This biomarker has also been measured in synovial fluid, and, at a cutoff level of 1.5 μg/l, 85% sensitivity and 62% specificity have been estimated [54]. However, the test is not widely available at the point of care, its measurement requires more time, and it is more expensive than CRP [55].

Imaging Studies

Imaging studies are not diagnostic for septic arthritis, but are helpful in detecting concomitant osteomyelitis and excluding other conditions. The initial radiographic examination is usually normal or may reveal soft tissue swelling, displacement of the muscles around the joint, widening of the joint space with or without luxation, or osteolytic changes suggesting contiguous osteomyelitis. However, it should be pointed out that plain radiographs have a low sensitivity for detecting joint effusions. When suppurative infection of the hip is suspected, radiographs should be obtained in the "frog-leg position" as well as with the legs extended at the knees and slightly internally rotated. Obliteration of the gluteal fat lines by accumulation of deep-seated edema and displacement of the femoral head laterally and upward and of the obturator internus muscle medially by a distended joint capsule would support the diagnosis [5].

Ultrasound has the advantage of being a noninvasive technique and is especially helpful for deep joints such as the hip, although its performance is operator-dependent. Ultrasound may detect accumulation of intra-articular fluid and guide performance of a diagnostic joint aspiration. Neither the size nor the echogenicity, however, allows one to firmly conclude whether the detected effusion is infected or not. Conversely, absence of joint fluid accumulation in the hip joint space helps to exclude septic arthritis.

Bone scans may be used to localize the joint affected when in doubt and detect unsuspected multifocal disease. The characteristic finding in septic arthritis consists of an increased uptake on both sides of the joint during the early phase, whereas in osteomyelitis, unilateral increased uptake is observed [56]. It should be pointed out that interpretation of bone scans is difficult in neonates.

Magnetic resonance imaging (MRI) provides high contrast and resolution and multiplanar imaging. These technical features result in superb definition of the extent of soft tissue and cartilage involvement and superior detection of an associated focus of osteomyelitis, especially in neonates and young infants. The examination may reveal effusion and periarticular edema, synovial thickening, joint debris, as well as extension of the process to the contiguous bones (erosions or abscesses). The technique may be also useful in differentiating suppurative arthritis from transient synovitis of the hip. Demonstration of contralateral (asymptomatic) joint effusion and signal intensity alterations and enhancement in surrounding soft tissues are associated with transient synovitis [57], while edema of the adjacent bone is exclusively associated with septic arthritis [58]. Use of MRI, however, has the limitations of costs and availability, and the need to sedate the young and uncooperative child [59].

Differential Diagnosis

The differential diagnosis of septic arthritis is wide and depends on the patient's age, underlying clinical conditions, and the joint involved (Table 5.3).

Table 5.3. Differential diagnosis of pediatric septic arthritis.

Osteomyelitis
Viral arthritis
Reactive arthritis
Transient synovitis
Rheumatic fever
Bone infarction
Autoimmune arthritis
Septic bursitis
Sickle cell anemia
Slipped capital femoral epiphysis
Perthes disease
Villonodular synovitis
Spondylodiscitis
Psoas abscess
Serum sickness
Henoch–Schönlein purpura
Hemophilia
Trauma
Familial Mediterranean fever
Chronic recurrent multifocal osteomyelitis
Leukemia
Neuroblastoma

Treatment

Because of the potential risk for long-term disabilities caused by the rapid destruction of the joint cartilage, septic arthritis should be considered a true emergency. Optimal management of suppurative arthritis of childhood requires a combination of medical and surgical interventions that is accomplished best with the joint efforts of pediatricians and experienced orthopedic surgeons. A high index of clinical suspicion, drawing blood and synovial fluid samples for bacteriological diagnosis, prompt joint drainage, and administration of adequate antimicrobial therapy are the cornerstones of an optimal treatment.

Joint Space Drainage

Evacuation of the joint space by close needle aspiration, arthroscopy, or surgical drainage provides synovial fluid samples for diagnostic purposes, reduces intracapsular pressure relieving pain, and removes bacteria and cartilage-damaging toxin products [60]. There is still controversy over the best mode of drainage, but most researchers have concluded that repeated aspirations of joints other than the hip are associated with a superior outcome as compared to arthrotomy [41, 61, 62]. However, because of the retrospective nature of most studies, a selection bias in which more clinically severe cases could have been selected for open surgery and, naturally, would have experienced a worst outcome, cannot be excluded. The customary indications for surgical drainage are summarized in Table 5.4.

Table 5.4. Indications for arthrotomy in pediatric septic arthritis.

Arthritis of the hip and shoulder
Presence of large amounts of fibrin, debris, or loculation within the joint space
Presence of an implant
Arthritis not responding to medical treatment within 3 days
Adjacent osteomyelitis

Because the femoral head receives its blood supply from a single intra-articular arterial branch, accumulation of pus within the hip joint may result in increased pressure and vascular occlusion, compromising bone tissue viability. Avascular necrosis of the femoral head, joint instability, premature physeal closure, and limb-length discrepancy are common complications of delayed or inadequate treatment of hip joint infections, necessitating complex corrective surgery, usually with dismal results [63]. Therefore, prompt open surgical drainage has been traditionally recommended [64]. In recent years, performance of ultrasound-guided aspiration of the joint space under local anesthesia alone or in combination with sedation has been advocated instead [65]. After the aspiration, the joint is irrigated using the same needle, and the procedure is repeated daily for 3–5 days. Only 4 of 28 patients treated with this modality did not improve and required surgical drainage, and no complications were detected after a mean follow-up of over 7 years [63]. In a second study, three-directional arthroscopic drainage and lavage were performed in children older than 6 years with staphylococcal hip arthritis, with excellent functional outcomes [66]. Despite these encouraging results, additional experience with this unorthodox treatment is needed before surgery-sparing approaches can be routinely recommended for pediatric suppurative arthritis of the hip.

Antibiotic Therapy

The choice of the initial antibiotic therapy (pending culture and/or nucleic acid amplification assays results) should be guided by Gram's stain examination of the fluid, patient's age, vaccination status, presence of specific risk factors (such as immunodeficiency), potential exposure to organisms such as *B. burgdorferi* or *Brucella* species, and the local prevalence of antibiotic resistance in relevant organisms such as *S. aureus* [67]. Guidelines for the initial antibiotic administration are provided in Table 5.5 and Figure 5.5.

Ceftriaxone offers the advantages of a wide antimicrobial spectrum, once-a-day dosage, good safety profile, and comparable results to those obtained with oxacillin in a retrospective study of adult skeletal infections [68]. However, because of its high serum protein binding (>90%) and relative high minimum inhibitory concentration (MIC) as compared to that of β-lactamase–resistant penicillins such as oxacillin or first-generation cephalosporins, there is reluctance to use it alone for suspected or culture-proven methicillin-susceptible *S. aureus* arthritis in children.

In patients under 4 years of age, a combination of an antistaphylococcal penicillin and a broad-spectrum cephalosporin (cefotaxime in the newborn and ceftriaxone or cefuroxime in older children) will provide adequate initial therapy for the most common bacterial pathogens. In premature babies and in those receiving intensive care, empiric therapy against nosocomial bacteria and yeasts should be considered. Over the age of 4 years, coverage against Gram-positive bacteria with narrow-spectrum antibiotics, such as a

Table 5.5. Guidelines for administration of initial antibiotic therapy, pending culture results.

Age group antibiotic		Regular dosage mg/kg/day	Regular dosage doses/day	Antibiotic	Increased dosage[a] mg/kg/day	Increased dosage[a] mg/kg/day
Neonate	BLRBL[b]	100	4			
	or					
	Vancomycin	30	2			
	or					
	Clindamycin	20–30	3			
	plus					
	Cefotaxime	100–150	3			
Child ≤4 years	BLRBL[b]	150	4	First-generation Cephalosporin *or* Clindamycin		
	or					
	Vancomycin	40–60	3			
	or					
	Clindamycin	30–40	3		≥150	4
	plus					
	Cefuroxime	150	3			
	or				≥40	4
	Ceftriaxone	100	1			
Child >4 years	BLRBL[a]	150	4			
	or					
	Vancomycin	45–60	3			
	or					
	Clindamycin	30–40	3			
Sexually active adolescent	BLRBL[b]	150	4			
	or					
	Vancomycin	40–60	3			
	or					
	Clindamycin	30–40	3			
	plus					
	Ceftriaxone	100	1			

[a]β-Lactamase–resistant β-lactam (nafcillin, cloxacillin, flucloxacillin, or dicloxacillin).
[b]From Refs. 52, 55, and 65.

β-lactamase–resistant penicillin, vancomycin, or clindamycin, is usually administered, unless an immunocompromising condition is present. It should be pointed out that in areas where CA-MRSA is prevalent (> 10% of staphylococcal isolates), vancomycin or clindamycin should be used instead of the β-lactamase–resistant penicillin, pending culture results. Co-trimoxazole (16 mg/kg/day, b.i.d.) is an alternative that is active against most CA-MRSA, provides adequate wide antimicrobial coverage, and has excellent oral bioavailability [67].

Because of the scarcity of well-designed, randomized, controlled, and sufficiently powered studies, no evidence-based recommendations on the duration of antibiotic therapy

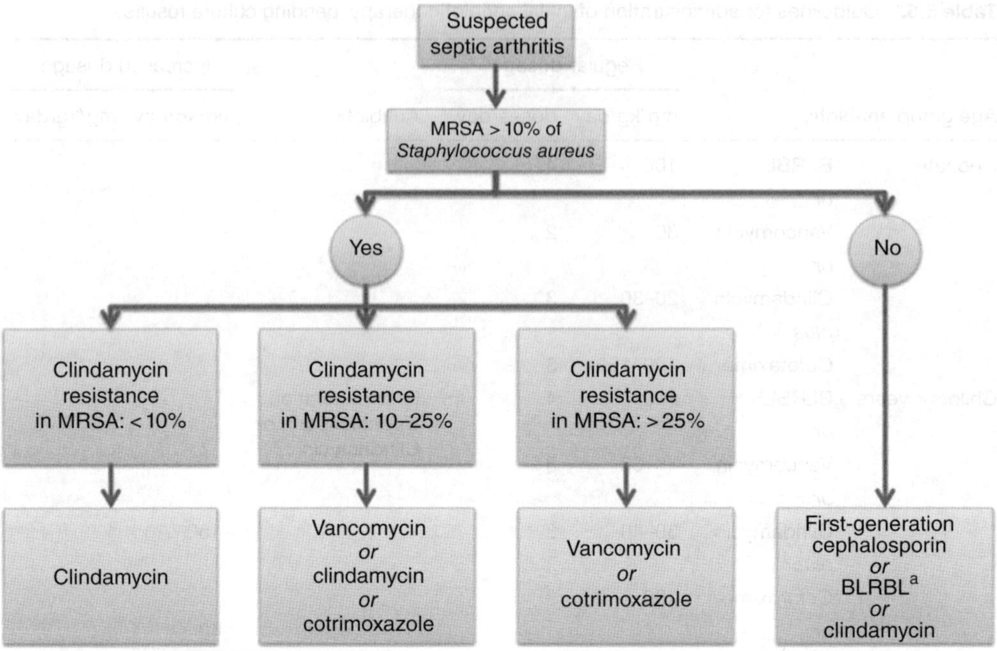

Figure 5.5. Initial administration of antibiotics based on the local prevalence of antibiotic resistance in *Staphylococcus aureus* [67]. [a] betalactamase-resistant betalactam. (*See insert for color representation of the figure.*)

for pediatric septic arthritis can be formulated. Most accepted treatment protocols are based on retrospective case analyses and personal experience, and, traditionally, prolonged in-hospital administration of IV antibiotics has been advocated. Usually, the total duration of the prescribed therapeutic courses varied between 3 and 6 weeks, depending on the patient's age, bacterial identity (longer for *S. aureus* and Gram-negative bacilli other than *H. influenzae* type b), and location of the infected site. This orthodox concept has been slowly evolving toward shorter antibiotic regimens, sequential parenteral–oral therapy, and early hospital discharge, if the following criteria are met: (i) the child is able to take oral medications; (ii) the identity of the causative organism is known; (iii) an oral agent with excellent bioavailability is available; (iv) the patient's compliance can be ascertained; (v) the family can be relied on to adhere to the antibiotic regimen at home; (vi) levels of CRP decrease and can be monitored [52, 68, 69]. The advantages of this approach are saving of hospitalization days, reduced treatment costs, lesser disruption of family life, shortened exposure to health care–associated infections, and avoidance of the untoward effects of prolonged parenteral antibiotic therapy [70]. Although in the past, peak serum bactericidal titers of the oral antibiotic ≥ 1:8 were required to switch to oral therapy [69], recent studies have further simplified patients' management by dispensing with the need for measurement of serum bactericidal levels, because of the adequate absorption of the oral drugs used in the treatment of pediatric skeletal system infections, their good penetration to joints, and their tolerance to high dosage [52, 55, 65, 67] (Figure 5.5) [67]. The initial IV antibiotic therapy is switched to oral therapy between days 2 and 4 after fever normalizes for 24 h, local findings and motion improve, and CRP levels decrease.

If concomitant osteomyelitis is present, antibiotic therapy should be continued for a total 20-day course [55]. According to the novel approach, surgical interventions should be limited to a minimum, and a single joint space aspiration is performed in the vast majority of patients, including those with hip or shoulder infections [67, 71]. Short-term results and 12-month follow-up have shown that the 10-day antibiotic regimen was not inferior to the traditional 30-day therapeutic course and all patients recovered with no significant orthopedic sequelae [52]. Despite the uncertainty regarding the etiology, children with bacteriologically unconfirmed septic arthritis can be treated like culture-positive cases, with comparable long-term results [36].

Adjunctive Anti-inflammatory Therapy

Because the host's immune response contributes to cartilage degradation [37], administration of anti-inflammatory drugs has been proposed in an attempt to limit joint tissues destruction and prevent disability. In experimental animal models of *S. aureus* and *H. influenzae* arthritis, administration of IV dexamethasone reduced the intra-articular concentration of polymorphonuclear leukocytes, cytokines, and stromelysin, as well as the extent of residual joint damage [72]. Two pioneer double-blind, randomized placebo-controlled studies have shown that an early 4-day course of dexamethasone significantly shortened the duration of symptoms in children with bacteriologically documented [73] and/or presumptive [74] pyogenic arthritis. One of the studies demonstrated the additional benefit of reduced incidence of long-term functional disabilities [73]. Despite these encouraging results, it should be realized that there is a subtle balance between an effective immune response aimed to eradicate the invading organism and the overexpression of the host's immune response that may harm the joint architecture and cause permanent functional damage. Attenuation of inflammation by administering corticosteroid therapy may reduce the effectiveness of the anti-infective host's reaction and add unnecessary and potentially dangerous side effects. Routine administration of adjuvant dexamethasone therapy cannot be recommended at this stage.

Prognosis

Long-term follow-up is required to assess the functional results of septic arthritis in children, because evaluation at the time of hospital discharge frequently fails to detect residual abnormalities. A variety of functional sequelae such as limping, decreased motion range or ankylosis, joint instability, permanent dislocation, or abnormal bone growth have been reported in 10–27% of children. Risk factors for a poor orthopedic outcome are listed in Table 5.6 [6, 75, 76].

Table 5.6. Risk factors for poor prognosis in pediatric septic arthritis.

Infection during the first 6 months of life
Involvement of hip or shoulder
Adjacent osteomyelitis
Delay in the diagnosis of >4 days
Infection with *S. aureus* or Gram-negative enteric organisms
Persistent positive culture after 1 week of appropriate antimicrobial therapy

Key Points

- The etiology of pediatric septic arthritis is age-dependent: *Staphylococcus aureus* predominates in neonates and children older than 4 years, whereas *Kingella kingae* is the most common etiology in the 6 months to 4 years age group.
- Children with *K. kingae* arthritis may present with low-grade fever and normal leukocyte counts, CRP levels, and erythrocyte sedimentation rates.
- A high index of clinical suspicion, drawing of synovial fluid and blood specimens for bacteriological diagnosis, prompt joint drainage, and administration of adequate antibiotic coverage are cornerstones of an optimal management of pediatric septic arthritis.
- Clinical improvement and decreasing CRP levels can be used to guide switching from parenteral to oral antibiotics and deciding on early home discharge.

References

[1] Riise OR, Handeland KS, Cvancarova M, *et al.* Incidence and characteristics of arthritis in Norwegian children: a population-based study. Pediatrics 2008; 121 (2): e299–306.

[2] Ike RW. Bacterial arthritis. Curr Opin Rheumatol 1998; 10 (4): 330–4.

[3] Grammatico-Guillon L, Maakaroun Vermesse Z, Baron S, *et al.* Pediatric bone and joint infections are more common in boys and toddlers: a national epidemiology study. Acta Pædiatrica 2013; 102 (3): e120–5.

[4] Weintraub A. Immunology of bacterial polysaccharide antigens. Carbohydr Res 2003; 338(23): 2539–47.

[5] Trujillo M, Nelson JD. Suppurative and reactive arthritis in children. Semin Pediatr Infect Dis 1997; 8 (4): 242–9.

[6] Welkon CJ, Long SS, Fisher MC, Alburger PD. Pyogenic arthritis in infants and children: a review of 95 cases. Pediatr Infect Dis 1986; 5 (6): 669–76.

[7] Martinez-Aguilar G, Avalos-Mishaan A, Hulten K, *et al.* Community-acquired, methicillin-resistant and methicillin-susceptible *Staphylococcus aureus* musculoskeletal infections in children. Pediatr Infect Dis J 2004; 23 (8): 701–6.

[8] Arnold SR, Elias D, Buckingham SC, *et al.* Changing patterns of acute hematogenous osteomyelitis and septic arthritis: emergence of community-associated methicillin-resistant *Staphylococcus aureus.* J Pediatr Orthop 2006; 26 (6): 703–8.

[9] Vardakas KZ, Kontopidis I, Gkegkes ID, *et al.* Incidence, characteristics, and outcome of patients with bone and joint infections due to community-associated methicillin-resistant *Staphylococcus aureus*: a systematic review. Eur J Clin Microbiol Infect Dis 2013; 32 (6): 711–21.

[10] Tyrrell GJ, Lovgren M, Kress B, Grimsrud K. Varicella-associated invasive group A streptococcal disease in Alberta, Canada 2000–2002. Clin Infect Dis 2005; 40 (7): 1055–7.

[11] Memon IA, Jacobs NM, Yeh TF, Lilien LD. Group B streptococcal osteomyelitis and septic arthritis. Its occurrence in infants less than 2 months old. Am J Dis Child 1979;133(9):921–3.

[12] Jacobs NM. Pneumococcal osteomyelitis and arthritis in children. A hospital series and literature review. Am J Dis Child 1991; 145 (1): 70–4.

[13] Ross JJ, Saltzman CL, Carling P, Shapiro DS. Pneumococcal septic arthritis: review of 190 cases. Clin Infect Dis 2003; 36 (3): 319–27.

[14] Howard AW, Viskontas D, Sabbagh C. Reduction in osteomyelitis and septic arthritis related to *Haemophilus influenzae* type b vaccination. J Pediatr Orthop 1999;19(6):705–9.

[15] Rotbart HA, Glode MP. *Haemophilus influenzae* type b septic arthritis in children: report of 23 cases. Pediatrics 1985;75(2):254–9.

[16] Le Quellec SN, Gaillot O, Chotel F, *et al.* Septic arthritis caused by nonencapsulated *Haemophilus influenzae*. J Clin Microbiol 2013; 51 (6): 1970–2.
[17] Yagupsky P, Porsch EA, St. Geme JW 3rd. *Kingella kingae*: an emerging pathogen in young children. Pediatrics 2011; 127 (3): 557–65.
[18] Chometon S, Benito Y, Boisset S, *et al.* Specific real-time polymerase chain reaction places *Kingella kingae* as the most common cause of osteoarticular infections in young children. Pediatr Infect Dis J 2007; 26 (5):377–81.
[19] Schaad UB. Arthritis in disease due to *Neisseria meningitidis*. Rev Infect Dis 1980; 2 (6): 880–8.
[20] Mathew S, Overturf GD. Complement and properidin deficiencies in meningococcal disease. Pediatr Infect Dis J 2006; 25 (3): 255–6.
[21] Rice PA. Gonococcal arthritis (disseminated gonococcal infection). Infect Dis Clin North Am 2005; 19 (4): 853–61.
[22] Aronoff SC, Scoles PV. Treatment of childhood skeletal infections. Pediatr Clin North Am 1983;30(2):271–80.
[23] Brancos MA, Peris P, Miro JM, *et al.* Septic arthritis in heroin addicts. Semin Arthritis Rheum 1991;21(2):81–7.
[24] Bosilkovski M, Urosevic VK, Cekovska Z, *et al.* Osteoarticular involvement of childhood brucellosis: experience with 133 cases in an endemic region. Pediatr Infect Dis J 2013; 32(8):815–9.
[25] Deanehan JK, Kimia AA, Tan Tanny SP, *et al.* Distinguishing Lyme from septic knee monoarthritis in Lyme disease-endemic areas. Pediatrics 2013;131(3):e695–e701.
[26] Franz A, Webster AD, Furr PM, Taylor-Robinson D. Mycoplasmal arthritis in patients with primary immunoglobulin deficiency: clinical features and outcome in 18 patients. Br J Rheumatol 1997;36(6):661–8.
[27] Brook I. Joint and bone infections due to anaerobic bacteria in children. Pediatr Rehabil 2002;5(1):11–9.
[28] Teo HE, Peh WC. Skeletal tuberculosis in children. Pediatr Radiol 2004;34(11):853–60.
[29] Flannery DD. Septic arthritis and concern for osteomyelitis in a child with rat bite fever. J Clin Microbiol 2013;51(6):1987–9.
[30] Offiah AC. Acute osteomyelitis, septic arthritis and discitis: differences between neonates and older children. Eur J Radiol 2006;60(2):221–32.
[31] Speiser JC, Moore TL, Osborn TG, *et al.* Changing trends in pediatric septic arthritis. Semin Arthritis Rheum 1985;15(2):132–8.
[32] Peltola H, Vahvanen V. Acute purulent arthritis in children. Scand J Infect Dis 1983;15(1): 75–80.
[33] Dubost JJ. Septic arthritis with no organism: a dilemma. Joint Bone Spine. 2006;73(4): 341–3.
[34] Lyon RM, Evanich JD. Culture-negative septic arthritis in children. J Pediatr Orthop. 1999; 19(5):655–9.
[35] Chang WS, Chiu NC, Chi H, *et al.* Comparison of the characteristics of culture-negative versus culture-positive septic arthritis in children. J Microbiol Immunol Infect 2005;38(3): 189–93.
[36] Pääkkönen M, Kallio MJT, Kallio PE, Peltola H. Significance of negative cultures in the treatment of acute hematogenous bone and joint infections in children. J Pediatr Infect Dis Soc 2013;2(2):119–25.
[37] Shirtliff ME, Mader JT. Acute septic arthritis. Clin Microbiol Rev 2012;15(4):527–44.
[38] Montgomery CO, Siegel E, Blasier RD, Suva LJ. Concurrent septic arthritis and osteomyelitis in children. J Pediatr Orthop 2013;33(4):464–7.
[39] Asnes RS, Arendar GM. Septic arthritis of the hip: a complication of femoral venipuncture. Pediatrics 1966;38(5):837–41.
[40] Ross JJ, Shamsuddin H. Sternoclavicular septic arthritis: review of 180 cases. Medicine (Baltimore) 2004;83(3):139–48.

[41] Goldenberg DL, Reed JI. Bacterial arthritis. N Engl J Med 1985;312(12):764–71.
[42] Goldenberg DL. Septic arthritis. Lancet 1998;351(9097):197–202.
[43] Wilson ML, Winn W. Laboratory diagnosis of bone, joint, soft tissue, and skin infections. Med Microbiol 2008;46(3):453–7.
[44] Press J, Peled N, Buskila D, Yagupsky P. Leukocyte count in the synovial fluid of children with culture-proven brucellar arthritis. Clin Rheumatol. 2002;21(3):191–3.
[45] Smith JW, Chalupa P, Shabaz Hasan M. Infectious arthritis: clinical features, laboratory findings and treatment. Clin Microbiol Infect 2006;12(4):309–14.
[46] Hughes JG, Vetter EA, Patel R, *et al.* Culture with BACTEC Peds Plus/F bottle compared with conventional methods for detection of bacteria in synovial fluid. J Clin Microbiol. 2001;39(12):4468–71.
[47] Bourbeau P, Riley J, Heiter BJ, *et al.* Use of the BacT/Alert blood culture system for culture of sterile body fluids other than blood. J Clin Microbiol 1998;36(11):3273–77.
[48] Fenollar F, Lévy PY, Raoult D. Usefulness of broad-range PCR for the diagnosis of osteoarticular infections. Curr Opin Rheumatol. 2008; 20 (4): 463–70.
[49] Okur M, Erbey F, Bektas MS, *et al.* Retrospective clinical and laboratory evaluation of children with brucellosis. Pediatr Int 2012;54(2):215–8.
[50] Hariharan P, Kabrhel C. Sensitivity of erythrocyte sedimentation rate and C-reactive protein for the exclusion of septic arthritis in emergency department patients. J Emerg Med 2011;40(4):428–31.
[51] Pääkkönen M, Kallio MJ, Kallio PE, Peltola H. Sensitivity of erythrocyte sedimentation rate and C-reactive protein in childhood bone and joint infections. Clin Orthop Relat Res 2010;468(3):861–6.
[52] Peltola H, Pääkkönen M, Kallio P, Kallio JT. Prospective, randomized trial of 10 days versus 30 days of antimicrobial treatment, including a short-term course of parenteral therapy, for childhood septic arthritis. Clin Infect Dis 2009;48(9):1201–10.
[53] Shen CJ, Wu MS, Lin WL, *et al.* The use of procalcitonin in the diagnosis of bone and joint infection: a systemic review and meta-analysis. Eur J Clin Microbiol Infect Dis 2013;32(6):817–14.
[54] Saeed K, Dryden M, Sitjar A, White G. Measuring synovial fluid procalcitonin levels in distinguishing cases of septic arthritis, including prosthetic joints, from other causes of arthritis and aseptic loosening. Infection 2013;14(4):47–54.
[55] Pääkkönen M, Peltola H. Treatment of acute septic arthritis. Pediatr Infect Dis J 2013; 32(6):684–5.
[56] Connolly LP, Connolly SA. Skeletal scintigraphy in the multimodality assessment of young children with acute skeletal symptoms. Clin Nucl Med 2003;28(9):746–54.
[57] Yang WJ, Im SA, Lim GY, *et al.* MR imaging of transient synovitis: differentiation from septic arthritis. Pediatr Radiol 2006;36(11):1154–8.
[58] Marin C, Sanchez-Alegre ML, Gallego C, *et al.* Magnetic resonance imaging of osteoarticular infections in children. Curr Probl Diagn Radiol 2004;33(2):43–59.
[59] Hambleton S, Berendt AR. Bone and joint infections in children. Hot Top Infect Immun Child 2004;549:47–62.
[60] Faust SN, Clark J, Pallett A, Clarke NM. Managing bone and joint infection in children. Arch Dis Child 2012;97(6):545–53.
[61] Nelson JD, Koontz WC. Septic arthritis in infants and children: a review of 117 cases. Pediatrics 1966;38(6):966–71.
[62] Morrey BF, Bianco AJ, Rhodes KH. Septic arthritis in children. Orthop Clin North Am 1975;6(4):923–34.
[63] Givon U, Liberman B, Schindler A, *et al.* Treatment of septic arthritis of the hip joint by repeated ultrasound-guided aspirations. J Pediatr Orthop 2004;24(3):266–70.
[64] Dunkele LM. Towards optimum management of serious focal infections: the model of suppurative arthritis. Pediatr Infect Dis J 1989;8(4):195–6.

[65] Pääkkönen M, Peltola H. Management of a child with suspected acute septic arthritis. Arch Dis Child 2012;97(3):287–92.
[66] Nusem I, McAlister A. Arthroscopic lavage for the treatment of septic arthritis of the hip in children. Acta Orthop Belg 2012;78(6):730–4.
[67] Pääkkönen M, Peltola H. Bone and joint infections. Pediatr Clin North Am 2013;60(2): 425–36.
[68] Wieland BW, Marcantoni JR, Warren DK, Marschall J. A retrospective comparison of ceftriaxone versus oxacillin for osteoarticular infections due to methicillin-susceptible *Staphylococcus aureus*. Clin Infect Dis 2012;54(5):585–90.
[69] Tetzlaff TR, McCracken GH, Nelson JD. Oral antibiotic therapy for skeletal infection of children. II. Therapy of osteomyelitis and suppurative arthritis. J Pediatr 1978;92(3):485–90.
[70] Faden D, Faden HS. The high rate of adverse drug events in children receiving prolonged outpatient parenteral antibiotic therapy for osteomyelitis. Pediatr Infect Dis 2009;28(6): 539–54.
[71] Pääkkönen M, Kallio M, Peltola H, Kallio PE. Pediatric septic hip with or without arthrotomy: a retrospective analysis of 62 consecutive nonneonatal culture-positive cases. J Pediartr Orthop B 2010;19(3):264–9.
[72] Jafari HS, Sáez-Llorens X, Paris M, *et al.* Dexamethasone attenuation of cytokine-mediated articular cartilage degradation in experimental lapine *Haemophilus* arthritis. J Infect Dis 1993;168(5):1186–93.
[73] Odio CM, Ramirez T, Arias G, *et al.* Double blind, randomized, placebo controlled study of dexamethasone therapy for hematogenous septic arthritis in children. Pediatr Infect Dis J 2003;22(10):883–8.
[74] Harel L, Prais D, Bar-On E, *et al.* Dexamethasone therapy for septic arthritis in children. Results of a randomized double-blind placebo-controlled study. J Pediatr Orthop 2011;31(2):211–5.
[75] Howard JB, Highgenboten CL, Nelson JD. Residual effects of septic arthritis in infancy and childhood. J Am Med Assoc 1976;236(8):932–5.
[76] Howard-Jones AR, Isaacs D, Gibbons PJ. Twelve-month outcome following septic arthritis in children. J Pediatr Orthop B 2013;22(5):486–490.

Chapter 6
Native Joint Arthritis

Ilker Uçkay, Mohamed Al-Mayahi, Domizio Suvà, Daniel Lew, and Pierre Vaudaux

Introduction

The presentation of different types of inflammatory arthritis, caused by autoimmunity, immune complexes, crystals, or cartilage damage, is clinically similar to native septic arthritis. Joint infection might be acute or chronic, community-acquired or nosocomial, hematogenously seeded as opposed to exogenously acquired, and monoarticular or polyarticular. It may be caused by a multitude of pathogens. This large microbiological spectrum is one of the particularities distinguishing arthritis from other orthopedic infections such as osteomyelitis or implant-related infections. Just to give an example, viral arthritis exists, while viral disease has never been proven for osteomyelitis or prosthetic joint–associated infections to the best of our knowledge. Moreover, immunological, postinfectious joint inflammations are well-known entities for native joints, but not other orthopedic infections. Finally, some pathogens causing synovial infections, such as gonococci, usually spare bone. The current literature often defines orthopedic infections as "osteoarticular," suggesting that diagnosis and treatment of bone and joint infection would be similar. However, septic arthritis, osteomyelitis, and orthopedic implant–associated infections [1] are different in nature, epidemiology, therapy, and outcome. Unfortunately, many physicians make analogies between these different entities, which may lead to inappropriate diagnostic and therapeutic approaches in daily clinical life. This chapter underlines the particularities of native joint arthritis with an emphasis on diagnosis, epidemiology, treatment, and some prevention aspects before planned joint interventions.

Pathogenesis, Epidemiology, and Microbiology

Septic arthritis harbors a high burden of morbidity and mortality. Most cases are primary native arthritis from presumed or proven hematogenous origin with an annual incidence of 2–10/100,000 persons [2–4]. The weight-bearing joints are the most frequently affected,

Bone and Joint Infections: From Microbiology to Diagnostics and Treatment, First Edition.
Edited by Werner Zimmerli.

involving the knee in half of all cases. A survey over 17 years from Northern Israel demonstrates that the knees are involved in 41.8% and the hips in 23.6% of cases [5]. Another 10-year survey from the United Kingdom reveals the knee joint as the most frequently affected (31%), followed in descending order by the hip, elbow, hand, ankle, wrist, and sternoclavicular and sacroiliac joints (see Chapter 7) [6]. The same authors report polyarticular infection in 15% of all episodes. Gram-positive cocci such as *Staphylococcus aureus* or streptococci are responsible for the vast majority of primary native arthritis among adults. Gram-negative rods account for 9–17% of all cases, and anaerobes are found in 1–3% [3]. This pathogen distribution is similar for postsurgical arthritis (SSI; surgical site infection). One important and particular exception is witnessed among young infants, with a high proportion of arthritis due to *Kingella kingae* [7], a Gram-negative rod that is almost exclusively seen in adults with endocarditis (see Chapter 5).

Septic arthritis can also result directly from a penetrating traumatic accidental injury. The exact proportion of primary versus posttraumatic arthritis has not been systematically investigated in the past. According to our literature review, the overall incidence of posttraumatic cases in terms of penetrating trauma among all septic arthritis episodes is between 4 and 22% (Table 6.1) [2, 3, 5, 6, 8–11]). Nevertheless, this main distinction of the pathophysiological pathway may have major implications in implicating the causal pathogen, and thus initial empirical antibiotic therapy postdrainage. Contrary to hematogenous arthritis, posttraumatic joint infection is predominantly observed in young healthy males (median age, 31 years) and involves the knee in 54% of cases [3]. Four distinct origins differ in pathogenesis and microbiology. The first pathogenic mechanism is related to bites; the second to thorn punctures; and the third and forth to trauma sustained in terrestrial and aquatic environments, respectively. Overall, causative microorganisms in posttraumatic arthritis are predominantly Gram-negative (51%), in contrast to primary native joint arthritis where *S. aureus* prevails. *Pantoea agglomerans* was the leading causative agent in thorn-related infections (6/11; 54%), while *Pseudomonas* spp. accounted for most injuries with instruments (7/28, 25%) or penetrating foot lesions. The latter

Table 6.1. Summary of reports with a proportional display of the origin of septic arthritis, 1945–2010.

Pathogenesis	Kaandorp [8]	Stutz [9]	Geirsson [2]	Morgan [10]	Weston [6]	Eder [5]	Gomez-Rodriguez [11]
Presumably hematogenous (*primary native joint arthritis*)	68%	54%	NA	72%	NA	NA	NA
Surgical site infection (total)	23%	42%	44%	NA	NA	16%	NA
Postoperative	21%	28%	26%	NA	NA	5%	NA
Intra-articular injection	2%	10%	NA	NA	3%	9%	NA
Arthrocentesis	NA	4%	18%	NA	NA	2%	NA
Penetrating trauma	7%	4%	NA	22%	9%	12%	8%

Adapted from Ref. 3.
NA, not available.

pathogen was equally the classical agent of nail-related wounds. *Clostridium* spp. were associated with trauma by metallic objects and soil (10/11, 91%). Fungi, mostly plant saprophytes of the *Scedosporioses* family, were retrieved in 36 cases (27%) in the context of injuries with soil contact. Nontuberculous mycobacteria were reported 21 times (16%), with three-quarters attributed to *Mycobacterium marinum* acquired in water-related injuries and all the rest to *Mycobacterium kansasii* in terrestrial injuries. Remaining pathogens were commensals of the natural environment (e.g., *Nocardia* spp.) or the zoonotic agent *Erysipelothrix rhusiopathiae* in the context of contact with either land mammals or fish. In case of empirical antibiotic treatment, a broader spectrum covering Gram-negative rods is more appropriate than simple anti–Gram-positive therapy [3]. Interestingly, similar observations have been reported for wound infections in the aftermath of terrestrial (earthquake) [12, 13] or aquatic (tsunami) [12] natural disasters, where usual skin organisms are less frequent than Gram-negative rods. First, soil, water, plants, and mammalian oral cavities are populated by Gram-negative microorganisms, in contrast to the human skin surface [3].

Diagnosis

There is no generally accepted definition of acute native joint infection, if blood and synovial cultures are negative. In such cases, clinical signs and symptoms, inflammatory blood parameters, and synovial cell counts have to be considered.

Signs and Symptoms

Most patients experience a short course of infection characterized by steadily increasing arthralgia, local heat and erythema, fever, and shivering. Similar signs and symptoms can be caused by crystal arthropathy or rheumatoid arthritis. Therefore, a general physical examination (signs of endocarditis?) should be performed in each case of suspected primary native joint infection. In addition, blood cultures should always be drawn, since they show growth in at least 50% of the cases [14].

Microbiological Diagnosis

The gold standard for the diagnosis of septic arthritis is the detection of an identical pathogen in two specimens from synovia, from a biopsy, or from blood. The presence of one pathogen, in only one specimen, with the others remaining negative, might indicate specimen contamination, especially when it is due to skin commensals such as *Corynebacterium* spp., *Bacillus* spp., or coagulase-negative staphylococci [1]. Culture growth may need to be extended beyond the standard incubation period of 5 days, especially in case of pretreatment with antibiotic agents. Cultures may be negative because of prior antimicrobial exposure, a low number of organisms, an inappropriate culture medium, fastidious organisms, or prolonged transport time to the microbiology laboratory. Culture-negative septic arthritis is then diagnosed clinically by the presence of pus, a high leukocyte count in the absence of crystal disease, and a context compatible with septic arthritis. Eubacterial polymerase chain reaction (PCR) yields a lower sensitivity and is still relatively expensive, which precludes its routine application. In polymicrobial infection, its interpretation may be difficult. Moreover, it does not provide information about antibiotic resistance, except for genes coding for methicillin or rifampin resistance. However, specific or multiplex PCR is beneficial in special circumstances in which slow-growing bacteria are suspected, such as *K. kingae* [7], *Brucella* spp., *Coxiella burnetii*, *Bartonella henselae*, *Mycobacterium tuberculosis*, or *Mycobacterium ulcerans*. Viral arthritis is typically not purulent and is often accompanied by cutaneous and systemic signs of generalized infection. Chronic septic arthritis may reveal a sinus tract or an open access to the intra-articular space. In these cases, the joint is contaminated and presumably infected, even if there are no local inflammatory signs. Chronic infectious arthritis occurs mainly in polyneuropathic foot infections with ulcerations of the toe joints (see Chapter 18). In case of suspicion of gonococcal disease (case history, cutaneous signs, oligoarthritis, tenosynovitis), the pathogen should be searched with a specific PCR in the urine and/or the joint sample.

Blood Chemistry

Elevation of the erythrocyte sedimentation rate suggests infection. However, this laboratory sign is neither sensitive nor specific, since it may be due to many other causes. Several studies mentioned the role of an elevated serum C-reactive protein (CRP) level for the diagnosis of septic arthritis. Ernst *et al.* [15] showed the good predictive value of a high CRP for septic joint. Harihanan and Kabrhel [16] reported a linear relationship between CRP level and the occurrence of septic arthritis, yielding a sensitivity of about 45% at a cutoff of 150 mg/l, in a group of 167 patients. Papanicolas *et al.* [17] identified a CRP >100 mg/l as predictive of septic arthritis with a sensitivity of 86%, with, however, a low specificity of only 48%. Other reports challenged the predictive value of serum inflammatory markers for the diagnosis of septic arthritis [18]. Regarding procalcitonin and primary septic arthritis, its diagnostic value at a cutoff of 0.25 μg/l is quite good, with a sensitivity of 93% and a specificity of 75% [19]. In the future, the use of other inflammatory markers in the diagnosis of infection will certainly emerge. As an example, serum leukocyte esterase warrants further investigation [20].

Synovial Fluid Analysis

Microbiological cultures take time to confirm infection. In addition, clinical judgment and serum laboratory parameters often do not allow correct initial decision making. A low concentration of microbial pathogens in the otherwise sterile synovial fluid is sufficient to trigger considerable inflammation that may lead to severe cartilaginous damage. In contrast to abscesses in soft tissue infections, this low bacterial concentration may remain undetected by microscopic examination. On the other hand, the responsible physicians should rapidly decide whether joint lavage has to be performed and whether empirical therapy should be started, while awaiting culture results. As cartilage damage occurs in most cases of nongonococcal septic arthritis, rapid drainage and start of empirical antibiotic treatment before receiving culture results is needed, unless there is a strong suspicion of recurrent crystal arthropathy [21–23]. The most cited tests are Gram or acridine orange staining, leukocyte differential counts, dark field microscopy for crystals, and ratio of synovial fluid glucose [18, 21, 24], proteins [21, 22], or lactic acid [19, 25]. These individual surrogate markers in synovia should be interpreted as a group and within the clinical context. The sensitivity of Gram staining is known to be suboptimal for the diagnosis of native joint septic arthritis [18, 21, 26–28]. Previous reports usually evaluated small size groups including less than 80 patients [22–30] and mainly focused on episodes of native joint septic arthritis [15, 24–31]. Few studies compared the efficacy of Gram with other staining procedures [32], for example, acridine orange, or evaluated different patient populations, such as those with prosthetic joint infections [16, 33], various immunocompromising conditions, or concomitant gout [28, 30] and other microcrystalline diseases [34, 35]. Episodes of monoarticular, concomitant septic, and crystalline arthritis were mostly reported in small case series [17, 28, 30, 36] or case reports. While many centers are still using Gram staining for a rapid, preliminary diagnosis of arthritis, the cost–benefit ratio of this approach has not been evaluated in detail. We investigated the performance of Gram and acridine orange staining in predicting positive culture results (unpublished results). Among 500 arthritis episodes, we analyzed 196 immunocompromised patients and 69 with gout or other crystal arthropathy. Gram staining revealed pathogens in 146 episodes (146/500, 29%) or in 146 of the 400 culture-positive episodes (37%). Overall, the sensitivity, specificity, and positive and negative predictive values of Gram stain for rapid diagnosis of septic arthritis were 0.37, 0.99, 0.99, and 0.28, respectively, with positive culture as gold standard. Quite similar values were recorded across the different patient subpopulations, in particular for sensitivity values that were 0.40 for immunocompromised patients, 0.36 for patients who were treated with antibiotics, and 0.52 for those with concomitant crystal arthropathy. A CRP level above 150 mg/l, bacteremia, and a synovial leukocyte count above 180,000/μl were also significantly linked with the outcome "positive Gram staining," but not the percentage of neutrophil counts in synovial fluid. In the literature, conflicting data are provided on the significance of synovial leukocyte counts. Most experts consider that a threshold of 50,000 cells/μl is particularly predictive of a septic origin [24, 27]. However, in the literature, different studies yielded different levels of sensitivity and used different cutoff levels. Some studies indicated that even 100,000/μl leukocytes could not reliably discriminate septic from nonseptic arthritis [22]. Regarding the significance of neutrophil differentials in synovial fluid, a literature review involving 6242 patients indicates that a neutrophil proportion above 90% has a threefold likelihood for septic arthritis, as compared to patients with less than 90% neutrophils in synovia [21]. This cutoff was set at 75% by other authors [24].

Radiology

In contrast to chronic osteomyelitis or periprosthetic joint infections, radiology is of low diagnostic value for native septic joint arthritis. A plain radiograph can display abnormal soft tissue swelling over the bone or may reveal hidden abscesses outside the infected joint. However, no imaging procedure is able to diagnose septic arthritis per se, especially not in acute disease, where concomitant osteomyelitis is still absent. Ultrasound allows detecting abscesses or abnormal synovial production, but it does not prove the nature of joint collection or inflammation.

Treatment

Treatment of joint infections is not standardized due to the variable clinical presentations and the lack of data from randomized controlled trials. Since septic arthritis represents a closed abscess collection, we recommend joint drainage, although no studies have been published comparing management with or without drainage procedures [14]. Bacterial nongonococcal infections or fungal disease requires some kind of joint drainage as the initial step of treatment, regardless of the cause and timing of the infection, and the condition of the host. In contrast, viral, mycobacterial, or gonococcal infections usually do not require surgical washout.

Surgical Drainage versus Arthrocentesis

Current literature suggests that any form of drainage is valid to reduce the infectious burden. It should be performed as soon and as meticulously as possible. In animal models, it has been shown that septic arthritis destroys cartilage within days [37]. Thus, rapid and effective drainage is needed, even though the exact duration until clinically significant destruction is unknown for humans, and might be influenced by several factors such as preexisting degeneration, rheumatological comorbidities, as well as the nature and inoculum size of pathogens. While surgeons usually prefer joint lavage by arthrotomy or arthroscopy [38], many rheumatologists and internists prefer joint drainage outside of the operating theater by (repeated) arthrocentesis. The discussion about the therapeutic value of each of these approaches is as old as the existence of the different medical and surgical specialities. To the best of our knowledge, there is no direct prospective head-to-head comparison in the literature, which might be difficult because of the large case mix of the arthritis population. As a rule, physicians prefer surgical drainage in life-threatening conditions such as sepsis or in joints, which are inconvenient for repeated arthrocentesis such as the hip. Conversely, many nonsurgical specialists prefer arthrocentesis for elbow and knee arthritis that is technically easier to aspirate. The literature does not indicate any significant differences in the outcome between arthroscopy and arthrotomy for the initial drainage of knee [38], hip [39], and other joints' septic arthritis [40], although arthrotomy usually allows a better intraoperative view and evaluation of the situation than arthroscopical approach.

Resection, Arthrodesis, or Amputation

When the infection has already destroyed the articular surface and the underlying bone, resection of the infected area becomes an option. This is mainly performed in severe hip arthritis by resection of the femoral head (Girdlestone hip), which subsequently requires joint reconstruction by arthroplasty. The mechanical removal of a nonfunctional joint and adjacent bone with cavernous infectious foci allows treating all hidden foci inside porous bone before a prosthetic joint can be implanted after 4–6 weeks of antibiotic therapy, and a 2-week free interval for observation. Antibiotic-loaded spacers may mechanically fill the gap until definitive arthroplasty [41], which is not the only mechanical solution for a destroyed articulation. Arthrodesis (the operative fusion of bones across the former synovial space) can provide a stable, generally painless, limb with some expected shortening. Arthrodesis is often performed after septic arthritis of the knee, ankle, or fingers, without major limitations in daily life, while a prosthesis approach is more applicable for the hip joint. The last and very rare surgical option is amputation. Amputation is indicated only for life-threatening infection or persistent local infection with massive bone loss not amenable to arthrodesis or prosthesis. Frequently, vascular and arterial insufficiencies may lead to the decision of amputation.

Antimicrobial Therapy

Antibiotic and surgical therapy is generally concomitant and requires in particular a good antimicrobial penetration into the synovial space.

Choice of Antibiotic Agents

Standard antimicrobial regimens for the most commonly encountered microorganisms are listed in Table 6.2 [44]. Intravenous β-lactam antibiotics can be used as long as the pathogen is susceptible. Two important drawbacks of this large class of antibiotics are their low oral bioavailability and limited synovial penetration [44]. Another cell wall synthesis inhibitor is the glycopeptide vancomycin, which has a serum half-life of 6 h. While it is believed that trough serum levels of 20 mg/ml are required for optimal treatment of bone infections [45], there is no similar consensus for optimal glycopeptide therapy of arthritis. Additional arguments against high-dose vancomycin are the nephrotoxic side effects. This issue may have been exaggerated in the past, since it was frequently confounded with the use of concomitant nephrotoxic medication in patients with chemotherapy. When used in continuous perfusion, the changes in vancomycin serum levels are less important than for intermittent application. The target concentrations are achieved faster with less adverse drug effects [46]. However, continuous perfusion does not automatically provide a better outcome in terms of remission. Vancomycin should be administered over at least 1 h to prevent a histamine-mediated "red man" syndrome, which should not be confounded with a true, but rare allergy. Another glycopeptide antibiotic is teicoplanin, which is available in Europe and several other countries outside Europe, but not in the United States. Teicoplanin, which has an extended serum half-life of 72 h, can be administered within 30 min, generally at a dose of 400 mg once a day parentally (after a loading dose of 2 × 400 mg the first day). Alternatively, teicoplanin can be administered three times a week or intramuscularly. For synovial infections, high serum levels may be required, although optimal trough levels and daily regimens remain to be established.

Another category of antibiotics includes daptomycin, which depolarizes bacterial membranes and yields a rapid, dose-dependent bactericidal effect. Daptomycin has a serum half-life of 9 h and is only available in parenteral form. It can be administered once a day at a dose of 6–8 mg/kg [47] in the absence of renal dysfunction, which makes it suitable for outpatient treatment. Some adverse events are known, in particular muscular toxicity detected by an elevation of creatinine phosphokinases.

Tigecycline, which belongs to the class of glycylcyclines, is a substance that has been developed from tetracyclines, but it exhibits a fivefold higher affinity to the target. Tigecycline inhibits ribosomal protein synthesis and is only available in parenteral form, using a loading IV dose of 100 mg, followed by 50 mg twice daily. In osteoarticular infections, tigecycline is still considered an experimental drug. Major adverse events include nausea and vomiting.

Among other antibiotics, aminoglycosides might be used in combination therapy for sustained bacteremia, but are not recommended for systemic monotherapy for osteoarticular infections. Aminoglycosides have a suboptimal activity in synovial fluid or bone [44], and their activity is known to be substantially reduced in a low-pH and low-oxygen environment.

Linezolid inhibits ribosomal protein synthesis and can be administered parenterally or orally at a dose of 600 mg twice daily without adjustment for renal insufficiency. It is an anti–Gram-positive bacteriostatic antibiotic displaying no cross-resistance to other

Table 6.2. Antibiotic treatment of bacterial native joint infections.

Microorganisms isolated	Parenteral therapy: Treatment of choice	Parenteral therapy: Alternatives	Oral therapy[a]
Methicillin-susceptible *S. aureus*	Nafcillin[b] (2 g 4×/day)	A cephalosporin II,[c] clindamycin	Clindamycin (600 mg every 8 h)
Methicillin-resistant *S. aureus*	Vancomycin (2 × 15 mg/kg/day)	Daptomycine (6–8 mg/kg/day), Clindamycin (600 mg every 8 h)	Trimethoprim/sulfamethoxazole (2 × 1 days tablet), clindamycin
Various streptococci	Penicillin G (3 million Units 4–6×/day)	A cephalosporin III[d]	Amoxicillin (750 mg 3×/day), Clindamycin (600 mg 3×/day)
Enteric Gram-negative rods	Cephalosporins II–III[d,e]	A cephalosporin III[d,e] or fluoroquinolone	Fluoroquinolone[f]
Serratia spp., *P. aeruginosa*	Ceftazidim (6 g IV/day)	Quinolone (with aminoglycosides)	Ciprofloxacin (750 mg 2×/day)[f]
	Piperacillin[e] (4 g 4×/day) and Gentamicin (5 mg/kg/day)		
Anaerobes	Clindamycin (600 mg 4×/day)	Metronidazole (500 mg 3×/day)	Clindamycin (600 mg 3×/day)
Mixed infection (aerobic and anaerobic microorganisms)	Amoxicillin–clavulanic acid (2.2 g 3×/day)	Imipenem[g] (500 mg 4×/day)	Amoxicillin–clavulanic acid (625 mg–1000 mg 3×/day)
Gonococcal arthritis[h]	Ceftriaxone (1–2 g 1×/day)		Fluoroquinolones
Empirical therapy	Cephalosporins II–III[d,e]	Amoxicillin-clavulanic acid (1.2 mg 3×/day)	Either IV or wait for results

Adapted from Refs. 42 and 43.
Regimens used at Geneva University Hospitals.
[a]Oral therapy is usually given after 4–6 weeks of parenteral therapy with exception of fluoroquinolones.
[b]Flucloxacillin in Europe.
[c]Second generation, such as cefuroxime (1500 mg 3–4×/day).
[d]Third generation, such as ceftazidime (2 g 3×/day).
[e]Depends on sensitivities; piperacillin/tazobactam and imipenem are useful alternatives.
[f]Fluoroquinolones, such as levofloxacin (500 mg 2×/day), ciprofloxacin (750 mg 2×/day).
[g]In cases of aerobic Gram-negative microorganisms resistant to amoxicillin–clavulanic acid.
[h]Total duration for only 1 week, at maximum 2 weeks.

antibiotics. It has an excellent bioavailability of 100%, which makes it a feasible and convenient choice for outpatient treatment [44]. Disadvantages of linezolid are its high cost, and some serious side effects include reversible bone marrow suppression (mainly thrombopenia), especially during prolonged administration for more than 2 weeks. Regular control for myelotoxicity is mandatory. Most serious adverse events are optic neuropathy and nonreversible peripheral neuropathy, which have been reported in 2–4% of patients with prolonged administration of linezolid. A severe serotonin syndrome in comedication with certain antidepressant drugs, such as monoamine oxidase inhibitors, has been described [48].

Trimethoprim/sulfamethoxazole is an inexpensive folate antagonist. In a study in IV drug addicts with *S. aureus* infection, it was only slightly less efficacious than vancomycin [49]. It may be less active for some severe infections, releasing significant amounts of thymidine from damaged host tissues and bacteria, which might antagonize its activity. Indeed, thymidine is known to antagonize the antistaphylococcal effects of both trimethoprim and sulfamethoxazole. Furthermore, it is also documented that *S. aureus* thermonuclease may release thymidine from DNA. Thus, failure with trimethoprim/sulfamethoxazole may well depend on the extent of tissue damage and organism burden [50]. Main adverse events during prolonged administration are nausea, rash, myelosuppression, allergy, and hepatitis.

Tetracyclines (doxycycline and minocycline; both 100 mg twice daily) are lipophilic, facilitating their uptake into tissues. Tetracyclines may be combined with rifampin, in order to get bactericidal activity against staphylococci [51]. Main adverse events include nausea and a risk of photosensitivity during summer. Oral fusidic acid administered 500 mg three times daily has demonstrated efficacy in osteoarticular infections [52] and inhibits protein synthesis. Most experts do not recommend fusidic acid monotherapy because of the development of (potentially reversible) resistance [53]. The time interval between onset of therapy and emergence of fusidic acid resistance is not defined and is quite variable. Hepatic failure has been reported when using fusidic acid and rifampin combinations, which may require a regular monitoring of liver function. Finally, streptogramins such as quinupristin–dalfopristin (IV) or pristinamycin (oral) inhibit protein synthesis by binding to bacterial ribosomes. Quinupristin–dalfopristin administration requires central venous access and dextrose infusion. However, adverse effects, such as myalgias, arthralgias, and nausea, limit their use [44].

In patients with anaerobic, streptococcal, and staphylococcal infection, bacterial protein synthesis inhibitors such as clindamycin (600–900 mg three times daily) may be an option. The clinical efficacy of clindamycin in arthritis can be explained by its excellent penetration despite its classification as a bacteriostatic agent [44]. However, a potential pitfall is the inducible clindamycin resistance. In isolates tested as susceptible to clindamycin but resistant to erythromycin in routine testing, resistance may be induced during ongoing clindamycin treatment [54]. Although staphylococci may be susceptible to fosfomycin and chloramphenicol, these antibiotics have not been recommended for osteoarticular infections and should be avoided, due to the potential risk of agranulocytosis under chloramphenicol administration. For anaerobic osteomyelitis, metronidazole is the drug of choice [44]. High-dose metronidazole may cause peripheral (irreversible) neuropathies.

Fluoroquinolones are the most important antimicrobial agents against Gram-negative infections. Because of their excellent bioavailability, they can be used by the oral route; practically from the start. A multicenter study in Switzerland proved that oral combination therapy with fluoroquinolones plus rifampicin was an alternative to standard parenteral therapy (cure rates; 86% versus 84%, respectively) for staphylococcal infections, including joint infections in 35 cases [55]. *Pseudomonas aeruginosa* and other nonfermenting Gram-negative rods may rapidly develop fluoroquinolone resistance in monotherapy. Therefore, combined therapy of fluoroquinolones with another antipseudomonal drug, or prolonged IV treatment of pseudomonal arthritis, would be wise. However, to the best of our knowledge, no antibiotic treatment adapted to this situation has been studied so far. Of note, the optimal oral dose of ciprofloxacin for synovial and bone [44] infections is set at 750 mg twice daily orally [44] for patients with good renal function.

Duration of Antibiotic Therapy and Duration of the Initial IV Part of Administration

The optimal duration of concomitant antibiotic treatment remains controversial and unknown since randomized controlled studies are lacking [56]. Different regimens have been recommended such as 2-week IV therapy for streptococci, 3–4 weeks IV for staphylococci and Gram-negative bacteria [57, 58], and more than 4 weeks for immunocompromised patients or damaged joints, for example, severe osteoarthritis [57]. Other authors recommend parenteral treatment for 2 weeks, followed by 2 additional weeks of oral treatment [56], or for 4 weeks without indicating the route of administration [57]. Noteworthy, many surgeons prescribe antimicrobials for longer periods without evidence-based arguments [40, 59]. The current opinion is that a standard course of 6 weeks, such as used for chronic osteomyelitis [44] or periprosthetic joint infections after implant removal, might be excessive for native joint infections. We performed a retrospective study of native joint infections among adults [14]. In multivariate analysis, total duration of antibiotic therapy (odds ratio [OR] 1.0, 95% confidence interval [CI] 0.95–1.05) or duration of IV antibiotic therapy (1.0, 0.95–1.05) were not associated with the risk of recurrence. Seven days of IV therapy had the same success rate as 8–21 days (0.4, 0.1–1.7) or <21 days (1.1, 0.4–3.1). In addition, 2 weeks of total antibiotic treatment had the same outcome as 2–4 weeks (0.4, 0.1–2.3) or >4 weeks (0.4, 0.1–1.6) [14].

Since prospective randomized trials are lacking, the current view of initial parenteral antibiotic treatment administered for 2 weeks for all severe orthopedic infections can hardly be modified, with the exception of fluroquinolone use, which enables oral therapy from day 1 or 2 [55]. Septic arthritis studies have been published only for children, which revealed a successful outcome of IV antibiotics administered for a short period of time (see Chapter 5). Prado *et al.* [60] reported that in 70 children with osteoarticular infection (60% septic arthritis), an initial regimen of IV antibiotic(s) for 7 days, followed by 3 to 5 weeks of oral therapy, was sufficient for cure. Peltola *et al.* [61] recommend a 10-day course of antimicrobial treatment in a randomized trial comparing 10 days (including a short-term IV therapy) to 30 days for childhood arthritis. In a prospective study, Jagodzinki *et al.* [62] showed that 3 days of high-dose IV therapy followed by 3 weeks of oral medication cured 70 children. Like in our study, an early switch from parenteral to oral antibiotics after a median duration of 7 days was equally effective in the treatment of pediatric arthritis among 186 children [63]. This cutoff of 7 days was supported by another trial where 7 versus 14 days of parenteral antibiotics yielded an equivalent outcome after surgical drainage in 130 cases with infectious arthritis [64]. For adults, Angly *et al.* [65] investigated 31 operated patients with finger arthritis. No recurrence of infection occurred after surgery combined with antibiotics administered for a median duration of 2 days IV and 17 days orally. Provided that the pathogen is susceptible to antibiotic agents with a good synovial penetration and oral bioavailability, IV antimicrobial therapy could also be limited to 7 days for adult patients, except for those with bacteraemia, endocarditis, sepsis with hypotension, or compromised intestinal absorption [55]. A reduced duration of antibiotic treatment for a total of 2–3 weeks may decrease not only antibiotic consumption, but also related costs, adverse effects, and selective pressure for antibiotic resistance among bacterial pathogens.

In general practice, the duration of antibiotic administration was rarely influenced by the nature of the pathogen with a few exceptions, that is, those proven to require long-lasting antibiotic treatment for eradication, such as tuberculosis, other mycobacteria, nocardiosis [66], or fungi. This uniform duration of concomitant antimicrobial treatment has remained constant for the last three decades.

Supportive Therapy

Besides surgical and antimicrobial chemotherapy, several adjunctive therapies are widely used or are under investigation for the treatment of septic arthritis. The most important aspect of arthritis therapy is to keep the joint mobility despite infection. Traditionally, many experts used to recommend bed rest or at least interdiction of weight bearing to patients with septic arthritis of the lower extremities. Although there are theoretical reasons for reducing the weight load, there is no scientific evidence for this practice. On the contrary, there is even increased cartilage damage if the infected joint is immobilized [67, 68]. Thus, patients should be allowed to move and stand, as long as the pain can be handled. This is especially important, because prolonged immobilization results in stiffening, which subsequently leads to additional morbidity and costs in the treatment of arthritis patients. A novel approach to the treatment of infectious arthritis, and a field of emerging research so far in children, has been the addition of systemic steroid therapy for native joint infections, especially to reduce the incidence of mechanical sequels due to infection [69]. Several animal studies indicate that injection of steroids in the infected joint (additional to systemic antibiotics) may reduce cartilaginous destruction without other deleterious effects [70]. Corticosteroids are supposed to diminish the number of T cells and macrophages in the synovial space, thus leading to improved cartilage preservation [70]. In the clinical situation, two randomized pediatric trials have examined this topic by randomizing in double-blind conditions 49 [71] or 123 infants [72] to receive antibiotics with or without systemic dexamethasone (the latter for 4 days) (see Chapter 5). Children on steroids showed a more rapid cure, a shorter hospital stay, and a significantly better functional outcome scale than those without steroids. Since there are no robust published data on corticosteroid therapy in septic arthritis of adults, we do not recommend their use in this population.

Outcome of Native Joint Septic Arthritis

The outcome of septic arthritis depends on the pathogenesis, the infecting agent, and the type of joint. In a literature review of posttraumatic cases, mortality equals zero and microbiological cure is achieved in 96%. In this particular population and according to our literature review, severe mechanical sequelae with arthrodesis, amputation, or invalidating pain were witnessed in 15% patients (12/82) [3]. There were considerable differences in outcome according to setting and the microorganism involved. Arthritis related to soil contact (*P. agglomerans*, *Nocardia*, *Actinomyces*, *Clostridium* spp.) showed the lowest functional impairment (2/16; 12%), as compared to water-related arthritis with three amputations for mixed bacterial and one for *M. marinum* infection. However, the worst functional outcome was seen in bite-inflicted cases with notably persistent pain in 5/6 of documented cases. Interestingly, fungal or mycobacterial infections resulted in relatively modest residual morbidity. In a patient population with a majority of primary cases and a high prevalence of *S. aureus*, we performed a retrospective case control study to evaluate the outcome recurrence of arthritis and sequels [14]. A total of 169 episodes of septic arthritis in 157 adult patients were studied. In 21 episodes (21/169, 12%), arthritis relapsed after the end of antibiotic treatment. Multivariate analysis showed that Gram-negative infection (OR 5.9, 95% CI 1.4–25.3) and immune suppression (5.3, 1.3–22.0) were associated with recurrence, but not the size of infected joints, the number of surgical drainages (1.3, 1.0–1.7), arthrotomy versus arthroscopic drainage (0.5, 0.2–1.8), or total duration of antibiotic therapy (1.0, 0.95–1.05). Steroid administration and organ transplantation were among the strongest predictors of recurrence [14]. Likewise, 7 days of IV therapy had the same efficacy as 8–21 days (OR 0.4, 0.1–1.8) and as 3 weeks or more of IV treatment (OR 1.6, 0.5–5.1) [14]. Sequels after microbiological cure of infection are another problem. The literature suggests that in the long term, roughly one-quarter of all patients witness moderate or severe sequels [14], leading to intermittent antalgic use or requiring orthopedic solutions for the damaged joint. In the aforementioned study, sequels in terms of functional impairment were observed in 26% of cases, as in the studies of Vispo-Seara *et al.* [40], with 20% sequels in patients with septic arthritis. Immune suppression (OR 3.6, 1.5–8.7) was associated with sequelae. No antibiotic-related parameters influenced sequels.

Key Points

- All septic arthritis cases require drainage to be cured.
- The recommended duration of concomitant antibiotic therapy is about 4 weeks.
- Most cases of primary septic arthritis and surgical site infections (SSIs) are due to Gram-positive pathogens, mainly *Staphylococcus aureus*. Posttraumatic cases show a greater variability with a higher proportion of Gram-negative and atypical pathogens.

Acknowledgments

We thank all colleagues of the Division of Orthopaedic and Trauma Surgery for clinical help and advice.

References

[1] Uçkay I, Pittet D, Vaudaux P, *et al.* Foreign body infections due to *Staphylococcus epidermidis*. Ann Med 2009;41(2):109–119.
[2] Geirsson AJ, Statkevicius S, Vikingsson A. Septic arthritis in Iceland 1990–2002: increasing incidence due to iatrogenic infections. Ann Rheum Dis 2008;67(5):638–643.
[3] Di Benedetto C, Hoffmeyer P, Lew I, Uçkay I. Post-traumatic septic arthritis. Eur Musculoskel Rev 2011;7(1):34–45.
[4] Favero M, Schiavon F, Riato L, *et al.* Septic arthritis: a 12 years retrospective study in a rheumatological university clinic. Reumatismo 2008;60(4):260–267.
[5] Eder L, Zisman D, Rozenbaum M, Rosner I. Clinical features and aetiology of septic arthritis in northern Israel. Rheumatology (Oxford, England) 2005;44(12):1559–1563.
[6] Weston VC, Jones AC, Bradbury N, *et al.* Clinical features and outcome of septic arthritis in a single UK Health District 1982–1991. Ann Rheumatol Dis 1999;58(4):214–219.

[7] Elyès B, Mehdi G, Kamel BH, *et al.* *Kingella kingae* septic arthritis with endocarditis in an adult. Joint Bone Spine 2006;73(4):472–473.
[8] Kaandorp CJ, Dinant HJ, van de Laar MA, *et al.* Incidence and sources of native and prosthetic joint infection: a community based prospective survey. Ann Rheum Dis 1997;56(8): 470–475.
[9] Stutz G, Kuster MS, Kleinstuck F, Gachter A. Arthroscopic management of septic arthritis: stages of infection and results. Knee Surg Sports Traumatol Arthrosc 2000;8(5):270–274.
[10] Morgan DS, Fisher D, Merianos A, Currie BJ. An 18 year clinical review of septic arthritis from tropical Australia. Epidemiol Infect 1996;117(3):423–428.
[11] Gomez-Rodriguez N, Ibanez-Ruan J, Gonzalez M, *et al.* Peripheral septic arthritis in adults. Epidemiologic study in a Galician health area. An Med Interna 2001;18(11):573–577.
[12] Uçkay I, Sax H, Harbarth S, *et al.* Multi-resistant infections in repatriated patients after natural disasters: lessons learned from the 2004 tsunami for hospital infection control. J Hosp Infect 2008;68(1):1–8.
[13] Tao C, Kang M, Chen Z, *et al.* Microbiologic study of the pathogens isolated from wound culture among Wenchuan earthquake survivors. Diagn Microbiol Infect Dis 2009;63(3): 268–270.
[14] Uçkay I, Tovmirzaeva L, Garbino J, *et al.* Short parenteral antibiotic treatment for adult septic arthritis after successful drainage. Int J Infect Dis 2013;17(3):199–205.
[15] Ernst AA, Weiss SJ, Tracy LA, Weiss NR. Usefulness of CRP and ESR in predicting septic joints. South Med J 2010;103(6):522–526.
[16] Hariharan P, Kabrhel C. Sensitivity of erythrocyte sedimentation rate and C-reactive protein for the exclusion of septic arthritis in emergency department patients. J Emerg Med 2011;40(4):428–431.
[17] Papanicolas LE, Hakendorf P, Gordon DL. Concomitant septic arthritis in crystal monoarthritis. J Rheumatol 2012;39(1):157–160.
[18] Carpenter CR, Schuur JD, Everett WW, Pines JM. Evidence-based diagnostics: adult septic arthritis. Acad Emerg Med 2011;18(8):781–796.
[19] Hugle T, Schuetz P, Mueller B, *et al.* Serum procalcitonin for discrimination between septic and non-septic arthritis. Clin Exp Rheumatol 2008;26(3):453–456.
[20] Parvizi J, Jacovides C, Antoci V, Ghanem E. Diagnosis of periprosthetic joint infection: the utility of a simple yet unappreciated enzyme. J Bone Joint Surg Am 2011;93(24):2242–2248.
[21] Margaretten ME, Kohlwes J, Moore D, Bent S. Does this adult patient have septic arthritis? JAMA 2007;297(13):1478–1488.
[22] Siva C, Velazquez C, Mody A, Brasington R. Diagnosing acute monoarthritis in adults: a practical approach for the family physician. Am Fam Physician 2003;68(1):83–90.
[23] Frischnecht J, Steigerwald JC. High synovial fluid white blood cell counts in pseudogout; possible confusion with septic arthritis. Arch Intern Med 1975;135(2):298–299.
[24] Brannan SR, Jerrard DA. Synovial fluid analysis. J Emerg Med 2006;30(3):331–339.
[25] Brook I, Reza MJ, Bricknell KS, *et al.* Synovial fluid lactic acid. A diagnostic aid in septic arthritis. Arthritis Rheum 1978;21(7):774–779.
[26] Faraj AA, Omonbude OD, Godwin P. Gram staining in the diagnosis of acute septic arthritis. Acta Orthop Belg 2002;68(4):388–391.
[27] McGillicuddy DC, Shah KH, Friedberg RP, *et al.* How sensitive is the synovial fluid white blood cell count in diagnosing septic arthritis? Am J Emerg Med 2007;25(7):749–752.
[28] Yu KH, Luo SF, Liou LB, *et al.* Concomitant septic and gouty arthritis—an analysis of 30 cases. Rheumatology (Oxford, England) 2003;42(9):1062–1066.
[29] Martinot M, Sordet C, Soubrier M, *et al.* Diagnostic value of serum and synovial procalcitonin in acute arthritis: a prospective study of 42 patients. Clin Exp Rheumatol 2005;23(3): 303–310.
[30] Gardner GC, Weisman MH. Pyarthrosis in patients with rheumatoid arthritis: a report of 13 cases and a review of the literature from the past 40 years. Am J Med 1990;88(5):503–511.

[31] Weng CT, Liu MF, Lin LH, *et al.* Rare coexistence of gouty and septic arthritis: a report of 14 cases. Clin Exp Rheumatol 2009;27(6):902.
[32] Madison BM. Application of stains in clinical microbiology. Biotech Histochem 2001;76(3):119–125.
[33] Morgan PM, Sharkey P, Ghanem E, *et al.* The value of intraoperative Gram stain in revision total knee arthroplasty. J Bone Joint Surg Am 2009;91(9):2124–2129.
[34] Ciancio G, Bortoluzzi A, Govoni M. Epidemiology of gout and chondrocalcinosis. Reumatismo 2012;63(4):207–220.
[35] Rosenthal AK. Crystals, inflammation, and osteoarthritis. Curr Opin Rheumatol 2011;23(2): 170–173.
[36] Archibeck MJ, Rosenberg AG, Sheinkop MB, *et al.* Gout-induced arthropathy after total knee arthroplasty: a report of two cases. Clin Orthop Relat Res 2001(392):377–382.
[37] Smith RL, Schurman DJ, Kajiyama G, *et al.* The effect of antibiotics on the destruction of cartilage in experimental infectious arthritis. J Bone Joint Surg Am 1987;69(7):1063–1068.
[38] Wirtz DC, Marth M, Miltner O, *et al.* Septic arthritis of the knee in adults: treatment by arthroscopy or arthrotomy. Int Orthop 2001;25(4):239–241.
[39] Nusem I, Jabur MK, Playford EG. Arthroscopic treatment of septic arthritis of the hip. Arthroscopy 2006;22(8):902 e1–902 e3.
[40] Vispo-Seara JL, Barthel T, Schmitz H, Eulert J. Arthroscopic treatment of septic joints: prognostic factors. Arch Orthop Trauma Surg 2002;122(4):204–211.
[41] Magnan B, Bondi M, Vecchini E, *et al.* A preformed antibiotic-loaded spacer for treatment for septic arthritis of the shoulder. Musculoskelet Surg 2013. DOI:10.1007/s12306-013-0268-x. [Epub ahead of print]
[42] Uçkay I, Pagani L, Gamulin A, Lew DP. Infections in skeletal prostheses and allografts. In: Jarvis WJ, editor. Bennet & Brachman's Hospital Infections 2013, 6th edition, Chapter 40. Philadelphia: Lippincott, Williams & Wilkins 2013, pp. 608–617.
[43] Rice PA. Gonococcal arthritis (disseminated gonococcal infection). Infect Dis Clin North Am 2005;19(4):853–861.
[44] Uçkay I, Jugun K, Gamulin A, *et al.* Chronic osteomyelitis. Curr Infect Dis Rep 2012; 14(5):566–575.
[45] Hidayat LK, Hsu DI, Quist R, *et al.* High-dose vancomycin therapy for methicillin-resistant *Staphylococcus aureus* infections: efficacy and toxicity. Arch Intern Med 2006;166(19): 2138–2144.
[46] Boffi El-Amari E, Vuagnat A, Stern R, *et al.* High versus standard dose vancomycin for osteomyelitis. Scand J Infect Dis 2004;36(10):712–717.
[47] Jugun K, Vaudaux P, Garbino J, *et al.* The safety and efficacy of high-dose daptomycin combined with rifampicin for the treatment of Gram-positive osteoarticular infections. Int Orthop 2013;37(7):1375–1380.
[48] Bernard L, Stern R, Lew D, Hoffmeyer P. Serotonin syndrome after concomitant treatment with linezolid and citalopram. Clin Infect Dis 2003;36(9):1197.
[49] Markowitz N, Quinn EL, Saravolatz LD. Trimethoprim-sulfamethoxazole compared with vancomycin for the treatment of *Staphylococcus aureus* infection. Ann Intern Med 1992; 117(5):390–398.
[50] Proctor RA. Role of folate antagonists in the treatment of methicillin-resistant Staphylococcus aureus infection. Clin Infect Dis 2008;46(4):584–593.
[51] Ruhe JJ, Monson T, Bradsher RW, Menon A. Use of long-acting tetracyclines for methicillin-resistant *Staphylococcus aureus* infections: case series and review of the literature. Clin Infect Dis 2005;40(10):1429–1434.
[52] Drancourt M, Stein A, Argenson JN, *et al.* Oral treatment of *Staphylococcus* spp. infected orthopaedic implants with fusidic acid or ofloxacin in combination with rifampicinJ Antimicrob Chemother 1997;39(2):235–240.
[53] Turnidge J, Collignon P. Resistance to fusidic acid. Int J Antimicrob Agents 1999;12(2): 35–44.

[54] Yılmaz G, Aydın K, Iskender S, *et al.* Detection and prevalence of inducible clindamycin resistance in staphylococci. J Med Microbiol 2007;56(3):342–345.
[55] Schrenzel J, Harbarth S, Schockmel G, *et al.*; Swiss Staphylococcal Study Group. A randomized clinical trial to compare fleroxacin-rifampicin with flucloxacillin or vancomycin for the treatment of staphylococcal infection. Clin Infect Dis 2004;39(9):1285–1292.
[56] Falagas ME, Giannopoulou KP, Kokolakis GN, Rafailidis PI. Fosfomycin: use beyond urinary tract and gastrointestinal infections. Clin Infect Dis 2008;46(7):1069–1077.
[57] Nade S. Septic arthritis. Best Pract Res Clin Rheumatol 2003;17(2):183–200.
[58] Syrogiannopoulos GA, Nelson JD. Duration of antimicrobial therapy for acute suppurative osteoarticular infections. Lancet 1988;1(8575–6):37–40.
[59] Matthews PC, Dean BJ, Medagoda K, *et al.* Native hip joint septic arthritis in 20 adults: delayed presentation beyond three weeks predicts need for excision arthroplasty. J Infect 2008;57(3):185–190.
[60] Prado SM, Lizama CM, Pena DA, *et al.* Short duration of initial intravenous treatment in 70 pediatric patients with osteoarticular infections. Rev Chilena Infectol 2008;25(1):30–36.
[61] Peltola H, Paakkonen M, Kallio P, Kallio MJ. Prospective, randomized trial of 10 days versus 30 days of antimicrobial treatment, including a short-term course of parenteral therapy, for childhood septic arthritis. Clin Infect Dis 2009;48(9):1201–1210.
[62] Jagodzinski NA, Kanwar R, Graham K, Bache CE. Prospective evaluation of a shortened regimen of treatment for acute osteomyelitis and septic arthritis in children. J Pediatr Orthop 2009;29(5):518–525.
[63] Ballock RT, Newton PO, Evans SJ, *et al.* A comparison of early versus late conversion from intravenous to oral therapy in the treatment of septic arthritis. J Pediatr Orthop 2009;29(6): 636–642.
[64] Jaberi FM, Shahcheraghi GH, Ahadzadeh M. Short-term intravenous antibiotic treatment of acute hematogenous bone and joint infection in children: a prospective randomized trial. J Pediatr Orthop 2002;22(3):317–320.
[65] Angly B, Steiger R, Zimmerli W. Septic arthritis of finger joints. Handchir Mikrochir Plast Chir 2007;39(2):118–123.
[66] Uçkay I, Bouchuiguir-Wafa K, Ninet B, *et al.* Posttraumatic ankle arthritis due to a novel *Nocardia* species. Infection 2010;38(5):407–412.
[67] Ando A, Suda H, Hagiwara Y, *et al.* Reversibility of immobilization-induced articular cartilage degeneration after remobilization in rat knee joints. Tohoku J Exp Med 2011;224(2): 77–85.
[68] Brandt KD. Response of joint structures to inactivity and to reloading after immobilization. Arthritis Rheum 2003;49(2):267–271.
[69] Kronig I, Schibler M, Rougemont M, Emonet S. Corticosteroids in the treatment of infectious diseases. Rev Med Suisse 2013;9:862–866.
[70] Wysenbeek AJ, Volchek J, Amit M, *et al.* Treatment of staphylococcal septic arthritis in rabbits by systemic antibiotics and intra-articular corticosteroids. Ann Rheum Dis 1998; 57(11):687–690.
[71] Harel L, Prais D, Bar-On E, *et al.* Dexamethasone therapy for septic arthritis in children: results of a randomized double-blind placebo-controlled study. J Pediatr Orthop 2011;31(2): 211–215.
[72] Odio CM, Ramirez T, Arias G, *et al.* Double blind, randomized, placebo-controlled study of dexamethasone therapy for hematogenous septic arthritis in children. Pediatr Infect Dis J 2003;22(10):883–888.

Chapter 7
Septic Arthritis of Axial Joints

Werner Zimmerli

In this chapter, arthritis in three different axial joints will be presented, namely, sternoclavicular arthritis, symphysitis pubis, and septic sacroiliitis. The common characteristic of these three joints is their small range of motion. Septic arthritis of these three joints is rare and generally occurs in patients with special risk factors. The sternoclavicular articulation is a gliding diarthrodial joint [1, 2]. Infection of this articulation can present as arthritis with a minimal amount of synovial empyema, as soft tissue abscess, or as concomitant osteomyelitis. In contrast, the pubic symphysis and the sacroiliac joint are amphiarthrotic, which means only slightly movable. The former is a synchondrosis, and the latter a syndesmosis. A synchondrosis is a cartilaginous joint connected by hyaline cartilage. A syndesmosis is a slightly movable articulation united by fibrous connective tissue forming an intraosseous membrane. Since both types of joints only have a virtual intraosseous space, infections of synchondrosis as well as syndesmosis manifest in most cases as osteomyelitis of the adjacent bones. Since the characteristics of these three types of arthritis are different, they will be presented separately.

Septic Arthritis of the Sternoclavicular Joint

Introduction

The sternoclavicular joint is a synovial joint connecting the axial skeleton with the upper extremity, and is involved in the movement of the upper extremities [3, 4]. It is composed of the anterior sternoclavicular ligament, the interclavicular ligament, the costoclavicular ligament, and two synovial cavities separated by a fibrocartilaginous disk [5, 6]. Septic arthritis can occur by the hematogenous route, exogenously by direct injury, or by continuous spread from adjacent infections. Hematogenous infection of the sternoclavicular joint is rare as compared to other joints, because this joint receives only a very small proportion of the total cardiac output. Its proximity to major vascular structures explains the exogenous infection in IV drug user (IVDU) using the internal jugular vein for

Bone and Joint Infections: From Microbiology to Diagnostics and Treatment, First Edition.
Edited by Werner Zimmerli.

injection [7, 8]. Similarly, it may occur as a complication after subclavian venous catheterization [6, 9]. In addition, due to the lack of substantial overlying soft tissue, exogenous infection by animal bites or scratches may occur. Continuous infection is transferred from adjacent lymph nodes or infections from the surroundings. Tuberculous sternoclavicular arthritis is reported to seed hematogenously during primary dissemination [10, 11]. However, since bilateral infection has been reported, continuous spread from sternoclavicular lymph nodes on both sides is also feasible [11].

Epidemiology

Sternoclavicular joint arthritis is a rare disease, especially in a population without risk factors. This may be one of the reasons why diagnosis is often delayed [8]. Table 7.1 shows the published data on the prevalence of sternoclavicular arthritis in 1336 patients with native joint arthritis for the last three decades [12–19]. In many additional case series, the exact number of patients with sternoclavicular arthritis has not been reported. The reported prevalence goes from 0.52 to 4.08% [13, 19]. This eightfold difference is due to distinct risk factors in the populations rather than the patients' origin in different countries. Indeed, in the Australian study no IVDU were reported, whereas in the Swiss study, 15% of the patients were IVDU. This is an obvious explanation for the large difference in the two series (see risk factors). Overall, the mean prevalence of septic sternoclavicular arthritis is 1.7% in an unselected population. However, in IVDU, axial joint involvement is observed in up to 67% of the cases [20]. In the study of Brancos *et al.* [21], 8/36 (22.2%) of the episodes of septic arthritis in IVDU occurred in the sternoclavicular joint.

Microbiology

Staphylococcus aureus is the most frequent microorganism in all types of septic arthritis. In a cumulative statistic reported by Ross *et al.* [17], 1066/2407 (44%) of the cases with each type of septic arthritis were caused by *S. aureus*, 8% by pyogenic streptococci, and 6% by *Streptococcus pneumoniae.* The same first author reported the microorganisms that

Table 7.1. Fraction of sternoclavicular arthritis in case series with septic arthritis.

References	Country	Sternoclavicular joint (SCJ) arthritis/total arthritis	Percent patients with SCJ arthritis (%)
Cooper *et al.* [12]	UK	1/75	1.33
Morgan *et al.* [13]	Australia	1/191	0.52
Kaandorp *et al.* [14]	NL	4/214	1.87
Kaandorp *et al.* [15]	NL	2/76	2.63
Weston *et al.* [16]	UK	2/243	0.82
Ross *et al.* [17]	USA	2/108[a]	1.85
Chanet *et al.* [18]	France	5/282	1.77
Clerc *et al.* [19]	Switzerland	6/147	4.08
Total	Three continents	23/1336	1.72

[a]Only patients with pneumococcal arthritis.

have been isolated in 176 cases of sternoclavicular septic arthritis [8] (Table 7.2). *S. aureus* is also by far the most frequent isolate (86/176 = 49%) in this type of arthritis. It is followed by *Pseudomonas aeruginosa*, which causes 10% of the sternoclavicular joint infections. Interestingly, this microorganism causes only 1% of unselected types of arthritis [17]. Thus, *Pseudomonas* is 10-fold overrepresented in patients with sternoclavicular arthritis. This may be due to several factors. First, *P. aeruginosa*, which is a common water contaminant, used to be a frequent microorganism in bloodstream infections of IVDU, before sterile paraphernalia became commonly available. Therefore, the risk of infection caused by this microorganism clearly increased. Second, according to Ross *et al.* [8], overrepresentation may be an artifact at least partially because of duplicate reporting of cases.

The overrepresentation of *Brucella melitensis* is less clear. According to a large series on brucellar arthritis, the sternoclavicular joint was involved in only 2/106 cases (1.9%), that is, not more frequently than in arthritis caused by other microorganisms [22]. Thus, the generally reported overrepresentation of *Brucella* spp. in axial arthritis is not well-documented.

Risk Factors

According to Ross *et al.* [8], only about one-quarter of the patients with septic sternoclavicular arthritis have no predisposing condition. However, even in the absence of risk

Table 7.2. Microorganisms isolated in 176 cases of sternoclavicular septic arthritis.

Microorganism	Isolates (*n* (%))
Stapylococcus aureus	86 (49%)
Pseudomonas aeruginosa	18 (10%)
Brucella melitensis	13 (7%)
Escherichia coli	8 (5%)
Group B streptococcus	6 (3%)
Mycobacterium tuberculosis	6 (3%)
Nonspecified streptococcus	5 (3%)
Streptococcus pneumoniae	4 (2%)
Anaerobes	2 (1%)
Group A streptococcus	2 (1%)
Haemophilus influenzae type b	2 (1%)
Group G streptococcus	2 (1%)
Streptococcus milleri group	2 (1%)
Neisseria gnorrhoeae	2 (1%)
Other enteric Gram-negative bacilli[a]	5 (3%)
Miscellaneous[b]	7 (4%)
Polymicrobial	6 (3%)

Data from Ross *et al.* [8].
[a]One each of *Acinetobacter anitratus, Burkholderia pseudomallei, Citrobacter diversus, Proteus mirabilis*, and *Serratia marcescens.*
[b]One each of *Candida albicans, Haemophilus aphrophilus, Mycobacterium avium* complex, *Pasteurella multocida, Propionibacterium acnes, Staphylococcus epidermidis, and Streptobacillus moniliformis.*

factors, sternoclavicular septic arthritis must be considered in patients with pertinent signs and symptoms. Bar-Natan *et al.* [23] reported a case series and review of 27 previously healthy adults. Since the publication of the comprehensive review by Ross *et al.* [8], the high prevalence of risk factors has been confirmed in several case series [6, 18, 24–27]. The following risk factors have been reported with various frequencies: IV drug use, diabetes mellitus, trauma, infected central line, chronic renal failure, hepatic dysfunction, alcohol abuse, corticosteroid, HIV infection, malignancy, and liver cirrhosis [8, 24, 27]. In addition, radiation therapy in the region of the sternoclavicular joint has also been reported as a risk factor. Chanet *et al.* [18] observed nine patients after radiation therapy for breast cancer. Six of them had septic shoulder arthritis and three suffered from sternoclavicular joint arthritis at the site of previous radiotherapy, after a median time interval of 16 years after irradiation. It is highly probable that there was a causal link between the two events, because the authors reported only 2/273 (0.65%) episodes of sternoclavicular arthritis without and 3/9 (33%) with previous radiotherapy.

In the review of Ross *et al.* [8], 25 patients had a diagnosed primary focus, the most frequent being pneumonia (10/180), cellulitis (8/180), and endocarditis (3/180). Cinquetti *et al.* [25] reported sternoclavicular arthritis after Lemierre's syndrome. It seems obvious that in this case, *Fusobacterium necrophorum* seeded continuously from the septic thrombosis of the jugular vein to the adjacent joint. Occasional cases caused by *Pasteurella multocida* were described in patients with licking dogs [28], and cats [29], respectively.

Venous catheter-related infections endanger the sternoclavicular joint by different pathogenic mechanisms. First, the infection may seed by the hematogenous route. Second, the joint may be directly inoculated during the puncture procedure. Finally, catheter-related infections may invade the adjacent capsule of the joint [6]. In IVDU, pathogenesis of infection is similar. These patients have an increased colonization by *S. aureus*, a higher risk of bacteremia, and occasionally attempt direct puncture of the internal jugular vein between the heads of the sternocleidomastoid muscle. During this trial, they may directly injure the adjacent joint with a contaminated needle.

Clinical and Laboratory Features

In the largest series of Ross *et al.* [8], 73% of the patients were men. This male preponderance has also been observed in other studies going from 60% [30] up to 86% [24]. This sex imbalance is at least partly due to the overrepresentation of IVDU in most series. Looking exclusively at this population, 91% were male [8]. However, this cannot be the only explanation, since in Bar-Natan *et al.*'s [23] series of 27 patients without underlying risk factors, 70% of the patients were male.

According to the most comprehensive case series of Ross *et al.* [8], the leading symptom is chest pain (78%), which is sometimes localized in the shoulder (24%). The clinical spectrum of signs ranges from pure arthritis to periarticular inflammation, to osteomyelitis, to frank abscess formation, and to mediastinitis. Only 65% of the patients have greater than 38°C fever. Functional impairment characterized by decreased range of shoulder motion is observed in less than 20% of the patients. The sternoclavicular joint is almost not distensible because of strong ligaments. Therefore, the leading clinical sign is a tender (90%), but not a swollen joint (4%). There is a slight preponderance of right-side arthritis (57%), in 5% both sternoclavicular joints are involved, and in 21% polyarticular septic arthritis is observed. This is especially typical for gonococcal arthritis. This etiology must be especially considered in patients with migratory tenosynovitis and oligoarthritis [31]. In

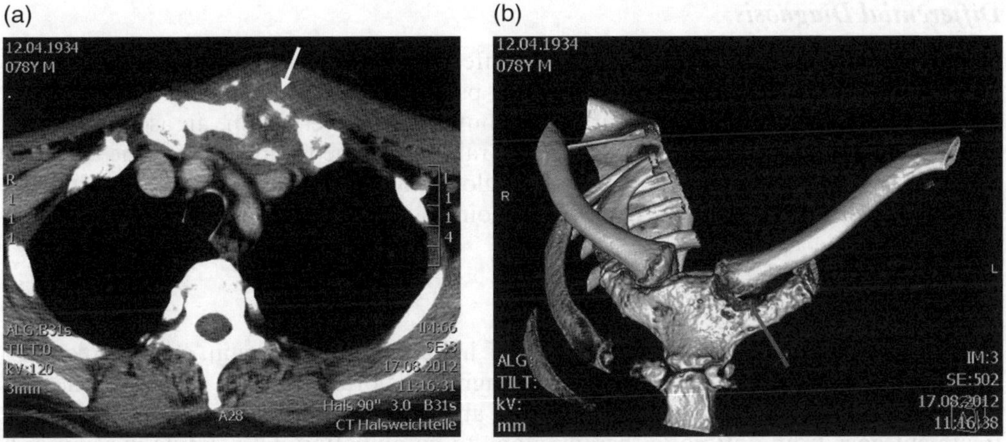

Figure 7.1. A 78-year-old man with monoclonal gammopathy of unknown significance presented with fever, cellulitis, and swelling on the left side of the upper thorax. Two blood cultures revealed *Staphylococcus aureus*. A first CT scan revealed a large abscess from the neck to the left nipple. Despite surgical drainage and adequate IV antibiotics, the infection progressed. (a) The CT scan control performed 5 weeks after initial drainage and start of antibiotic therapy shows a pyarthrosis of the left sternocalvicular joint, osteomyelitis of the proximal left clavicula and manubrium sterni with sequestrated bone (arrow), as well as a retrosternal abscess. (b) The reconstruction shows a dislocated joint (arrow) as well as erosions of the manubrium sterni and the first rib.

patients with central venous catheter-related infection, the diagnosis of sternoclavicular arthritis may be difficult. The typical leading sign is cellulitis around the insertion site [6].

Since the sternoclavicular joint has only a minimal synovial space, arthritis rapidly progresses to osteomyelitis, by invasion of medial parts of the clavicle, sternum, and ribs. Sometimes, pyogenic complications (chest wall abscess, mediastinitis) are the first signs of sternoclavicular arthritis. Overall, pyogenic complications such as abscess formation, septic thrombosis of the subclavian or internal jugular vein, mediastinitis, and empyema are frequent [26, 32–35]. According to Wohlgethan *et al.* [33], 20% of the patients suffer from an abscess (Figure 7.1). In most series, mortality is less than 5%. It mainly depends on the comorbidity of the patient population.

As in most types of arthritis, systemic laboratory parameters are diagnostically not helpful, due to their low sensitivity and/or specificity. In the case series of Ross *et al.* [8], only 56% of the patients had leukocytosis greater than 11 giga/l, and only 62% had bacteremia. Thus, rapid diagnosis needs imaging procedures followed by joint puncture or biopsy.

Imaging Procedures

Plain radiographs have a low sensitivity to reveal sternoclavicular arthritis. Three-phase bone scan has a good sensitivity, but a low specificity for the diagnosis of osteomyelitis [36]. In view of the frequency of pyogenic complications, computed tomography (CT) scan or magnetic resonance imaging (MRI) should be performed in all cases. Gas within the joint space does not prove arthritis. It is also known as vacuum phenomenon in normal individuals [4]. MRI should be preferred to CT scan in case osteomyelitis is suspected.

Differential Diagnosis

Septic sternoclavicular arthritis has a broad differential diagnosis including osteoarthritis, rheumatoid arthritis, ankylosing spondylitis, psoriasis, gout, and chondrocalcinosis. If clinical symptoms of arthritis occur in the context of skin eruption, an aseptic neutrophilic dermatosis should be considered. This rare syndrome is labeled with the acronym SAPHO (synovitis, acne, palmoplantar pustulosis, hyperostosis, osteitis). In most cases (65–90%) with SAPHO, the sternoclavicular joint is involved [37].

Management

The treatment aim is complete eradication of infection, relief of pain, and the recovery of function of the upper extremity. Since diagnosis is often delayed, this aim generally needs a combined noninvasive (antibiotics) and surgical approach. The spectrum of treatment goes from antibiotic therapy alone to combination with simple incision and drainage to invasive surgery, such as en bloc resection [30]. Antibiotics should be withheld until appropriate microbiological diagnosis has been performed. The choice of antibiotics does not differ from the one in all other types of native arthritis (see Chapter 6). In general, treatment is started with intravenous (IV) antibiotics. If the infecting agent is susceptible to drugs with excellent bioavailability, oral treatment can be started after a few days [38]. The duration has not been tested in comparative studies. In general, a 6-week course is proposed. In case of pyogenic complications, which are not completely removed by surgery, a longer treatment may be required.

Table 7.3 shows a classification that has been proposed by Abu Arab [24]. Patients with grade I–III can be treated either with antibiotics alone or combined with incision and drainage in case of effusion or local abscesses. Some patients with grade III may require debridement of necrotic bone. In patients with grade IV or V, sternoclavicular joint resection is generally needed [24, 27, 30, 36, 39–43]. A combined approach by a thoracic and an orthopedic surgeon may be favorable in case of resection arthroplasty. In case of large soft tissue defects, a plastic reconstructive surgeon should be consulted in addition.

Key Points

- Low prevalence of 0.5–4% in general population with septic arthritis, but much more frequent (15–22%) in septic arthritis of IVDU.
- Three-quarters of patients are men.
- In most cases, combined antibiotic and surgical management is needed.

Septic Arthritis of the Symphysis Pubis

Introduction

The symphysis pubis is a synchondrosis allowing minimal motion, but undergoing big sheering forces in pregnant women, especially during delivery. In addition, athletes including runners, and football, soccer, and ice hockey players suffer from repetitive

Table 7.3. Classification of sternoclavicular joint (SCJ) modified according to Abu Arab [24].

Grade I	Signs: Inflammation over SCJ with intact overlying skin, no systemic signs of infection Symptoms: Mild pain Imaging (X-ray, CT, MRI): Minimal swelling, no signs of osteomyelitis (intact clavicle, sternum and first rib), minimal or no effusion at SCJ
Grade II	Signs: Moderate to large swelling of SCJ, ± inflammation, ± systemic signs of infection Symptoms: Pain Imaging: Moderate swelling, moderate to large effusion at SCJ, no signs of osteomyelitis (intact clavicle, sternum, and first rib)
Grade III	Signs and symptoms: Any of the criteria mentioned in Grade II SCJ Imaging: Any of the criteria mentioned in Grade II SCJ plus minimal radiological signs of osteomyelitis
Grade IV	Any of the clinical or imaging criteria of grade I–II plus any of the following: Severe osteomyelitis Sinus tract Persistence or recurrence of infection
Grade V	Any of the clinical or imaging criteria of grade I–IV plus evidence of mediastinitis

Modified from Abu Arab W, Khadragui I, Echave V, *et al*. Surgical management of sternoclavicular joint infection. *Eur J Cardiothorac Surg*. 2011;40(3):630–4.

traumatisms at the insertion of the adductor muscles, leading to tendinitis, as well as inflammation and sclerosis of the symphysis pubis [44]. The diagnosis of septic arthritis of the pubic symphysis is difficult, since the clinical pictures of noninfectious osteitis pubis and septic arthritis of the symphysis are similar. In addition, in a joint lacking synovial fluid, arthritis cannot be clinically diagnosed. Therefore, it is important to maintain a high level of suspicion in case of typical risk factors, signs, and symptoms.

Epidemiology

The incidence of septic arthritis of the pubic symphysis is very low. In seven studies on septic arthritis reporting the localization of 1051 episodes, none reported a single case [12–17, 19]. This may be an artifact in part, because some of the cases are probably reported as osteomyelitis and not as arthritis, since greater than 95% of the cases have not only arthritis, but also adjacent osteomyelitis [45]. These cases may be missed with a PubMed search using the keywords "septic AND arthritis AND symphysis." Indeed, in the review of Brancos *et al.* [21], reporting 217 episodes of septic arthritis in heroin addicts, 19 (8.8%) were localized in the symphysis pubis. In IVDU each type of septic axial arthritis is more frequent than in a general population of patients with septic arthritis [21].

Microbiology

According to Ross and Hu [45], *S. aureus* causes one-third of septic arthritis of the symphysis pubis. It is the predominant pathogen in athletes. One-fourth is caused by *P. aeruginosa*. This microorganism was cultured from 87% of the patients with IV drug use. In the 1970s, in the United States most IVDU used pentazocine, an illicit drug, which was desolved in nonsterile water [8, 46]. The preponderance of this microorganism in drug addicts is explained by the fact that tap water is routinely contaminated with *P. aeruginosa* [47]. Interestingly, *Pseudomonas* infection in IVDU has been rarely observed in Europe, where most drug addicts inject heroin, which must be solved in acid. In the 1980s, lemon juice, occasionally containing *Candida* spp., was used for this purpose [48]. However, since at least 20 years, sterile paraphernalia are offered in so-called shooting galleries. As a result, this pathogen disappeared in more recent case series of infections in IVDU.

Most patients with pelvic malignancy have anaerobic/aerobic mixed infection, mainly due to the presence of a sinus tract [45]. After incontinence surgery, different enterobacteriaceae (*Escherichia coli*, *Proteus* spp., *Klebsiella pneumonia*, etc.), enterococci, or group B streptococci are observed. Other microorganisms such as *S. pneumoniae*, *Salmonella* spp., or *Brucella* spp. are found in rare cases [45, 49–51]. Bali *et al.* [52] reviewed nine cases with tuberculous pubic arthritis published during the last three decades. In addition, they cited studies published between 1888 and 1974, in which more than 100 cases suffering from tuberculosis of the symphysis pubis have been reported. Thus, before the availability of antituberculous drugs, apparently, mycobacteria preferentially seeded in the symphysis.

Risk Factors

Only a minority of patients have no predisposing condition. There are five main risk groups. Patients with previous incontinence surgery (Marshall–Marchetti–Krantz urethropexy) or pelvic malignancy (often with sinus tract) have the highest risk. Bouza *et al.* [51] reported three male patients suffering from septic arthritis of the symphysis pubis after implantation of an anti-incontinence device. Interestingly, two of them had previous local irradiation for prostate cancer, which may also be a predisposing factor (confer sternoclavicular arthritis). In addition, athletes, IVDU, and postpartal women are also at risk [45, 53]. Interestingly, IVDU are prone to all types of axial septic arthritis for unknown reasons. Part of the explanation could be their relatively low age, since the risk for axial septic arthritis increases with younger age. Indeed, in the series of Ross and Hu [45], the mean age was only 48 years, as compared to 65 years in a review of adult patients with all types of arthritis [54].

Clinical and Laboratory Features

According to Ross and Hu [45], as well as many case reports, pain is the leading symptom [46, 49–53, 55–62]. Two-thirds of patients have a localized pubic pain, which allows a straightforward workup. However, if the pain is localized in the groin (41%), thigh (15%), or hip (12%), more frequent alternative diagnoses have to be excluded. In individual case reports, pain has also been reported in the buttock [57] or in the testis [58]. Leading signs are pubic tenderness (88%), fever greater than 38°C (74%), painful (waddling) gait (59%), and pain with hip motion (45%) [45].

Laboratory signs are either not sensitive or nonspecific, or both. Leukocytosis (>11 giga/l) is observed in only one-third of the patients. In contrast, increased sedimentation rate and/or C-reactive protein (CRP) have been mentioned in most case reports, but are also nonspecific.

Imaging Procedures

Plain radiograph findings lag behind clinical symptoms about 2–4 weeks. In the series of Ross and Hu [45], 24/76 (32%) of the pelvic radiographs were normal at the time of the first examination. Later, signs of arthritis (symphyseal widening) or osteomyelitis (marginal erosions, bone destruction) can be observed. ^{99m}TC-MDP bone scan is positive within a few days, but has a low specificity [57]. It documents bone turnover, which is observed in each type of inflammation, including sterile osteitis. A CT scan shows abnormalities at an early time point. It documents soft tissue inflammation (phlegmon, abscess, or sinus tract) and symphyseal widening as signs of arthritis and bone erosion and sequester as signs of osteomyelitis (Figure 7.2). The presence of bone sequesters differentiates osteomyelitis from athletes' osteitis. Except during pregnancy, and in the immediate postpartum period, there is no gas observed within the normal pubic symphysis. Thus, this may be a sign of septic arthritis of the symphysis pubis [63]. MRI is the most sensitive and specific imaging technique in patients with arthritis of the pubic symphysis. It shows inflammation of the skin and muscles, intra-articular synovial fluid or abscesses, and early signs of osteomyelitis, which complicates arthritis in 97% of the patients [45, 53].

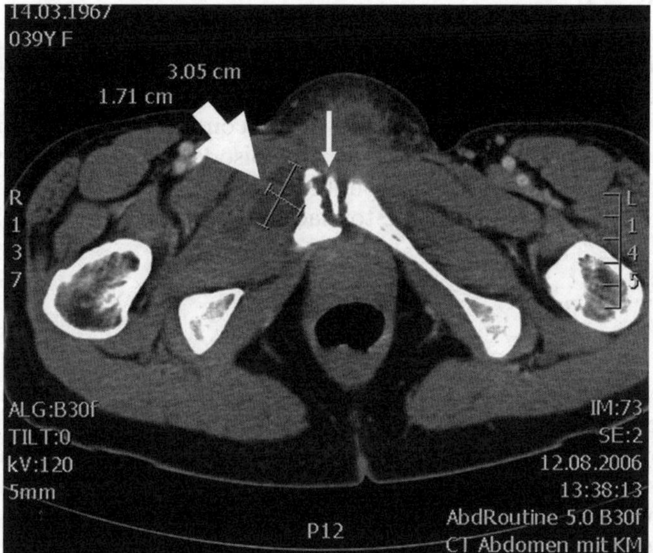

Figure 7.2. A 39-year-old woman with IV drug use (heroin and cocaine) since age 15. The patient suffered from HIV infection CDC B2 and chronic hepatitis C. During 5 years, she had three episodes of *Staphylococcus aureus* endocarditis of the mitral valve. After a free interval of 1 year with occasional IV drug use, she suffered from severe pubic pain. Septic symphysitis pubis was suspected on a computed tomography (CT) scan and proven with a biopsy of the symphysis. Bone biopsies and blood cultures showed growth of an identical *Staphylococcus epidermidis.* At the end of a correct 6-week course of antibiotics, bone erosions (thin arrow), sequesters, and prepubic abscesses (thick arrow) were still visible.

Differential Diagnosis

The clinical differential diagnosis includes syndromes characterized by pain as in pubic arthritis. It may imitate acute appendicitis (abdominal pain and fever), groin hernia (groin pain), slipped disk (backache and radiating pain to the thigh), prostatitis, and cystitis. Most of these entities can be easily excluded either by appropriate laboratory workup or by imaging procedures.

Pathological findings in different imaging procedures (plain radiograph, CT scan, bone scintigraphy) have a broad differential diagnosis including arthritis of the pubic symphysis, noninfectious osteitis pubis, seronegative spondyloarthropathies, posttraumatic changes in the context of pregnancy or delivery, overuse stress injury in athletes (soccer, football, running, ice hockey, etc.), osteoporotic fracture, osteoarthritis and idiopathic hyperostosis, metabolic (hemochromatosis etc.) and crystal-induced diseases (gout, chondrocalcinosis), as well as malignancies.

Osteitis pubis in athletes is generally hard to differentiate from septic pubic arthritis. In addition, the presence of one diagnosis does not exclude the other, since osteitis pubis predisposes to subsequent infection. Consecutive appearance of both diagnoses has been reported [49]. Radiographic changes of the symphysis are frequent in soccer players. Harris and Murray [64] reported radiological signs of osteitis pubis in 76% of players routinely examined. In addition, in 58% it correlated with symptoms such as groin and lower back pain.

Patients with ankylosing spondylitis have signs of not only sacroiliitis, but frequently also radiographic abnormalities of the pubic symphysis [65]. Less frequently, such changes are also observed in patients with psoriatic arthritis [66].

Calcification of the symphysis pubis is a rather common incidental manifestation of chondrocalcinosis [63]. However, most of these episodes obviously occur asymptomatically, because there is only one single publication of clinical pseudogout in the pubic symphysis [67]. Similarly, symptomatic tophaceous gout in the symphysis is also a very rare event [68].

Management

All patients need antimicrobial therapy, as soon as the microbiological testing shows growth of a relevant microorganism. Since septic arthritis can only be differentiated from osteitis pubis by the presence of a positive culture, antibiotics should be withheld until formal proof of infection. If blood cultures remain negative and cultures of the needle aspirates do not show any microbial growth, open debridement surgery for diagnostic purposes is needed before starting antimicrobial therapy.

Since the diagnosis of septic pubic arthritis is often delayed, most patients suffer from concomitant osteomyelitis. In addition, in many patients retropubic abscesses are observed. Therefore, in at least half of the patients, surgery is required. This includes incision and drainage of an abscess, debridement of dead bone, or both.

The choice of the antimicrobial agent does not differ from that in other types of native arthritis or osteomyelitis (see Chapters 6 and 15). In general, a duration of 6 weeks is suggested [45]. Antimicrobial therapy is generally started by the IV route. If possible, treatment can be switched to the oral route, if drugs with excellent bioavailability are available.

Key Points

- Most patients have a predisposing condition, such as urogenital surgery, IV drug use, participation in heavy sports.
- Pain and pubic tenderness are leading symptoms and signs.
- Surgery is needed in case of chronic osteomyelitis or pyogenic complications.

Septic Arthritis of the Sacroiliac Joint

Introduction

The sacroiliac joint articulates between the sacrum and the innominate bone of the pelvis. Whereas the posterosuperior portion is a syndesmosis, the anterior portion is synovial [69]. This joint is typically affected by inflammatory rheumatic diseases, such as ankylosing spondylitis or psoriatic spondyloarthritis [70]. In addition, like pubic symphysis, it undergoes mechanical stress during pregnancy, delivery, and heavy sport exposure [71, 72]. Thus, the clinical differentiation between inflammatory rheumatic disease, mechanical stress, and septic arthritis may be difficult.

Epidemiology

The incidence of septic sacroiliitis is low. In seven studies reporting the localization of 1051 episodes of septic arthritis, only five (0.48%) were localized in the sacroiliac joint [12–17, 19]. The incidence is about three times higher in children than in adults. In children, 55–60% of the episodes are observed in males, whereas in adults, there is a female preponderance of about 60–65%, probably due to the increased risk during pregnancy [73–76]. Adult septic sacroiliitis generally occurs in adolescents and during child-bearing age [76]. The median/mean age of the adult patients has been reported to be around 29 years [77], 37 years [78], and 50 years [73]. Adults with septic sacroiliitis are about 10 years younger as compared to adults with other types of septic arthritis [19, 73]. In IVDU, the incidence of septic arthritis of the sacroiliac joint is very high. In a compiled statistic of 217 heroin addicts with arthritis, 67(30.9%) were localized in the sacroiliac joint [21].

Microbiology

According to a large review of cases with septic sacroiliitis, the three most frequent causal agents were *S. aureus* (70%), streptococci (mainly Group A and B) (8.9%), and *P. aeruginosa* (5.2%) [77]. Any microorganism causing bacteremia can seed in the sacroiliac joint. As an example, a double infection with *Campylobacter rectus* and *Actinomyces odontolyticus* has been reported in a patient with severe periodontitis [79]. Furthermore, in a young HIV-infected patient, Chlamydia DNA was detected in a CT-guided puncture from the sacroiliac joint, 3 weeks after an episode of urethritis [80]. Thus, detection of the infectious focus may give a hint regarding the microorganism causing septic sacroiliitis. Interestingly, *Salmonella* spp. sacroiliitis has been repeatedly reported in children as well as in adults [81]. In the review of Zimmermann *et al.* [77], 4.9% of the episodes were caused by *Salmonella* spp. Feldman [82] reviewed 24 *Salmonella* cases reported between 1977 and 2006. Most patients were adolescents (mean age 18.8 years), none of them had a known immunodeficiency, and none had sickle cell anemia. In endemic zones for *Brucella* (Mediterranean countries of Europe, Northern and Eastern Africa, Near East countries, India, Central Asia, Mexico, and Central and South America), brucellar sacroiliitis has to be considered, especially in case of subacute or chronic presentation [83]. Tuberculous sacroiliitis has been observed in about 10% of the patients with bone and joint tuberculosis. It is also mainly reported from endemic areas. Gao *et al.* [84] reported 15 patients observed between 1997 and 2007 in China. In a very large original series from France, comprising 214 episodes of septic sacroiliitis, 65% were due to pyogenic microorganisms, 25% to *Mycobacterium tuberculosis*, and 10% to *Brucella* spp. [78].

Risk Factors

IV drug use is the most frequent associated factor, reported in 14% of the cases [85]. About 10% of children and 6.6% of adults suffering from septic sacroiliitis have a history of previous pelvic trauma [74, 85]. Other important predisposing conditions are abortion or pregnancy [77, 86]. In addition, sacroiliac joint steroid injection, which is an increasingly used intervention, can also result in septic sacroiliitis [87].

Since septic sacroiliitis is generally a hematogenous arthritis, some patients have a history of primary infection. According to Vyskocil *et al.* [85], 7.2% of the patients suffered from skin infection, followed by upper and lower respiratory tract infection (3.6% each), gynecological infection (3%), and urinary tract (1.2%) infection.

Clinical and Laboratory Features

The clinical diagnosis of septic sacroiliitis is difficult due to the nonspecific signs and symptoms. In addition, fever is often lacking, probably due to the frequent use of analgesics lowering the temperature. In a large French study, only 16 out of 39 patients (41%) were febrile greater than 37.8°C [76]. In contrast, in a literature review comprising 103 cases, fever was more common (75%) [85]. Similarly, in the largest original study on 138 patients with pyogenic sacroiliitis, 82% had a temperature of at least 38°C [78]. The classical triad of fever, low back pain, and difficulty in weight bearing is more frequently seen in pediatric (82%) than in adult (64%) patients [73].

The leading symptom is pain, described as low back, buttock, abdominal, pubic, hip, coxofemoral, thigh, and calf pain. The radiating pain can be explained by the fact that the first two sacral nerves (superior gluteal and obturator nerves) cross anterior to the sacroiliac joint [88]. Clinical signs can be summarized as fever, tenderness on pressure on the sacroiliac joint, tenderness on rectal palpation, pain exacerbation by weight bearing or by moving the sacroiliac joint, painful limping, pain on hip mobilization, psoas sign, and painful abdomen [70, 73, 75, 76, 78, 79, 81–100]. In addition, provocation maneuvers such as Mennell test (also labeled as Gaenslen's test in the US literature) or the FABER (flexion, abduction, external rotation) test are generally positive.

Leukocyte count is not a sensitive parameter. In the study of Wu *et al.* [73], only 46% had leukocytosis, whereas a CRP greater than 60 mg/l was observed in 73% of the patients.

Blood cultures are positive in a minority of cases, namely, 23 and 58%, respectively, in two different case series [73, 85]. This low yield may be due to previous antibiotic therapy.

We observed eight non-pretreated patients, all with positive blood cultures [100]. Synovial fluid aspiration is rarely performed without evidence of fluid in an imaging procedure. However, the rate of positive synovial fluid aspiration is as high as 75%, if an abscess is visible on CT or MRI [73]. If both blood cultures and synovial fluid do not show microbial growth, a biopsy of the sacroiliac joint should be performed before starting antimicrobial therapy.

Imaging Procedures

Plain radiographs have a low sensitivity, and are usually normal until several weeks after infection [85, 101–103]. In contrast, ^{99m}Tc-MDP bone scintigraphy is very sensitive for bone and joint infection and can be positive as early as 2 days after onset of symptoms [100]. After the first week, the majority is positive [88, 100]. However, bone scintigraphy is nonspecific, that is, it shows the same pathology in all types of bony inflammation. In addition, due to the high bone turnover, there is a physiological high uptake of technetium methyldiphosphonate in the sacroiliac joints. A pathological accumulation of the tracer is easier to detect in unilateral arthritis. Thus, bone scintigraphy is still useful as a screening test, if unilateral septic sacroiliitis is suspected. The hybrid procedure (single-photon emission computed tomography/computed tomography: SPECT–CT) allows a better localization of the inflammatory process [95]. However, this procedure is more specific only if it is combined with labeled antigranulocyte antibodies or granulocyte scintigraphy [104].

CT allows the evaluation of soft tissue abnormalities, bone irregularities, joint erosions, and abscesses (Figure 7.3) [103]. Thus, it is especially useful as follow-up examinations in patients with an unsatisfactory response to treatment, in order to look for pyogenic complications (e.g., *Musculus iliacus* abscess).

As in all types of osteomyelitis, MRI is the most sensitive and specific imaging procedure. Bone marrow edema and inflammatory changes of the adjacent muscle tissue differentiate septic sacroiliitis from arthritis in patients with inflammatory rheumatoid disease [94, 101, 102, 105].

With F-18 FDG PET/CT, hypermetabolic activity cannot only be shown in tumor tissue, but also in infection [106]. However, its role in the diagnosis of septic sacroiliitis has to be confirmed.

Differential Diagnosis

Since the sacroiliac joint can be affected by inflammatory rheumatic disease, trauma, and infection, the differential diagnosis is rather broad. It includes different types of seronegative spondyloarthropathies (e.g., psoriatic arthritis, ankylosing spondyloarthropathy), crystal arthropathy, rheumatoid arthritis, familial Mediterranean fever, Behcet's disease, Whipple's disease, hyperparathyroidism, and traumatic lesions [77, 107, 108]. Occasionally, metastatic carcinoma or sarcoma may also mimic sacroiliitis [109, 110].

In a French case series of 39 adults with septic sacroiliitis, the suspected clinical diagnosis at admission included lumbar disk herniation, vertebral osteomyelitis, mechanical low back pain, septic arthritis of the hip, and inflammatory sacroiliitis [76, 86]. According to a review of 191 cases published between 1929 and 1983, abdominal pain is a frequent symptom occurring in 24/191 (12.6%) of the cases [111]. Thus, septic sacroiliitis is an imitator of the acute abdomen, which frequently led to laparotomy before CT scan and MRI were available.

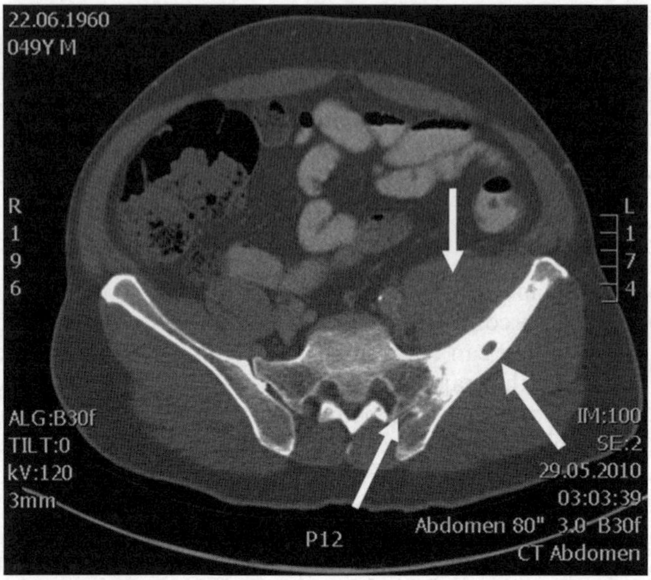

Figure 7.3. A 49-year-old man, originally from Sri Lanka, with a 3-week history of pain in the left leg and sacrum. No trauma, no injections, intermittent fever, and shivering. At hospitalization, the patient had 38.2°C fever and was unable to stay on his left leg. In addition, his sacroiliac joint was painful on palpation, and the provocation maneuver (Mennel sign) was positive on the left side. CRP 261 mg/l, leukocytes 11.9 G/l, growth of *Staphylococcus aureus* in 3/3 biopsies. The computed tomography (CT) scan was performed after a 3-week history of symptoms at the time of hospitalization. It shows signs of spontaneous arthrodesis (thin arrow from dorsal) after septic arthritis of the left sacroiliac joint with osteomyelitis of the ileum (thick arrow) and a large abscess of the *Musculus iliacus* (thin arrow from ventral). Open surgical revision was performed.

Management

Rapid IV antimicrobial therapy is the cornerstone of treatment. However, as in all types of bone and joint infection, antibiotics should only be started when infection is microbiologically documented. As an exception, in patients with a sepsis syndrome, empirical therapy should be started immediately after sampling of blood cultures and CT-guided puncture of possible abscesses. Such an initial therapy has to be guided according to the most frequent spectrum of microorganisms, that is, it is an educated guess. In IVDU, it should include the spectrum of *S. aureus* and *P. aeruginosa* (e.g., piperacillin/tazobactam). In pregnant or postpartal women Group B streptococci (e.g., ceftriaxone), and in patients with previous episodes of diarrhea, *Salmonella* spp. (e.g., ceftriaxone) should be covered by the empirical antibiotic. If the microorganism and its susceptibility are known, the choice of the antimicrobial agent does not differ from the one in other types of native arthritis or osteomyelitis (see Chapters 6 and 15). In general, a 6-week course is suggested [45]. Antimicrobial therapy is generally started by the IV route. If possible, treatment can be switched to the oral route, if drugs with excellent bioavailability are available.

In case of rapid diagnosis, open surgical intervention is rarely required [86, 100]. However, if imaging procedures reveal an abscess, CT-guided drainage should be urgently performed for diagnostic and therapeutic reasons. Open surgery is needed in case of large abscesses, lack of response to antibiotics and sequesters, or late instability surgical debridement and arthrodesis may be required [77, 97, 99]. For this purpose, a minimally invasive technique has been described [97].

Key Points

- Mean age of adults with sacroiliitis is about 10 years younger than in other types of septic arthritis. There is a female preponderance.
- Classical triad of fever, low back pain, and difficulty in weight bearing is present in only two-thirds of the patients.
- Seronegative spondyloarthropathy is the most frequent differential diagnosis.

References

[1] Yood RA, Goldenberg DL. Sternoclavicular joint arthritis. Arthritis Rheum 1980;23(2): 232–239.
[2] Renfree KJ, Wright TW. Anatomy and biomechanics of the acromioclavicular and sternoclavicular joints. Clin Sports Med 2003;22(2):219–237.
[3] Stark P, Jaramillo D. CT of the sternum. AJR Am J Roentgenol 1986;147(1):72–77.
[4] Restrepo CS, Martinez S, Lemos DF, *et al.* Imaging appearances of the sternum and sternoclavicular joints. Radiographics 2009;29(3):839–859.
[5] Barbaix E, Lapierre M, Van Roy P, *et al.* The sternoclavicular joint: variants of the discus articularis. Clin Biomech (Bristol, Avon) 2000;15(Suppl 1):S3–S7.
[6] Pradhan C, Watson NF, Jagasia N, *et al.* Bilateral sternoclavicular joint septic arthritis secondary to indwelling central venous catheter: a case report. J Med Case Rep 2008;2:131.
[7] Ghasemi Barghi R, Mirakbari SM. Septic arthritis of sternoclavicular joint: a case report of a rare finding in injecting drug users. Arch Iran Med 2010;13(3):248–250.
[8] Ross JJ, Shamsuddin H. Sternoclavicular septic arthritis: review of 180 cases. Medicine (Baltimore) 2004;83(3):139–148.
[9] Aglas F, Gretler J, Rainer F, *et al.* Sternoclavicular septic arthritis: a rare but serious complication of subclavian venous catheterization. Clin Rheumatol 1994;13(3):507–512.
[10] Khan SA, Zahid M, Asif N, *et al.* Tuberculosis of the sternoclavicular joint. Indian J Chest Dis Allied Sci 2002;44(4):271–273.
[11] Pandita KK, Sharma R, Dogra S, *et al.* Bilateral sternoclavicular joint tubercular cold abscess. Ann Thorac Med 2010;5(1):56–57.
[12] Cooper C, Cawley MI. Bacterial arthritis in an English health district: a 10 year review. Ann Rheum Dis 1986;45(6):458–463.
[13] Morgan DS, Fisher D, Merianos A, *et al.* An 18 year clinical review of septic arthritis from tropical Australia. Epidemiol Infect 1996;117(3):423–428.
[14] Kaandorp CJ, Dinant HJ, van de Laar MA, *et al.* Incidence and sources of native and prosthetic joint infection: a community based prospective survey. Ann Rheum Dis 1997;56(8):470–475.
[15] Kaandorp CJ, Krijnen P, Moens HJ, *et al.* The outcome of bacterial arthritis: a prospective community-based study. Arthritis Rheum 1997;40(5):884–892.
[16] Weston VC, Jones AC, Bradbury N, *et al.* Clinical features and outcome of septic arthritis in a single UK Health District 1982-1991. Ann Rheum Dis 1999;58(4):214–219.
[17] Ross JJ, Saltzman CL, Carling P, *et al.* Pneumococcal septic arthritis: review of 190 cases. Clin Infect Dis 2003;36(3):319–327.

[18] Chanet V, Soubrier M, Ristori JM, *et al.* Septic arthritis as a late complication of carcinoma of the breast. Rheumatology (Oxford) 2005;44(9):1157–1160.

[19] Clerc O, Prod'hom G, Greub G, *et al.* Adult native septic arthritis: a review of 10 years of experience and lessons for empirical antibiotic therapy. J Antimicrob Chemother 2011; 66(5):1168–1173.

[20] Belzunegui J, Rodriguez-Arrondo F, Gonzalez C, *et al.* Musculoskeletal infections in intravenous drug addicts: report of 34 cases with analysis of microbiological aspects and pathogenic mechanisms. Clin Exp Rheumatol 2000;18(3):383–386.

[21] Brancos MA, Peris P, Miro JM, *et al.* Septic arthritis in heroin addicts. Semin Arthritis Rheum 1991;21(2):81–87.

[22] Geyik MF, Gur A, Nas K, *et al.* Musculoskeletal involvement of brucellosis in different age groups: a study of 195 cases. Swiss Med Wkly 2002;132(7–8):98–105.

[23] Bar-Natan M, Salai M, Sidi Y, *et al.* Sternoclavicular infectious arthritis in previously healthy adults. Semin Arthritis Rheum 2002;32(3):189–195.

[24] Abu Arab W, Khadragui I, Echave V, *et al.* Surgical management of sternoclavicular joint infection. Eur J Cardiothorac Surg 2011;40(3):630–634.

[25] Cinquetti G, Banal F, Mohamed S, *et al.* Rare complications of Lemierre's syndrome: septic sternoclavicular joint arthritis and cavitating pneumonia. Rev Med Interne 2009;30(12): 1061–1063.

[26] El Ibrahimi A, Daoudi A, Boujraf S, *et al.* Sternoclavicular septic arthritis in a previously healthy patient: a case report and review of the literature. Int J Infect Dis 2009;13(3): e119 –e121.

[27] Bakaeen FG, Huh J, Fagan SP, *et al.* Surgical treatment of sternoclavicular joint infections in cirrhotic patients. Am J Surg 2008;195(1):130–133.

[28] Blanchais A, Cormier G, Varin S, *et al. Pasteurella multocida* septic oligoarthritis. Med Mal Infect 2010;40(2):123–125.

[29] Nitsche JF, Vaughan JH, Williams G, *et al.* Septic sternoclavicular arthritis with *Pasteurella multocida* and *Streptococcus sanguis*. Arthritis Rheum 1982;25(4):467–469.

[30] Chun JM, Kim JS, Jung HJ, *et al.* Resection arthroplasty for septic arthritis of the sternoclavicular joint. J Shoulder Elbow Surg 2012;21(3):361–366.

[31] Guillot X, Delattre E, Prati C, *et al.* Destructive septic arthritis of the sternoclavicular joint due to Neisseria gonorrhoeae. Joint Bone Spine 2010;79(5):519–520.

[32] Pollack MS. Staphylococcal mediastinitis due to sternoclavicular pyarthrosis: CT appearance. J Comput Assist Tomogr 1990;14(6):924–927.

[33] Wohlgethan JR, Newberg AH, Reed JI. The risk of abscess from sternoclavicular septic arthritis. J Rheumatol 1988;15(8):1302–1306.

[34] Moreno Martinez MJ, Moreno Ramos MJ, Linares Ferrando LF, *et al.* Sternoclavicular septic arthritis and empyema. Reumatol Clin 2012;8(2):102–103.

[35] Asnis DS, Dhaliwal GS. Bilateral sternoclavicular joint septic arthritis presenting as cutaneous abscesses. Clin Infect Dis 1994;19(5):964–966.

[36] Song HK, Guy TS, Kaiser LR, *et al.* Current presentation and optimal surgical management of sternoclavicular joint infections. Ann Thorac Surg 2002;73(2):427–431.

[37] Earwaker JW, Cotten A. SAPHO: syndrome or concept? Imaging findings. Skeletal Radiol 2003;32(6):311–327.

[38] Schrenzel J, Harbarth S, Schockmel G, *et al.* A randomized clinical trial to compare fleroxacin-rifampicin with flucloxacillin or vancomycin for the treatment of staphylococcal infection. Clin Infect Dis 2004;39(9):1285–1292.

[39] Joethy J, Lim CH, Koong HN, *et al.* Sternoclavicular joint infection: classification of resection defects and reconstructive algorithm. Arch Plast Surg 2012;39(6):643–648.

[40] Lim CH, Lim J, Naik M, *et al.* Surgical management of sternoclavicular joint infection. Asian Cardiovsc Thorac Ann 2001;9:291–295.

[41] Nusselt T, Klinger HM, Freche S, *et al.* Surgical management of sternoclavicular septic arthritis. Arch Orthop Trauma Surg 2011;131(3):319–323.

[42] Puri V, Meyers BF, Kreisel D, *et al.* Sternoclavicular joint infection: a comparison of two surgical approaches. Ann Thorac Surg 2011;91(1):257–261.
[43] Burkhart HM, Deschamps C, Allen MS, *et al.* Surgical management of sternoclavicular joint infections. J Thorac Cardiovasc Surg 2003;125(4):945–949.
[44] Paajanen H, Ristolainen L, Turunen H, *et al.* Prevalence and etiological factors of sport-related groin injuries in top-level soccer compared to non-contact sports. Arch Orthop Trauma Surg 2011;131(2):261–266.
[45] Ross JJ, Hu LT. Septic arthritis of the pubic symphysis: review of 100 cases. Medicine (Baltimore) 2003;82(5):340–345.
[46] Magarian GJ, Reuler JB. Septic arthritis and osteomyelitis of the symphysis pubis (osteitis pubis) from intravenous drug use. West J Med 1985;142(5):691–694.
[47] Fujitani S, Sun HY, Quintiliani R, *et al.* Pneumonia due to *Pseudomonas aeruginosa*: part II: antimicrobial resistance, pharmacodynamic concepts, and antibiotic therapy. Chest 2011;139(5):1172–1185.
[48] Miro JM, Puig de la Bellacasa J, Odds FC, *et al.* Systemic candidiasis in Spanish heroin addicts: a possible source of infection. J Infect Dis 1987;156(5):857–858.
[49] Pauli S, Willemsen P, Declerck K, *et al.* Osteomyelitis pubis versus osteitis pubis: a case presentation and review of the literature. Br J Sports Med 2002;36(1):71–73.
[50] Qassemyar Q, Smail A, Perignon D, *et al.* Cold abscess and Bazin's indurated erythema associated to pubic symphysis tuberculosis. Med Mal Infect 2010;40(9):552–554.
[51] Bouza E, Winston DJ, Hewitt WL. Infectious osteitis pubis. Urology 1978;12(6):663–669.
[52] Bali K, Kumar V, Patel S, *et al.* Tuberculosis of symphysis pubis in a 17 year old male: a rare case presentation and review of literature. J Orthop Surg Res 2010;5:63.
[53] Viceneux P, Rist S, Bosquet A. Arthrites septiques des sacro-iliaques et de la symphyse pubienne. Septic arthritis of the sacroiliac joint and pubic symphysis. Revue du Rhumatisme 2005;73:177–182.
[54] Kaandorp CJ, Van Schaardenburg D, Krijnen P, *et al.* Risk factors for septic arthritis in patients with joint disease. A prospective study. Arthritis Rheum 1995;38(12):1819–1825.
[55] Brewer M. Acute pubic osteomyelitis. Infect Dis Newslett 1993;12:7–8.
[56] Charles P, Ackermann F, Brousse C, *et al.* Spontaneous streptococcal arthritis of the pubic symphysis. Rev Med Interne 2011;32(7):e88–e90.
[57] Cheer K, Pearce S. Osteoarticular infection of the symphysis pubis and sacroiliac joints in active young sportsmen. BMJ 2009;339:b5019.
[58] Andole SN, Gupta S, Pelly M. Septic arthritis affecting pubic symphysis. BMJ Case Rep 2011. DOI:10.1136/bcr.042011.4089
[59] To F, Tam P, Villanyi D. Septic arthritis of the pubic symphysis from *Pseudomonas aeruginosa*: reconsidering traditional risk factors and symptoms in the elderly patient. BMJ Case Rep 2012. pii:bcr201200677. DOI:1136/bcr-2012-006777.
[60] Garcia-Porrua C, Picallo JA, Gonzalez-Gay MA. Osteitis pubis after Marshall-Marchetti-Krantz urethropexy. Joint Bone Spine 2003;70(1):61–63.
[61] Ferraro K, Cohen MA. Acute septic sacroiliitis in an injection drug user. Am J Emerg Med 2004;22(1):60–61.
[62] McPhee E, Eskander JP, Eskander MS, *et al.* Imaging in pelvic osteomyelitis: support for early magnetic resonance imaging. J Pediatr Orthop 2007;27(8):903–909.
[63] Sequeira W. Diseases of the pubic symphysis. Semin Arthritis Rheum 1986;16(1):11–21.
[64] Harris NH, Murray RO. Lesions of the symphysis in athletes. Br Med J 1974;4(5938): 211–214.
[65] Resnick D, Dwosh IL, Goergen TG, *et al.* Clinical and radiographic abnormalities in ankylosing spondylitis: a comparison of men and women. Radiology 1976;119(2):293–297.
[66] Maldonado-Cocco JA, Porrini A, Garcia-Mortoe O. Prevalence of sacroiliitis and ankylosing spondylitis in psoriasis patients. J Rheumatol 1978;5(3):311–313.
[67] Kenzaka T, Wakabayashi T, Morita Y. Acute crystal deposition arthritis of the pubic symphysis. BMJ Case Rep 2013. pii:bcr2013009239. DOI:10.1136/bcr-2013-009239.

[68] Justiniano ME, Colmegna I, Gimenez CR, *et al.* Tophaceous gout of the symphysis pubis. Arthritis Rheum 2005;52(12):4052.
[69] Sgambati E, Stecco A, Capaccioli L, *et al.* Morphometric analysis of the sacroiliac joint. Ital J Anat Embryol 1997;102(1):33–38.
[70] Ehrenfeld M. Spondyloarthropathies. Best Pract Res Clin Rheumatol 2012;26(1):135–145.
[71] Keriakos R, Bhatta SR, Morris F, *et al.* Pelvic girdle pain during pregnancy and puerperium. J Obstet Gynaecol 2011;31(7):572–580.
[72] Ruhe A, Bos T, Herbert A. Pain originating from the sacroiliac joint is a common non-traumatic musculoskeletal complaint in elite inline-speedskaters—an observational study. Chiropr Man Therap 2012;20(1):5.
[73] Wu MS, Chang SS, Lee SH, *et al.* Pyogenic sacroiliitis—a comparison between paediatric and adult patients. Rheumatology (Oxford) 2007;46(11):1684–1687.
[74] Schaad UB, McCracken GH, Jr., Nelson JD. Pyogenic arthritis of the sacroiliac joint in pediatric patients. Pediatrics 1980;66(3):375–379.
[75] Almoujahed MO, Khatib R, Baran J. Pregnancy-associated pyogenic sacroiliitis: case report and review. Infect Dis Obstet Gynecol 2003;11(1):53–57.
[76] Hermet M, Minichiello E, Flipo RM, *et al.* Infectious sacroiliitis: a retrospective, multicentre study of 39 adults. BMC Infect Dis 2012;12:305.
[77] Zimmermann B, 3rd, Mikolich DJ, Lally EV. Septic sacroiliitis. Semin Arthritis Rheum 1996;26(3):592–604.
[78] Feldmann JL, Menkes CJ, Weill B, *et al.* Infectious sacroiliitis. Multicenter study of 214 cases. Rev Rhum Mal Osteoartic 1981;48(1):83–91.
[79] Harvey P, Bayardelle P, Belanger R, *et al.* Sacroiliitis and septicemia caused by Campylobacter rectus and Actinomyces odontolyticus. Can J Infect Dis 1994;5(3):133–136.
[80] Rihl M, Wagner AD, Bakhsh KA, *et al.* Detection of chlamydial DNA in the inflamed sacroiliac joint of a patient with multiple infections. J Clin Rheumatol 2009;15(4):195–197.
[81] Schmidt DB, Meier R, Ochsner PE, *et al.* Acute bacterial sacroiliitis caused by Salmonella cholerae-suis. Dtsch Med Wochenschr 1988;113(38):1474–1477.
[82] Feldman LS. Salmonella septic sacroiliitis: case report and review. Pediatr Infect Dis J 2006;25(2):187–189.
[83] Dayan L, Deyev S, Palma L, *et al.* Long-standing, neglected sacroiliitis with remarked sacroiliac degenerative changes as a result of *Brucella* spp. infection. Spine J 2009;9(3):e1–e4.
[84] Gao F, Kong XH, Tong XY, *et al.* Tuberculous sacroiliitis: a study of the diagnosis, therapy and medium-term results of 15 cases. J Int Med Res 2011;39(1):321–335.
[85] Vyskocil JJ, McIlroy MA, Brennan TA, *et al.* Pyogenic infection of the sacroiliac joint. Case reports and review of the literature. Medicine (Baltimore) 1991;70(3):188–197.
[86] Doita M, Yoshiya S, Nabeshima Y, *et al.* Acute pyogenic sacroiliitis without predisposing conditions. Spine (Phila Pa 1976) 2003;28(18):E384–E389.
[87] McHugh RC, Tiede JM, Weingarten TN. Clostridial sacroiliitis in a patient with fecal incontinence: a case report and review of the literature. Pain Physician 2008;11(2): 249–252.
[88] Carlson SA, Jones JS. Pyogenic sacroiliitis. Am J Emerg Med 1994;12(6):639–641.
[89] Liu XQ, Li FC, Wang JW, *et al.* Postpartum septic sacroiliitis misdiagnosed as sciatic neuropathy. Am J Med Sci 2010;339(3):292–295.
[90] Le Bars H, Lamini N, Brunet JF, *et al.* Sacroiliitis due to Kingella kingae in an adult: updates on this pathogen. Ann Biol Clin (Paris) 2010;68(3):341–345.
[91] McKenna T, O'Brien K. Case report: group B streptococcal bacteremia and sacroiliitis after mid-trimester dilation and evacuation. J Perinatol 2009;29(9):643–645.
[92] Bindal M, Krabak B. Acute bacterial sacroiliitis in an adult: a case report and review of the literature. Arch Phys Med Rehabil 2007;88(10):1357–1359.
[93] Cinar M, Sanal HT, Yilmaz S, *et al.* Radiological followup of the evolution of inflammatory process in sacroiliac joint with magnetic resonance imaging: a case with pyogenic sacroiliitis. Case Rep Rheumatol 2012;2012:509136.

[94] Bellussi A, Busi Rizzi E, Schinina V, *et al.* STIR sequence in infectious sacroiliitis in three patients. Clin Imaging 2002;26(3):212–215.

[95] Moros ML, Rodrigo C, Villacampa A, *et al.* Septic shock in pregnancy due to pyogenic sacroiliitis: a case report. J Med Case Rep 2009;3:6505.

[96] Roca B, Torres V. Pyomyositis of the iliacus muscle complicated with septic sacroiliitis. QJM 2008;101(12):983–984.

[97] Giannoudis PV, Tsiridis E. A minimally-invasive technique for the treatment of pyogenic sacroiliitis. J Bone Joint Surg Br 2007;89(1):112–114.

[98] Kanakaris NK, Psarakis S, Chalidis B, *et al.* Management of pelvic instability secondary to chronic pyogenic sacroiliitis: case report. Surg Infect (Larchmt) 2009;10(4):353–358.

[99] Ebraheim NA, Ramineni SK, Alla SR, *et al.* Sacroiliac joint fusion with fibular bone graft in patients with failed percutaneous iliosacral screw fixation. J Trauma 2012;69(5): 1226–1229.

[100] Spoendlin M, Zimmerli W. Pyogenic sacroiliitis. Review of 8 personal cases and 200 cases from the literature. Schweiz Med Wochenschr 1988;118(21):799–805.

[101] Sturzenbecher A, Braun J, Paris S, *et al.* MR imaging of septic sacroiliitis. Skeletal Radiol 2000;29(8):439–446.

[102] Braun J, Sieper J, Bollow M. Imaging of sacroiliitis. Clin Rheumatol 2000;19(1):51–57.

[103] Ford LS, Ellis AM, Allen HW, *et al.* Osteomyelitis and pyogenic sacroiliitis: a difficult diagnosis. J Paediatr Child Health 2004;40(5–6):317–319.

[104] Abbott GT, Carty H. Pyogenic sacroiliitis, the missed diagnosis? Br J Radiol 1993;66(782):120–122.

[105] Puhakka KB, Melsen F, Jurik AG, *et al.* MR imaging of the normal sacroiliac joint with correlation to histology. Skeletal Radiol 2004;33(1):15–28.

[106] Ho CL, Wu WC, Chen S, *et al.* F-18 FDG PET/CT in an adult case of group B streptococcal sacroiliitis. Clin Nucl Med 2010;35(10):834–835.

[107] Kerr R. Radiologic case study. Sacroiliac joint involvement by gout and hyperparathyroidism. Orthopedics 1988;11(1):185, 7–185, 90.

[108] Garland TA, Innes DJ, Jr., Stewart FM. Calcium pyrophosphate dihydrate crystal deposition in the sacroiliac joint. Arch Pathol Lab Med 1983;107(8):441.

[109] Humphrey SM, Inman RD. Metastatic adenocarcinoma mimicking unilateral sacroiliitis. J Rheumatol 1995;22(5):970–972.

[110] Adelman HM, Wallach PM, Flannery MT. Ewing's sarcoma of the ilium presenting as unilateral sacroiliitis. J Rheumatol 1991;18(7):1109–1111.

[111] Cohn SM, Schoetz DJ, Jr. Pyogenic sacroiliitis: another imitator of the acute abdomen. Surgery 1986;100(1):95–98.

Chapter 8

Periprosthetic Joint Infection: General Aspects

Werner Zimmerli

Introduction

Prosthetic joints are used to replace articulations, which are damaged because of degeneration, trauma, or inflammation. Since damaged cartilage can still not be replaced to a sufficient extent, arthroplasty remains the only treatment, which is able to completely relieve pain and restore function. In case of an optimal treatment result, the prosthetic joint can be functionally used similar to a native joint. Due to the increasing life expectancy, the number of patients with osteoarthritis is steadily rising [1–4]. In contrast, the need for joint replacement in patients with rheumatoid arthritis is decreasing, since efficacious disease-modifying drugs are widely available [5].

Several complications can be observed after joint replacement, including mechanical problems such as luxation, heterotopic ossification [6], aseptic loosening because of wear particles [7], or insufficient adaptation of the implant shape to the bone [8]. The most feared complication, however, is periprosthetic joint infection (PJI), because it may lead to the loss of the device [3, 9–11]. Furthermore, it has big economical consequences [2]. Kurtz *et al.* [1] projected a total cost of US$1 billion for surgical procedures in patients with PJI in 2014. Thus, it is of paramount interest for patients and society to avoid PJI, or, if it occurs, to manage this complication in an optimal way.

Regardless of the type of device, implant-associated infections have some common features: They are highly susceptible to infection, and cannot be cured with antibiotics alone [10–14]. In addition, prosthetic devices are endangered for infection not only during implantation, but as long as they remain in the body [15–18]. In this chapter, common features of PJI will be presented. Specific problems that are associated with each type of artificial joint are discussed in separate chapters dealing with the different joint replacements (Chapters 9–12).

Bone and Joint Infections: From Microbiology to Diagnostics and Treatment, First Edition.
Edited by Werner Zimmerli.

Definition

There is no unanimously accepted gold standard for the diagnosis of PJI. Therefore, various diagnostic criteria have been used in different publications [9, 11, 19]. For clinical studies, the definition should have a very high specificity for a meaningful comparison of published results. In contrast, in clinical practice, the sensitivity should be as high as possible, in order not to miss any PJI. If the diagnosis is delayed by 3–4 weeks, cure with implant retention has a very low chance [9–11]. Therefore, in such cases, implant removal is generally needed. As a consequence, even if a patient does not strictly fulfill the definition of PJI, diagnostic surgical debridement should be considered.

Surgical site infections can be classified as superficial incisional, deep incisional, and organ/space infections [20]. However, this classification is not reliable after joint replacement. Superficial infection rapidly progress to deep infection, and the differentiation is clinically not possible [21]. With the wrong diagnosis of superficial surgical site infection, a short course of antibiotics is given, which typically delays the diagnosis of PJI. Since early postoperative evaluation heavily underestimates the rate of PJI, the Centers for Disease Control's (CDC's) National Healthcare Safety Network (NHSN) has proposed special surgical site infection definitions for procedures involving prosthetic devices [22].

The Infectious Disease Society of America (IDSA) has recently proposed a definition, which is summarized in Table 8.1 [9, 11]. Since synovia and tissue cultures cannot be considered as gold standard, microbiological proof is not required in the IDSA case definition. This is important, since culture-negative PJIs are quite frequent [23]. In the presence of a sinus tract communicating with the implant, the diagnosis of PJI is proven. Purulence around an implant proves PJI, provided that it cannot be explained by an alternative pathology such as crystal arthropathy [24] or a large amount of wear particles [7, 25]. Criteria for histopathology are not clearly defined. Generally, the cutoff for a positive result is set at greater than 5 granulocytes per high power field (see later). The number of leukocytes in synovia, which is required as criterion for PJI, depends on the joint, the postoperative interval, and the type of underlying disease [26–28] (see later).

There are alternative propositions of definition, which include parameters of generalized inflammation such as erythrocyte sedimentation rate (ESR; >30 mm/h) and C-reactive protein (CRP; >10 mg/l) [19]. Since these parameters are neither specific nor sensitive for PJI, they are only useful in combination with other parameters [29].

Table 8.1. Definition of periprosthetic joint infection[a] [9, 11].

Presence of a sinus tract communicating with the prosthetic joint
Presence of purulence without another known etiology surrounding the prosthetic device
Acute inflammation consistent with infection at histopathological examination of periprosthetic tissue
Elevated leukocyte count in the synovial fluid and/or predominance of neutrophils [26, 27, 28]
Growth of identical microorganisms in at least two intraoperative cultures or combination of preoperative aspiration and intraoperative cultures in case of a low-virulence microorganism (coagulase-negative staphylococci, *Propionibacterium acnes*, etc.). In case of a virulent microorganism (e.g., *S. aureus, E. coli*, etc.), growth in a single specimen from synovial fluid and/or periprosthetic tissue and/or sonication fluid may also represent PJI. However, growth in a single specimen must always consider other criteria and constellation of diagnostic procedures (e.g., previous antimicrobial treatment)

[a] For the diagnosis of PJI at least one of the five criteria is required.

Table 8.2. Novel classification of periprosthetic joint infection [9].

Type of infection	Characteristics
Acute hematogenous PJI	Infection with 3 weeks or less of duration of symptoms after an uneventful postoperative period
Early postinterventional PJI	Infection that manifests within 1 month after an invasive procedure such as surgery or arthrocentesis
Chronic PJI	Infection with symptoms that persist for >3 weeks and are beyond the early postinterventional period

Classification

Traditionally, PJIs are classified as early (<3 months after surgery), delayed (3–24 months after surgery), and late infections (>2 years after surgery) [9–11]. Early and delayed infections are mainly exogenously acquired in the perioperative period, whereas most late PJIs are hematogenously acquired. For clinical purposes, a classification considering the surgical treatment concepts is more useful. Table 8.2 shows the three types of PJIs. Acute hematogenous PJIs of less than 3 weeks' duration and early postinterventional PJIs (<1 month after surgery) can generally be treated with implant retention (see later) [9–11]. In contrast, in patients with chronic PJI, the biofilm on implant material can generally not be eliminated by antimicrobial agents [30–32]. Therefore, all foreign material has to be removed.

Pathogenesis

Implanted devices are highly susceptible to bacterial and fungal infections [4, 10, 33]. Elek and Conen [34] demonstrated in human volunteers that an inoculum of 100 CFU *Staphylococcus aureus* was enough to get a stitch abscess, whereas greater than 10,000-fold more staphylococci were required to get a subcutaneous abscess in the absence of a foreign body. We reproduced the same phenomenon in a guinea pig model with subcutaneous tissue cages [12, 14, 35]. These experimental observations point toward a locally acquired granulocyte defect, since staphylococcal infection is controlled by phagocytes.

Once established, device-related infections are difficult to eradicate. This indicates a locally impaired host defense on the one side and a phenotypic resistance of adhering microorganisms on the other side. The reason for this is an obvious unfavorable interaction between the local defense mechanism (e.g., granulocytes, complement) with the implanted foreign body and a transition from the susceptible planktonic microorganism to a resistant biofilm. Interestingly, the type of material plays only a minor role regarding the susceptibility of a foreign body to infection [36]. Ha *et al.* [37] demonstrated that biofilm-forming *Staphylococcus epidermidis* adhere to a higher degree on pure titanium than on stainless steel. Nevertheless, this difference could not be reproduced in vivo. The reason for this paradox could be an immediate coating of the implant by host proteins, which are more relevant for bacterial adherence than the type of material. The main argument for this mechanism is the observation that fibronectin and other proteins act as receptors for staphylococci [38, 39].

The Role of the Host

Innate or nonspecific host defense mechanisms are responsible for rapid and efficacious elimination of microorganisms, as soon as the device is placed in the body [13]. As a first step, microorganisms have to be opsonized for rapid ingestion by granulocytes or mononuclear leukocytes [40]. This process involves nonspecific (complement, bacterial remnants) and specific (antibodies) soluble components in the humoral phase and intact corresponding receptors on phagocytes. If one component of this process is impaired, the susceptibility to infection is increased [41]. Various possible mechanisms for the impaired bacterial clearance have been hypothesized. In addition, the paradox of microbial persistence in the presence of abundant granulocytes around the implant has been studied.

Interaction of the Implant with Granulocytes

The role of granulocytes is killing microbes and elimination of foreign bodies by phagocytosis (Figure 8.1) [13]. If the foreign material is too large for phagocytosis, granulocytes interact with the nonphagocytosable surface, a process that is called frustrated phagocytosis [42, 43]. This concept has been tested in an animal model [12, 14]. The guinea pig tissue cage model perfectly simulates the clinical situation, that is, very low numbers of microorganisms lead to permanent implant-associated infection, which never spontaneously heals [4, 10, 44]. Even so-called apathogenic bacteria such as *Propionibacterium acnes* or *S. epidermidis* cause infection in this model [45, 46]. Granulocytes purified from the interstitial fluid accumulating around subcutaneous implants indeed revealed a severe defect in ingestion, staphylococcal killing, and superoxide production [12, 14]. In vitro experiments indicate that this defect is due to the interaction with the nonphagocytosable surface. Granulocytes partially degranulate during interaction with the implant [14]. Since the liberated granulocytes also contain collagenase, this phenomenon may also be responsible for loosening of the device during infection [47].

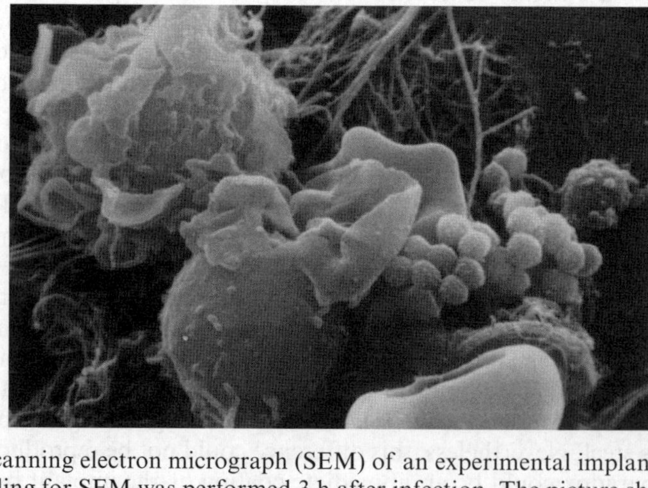

Figure 8.1. Scanning electron micrograph (SEM) of an experimental implant–associated infection. Sampling for SEM was performed 3 h after infection. The picture shows two granulocytes facing an aggregate of *Staphylococcus aureus*. The irregular surface of *S. aureus* shows exopolysaccharides of the young biofilm. Reprinted with permission from Zimmerli and Sendi [13].

Interaction of Wear Particles with Phagocytes

After arthroplasty, wear particles are produced in variable amounts depending on the biomechanical situation. Abundant wear particles are supposed to be a risk factor for infection [48]. Therefore, not only the implant, but also wear particles may interact with granulocytes. Indeed, Bernard *et al.* [49, 50] described an impaired bactericidal activity of neutrophils after interaction with wear particles in vitro. Similarly, in the tissue surrounding an implant with wear particles, cytokines are liberated by local macrophages [51]. Some of these cytokines, such as M-CSF and TGF-alpha, directly stimulate osteoclastogenesis, which favors implant loosening by bone resorption.

The Role of the Microorganism (Biofilm)

Microorganisms, which are not immediately eliminated during insertion of a device, rapidly adhere to the implant and resist elimination by host defense mechanisms [30]. Therefore, preventive measures are of paramount importance in implant surgery. Adherence to the surface involves nonspecific physical factors (e.g., surface tension, hydrophobicity, and electrostatic interaction) or specific adhesins such as fibronectin. This initial process is followed by biofilm formation, which is mediated in part by the polysaccharide intercellular adhesion encoded by the intercellular adhesion (*ica*) operon [52]. Within biofilms, microbes are enclosed in a polymeric matrix and develop into organized, complex communities, resembling multicellular organisms. At high microbial density, so-called quorum sensing genes are activated in order to control the size of the biofilm [53]. Biofilm bacteria are up to 1000-fold more resistant to antibiotics than planktonic bacteria [54]. This explains the persistence of implant-associated infection once it is established. There is a transformation of the biofilm over time. Clinically, it can be observed that in case of implant retention, the chance of cure dramatically falls from about 80–90% to 30–60%, if treatment is started later than 3–4 weeks after infection [55–60].

The Route of Infection

Inoculation of PJI occurs either by the exogenous or by the hematogenous route [9, 10, 61]. Exogenous infections are mainly acquired during the perioperative period. They manifest themselves during the first 2 years after surgery. The less virulent the infecting agent, the more delayed is the infection. As an example, the diagnosis of PJI due to *P. acnes* is often delayed by months or even years [62, 63]. Microorganisms reach the implant not only during but also after surgery, as long as the wound is not completely dry. Microorganisms penetrate along drainage tubes or directly through the wound. The risk is especially high in patients with large hematoma. Thus, the primary aim in the early postoperative period is rapid wound healing. Therefore, in case of wound healing disturbance due to high volume secretion, diagnostic and therapeutic debridement surgery should not be delayed. Rarely, exogenous infection occurs after arthrocentesis or after spontaneous or traumatic skin perforation from outside or from inside the device. Exogenous PJI is more frequent in joints with poor soft tissue coverage such as knee, elbow or ankle arthroplasty. Indeed, 85% of periprosthetic ankle joint [64] but only 57% of hip joint infections occurred by the exogenous route [57]. Looking exclusively at patients with *S. aureus* PJI, the difference was even more impressive. PJI occurred by the exogenous route in 45% of the knee, but only 22% of the hip arthroplasties [61].

Hematogenous PJI are acquired via bloodstream at any time after surgery. The incidence rate of knee and hip PJI is estimated as 5.9 episodes per 1000 joint-years during the first 2 years, and 2.3 episodes per 1000 joint-years thereafter [16]. This difference in susceptibility reflects an early predominance of perioperative infections on the one hand, and higher initial risk for hematogenous PJI on the other side. Due to the earlier-mentioned locally acquired granulocyte defect around the implant, the device is a *locus minoris resistentiae*, which is prone to hematogenous seeding. Indeed, Blomgren *et al.* [65, 66] showed that knee prostheses could be infected by the hematogenous route in a rabbit model. We quantified the risk for bacterial seeding in the guinea pig tissue cage infection model [67]. With a bacteremia of 1000 CFU *S. aureus* per ml blood, 42% subcutaneous implants could be selectively infected. With lower bacterial density in the bloodstream, no extravascular devices were infected. With a higher bacterial load, bacterial seeding was not selective, because not only implants, but also different organs were infected. Taken together, implants are favorite sites of bacterial seeding at the high bacterial density occurring during *S. aureus* bacteremia (> 1000 CFU/ml blood).

Diagnostic Algorithm

If a patient has a sinus tract communicating with the prosthetic device, PJI is confirmed (see Table 8.1), and treatment should be planned immediately. Similarly, in patients with prolonged or new-onset wound secretion after implantation, diagnostic debridement surgery is required regardless of whether the patient has clinical or laboratory signs of infection. If such a patient is already in a rehabilitation center, he/she should be referred back for evaluation by the orthopedic surgeon and an infectious disease specialist. In all these situations, antibiotics should not be given before the microbiological workup. If the patient has already been treated empirically with antibiotics without etiologic diagnosis, debridement surgery should be postponed whenever possible, in order to improve the chance for finding the responsible microorganism.

In patients with an acute or persistently painful prosthetic joint, the diagnosis of PJI should be considered. Figure 8.2 shows the diagnostic algorithm for these patients [11]. The first step is the case history, which includes questions about previous symptoms of local inflammation, postoperative wound disturbance, bridging symptoms between implantation and acute onset, and recent focal or systemic infections (especially skin/soft tissue infection, febrile diarrhea, pneumonia, pyelonephritis, sepsis syndrome). In the clinical examination, not only local signs of infection, but also possible primary foci should be looked for. The laboratory workup includes CRP, which has a sensitivity of about 90% to detect PJI according to a meta-analysis [29]. Drawing blood cultures is suggested in patients with sepsis syndrome. In all patients, a plain radiograph and/or a bone scintigraphy should be performed, mainly to detect either signs of loosening or mechanical reasons for pain. The diagnostic value of other imaging procedures is presented in Chapter 9. If PJI is still suspected after this workup, arthrocentesis should be performed. Synovial cells allow rapid confirmation of PJI (see later). If PJI is confirmed with positive synovial cell counts and/or culture, surgical management should be planned without delay (see later) [10, 11, 68]. If PJI is only suspected, diagnostic debridement surgery should be considered. In case of a short duration of infection, this intervention is a therapeutic intervention in many cases. Removal and sonication of the modular parts allows a more sensitive diagnosis in patients with previous antibiotic therapy [69]. If PJI

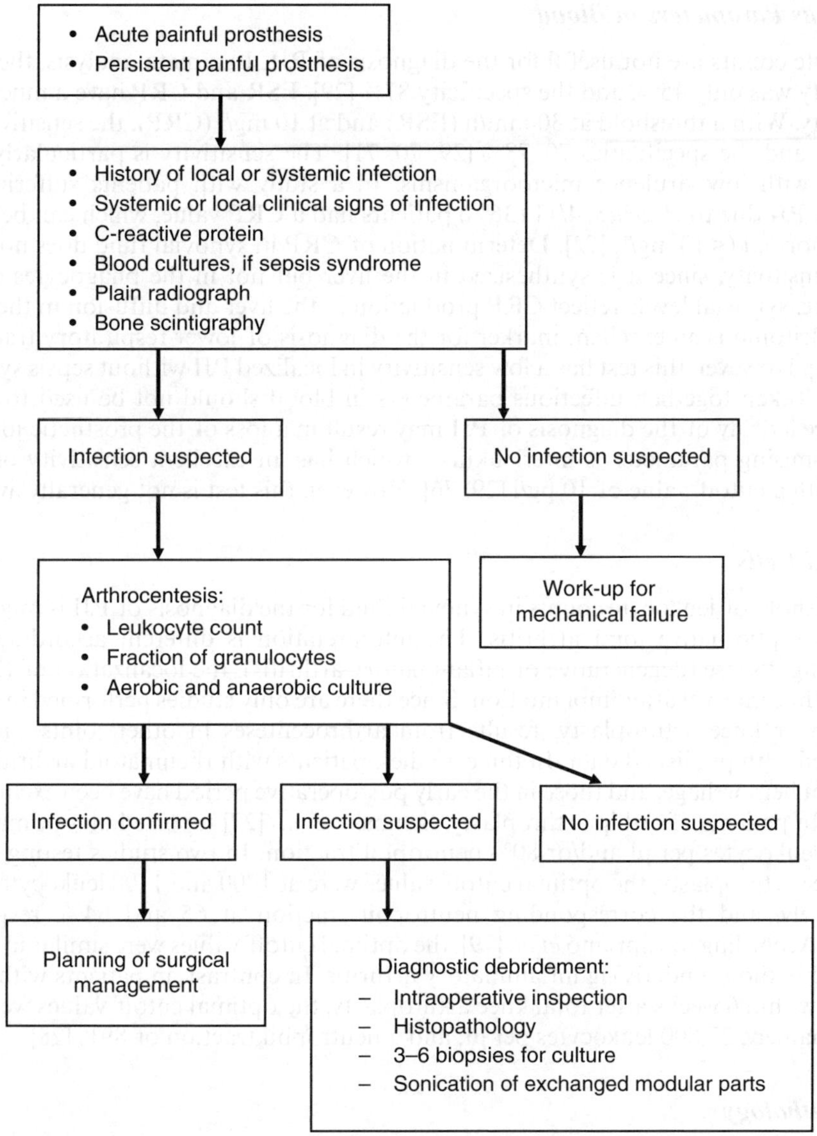

Figure 8.2. Diagnostic algorithm for patients with suspected but not confirmed PJI.

is not suspected after arthrocentesis, other reasons, such as mechanical failure of the device, should be searched.

Laboratory Investigation

Several laboratory parameters have been used to look for and to confirm PJI. Some of them are not sensitive enough and can therefore only be used as supportive arguments for PJI.

Infectious Parameters in Blood

Leukocyte counts are not useful for the diagnosis of PJI. In a meta-analysis, the pooled sensitivity was only 45%, and the specificity 87% [29]. ESR and CRP have a much better sensitivity. With a threshold at 30 mm/h (ESR) and at 10 mg/l (CRP), the sensitivities are 91–97%, and the specificities 70–78% [29, 70, 71]. The sensitivity is particularly low in patients with low-virulence microorganisms. In a study with patients suffering from shoulder PJI due to *P. acnes*, 4/11 (36%) patients had a CRP value, which can be considered as normal (< 10 mg/l) [72]. Determination of CRP in synovial fluid does not have a better sensitivity, since it is synthesized in the liver but not in the phagocytes [73, 74]. Therefore, synovial levels reflect CRP production in the liver and diffusion in the joint.

Procalcitonin is an excellent marker for the diagnosis of lower respiratory tract infections [75]. However, this test has a low sensitivity in localized PJI without sepsis syndrome [76, 77]. Taken together, infectious parameters in blood should not be used to exclude PJI, since a delay of the diagnosis of PJI may result in a loss of the prosthetic joint. The only promising parameter is interleukin-6, which has an excellent sensitivity of nearly 100% with a cutoff value of 10 pg/l [29, 76]. However, this test is not generally available.

Synovial Cells

The threshold of leukocyte counts in synovial fluid for the diagnosis of PJI is much lower than for septic native joint arthritis. The interpretation is different according to the underlying disease (degenerative or inflammatory arthritis), the localization of the joint, and the time interval after implantation. Since there are only studies performed in patients with hip or knee arthroplasty, results from arthrocenteses in other joints cannot be compared with published data. In three studies, patients with rheumatoid arthritis, those with joint hemorrhage, and those in the early postoperative period have been excluded [26, 27, 78]. In patients after hip arthroplasty, Schinsky *et al.* [27] reported an optimal cutoff at 4200 leukocytes per μl and/or 80% neutrophil fraction. In two studies testing patients with knee arthroplasty, the optimal cutoff values were at 1700 and 1100 leukocytes per μl, respectively, and the corresponding neutrophil fraction at 65 and 64%, respectively [26, 78]. According to Cipriano *et al.* [79], the optimal cutoff values were similar in patients with and without underlying inflammatory arthritis. In contrast, in patients with arthrocentesis within 6 weeks after total knee arthroplasty, the optimal cutoff values were much higher, namely, 27,800 leukocytes per μl, and a neutrophil fraction of 89% [28].

Histopathology

Intraoperative frozen sections should only be used in centers with experienced pathologists, who are able to differentiate between mechanical failure and infection [80]. Biopsies, which are sampled during surgery, should be divided in two parts: one for microbiology and the other for conventional histopathology. Comparing pairs of biopsies allows a better interpretation of culture results (contamination). In addition, in case of negative culture, the presence of granulocytes in biopsy specimens indicates culture-negative PJI (Table 8.1).

Culture

Swab cultures have a low sensitivity and should therefore be avoided [81]. Culture of synovial fluid has a sensitivity of about 85% and a specificity of at least 95% [71].

The sensitivity is better when using polymerase chain reaction (PCR) [82] or by culturing synovial fluid in blood culture flasks [83]. For the diagnosis of PJI during surgery, at least three but optimally six biopsies should be obtained [11].

Sonication

Microorganisms persist as a biofilm on the implant. Sonication has been used, in order to detach microbes from the surface. In many centers, this technique is used for examination of explanted prostheses or modular parts. Since sonication may harm the viability of microorganisms, only evaluated technical protocols should be used [69, 84]. In the study of Trampuz *et al.* [69], sonication was significantly more sensitive only in patients treated with antibiotics within 2 weeks before sampling (75% versus 45%, $P < 0.001$). The appropriate cutoff for differentiating infection from contamination depends on the sonication technique (amount of fluid, concentration step, etc.). Thus, culture of sonication fluid is useful in patients with recent antibiotic therapy before surgery.

Molecular Diagnosis

The role of molecular diagnosis in patients with suspected PJI is still not clear. It can be performed in synovial fluid, biopsy specimens, and sonication fluid (see earlier). Different techniques have been used, either broad-range PCR or specific multiplex PCR [84–86]. The main drawback of broad-range PCR is its low sensitivity. In the commercially available specific multiplex PCR, the lack of some primers, which are relevant for detecting PJI (e.g., *P. acnes*), limits its clinical use [86].

Therapeutic Management

In contrast to many other infections, such as pneumonia, sinusitis, or pyelonephritis, PJIs never spontaneously heal. The lack of clinical symptoms does not necessarily indicate cure, but could also indicate asymptomatic persistence. This situation is mainly observed after antimicrobial therapy without debridement surgery or in patients with PJI due to low-virulence microorganisms such as *P. acnes*. In order to get an optimal treatment result, antibiotics should only be started after appropriate diagnostic procedures, and antimicrobial therapy should always be combined with a surgical intervention [9–11]. The most appropriate surgical management should be chosen for each individual patient. Choosing the least invasive procedure, regardless of its risk for failure, is not a good option. Cure by the first treatment attempt avoids soft tissue damage and loss of functional integrity of the joint. Therefore, early referral to a specialized center is advised.

As the first step, it should be decided whether a curative or a palliative management is indicated. The palliative option should only be chosen if a specialized team argues against a curative procedure in a particular patient. Cure is defined as a long-term pain-free functional joint with complete eradication of infection. This requires a combination of both an appropriate surgical procedure and long-term antimicrobial therapy. In contrast, the aim of palliative therapy is suppression of symptoms, regardless of the functional outcome. This approach requires either crude surgery (resection arthroplasty or amputation) or no surgery but lifelong suppressive antimicrobial therapy in patients at very high risk for surgery.

Surgical Interventions

The traditional surgical treatment of PJI is two-stage exchange of the device. The first intervention includes removal of all necrotic tissue and foreign material. During the implant-free period, the patient is treated with antibiotics before reimplantation [87]. This procedure is expensive and invasive and generally compromises function [88]. Less invasive procedures have lower cure rates, if the selection of patients is not appropriate [55, 58, 89–92]. During the last two decades, an algorithm for the optimal surgical treatment of the different presentations has been developed at our institution [10, 68]. Four curative procedures can be performed: debridement with implant retention, one-stage exchange, two-stage exchange with a short interval, and two-stage exchange with a long interval. The algorithm indicates the least invasive procedure with the highest cure rate for each patient. In brief, only patients with acute hematogenous or early postoperative infection (Table 8.2) can be successfully treated with implant retention. In all other patients, the implant has to be removed or exchanged in order to get a high chance of cure. The detailed treatment concepts are integrated in the chapters dealing with specific prosthetic joints (see Chapters 9–12). If the appropriate intervention is selected for each patient, the cure rate is above 80% for all four procedures [56, 57]. If for any reason a palliative procedure is preferred, the main options are implant removal without replacement [93], or long-term suppressive antimicrobial therapy with or without surgery [94]. In patients with periprosthetic knee infection, resection arthroplasty is generally followed by arthrodesis after healing of infection [95]. Amputation should be considered in life-threatening situations, such as PJI with necrotizing fasciitis, or after multiple treatment failures [96, 97].

Antimicrobial Therapy

Biofilm bacteria are in a stationary phase of growth, because of oxygen and glucose limitation [98]. Therefore, successful treatment of PJI should consider the biofilm status of the microbes. In vitro studies revealed that most antimicrobial agents have a minimal bactericidal concentration (MBC), which is up to greater than 100-fold higher in the stationary phase of growth [99–109]. It has been shown that antibiotics with a high stationary-phase MBC are not able to clear bacteria adhering on sinter glass beads [100, 103]. The high stationary-phase MBC, and the lack of efficacy on adhering bacteria, is predictive of failure of antimicrobial therapy in orthopedic implant–associated infections [60, 107, 110–112]. Unfortunately, up to now, only two classes of drugs have shown the properties that are needed for efficacious elimination of biofilm bacteria, namely, rifampin and other rifamycins against staphylococci [60, 110, 113] and fluoroquinolones against Gram-negative bacilli [107, 111, 112, 114].

Detailed information on the type and dose of antimicrobial drugs, the route of administration, and the duration of treatment is given in the chapters on specific prosthetic joints (see Chapters 9–12).

Prophylaxis

Perioperative prophylaxis

Primary arthroplasty is clean surgery, which, nevertheless, requires antibiotic prophylaxis, since any single PJI is an extremely severe adverse event. This is based on studies from the 1970s [115]. The antimicrobial agent should be effective against the majority of microorganisms introduced from skin. In addition, the drug should be selected according to the local epidemiology regarding the resistance pattern. In general, a first- or second-generation cephalosporin is chosen [115]. Tissue levels above the minimal inhibitory concentration (MIC) of skin microorganisms should be reached at the time or very close to the time of incision [116]. These levels should be maintained during surgery. The duration of prophylaxis should generally not exceed the time of surgery [115].

In case of intervention because of PJI, the first dose of antibiotics can be postponed to the time after sampling of the biopsies for culture, since the patient does not get single-dose prophylaxis, but long-term antibiotic treatment after surgery.

The role of antibiotic-impregnated cement has never been tested in a large controlled trial. However, according to the Finnish Arthroplasty Registry, reporting on greater than 40,000 patients, the risk for knee PJI was significantly lower in patients receiving antibiotic-impregnated cement in addition to IV prophylaxis. This was observed not only in patients with revision arthroplasty but also in those undergoing primary arthroplasty [117].

Prevention of Hematogenous Infection

As mentioned earlier, prosthetic joints are at risk for hematogenous infection as long as they remain in the body [16]. The risk is especially high (29–39%) during *S. aureus* bacteremia [15, 17, 18], but relatively low (1.3%) during exposure to a remote infection [118]. Thus, patients with prosthetic joints should be informed about this risk, and a physician should be contacted as soon as possible in case of an infection. Rapid antimicrobial therapy of bacterial infections (e.g., skin and soft tissue infection, febrile diarrhea, pneumonia) is important for prevention of bacteremia, and hence seeding on the prosthetic device. In addition, oro-dental hygiene and regular dental treatment is important for prevention of PJI of dental origin. However, antibiotic prophylaxis during routine dental work, including extraction of noninfected teeth, is not needed [119, 120]. The bacterial density during bacteremia provoked by low- and high-risk dental procedures (< 50 CFU/ml blood) is too low to result in hematogenous seeding of bacteria on the prosthetic device [119, 121].

Errors in the Management of PJI

Several errors in the management of PJI jeopardize the outcome. First, if PJI is suspected, the decision for a diagnostic procedure should not be postponed, because the time for successful implant retention is limited. Second, treating with antibiotics, without previous culturing appropriate samples, increases the risk for culture-negative PJI, which cannot be treated in a rational way. Third, not considering the biofilm status of the implant-adhering microorganisms results in poorer outcome. Fourth, not combining antimicrobial therapy with an appropriate surgical intervention and vice versa is only palliative but not curative management. Fifth, inadequate sampling (e.g., swabs instead of biopsies) and/or insufficient microbiological techniques (e.g., short incubation time, no anaerobic incubation) increase the risk for false-negative cultures. Sixth, one single surgical technique for everybody (e.g., two-stage exchange) should be replaced by a differentiated procedure according to a rational algorithm, in order to get the best functional result with the highest microbiological cure rate. Seventh, implant retention in patients who do not qualify for this procedure may be the beginning of repetitive treatment failures resulting in a poor functional outcome.

Key Points

- Classification of PJI should consider the time interval from infection to diagnosis, in order to be useful for planning the surgical management.
- Prosthetic devices are highly susceptible to infection due to an impaired local host defense.
- The risk for exogenous infection is higher in patients with prosthetic joints with poor soft tissue coverage (knee, elbow, ankle), than in those with hip arthroplasty.
- Arthrocentesis is the most important single test for diagnosis of PJI.
- Acute hematogenous and early postinterventional PJI can generally be treated with implant retention.
- Whenever possible, antibiotics with efficacy on biofilm bacteria should be chosen.

References

[1] Kurtz SM, Lau E, Watson H, *et al.* Economic burden of periprosthetic joint infection in the United States. J Arthroplasty 2012;27(8 Suppl):61–5 e1.

[2] Kurtz SM, Lau E, Schmier J, *et al.* Infection burden for hip and knee arthroplasty in the United States. J Arthroplasty 2008;23(7):984–991.

[3] Aslam S, Darouiche RO. Prosthetic joint infections. Curr Infect Dis Rep 2012;14(5):551–557.

[4] Del Pozo JL, Patel R. Clinical practice. Infection associated with prosthetic joints. N Engl J Med 2009;361(8):787–794.

[5] Jamsen E, Virta LJ, Hakala M, *et al.* The decline in joint replacement surgery in rheumatoid arthritis is associated with a concomitant increase in the intensity of anti-rheumatic therapy: a nationwide register-based study from 1995 through 2010. Acta Orthop 2013;84(4):331–337.

[6] Zeckey C, Hildebrand F, Frink M, *et al.* Heterotopic ossifications following implant surgery—epidemiology, therapeutical approaches and current concepts. Semin Immunopathol 2011;33(3):273–286.

[7] Catelas I, Wimmer MA, Utzschneider S. Polyethylene and metal wear particles: characteristics and biological effects. Semin Immunopathol 2011;33(3):257–271.

[8] Ochsner PE. Osteointegration of orthopaedic devices. Semin Immunopathol 2011; 33(3):245–256.

[9] Zimmerli W, Sendi P. Orthopedic implant-associated infections. In: Bennett JE, Dolin R, Blaser M, editors. Mandell, Douglas, and Bennett's Principles and Practice of Infectious Diseases, 8th edition. Philadelphia, PA: Elsevier; 2014.

[10] Zimmerli W, Trampuz A, Ochsner PE. Prosthetic-joint infections. N Engl J Med 2004;351(16):1645–1654.

[11] Osmon DR, Berbari EF, Berendt AR, *et al.* Executive summary: diagnosis and management of prosthetic joint infection: clinical practice guidelines by the Infectious Diseases Society of America. Clin Infect Dis 2013;56(1):1–10.

[12] Zimmerli W, Waldvogel FA, Vaudaux P, *et al.* Pathogenesis of foreign body infection: description and characteristics of an animal model. J Infect Dis 1982;146(4):487–497.

[13] Zimmerli W, Sendi P. Pathogenesis of implant-associated infection: the role of the host. Semin Immunopathol 2011;33(3):295–306.

[14] Zimmerli W, Lew PD, Waldvogel FA. Pathogenesis of foreign body infection. Evidence for a local granulocyte defect. J Clin Invest 1984;73(4):1191–1200.

[15] Lalani T, Chu VH, Grussemeyer CA, *et al.* Clinical outcomes and costs among patients with *Staphylococcus aureus* bacteremia and orthopedic device infections. Scand J Infect Dis 2008;40(11–12):973–977.

[16] Steckelberg J, Osmon DR. Prosthetic joint infection. In: Waldvogel FA, Bisno AL, editors. Infections Associated with Indwelling Medical Devices. Washington, DC: ASM Press; 2000, pp. 173–209.

[17] Murdoch DR, Roberts SA, Fowler Jr VG, Jr., *et al.* Infection of orthopedic prostheses after *Staphylococcus aureus* bacteremia. Clin Infect Dis 2001;32(4):647–649.

[18] Sendi P, Banderet F, Graber P, *et al.* Periprosthetic joint infection following *Staphylococcus aureus* bacteremia. J Infect 2011;63(1):17–22.

[19] Parvizi J, Jacovides C, Zmistowski B, *et al.* Definition of periprosthetic joint infection: is there a consensus? Clin Orthop Relat Res 2011;469(11):3022–3030.

[20] Horan TC, Gaynes RP, Martone WJ, *et al.* CDC definitions of nosocomial surgical site infections, 1992: a modification of CDC definitions of surgical wound infections. Am J Infect Control 1992;20(5):271–274.

[21] Berbari EF, Hanssen AD, Duffy MC, *et al.* Risk factors for prosthetic joint infection: case-control study. Clin Infect Dis 1998;27(5):1247–1254.

[22] Yokoe DS, Avery TR, Platt R, *et al.* Reporting surgical site infections following total hip and knee arthroplasty: impact of limiting surveillance to the operative hospital. Clin Infect Dis 2013;57(9):1282–1288.

[23] Berbari EF, Marculescu C, Sia I, *et al.* Culture-negative prosthetic joint infection. Clin Infect Dis 2007;45(9):1113–1119.

[24] Zadaka A, Gioe T, Gertner E. Acute crystal-induced arthritis following arthroplasty. J Knee Surg 2010;23(1):17–20.

[25] Pajarinen J, Kouri VP, Jamsen E, *et al.* The response of macrophages to titanium particles is determined by macrophage polarization. Acta Biomater 2013;9(11):9229–9240.

[26] Trampuz A, Hanssen AD, Osmon DR, *et al.* Synovial fluid leukocyte count and differential for the diagnosis of prosthetic knee infection. Am J Med 2004;117(8):556–562.

[27] Schinsky MF, Della Valle CJ, Sporer SM, *et al.* Perioperative testing for joint infection in patients undergoing revision total hip arthroplasty. J Bone Joint Surg Am 2008;90(9):1869–1875.

[28] Bedair H, Ting N, Jacovides C, *et al.* The Mark Coventry Award: diagnosis of early postoperative TKA infection using synovial fluid analysis. Clin Orthop Relat Res 2011;469(1):34–40.

[29] Berbari E, Mabry T, Tsaras G, *et al.* Inflammatory blood laboratory levels as markers of prosthetic joint infection: a systematic review and meta-analysis. J Bone Joint Surg Am 2010;92(11):2102–2109.

[30] Gristina AG. Biomaterial-centered infection: microbial adhesion versus tissue integration. Science 1987;237(4822):1588–1595.

[31] Costerton JW, Stewart PS, Greenberg EP. Bacterial biofilms: A common cause of persistent infections. Science 1999;284:1318–1322.

[32] Hoiby N, Ciofu O, Johansen HK, *et al.* The clinical impact of bacterial biofilms. Int J Oral Sci 2010;3(2):55–65.

[33] Darouiche RO. Treatment of infections associated with surgical implants. N Engl J Med 2004;350(14):1422–1429.

[34] Elek SD, Conen PE. The virulence of *Staphylococcus pyogenes* for man; a study of the problems of wound infection. Br J Exp Pathol 1957;38(6):573–586.

[35] Zimmerli W. Experimental models in the investigation of device-related infections. J Antimicrob Chemother 1993;31(Suppl D):97–102.

[36] Rochford ET, Richards RG, Moriarty TF. Influence of material on the development of device-associated infections. Clin Microbiol Infect 2012;18(12):1162–1167.

[37] Ha KY, Chung YG, Ryoo SJ. Adherence and biofilm formation of *Staphylococcus epidermidis* and Mycobacterium tuberculosis on various spinal implants. Spine (Phila Pa 1976) 2005;30(1):38–43.

[38] Herrmann M, Vaudaux PE, Pittet D, *et al.* Fibronectin, fibrinogen, and laminin act as mediators of adherence of clinical staphylococcal isolates to foreign material. J Infect Dis 1988;158(4):693–701.

[39] Lopes JD, dos Reis M, Brentani RR. Presence of laminin receptors in *Staphylococcus aureus*. Science 1985;229(4710):275–277.
[40] Verhoef J. The phagocytic process and the role of complement in host defense. J Chemother 1991;3(Suppl 1):93–97.
[41] Zimmerli W, Zarth A, Gratwohl A, *et al.* Neutrophil function and pyogenic infections in bone marrow transplant recipients. Blood 1991;77(2):393–399.
[42] Klock JC, Stossel TP. Detection, pathogenesis, and prevention of damage to human granulocytes caused by interaction with nylon wool fiber. Implications for filtration leukapheresis. J Clin Invest 1977;60(5):1183–1190.
[43] Wright DG, Gallin JI. Secretory responses of human neutrophils: exocytosis of specific (secondary) granules by human neutrophils during adherence in vitro and during exudation in vivo. J Immunol 1979;123(1):285–294.
[44] Raad I, Alrahwan A, Rolston K. *Staphylococcus epidermidis*: emerging resistance and need for alternative agents. Clin Infect Dis 1998;26(5):1182–1187.
[45] Furustrand Tafin U, Corvec S, Betrisey B, *et al.* Role of rifampin against *Propionibacterium acnes* biofilm in vitro and in an experimental foreign-body infection model. Antimicrob Agents Chemother 2012;56(4):1885–1891.
[46] Widmer AF, Frei R, Rajacic Z, *et al.* Correlation between in vivo and in vitro efficacy of antimicrobial agents against foreign body infections. J Infect Dis 1990;162(1):96–102.
[47] Syggelos SA, Eleftheriou SC, Giannopoulou E, *et al.* Gelatinolytic and collagenolytic activity in periprosthetic tissues from loose hip endoprostheses. J Rheumatol 2001; 28(6):1319–1329.
[48] Hosman AH, van der Mei HC, Bulstra SK, *et al.* Effects of metal-on-metal wear on the host immune system and infection in hip arthroplasty. Acta Orthop 2010;81(5):526–534.
[49] Bernard L, Vaudaux P, Merle C, *et al.* The inhibition of neutrophil antibacterial activity by ultra-high molecular weight polyethylene particles. Biomaterials 2005;26(27):5552–5557.
[50] Bernard L, Vaudaux P, Huggler E, *et al.* Inactivation of a subpopulation of human neutrophils by exposure to ultrahigh-molecular-weight polyethylene wear debris. FEMS Immunol Med Microbiol 2007;49(3):425–432.
[51] Tuan RS, Lee FY, Konttinen Y, *et al.* What are the local and systemic biologic reactions and mediators to wear debris, and what host factors determine or modulate the biologic response to wear particles? J Am Acad Orthop Surg 2008;16(Suppl 1):S42–S48.
[52] Laverty G, Gorman SP, Gilmore BF. Biomolecular mechanisms of staphylococcal biofilm formation. Future Microbiol 2013;8(4):509–524.
[53] Simonetti O, Cirioni O, Mocchegiani F, *et al.* The efficacy of the quorum sensing inhibitor FS8 and tigecycline in preventing prosthesis biofilm in an animal model of staphylococcal infection. Int J Mol Sci 2013;14(8):16321–16332.
[54] Stewart PS, Costerton JW. Antibiotic resistance of bacteria in biofilms. Lancet 2001;358(9276):135–138.
[55] Deirmengian C, Greenbaum J, Lotke PA, *et al.* Limited success with open debridement and retention of components in the treatment of acute *Staphylococcus aureus* infections after total knee arthroplasty. J Arthroplasty 2003;18(7 Suppl 1):22–26.
[56] Laffer RR, Graber P, Ochsner PE, *et al.* Outcome of prosthetic knee-associated infection: evaluation of 40 consecutive episodes at a single centre. Clin Microbiol Infect 2006;12(5):433–439.
[57] Giulieri SG, Graber P, Ochsner PE, *et al.* Management of infection associated with total hip arthroplasty according to a treatment algorithm. Infection 2004;32(4):222–228.
[58] Marculescu CE, Berbari EF, Hanssen AD, *et al.* Outcome of prosthetic joint infections treated with debridement and retention of components. Clin Infect Dis 2006;42(4): 471–478.
[59] Sendi P, Zimmerli W. Antimicrobial treatment concepts for orthopaedic device-related infection. Clin Microbiol Infect 2012;18(12):1176–1184.

[60] Zimmerli W, Widmer AF, Blatter M, *et al.* Role of rifampin for treatment of orthopedic implant-related staphylococcal infections: a randomized controlled trial. Foreign-Body Infection (FBI) Study Group. JAMA 1998;279(19):1537–1541.
[61] Sendi P, Banderet F, Graber P, *et al.* Clinical comparison between exogenous and haematogenous periprosthetic joint infections caused by *Staphylococcus aureus*. Clin Microbiol Infect 2011;17(7):1098–1100.
[62] Zappe B, Graf S, Ochsner PE, *et al.* Propionibacterium spp. in prosthetic joint infections: a diagnostic challenge. Arch Orthop Trauma Surg 2008;128(10):1039–1046.
[63] Lutz MF, Berthelot P, Fresard A, *et al.* Arthroplastic and osteosynthetic infections due to *Propionibacterium acnes*: a retrospective study of 52 cases, 1995–2002. Eur J Clin Microbiol Infect Dis 2005;24(11):739–744.
[64] Kessler B, Sendi P, Graber P, *et al.* Risk factors for periprosthetic ankle joint infection: a case-control study. J Bone Joint Surg Am. 2012;94(20):1871–1876.
[65] Blomgren G, Lindgren U. Late hematogenous infection in total joint replacement: studies of gentamicin and bone cement in the rabbit. Clin Orthop Relat Res. 1981;(155):244–248.
[66] Blomgren G, Lundquist H, Nord CE, *et al.* Late anaerobic haematogenous infection of experimental total joint replacement. A study in the rabbit using *Propionibacterium acnes*. J Bone Joint Surg Br 1981;63B(4):614–618.
[67] Zimmerli W, Zak O, Vosbeck K. Experimental hematogenous infection of subcutaneously implanted foreign bodies. Scand J Infect Dis. 1985;17(3):303–310.
[68] Zimmerli W, Ochsner PE. Management of infection associated with prosthetic joints. Infection. 2003;31(2):99–108.
[69] Trampuz A, Piper KE, Jacobson MJ, *et al.* Sonication of removed hip and knee prostheses for diagnosis of infection. N Engl J Med. 2007;357(7):654–663.
[70] Sendi P, Zimmerli W. Diagnosis of periprosthetic joint infections in clinical practice. Int J Artif Organs. 2012;35(10):913–922.
[71] Widmer AF. New developments in diagnosis and treatment of infection in orthopedic implants. Clin Infect Dis 2001;33(Suppl 2):S94–S106.
[72] Dodson CC, Craig EV, Cordasco FA, *et al.* *Propionibacterium acnes* infection after shoulder arthroplasty: a diagnostic challenge. J Shoulder Elbow Surg 2010;19(2):303–307.
[73] Vanderstappen C, Verhoeven N, Stuyck J, *et al.* Intra-articular versus serum C-reactive protein analysis in suspected periprosthetic knee joint infection. Acta Orthop Belg 2013;79(3):326–330.
[74] Parvizi J, McKenzie JC, Cashman JP. Diagnosis of periprosthetic joint infection using synovial C-reactive protein. J Arthroplasty 2012;27:12–16.
[75] Schuetz P, Christ-Crain M, Thomann R, *et al.* Effect of procalcitonin-based guidelines vs standard guidelines on antibiotic use in lower respiratory tract infections: the ProHOSP randomized controlled trial. JAMA 2009;302(10):1059–1066.
[76] Worthington T, Dunlop D, Casey A, *et al.* Serum procalcitonin, interleukin-6, soluble intercellular adhesin molecule-1 and IgG to short-chain exocellular lipoteichoic acid as predictors of infection in total joint prosthesis revision. Br J Biomed Sci 2010;67(2):71–76.
[77] Drago L, Vassena C, Dozio E, *et al.* Procalcitonin, C-reactive protein, interleukin-6, and soluble intercellular adhesion molecule-1 as markers of postoperative orthopaedic joint prosthesis infections. Int J Immunopathol Pharmacol 2011;24(2):433–440.
[78] Ghanem E, Parvizi J, Burnett RS, *et al.* Cell count and differential of aspirated fluid in the diagnosis of infection at the site of total knee arthroplasty. J Bone Joint Surg Am 2008;90(8):1637–1643.
[79] Cipriano CA, Brown NM, Michael AM, *et al.* Serum and synovial fluid analysis for diagnosing chronic periprosthetic infection in patients with inflammatory arthritis. J Bone Joint Surg Am 2012;94(7):594–600.
[80] Tohtz SW, Muller M, Morawietz L, *et al.* Validity of frozen sections for analysis of periprosthetic loosening membranes. Clin Orthop Relat Res 2010;468(3):762–768.

[81] Aggarwal VK, Higuera C, Deirmengian G, *et al.* Swab cultures are not as effective as tissue cultures for diagnosis of periprosthetic joint infection. Clin Orthop Relat Res 2013;471(10):3196–3203.
[82] Gallo J, Kolar M, Dendis M, *et al.* Culture and PCR analysis of joint fluid in the diagnosis of prosthetic joint infection. New Microbiol 2008;31(1):97–104.
[83] Font-Vizcarra L, Garcia S, Martinez-Pastor JC, *et al.* Blood culture flasks for culturing synovial fluid in prosthetic joint infections. Clin Orthop Relat Res 2010;468(8):2238–2243.
[84] Portillo ME, Salvado M, Sorli L, *et al.* Multiplex PCR of sonication fluid accurately differentiates between prosthetic joint infection and aseptic failure. J Infect 2012;65(6):541–548.
[85] Vandercam B, Jeumont S, Cornu O, *et al.* Amplification-based DNA analysis in the diagnosis of prosthetic joint infection. J Mol Diagn 2008;10(6):537–543.
[86] Achermann Y, Vogt M, Leunig M, *et al.* Improved diagnosis of periprosthetic joint infection by multiplex PCR of sonication fluid from removed implants. J Clin Microbiol 2010;48(4):1208–1214.
[87] Westrich GH, Bornstein L, Brause BD, *et al.* Historical perspective on two-stage reimplantation for infection after total hip arthroplasty at Hospital for Special Surgery, New York City. Am J Orthop (Belle Mead NJ) 2011;40(11):E236–E240.
[88] De Man FH, Sendi P, Zimmerli W, *et al.* Infectiological, functional, and radiographic outcome after revision for prosthetic hip infection according to a strict algorithm. Acta Orthop 2011;82(1):27–34.
[89] Brandt CM, Sistrunk WW, Duffy MC, *et al.* *Staphylococcus aureus* prosthetic joint infection treated with debridement and prosthesis retention. Clin Infect Dis 1997;24(5):914–919.
[90] Krause R, Hoenigl M, Valentin T, *et al.* Debridement antibiotic therapy and implant retention in *Staphylococcus aureus* prosthetic joint infections. Clin Infect Dis 2013;56:1843.
[91] Lora-Tamayo J, Murillo O, Iribarren JA, *et al.* A large multicenter study of methicillin-susceptible and methicillin-resistant *Staphylococcus aureus* prosthetic joint infections managed with implant retention. Clin Infect Dis 2013;56(2):182–194.
[92] Lora-Tamayo J, Murillo O, Ariza J. Reply to Krause et al. Clin Infect Dis 2013; 56(12):1843–1844.
[93] Bittar ES, Petty W. Girdlestone arthroplasty for infected total hip arthroplasty. Clin Orthop Relat Res 1982;(170):83–87.
[94] Lentino JR. Prosthetic joint infections: bane of orthopedists, challenge for infectious disease specialists. Clin Infect Dis 2003;36(9):1157–1161.
[95] Rand JA, Bryan RS, Chao EY. Failed total knee arthroplasty treated by arthrodesis of the knee using the Ace-Fischer apparatus. J Bone Joint Surg Am 1987;69(1):39–45.
[96] Krijnen MR, Wuisman PI. Emergency hemipelvectomy as a result of uncontrolled infection after total hip arthroplasty: two case reports. J Arthroplasty 2004;19(6):803–808.
[97] Isiklar ZU, Landon GC, Tullos HS. Amputation after failed total knee arthroplasty. Clin Orthop Relat Res 1994;(299):173–178.
[98] Anderl JN, Zahller J, Roe F, *et al.* Role of nutrient limitation and stationary-phase existence in Klebsiella pneumoniae biofilm resistance to ampicillin and ciprofloxacin. Antimicrob Agents Chemother 2003;47(4):1251–1256.
[99] Baldoni D, Haschke M, Rajacic Z, *et al.* Linezolid alone or combined with rifampin against methicillin-resistant *Staphylococcus aureus* in experimental foreign-body infection. Antimicrob Agents Chemother 2009;53(3):1142–1148.
[100] Blaser J, Vergeres P, Widmer AF, *et al.* In vivo verification of in vitro model of antibiotic treatment of device-related infection. Antimicrob Agents Chemother 1995;39(5):1134–1139.
[101] Furustrand Tafin U, Majic I, Zalila Belkhodja C, *et al.* Gentamicin improves the activities of daptomycin and vancomycin against *Enterococcus faecalis* in vitro and in an experimental foreign-body infection model. Antimicrob Agents Chemother 2011;55(10):4821–4827.
[102] John AK, Baldoni D, Haschke M, *et al.* Efficacy of daptomycin in implant-associated infection due to methicillin-resistant *Staphylococcus aureus*: importance of combination with rifampin. Antimicrob Agents Chemother 2009;53(7):2719–2724.

[103] Schwank S, Rajacic Z, Zimmerli W, *et al.* Impact of bacterial biofilm formation on in vitro and in vivo activities of antibiotics. Antimicrob Agents Chemother 1998;42(4):895–898.
[104] Corvec S, Furustrand Tafin U, Betrisey B, *et al.* Activities of fosfomycin, tigecycline, colistin, and gentamicin against extended-spectrum-beta-lactamase-producing *Escherichia coli* in a foreign-body infection model. Antimicrob Agents Chemother 2013;57(3):1421–1427.
[105] Widmer AF, Wiestner A, Frei R, *et al.* Killing of nongrowing and adherent *Escherichia coli* determines drug efficacy in device-related infections. Antimicrob Agents Chemother 1991;35(4):741–746.
[106] Zimmerli W, Frei R, Widmer AF, *et al.* Microbiological tests to predict treatment outcome in experimental device-related infections due to *Staphylococcus aureus*. J Antimicrob Chemother 1994;33(5):959–967.
[107] Widmer AF, Colombo VE, Gachter A, *et al.* Salmonella infection in total hip replacement: tests to predict the outcome of antimicrobial therapy. Scand J Infect Dis 1990;22(5):611–618.
[108] Trampuz A, Murphy CK, Rothstein DM, *et al.* Efficacy of a novel rifamycin derivative, ABI-0043, against *Staphylococcus aureus* in an experimental model of foreign-body infection. Antimicrob Agents Chemother 2007;51(7):2540–2545.
[109] Niska JA, Shahbazian JH, Ramos RI, *et al.* Vancomycin-rifampin combination therapy has enhanced efficacy against an experimental *Staphylococcus aureus* prosthetic joint infection. Antimicrob Agents Chemother 2013;57(10):5080–5086.
[110] Widmer AF, Gaechter A, Ochsner PE, *et al.* Antimicrobial treatment of orthopedic implant-related infections with rifampin combinations. Clin Infect Dis 1992;14(6):1251–1253.
[111] Hsieh PH, Lee MS, Hsu KY, *et al.* Gram-negative prosthetic joint infections: risk factors and outcome of treatment. Clin Infect Dis 2009;49(7):1036–1043.
[112] Aboltins CA, Dowsey MM, Buising KL, *et al.* Gram-negative prosthetic joint infection treated with debridement, prosthesis retention and antibiotic regimens including a fluoroquinolone. Clin Microbiol Infect 2010;17(6):862–867.
[113] Peel TN, Buising KL, Dowsey MM, *et al.* Outcome of debridement and retention in prosthetic joint infections by methicillin-resistant staphylococci, with special reference to rifampin and fusidic acid combination therapy. Antimicrob Agents Chemother 2013; 57(1):350–355.
[114] Rodriguez-Pardo D, Pigrau C, Lora-Tamayo J, et al. Gram-negative prosthetic joint infection: outcome of a debridement, antibiotics, and implant retention approach. A large multicenter study. Clin Microbiol Infect 2014. DOI: 10.1111/1469-0691.12649 [Epub ahead of print].
[115] Zimmerli W. Antibiotic prophylaxis. In: Rüedi T *et al.*, editors. AO Principles of Fracture Management. New York: Thieme; 2007, pp. 425–433.
[116] Weber WP, Marti WR, Zwahlen M, *et al.* The timing of surgical antimicrobial prophylaxis. Ann Surg 2008;247(6):918–926.
[117] Jamsen E, Huhtala H, Puolakka T, *et al.* Risk factors for infection after knee arthroplasty. A register-based analysis of 43,149 cases. J Bone Joint Surg Am 2009;91(1):38–47.
[118] Uckay I, Bernard L. Gram-negative versus gram-positive prosthetic joint infections. Clin Infect Dis 2010;50(5):795.
[119] Berbari EF, Osmon DR, Carr A, *et al.* Dental procedures as risk factors for prosthetic hip or knee infection: a hospital-based prospective case-control study. Clin Infect Dis 2010;50(1):8–16.
[120] Legout L, Beltrand E, Migaud H, *et al.* Antibiotic prophylaxis to reduce the risk of joint implant contamination during dental surgery seems unnecessary. Orthop Traumatol Surg Res 2012;98(8):910–914.
[121] Zimmerli W, Sendi P. Antibiotics for prevention of periprosthetic joint infection following dentistry: time to focus on data. Clin Infect Dis 2010;50(1):17–19.

Chapter 9
Periprosthetic Joint Infection after Total Hip and Knee Arthroplasty

Werner Zimmerli and Martin Clauss

Introduction

Total hip arthroplasty (THA) and total knee arthroplasty (TKA) have been used since about 50 years [1, 2]. Both procedures are equally successful regarding alleviation of pain, implant function, and device survival. Between 1990 and 2007, in the United States, the number of THA increased twofold to about 200,000 and the number of TKA increased almost fivefold to about 550,000 [3, 4]. In Finland, the number of THA increased from 5000 to 9200, and the number of TKA increased from 3000 to 9100 between 1995 and 2009 [5]. The comparison of these figures shows a significant increase on both sides of the Atlantic. However, it clearly indicates a less strict indication of TKA in the United States, as opposed to Finland, where an equal number of hip and knees are replaced.

Prosthetic joints are highly susceptible to infection, due to an impaired host defense around the implant (see Chapter 8) [6–8]. PJIs are acquired during the perioperative period mainly by the exogenous route and lifelong by the hematogenous route. Kurtz *et al.* [9] projected that 65,000 PJIs associated with THA and TKA will be observed in the year 2020 in the United States. Thus, the number of PJIs will increase heavily. Therefore, the indication for THA and TKA must be strict. In addition, rapid diagnosis and correct management of PJI is important not only for the individual patient, but economically for the whole society.

Risk Factors

Risk factors for PJI have mainly been analyzed in cohort studies [10–13]. Most studies revealed patient characteristics, intervention-related factors, and remote infection/bacteremia as risk factors. Comorbidities such as diabetes mellitus, obesity (body mass index [BMI] > 30 kg/m^2), inflammatory rheumatic diseases, and immunosuppression impair

Bone and Joint Infections: From Microbiology to Diagnostics and Treatment, First Edition.
Edited by Werner Zimmerli.

wound healing, and therefore increase the risk for PJI. Previous joint surgery, prolonged operative time, and multiple simultaneous joint implantations are intervention-related risk factors. During the early postoperative period, delayed wound healing, prolonged or severe wound secretion, as well as large hematoma increase the risk for PJI.

In a large case control study, Berbari *et al.* [14] defined four independent risk factors for PJI, namely, superficial surgical site infection (odds ratio [OR] 35.9; 95% confidence interval [CI] 8.3–154.6), a NNIS (National Nosocomial Infections Surveillance) surgical patient risk index score of 2 (OR 3.9; 95% CI 1.3–7.5%), the presence of a malignancy (OR 3.1; 95% CI 1.3–7.2), and a history of prior arthroplasty (OR 2.0; 95% CI 1.4–3.0%).

Remote infections and bacteremia can be the source of hematogenous PJI, as long as the artificial joint remains in the body. The risk is very high (> 30%) during documented *Staphylococcus aureus* bacteremia [15–17], but relatively low (< 2%) during exposure to a remote infection. In the Geneva cohort, only one PJI occurred per 79 exposures to remote infections (1.25%) [18].

Microbiology

Almost all microorganisms, including *Mycobacterium tuberculosis* [19], nontuberculous mycobacteria [20], *Mycoplasma* [21], *Legionella* [22], and fungal agents [23], have been reported in PJI. The relative frequency of the microorganisms differs in different joints. However, in all types of PJI, staphylococci predominate.

Table 9.1 summarizes the microorganisms that have been found in patients with PJI associated with THA and TKA. Data from five studies with a nonselected population, reporting only cases from a defined joint, and giving the type of microorganisms in sufficient detail, are summarized [24–28]. After THA, the predominant microorganism responsible for PJI was *S. aureus*, causing 43.2% of the episodes. In contrast, in TKA PJI, *S. aureus* and coagulase-negative staphylococci (CNS) were about equally frequent, causing 30.4 and 28.4% of the episodes, respectively. For other microorganisms, there were no

Table 9.1. Microorganisms in periprosthetic hip and knee infection.

Microorganism	Total hip arthroplasty *n* = 118[a]	Total knee arthroplasty *n* = 500[b]
Staphylococcus aureus	51/118 = 43.2%	152/500 = 30.4%
Coagulase-negative staphylococci	22/118 = 18.6%	142/500 = 28.4%
Streptococcus spp.	11/118 = 9.3%	43/500 = 8.6%
Enterococcus spp.	4/118 = 3.9%	37/500 = 7.4%
Propionibacterium spp.	1/118 = 0.8	11/500 = 2.2%
Gram-negative bacilli	7/118 = 5.9%	33/500 = 6.6%
Miscellaneous	2/118 = 1.7%	9/500 = 1.8%
Polymicrobial	9/118 = 7.6%	29/500 = 5.8%
No growth	11/118 = 9.3%	44/500 = 8.8%

[a] Giulieri *et al.* [24], Schinsky *et al.* [25].
[b] Laffer *et al.* [26], Trampuz *et al.* [27], Stefansdottir *et al.* [28].
Adapted from Refs. 30, 46 and 48.

remarkable differences between the two different joints. Polymicrobial infections were observed in 7.6% of the patient with THA and 5.8% of those with TKA. In 5–12.7%, the cultures did not reveal any microorganism.

Clinical Features

The clinical presentation of PJI depends on the pathophysiology (exogenous versus hematogenous), the type of joint, and the duration of infection. The hallmarks of acute exogenous PJI are local signs of inflammation, such as skin erythema, hyperthermia, wound healing disturbance, drainage of fluid through the open wound, and purulent discharge [15, 29]. High-grade fever is rather the exception [15]. Acute-onset PJI beyond the perioperative period is generally of hematogenous origin [15, 17, 30–32]. It is mainly caused by virulent microorganisms such as *S. aureus*, β-hemolytic streptococci, and less frequently Gram-negative bacilli [18, 28, 31, 33, 34]. In hematogenous PJI caused by *S. aureus,* less than 50% of the patients have an identified focus [15]. The most frequent sources are skin and soft tissue as well as respiratory tract infections. Systemic signs of infection (sepsis syndrome) are frequent, whereas local inflammation is lacking in the early stage. The most important diagnostic clue is a profound pain at the site of the implant. In a series comparing PJIs caused by *S. aureus*, the leading sign in hematogenous infection was pain, occurring in 73% of the patients [15].

Chronic PJI is mostly acquired during the perioperative period. Typically, low-virulence microorganisms such as CNS or *Propionibacterium acnes* cause chronic infections because the early symptoms are subtle. Key symptoms are pain caused by local inflammation, chronic joint effusion, or implant loosening. If the diagnosis is missed for a long time, a sinus tract with spontaneous drainage of pus can be observed [24]. In our cohorts, abscesses and sinus tracts were more frequent in PJIs after THA (51%) than in those after TKA (25%) [24, 26]. This is probably due to the fact that PJIs after THA are diagnosed later than after TKA. Indeed, in our cohorts 41% of the PJI episodes after THA, but only 22.5% of those after TKA were chronic episodes, diagnosed more than 3 months after implantation [24, 26].

Laboratory Investigation

Data on the different laboratory tests for diagnosing PJI are presented in Chapter 8. Therefore, in this chapter only data that are reported specifically for patients with PJI after THA and TKA are summarized.

C-Reactive Protein and Erythrocyte Sedimentation Rate

C-reactive protein (CRP) is the most frequently used laboratory test for detection of infection [35]. Unfortunately, its specificity for detecting infection is limited, and various cutoffs have been proposed for different types of infections. Erythrocyte sedimentation rate (ESR) is rarely used in European countries for detection of infection. However, it is still a standard test in the United States. There are two main indications for testing parameters of inflammation in patients after arthroplasty. First, postoperative healing can be followed in a longitudinal fashion. Second, CRP and ESR may be useful for differentiating infectious versus mechanical failure in patients undergoing revision arthroplasty. Piper *et al.* [36] analyzed preoperative CRP and ESR values in 297 patients undergoing revision TKA and 221 undergoing revision THA. Patients with underlying inflammatory arthritis were excluded. Patients with PJI after TKA (n = 82) had a median ESR of 54 mm/h (range 6–128) and those with aseptic failure (n = 215) 11 mm/h (0–68). The corresponding values for CRP were 51 mg/l (3–444) versus 4 mg/l (0.1–174). ESR with a cutoff at greater than 30 mm/h had a sensitivity of 71% in patients with TKA and 47% with THA. The corresponding specificities were 89 and 84%, respectively. Thus, the sensitivity was clearly lower for differentiating PJI from aseptic failure after THA than after TKA. This difference was smaller for CRP. With a cutoff at greater than 10 mg/l, the sensitivity for detecting PJI after TKA was 83% and for PJI after THA 74%. The specificities were 79 and 78%, respectively. Considering both parameters together, the negative predictive values were 94% after both TKA and THA. Thus, it may be useful to determine both ESR and CRP. However, despite the high negative predictive value, PJI cannot be excluded with certitude in patients undergoing revision arthroplasty.

Synovial Cells

Leukocyte and differential counts in synovial fluid are the most important diagnostic tests in patients with suspected PJI. They allow differentiating PJI from aseptic failure with high sensitivity and specificity. In patients with painful THA, a synovial leukocyte count greater than 4200/µl had a sensitivity of 84%, and a specificity of 93%. Similarly, a neutrophil fraction greater than 80% had a sensitivity of 84% and a specificity of 82% [25]. In patients undergoing revision TKA, a synovial leukocyte count greater than 1700/µl had a sensitivity of 94% and a specificity of 88%. With a cutoff greater than 65% neutrophils, the sensitivity for detecting PJI was 97% and the specificity 98% [27]. Taken together, cell counts in synovial fluid allow detecting PJI with high probability, even if synovial cultures remain negative.

Imaging Procedures

The first diagnostic step in the imaging workup of PJI is the conventional radiograph. It allows detecting radiolucency, osteolysis, and migration, which, however, are signs of not only infection, but also aseptic loosening [37]. Ultrasonography can be used for guidance for joint aspiration, which is mainly useful in the hip.

Computed tomography (CT) is superior for detecting soft tissue infection (abscesses, sinus tracts), prosthetic loosening, and bone erosion. With the use of special techniques, metallic artifacts can be minimized [38]. Magnetic resonance imaging (MRI) can be used, if the device is not ferromagnetic (e.g., titanium and tantalum) [39]. Using techniques reducing metal artifacts, MR images excellently visualize not only soft tissue distant from the device but also bone and tendons directly adjacent to the prosthetic joint [40].

With radionuclide imaging, signs of infection are visible before anatomical changes. The three-phase bone scan using a bone-seeking tracer (e.g., technetium-99m-labeled methylene diphosphonate [^{99m}Tc-MDP]) is very sensitive for detecting infection, but has poor specificity [41]. Bone remodeling, and hence marker uptake, is increased for at least 1 year after implantation [42]. Using a specific radiotracer, such as ^{99m}Tc antigranulocyte monoclonal antibodies, improves the specificity to 68% in patients with TKA and THA, which is still insufficient for a reliable diagnosis [43]. The spatial resolution of nuclear scanning techniques is limited. Therefore, single-photon emission computed tomography plus conventional CT (SPECT/CT), which is performed with an integrated hybrid machine, should be preferred. This technique offers a more precise localization of the radiotracer. It is mainly used with ^{99m}Tc-MDP, labeled leukocytes, or labeled antigranulocyte monoclonal antibodies. It improves the sensitivity and specificity as compared to the planar image. In 31 consecutive patients with suspected low-grade PJI, sensitivity, specificity, and accuracy improved from 66, 60, and 61%, respectively, to 89, 73, and 77%, respectively. The prevalence of PJI in this population was 29% [44].

Positron emission tomography (PET) plays an important role in the detection of malignant tumors. ^{18}F-fluorodeoxyglucose (FDG) accumulates in cells such as neutrophils and is intracellularly phosphorylated to a stable molecule. Therefore, accumulation occurs not only in tumors, but also in inflammatory foci. According to a recent systematic review of 11 studies, including a total sample size of 635 prostheses, FDG–PET has good sensitivity (84.6%; 95% CI 71–92%) and specificity (84.0%; 95% CI 68–93%) for the detection of PJIs in hip and knee arthroplasties [45]. Nevertheless, it cannot yet be recommended for routine clinical practice. In many countries, costs are not covered for these novel imaging techniques.

Management

The therapeutic management of PJI includes surgical debridement with or without exchange of the device, antimicrobial therapy, and plastic-reconstructive surgery in cases with severely damaged skin and soft tissue. The different specialists should coordinate their respective interventions. The first question is whether a curative or only a palliative management is planned (see Chapter 8). Cure means complete eradication of infection and restoring a reasonable joint function. In contrast, the aim of palliative treatment is suppression of symptoms either by lifelong oral antibiotic therapy or by surgical removal of the infectious focus without considering function (removal without replacement or amputation).

Antibiotic Treatment

After microbiological sampling, antibiotic therapy is immediately started, initially by the intravenous (IV) route. This guarantees the highest tissue levels possible and overcomes possible perioperative disturbance of enteral resorption. Table 9.2 summarizes pathogen-specific therapy [30, 46, 47]. The correct duration of treatment has never been tested in a randomized controlled trial. Long-term treatment is based on the concept that remaining bacteria cannot be killed by host defense, if the microorganisms persist as biofilm. Therefore, we suggest a 3-month course in case of debridement with implant retention, one-stage exchange, or two-stage exchange with a short interval (2–3 weeks) [30, 47, 48]. For knee PJI, even a 6-month course is suggested [30, 47, 49]. However, in a recent observational study, patients undergoing debridement with implant retention were successfully treated for only 2 months for THA PJI and 3 months for TKA PJI [50]. Since this was not a comparative study, the length of therapy remains a matter of debate. In case of two-stage exchange with a long interval (≥8 weeks), a 6-week course is long enough, because the site of the device is supposed to be sterile at the time of implantation.

Table 9.2. Treatment of implant-associated infections.

Microorganism	Antimicrobial agent[a]	Dose	Route
Staphylococcus spp.	Recommendation for the initial treatment phase (for 2 weeks)		
Methicillin susceptible	Rifampin plus	450 mg q12 h[b]	PO/IV
	Nafcillin *or* oxacillin[c]	2 g q6 h	IV
Methicillin resistant	Rifampin plus	450 mg q12 h[b]	PO/IV
	Vancomycin *or*	15 mg/kg q12 h[d]	IV
	Daptomycin	6 mg/kg q24 h[e]	IV
Staphylococcus spp.	Recommendation after completion of the initial treatment phase		
	Rifampin plus	450 mg q12 h[b]	PO
	Levofloxacin *or*	750 mg q24 h or 500 mg q12 h	PO
	Ciprofloxacin *or*	750 mg q12 h	PO
	Teicoplanin *or*	400 mg q24 h[f]	IV
	Fusidic acid *or*	500 mg q8 h	PO
	Trimethoprim–sulfamethoxazole *or*	1 DS tablet q8 h[g]	PO
	Minocycline[h] *or*	100 mg q12 h	PO
	Linezolid *or*	600 mg q12 h	PO
	Clindamycin[i]	1200–1350 mg/day divided in 3–4 doses	
Streptococcus spp.[j]	Penicillin G[c] *or*	18–24 million Units/day divided in 4–6 doses	IV
	Ceftriaxone	2 g q24 h	IV
	for 4 weeks, followed by		
	Amoxicillin *or*	750 or 1000 mg q6 h to q8 h	PO
	Clindamycin[i]	1200 or 1350 mg/day divided in 3–4 doses	PO
Enterococcus spp.[k]			
Penicillin susceptible	Penicillin G[l] *or*	24 million Units/day divided in 6 doses	IV
	Ampicillin or amoxicillin[l]	2 g q6 h to q4 h	IV
Penicillin resistant	Vancomycin *or*	15 mg/kg q12 h[d]	IV
	Daptomycin *or*	6 mg/kg q24 h[e]	IV
	Linezolid	600 mg q12 h	IV/PO
Enterobacteriaceae	β-lactam based on in vitro susceptibilities for 2 weeks[m] followed by		IV
	Ciprofloxacin	750 mg q12 h	PO
Enterobacter spp.[n] and nonfermenters[o] (e.g., *Pseudomonas aeruginosa*)	Cefepime *or*	2 g q8 h	IV
	Ceftazidime *or*	2 g q8 h	IV
	Meropenem	1 g q8 h[p]	IV
	For 2–4 weeks, followed by		
	Ciprofloxacin	750 mg q12 h	PO
Propionibacterium spp.	Penicillin G *or*	18–24 million Units/day divided in 6 doses	IV

Table 9.2. (*Continued*)

Microorganism	Antimicrobial agent[a]	Dose	Route
	Clindamycin[i]	600 or 900 mg q8 h	IV
	For 2–4 weeks followed by		
	Amoxicillin *or*	750 or 1000 mg q8 h to q6 h	PO
	Clindamycin[i]	1200 or 1350 mg/day divided in 3–4 doses	PO
Gram-negative anaerobes (e.g., *Bacteroides* spp.)	Metronidazole	500 mg q8 h	IV/PO
Mixed infections (without methicillin-resistant staphylococci)	Ampicillin/sulbactam *or*	3 g q6 h	IV
	Amoxicilin/clavulate[q] *or*	2.2 g q6 h	IV
	Piperacillin/tazobactam *or*	4.5 g q8 h	IV
	Imipenem *or*	500 mg q6 h	IV
	Meropenem	1 g q8 h[p]	IV
	For 2–4 weeks, followed by individual regimens according to antimicrobial susceptibility		

Adapted from Refs. 30, 46 and 48.

Antimicrobial dosage recommendations are based on normal renal and hepatic function. Antimicrobials should be chosen based on in vitro susceptibility as well as patient drug allergies, intolerances, and potential drug interactions or contraindications to a specific antimicrobial.

PO, oral; IV, intravenous; DS, double strength.

[a]For total duration of antimicrobial treatment, see text.

[b]Other dosages and intervals of administration have been reported with equivalent success rates [46].

[c]In patients with delayed hypersensitivity, cefazolin (2 g every 8 h IV) can be administered. In patients with immediate hypersensitivity, penicillin should be replaced by vancomycin.

[d]Recommended doses are based on AUC_{0-24}/MIC and trough levels. Trough levels should be monitored for nephrotoxicity.

[e]Recommended dose according to IDSA guidelines [47]. However, dosages up to 10 mg/kg q 24 h are reported [46].

[f]Teicoplanin: Loading dose (day 1–3 of treatment) 800 mg q24 h is recommended. Not available in the United States.

[g]Double strength = trimethoprim 160 mg plus sulfamethoxazole 800 mg.

[h]Lack of data on bone penetration. Alternatively, doxycycline 100 mg q12 h PO is possible [46].

[i]Higher doses PO (e.g., up to 2400 mg/day) are possible, but frequently not tolerated due to side effects.

[j]We recommend determination of minimal inhibitory concentration for penicillin.

[k]Combination therapy with an aminoglycoside is optional, since its superiority to monotherapy in PJI is unproven. When using combination therapy, monitor signs of ototoxicity and nephrotoxicity of aminoglycosides; the latter is potentiated with other nephrotoxic agents (e.g., vancomycin).

[l]In patients with hypersensitivity to penicillin, see treatment options for penicillin-resistant enterococci.

[m]In patients with hypersensitivity to β-lactam, ciprofloxacin (PO or IV) can be administered.

[n]Cetriaxone and ceftazidime should not be administered for *Enterobacter* spp., even when tested to be susceptible in the laboratory. ESBL-producing strains should not be treated with any cephalosporin, including cefepime. In infections due to *Enterobacter* spp. ertapenem 1 g q24 h can be administered alternatively. However, ertapenem is not effective against *Pseudomonas* spp. and other nonfermenters.

[o]Addition of aminoglycoside optional. Use of two active drugs could be considered based on clinical circumstance of the patient. If aminoglycoside is used in the spacer, and the organism is susceptible to aminoglycoside, then double coverage should be provided with recommended IV or oral monotherapy [47].

[p]Dosage recommended according to IDSA guidelines [47]. In Europe, 2 g q8 h is suggested in case of *P. aeruginosa* infection [46].

[q]Not available in the United States as an IV drug.

Empirical therapy should be optimized as soon as the susceptibility pattern of the microorganism(s) is known. After the initial 1- to 2-week course of IV therapy, switch to oral therapy can be considered. The decision should be based on clinical and laboratory signs of inflammation, formation of hematoma, wound healing, and wound secretion [46]. In addition, an appropriate drug with excellent bioavailability should be available, otherwise IV therapy has to be continued. In case of staphylococcal infection, the oral combination regimen should include rifampin [49], and in PJI caused by Gram-negative bacilli, a fluoroquinolone [51, 52, 52a] is the best choice (see Chapter 8). The role of rifampin for treatment of pathogens other than staphylococci has been reviewed recently. There are not enough data to suggest rifampin combination therapy in streptococcal, enterococcal, *P. acnes*, or Gram-negative PJI [46].

In patients on rifampin therapy, drug interactions should be considered. An additional problem is the rapid emergence of resistance, if rifampin is inappropriately used [53]. According to a case control study, risk factors for emergence of resistance are multiple previous surgical revisions, treatment with high initial bacterial load (inadequate debridement surgery, <2 weeks of initial IV combination therapy), and inadequate use (monotherapy or combination with drug with low bioavailability) [54].

Surgical Interventions

Figures 9.1 and 9.2 show algorithms for the surgical management of PJI. They have been developed for patients with PJI after hip or knee arthroplasty [24, 26, 30, 55].

Debridement and Implant Retention

Debridement with implant retention is an established procedure for either (i) early postoperative infection within 4 weeks after implantation or (ii) acute hematogeneous infections within 3 weeks after onset of symptoms, provided that the implant is stable, the pathogen is susceptible to a biofilm-active antimicrobial agent, and skin and soft tissue are intact [30, 47, 55] (Figure 9.1). Debridement aims to reduce the bacterial burden around a stable implant. For a successful treatment, it has to be combined with long-term (3 months THA and 6 months TKA) antimicrobial treatment [30, 46, 47].

The surgical procedure starts with (i) blunt opening of the healing wound, or (ii) excision of the scar, whenever possible, without disturbing later wound closure. Afterward, preparation follows the established surgical approach. All potentially infected loose foreign bodies such as suture material, wires [56], and bone grafts [57, 58], should be removed. The next step is a complete synovectomy. Removing all modular parts of the prosthesis to complete synovectomy of the postero-medial capsule (THA) or the poplitea (TKA) is mandatory. Afterward, the joint should be rinsed with a large amount (>3–5 l) of either sterile saline or a disinfectant (e.g., polyhexanid/Lavasept®). Jet lavage should be used only on fixed prosthetic components, while it is not recommended on soft tissue, as it spreads bacteria in deeper layers of the tissue instead of clearing the wound [59]. Then mobile parts are replaced. It is important to assess joint stability (e.g., with a thicker polyethylene (PE) inlay or longer femoral head) as this might have changed during synovectomy. In THA, metal heads or special revision heads should be used to avoid later breakage of the head as the cone might have been damaged during the procedure. In the final step, the approach is closed using several drains to drain potential hematoma formation.

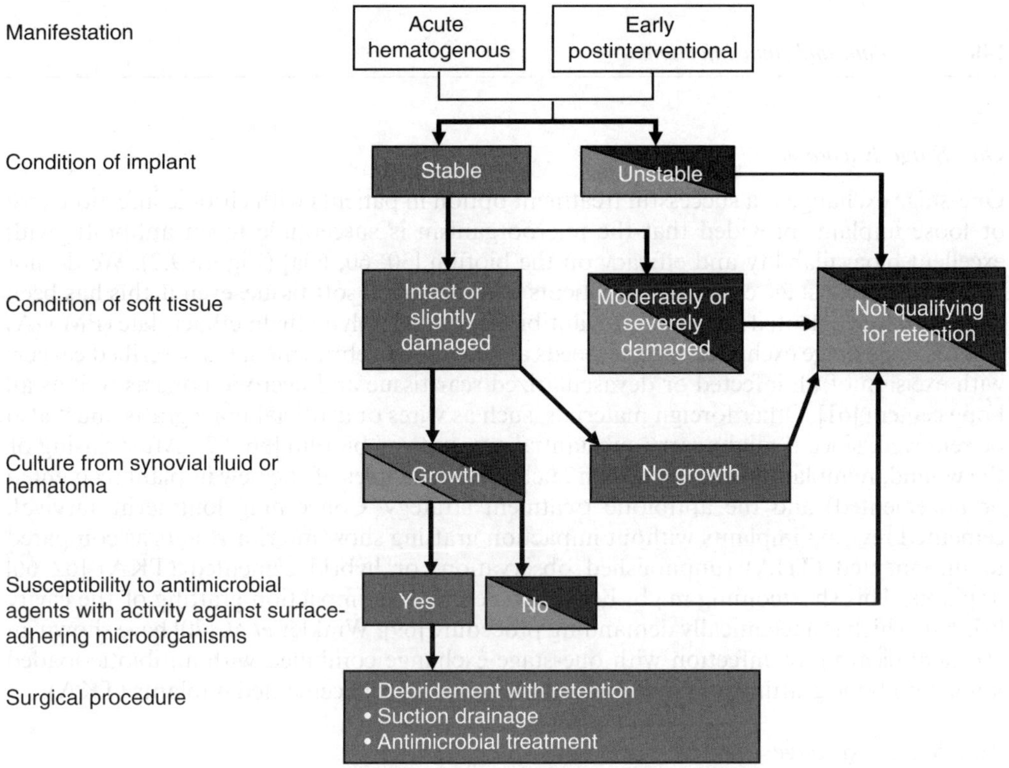

Figure 9.1. Treatment algorithm of acute hematogenous and early postinterventional PJI. Modified from Ref. 30. (*See insert for color representation of the figure.*)

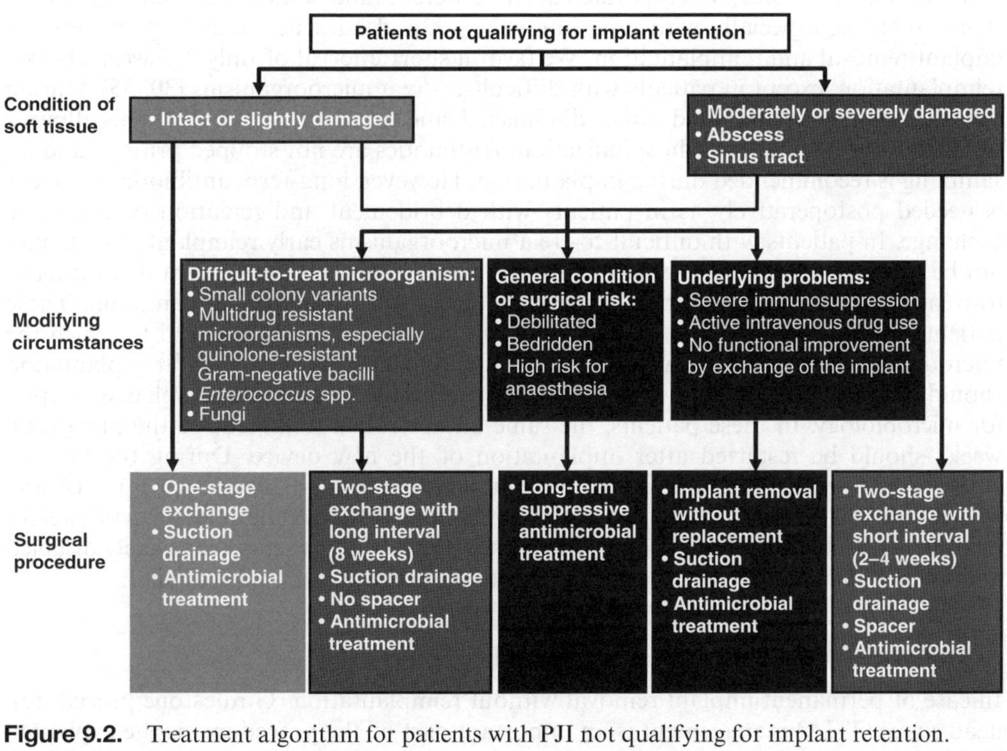

Figure 9.2. Treatment algorithm for patients with PJI not qualifying for implant retention. Modified from Ref. 30. (*See insert for color representation of the figure.*)

One-Stage Exchange

One-stage exchange is a successful treatment option in patients with chronic infection and/or loose implant, provided that the microorganism is susceptible to an antibiotic with excellent bioavailability and efficacy on the biofilm [30, 60, 60a] (Figure 9.2). We do not recommend one-stage exchange in patients with damaged soft tissue, even if this has been suggested for cemented implants with antibiotic-loaded polymethylmethacrylate (PMMA) [58, 61]. One-stage exchange surgery needs a meticulous debridement (as described earlier) with excision of all infected or devascularized scar tissue and necrotic bone as well as all bone cement [61]. Other foreign materials, such as wires or artificial bone grafts, must also be removed, since it might carry substantial amounts of biofilm [56, 57]. After rinsing of the wound, reimplantation depends on anchoring principles of the new implant (cemented or uncemented) and the antibiotic treatment strategy. Concerning long-term survival, cemented revision implants without impaction grafting show inferior results as compared to uncemented (THA) (unpublished observation) or hybrid cemented (TKA) [62–66] implants. This shortcoming might be overcome by using impaction grafting of the cavity [67, 68], which is a technically demanding procedure [69]. Winkler *et al.* [70] have shown an excellent clearing of infection with one-stage exchange combined with antibiotic-loaded impaction bone grafting and uncemented (THA) or hybrid cemented implants (TKA).

Two-Stage Exchange

Two-stage exchange was the standard procedure in the United States before publication of the Infectious Diseases Society of America (IDSA) guidelines [47, 71] (Figure 9.2). The surgical steps are the same as in one-stage exchanges (meticulous debridement and reimplantation of the device), but separated at two different time points. A crucial step in two-stage exchange, especially with long intervals, is the dead space management between implant removal and reimplantation. We favor a short interval of only 2–3 weeks before reimplantation, except in patients with difficult-to-treat microorganisms [30, 55]. During this interval the bacterial load can be diminished, and the soft tissue healed. This allows a treatment during one single hospitalization. Antibiotics are not stopped prior to and no sampling is recommended during implantation. However, long-term antibiotic treatment is needed postoperatively as in patients with debridement and retention or one-stage exchange. In patients with difficult-to-treat microorganisms early reimplantation should not be chosen [30, 46, 55]. The rationale for the long interval is the concept that difficult-to-treat microorganisms must be completely eradicated prior to reimplantation. These patients should be treated for 6 weeks without any foreign body material to which the microorganism could potentially adhere again and form a biofilm. Then, reimplantation should be delayed for two more weeks free of antibiotics in order to get reliable samples for microbiology. In these patients, the same antimicrobial treatment of the previous 6 weeks should be restarted after implantation of the new device. During the interval without prosthetic device, limb control can be achieved with a PMMA spacer (THA and TKA), external fixation (TKA), or traction (TKA and THA). While at least partial weight bearing is possible with the use of a PMMA spacer [72–74], patients are severely disabled with a joint spanning fixation device or bedridden during a traction treatment.

Resection Arthroplasty

In case of permanent implant removal without reimplantation (Girdlestone procedure), dead space (THA) management is an important part of the procedure, as the explanted

THA removal leaves a huge cavity behind. Excellent infection clearing and dead space control using a *Musculus vastus lateralis* transfer in the cavity of an explanted hip has been proposed [75]. However, the functional results are inferior to one- or two-stage exchanges. Therefore, it should only be used as a salvage procedure.

Specific Surgical Problems in Hip PJI

Despite continuous monitoring, education, and surgical training, there has been an increasing incidence of PJI during the last decade in all Scandinavian registers without finding clear risk factors [76]. A shortcoming of register data is the lack of surgery-related risk factors, such as duration of surgery or timing of antibiotic prophylaxis [77, 78]. Other surgical factors that might influence the rate of PJI are the surgical approach to the hip (anterior, anterio-lateral, lateral, posterior) and positioning of the patient (supine or lateral). We could not find any published data advocating a specific approach or positioning of the patient as a risk factor. Minimal invasive surgery has become popular during the last decade. There are various studies describing increasing surgical complications like femoral fractures, implant malpositioning, or increased duration of surgery, but none of these studies has shown an increased risk of PJI [79–83].

The first step of a two-stage exchange of infected THA is implant removal, which leaves a considerable cavity in the hip area. Usually, this cavity is filled with a PMMA spacer [73, 84–89]. Spacers can be obtained preformed and industrially loaded with antibiotics, or they can be produced custom-made with an individual antibiotic loading without differences concerning infection control [89, 90]. The use of antibiotic-loaded PMMA is well-established from the treatment of osteomyelitis [91, 92]. The use of antibiotics in spacers is controversial. The delegates from the International Consensus Meeting on PJI strongly recommend the addition of antibiotics [89, 90]. This is in contrast to a recent review, in which no benefit could be observed by adding antibiotics to temporary PMMA spacers in 824 patients with PJI treated with systemic antibiotics [93]. The creation of a so-called bead pouch is a variation to fill cavities in the first interval of a two-stage exchange. This would increase the surface area/volume of the bone graft substitute [94], but will not allow weight bearing of the joint. Thus, this might be an option in the exchange of, for example, infected shoulder and elbow arthroplasties where weight bearing is not needed. We found only one study dealing with infected total hip arthroplasties using a bead pouch in the first step and impaction grafting in the second step [95]. Without using any IV antibiotics, the success rate was 86%. Because of various limitations of this study, the procedure cannot be considered as standard. Other options for dead space management are the use of antibiotic-loaded collagen fleeces [94] or antibiotic-loaded calcium sulfate [96]. There are various other synthetic materials, which are currently being tested in vitro and in vivo, but clinical data are still lacking.

Specific Surgical Problem in Knee PJI

In general, TKA-associated infection is rapidly diagnosed due to the sparse soft tissue mantle. Indeed, Laffer *et al.* [26] reported in a series of 40 episodes of TKA PJI that 77% of the patients had either early postoperative or late hematogenous infection. Correspondingly, 53% of the patients could be treated with debridement and retention, with 95.2% success. The optimal surgical technique for this procedure has not yet been established. Arthroscopic as well as open debridement have been suggested. In published studies, the success rate of open debridement was reported to be 83–100% [26, 97], whereas the one of arthroscopic debridement was only 38–88% [26, 98–100]. Unfortunately, these data are hard to interpret, because in no study were comparable cases randomized to either arthroscopic or open debridement. Thus, it can be speculated that more severe cases were treated with open debridement. Therefore, the tendency for worse outcome with arthroscopic debridement indicates that this is not an optimal procedure. We prefer open debridement, not least because modular parts of the device can be exchanged in case of PJI treated with implant retention.

Instructive Cases

Case 1: Acute Hematogenous PJI due to Streptococcus dysgalactiae *8 Months after THA*

A 66-year-old man had a right total THA due to osteoarthritis. His case history revealed spinal stenosis with right-sided pseudo-radicular pain (L4), and obesity (BMI 45 kg/m^2). He was discharged 5 days after surgery, but presented 6 days later at the Emergency Department with severe pain from the back crossing the hip down until the lateral aspect of the knee, caused by a massive hematoma in his right thigh. Leukocyte counts were normal, and CRP was declining as compared to the discharge parameters. Six weeks after surgery, the patient consulted again because of persisting pain and swelling. Due to a suspected early infection, his THA was punctured. The puncture revealed 10 ml of dark liquid with a leukocyte count of 6000/µl (76% neutrophils). Synovial fluid sampled in blood culture bottles did not show any growth.

Eight months after surgery, the patient suddenly suffered from fatigue. One day later, he developed fever and chills, as well as acute reddening and swelling over his right THA. The patient was hospitalized, and treated with penicillin (5 Mio IU 5 times daily i.v.) for 5 days, followed by cefuroxime axetil (500 mg 4 times daily) plus clindamycin (600 mg 4 times daily) during 7 days. Initial blood cultures revealed *Staphylococcus epidermidis*.

He was referred to our clinic 7 days later. As the patient was free of symptoms, he was scheduled for arthrocentesis 5 days after stopping antibiotics. The puncture revealed 56,000 leukocytes/µl (96% neutrophils), and showed growth of *Streptococcus dysgalactiae*.

Two days later, open debridement and exchange of the femoral head was performed. Intraoperative biopsies again showed *S. dysgalactiae*. Antibiotic therapy was started with penicillin (6×5 Mio IU/day), followed by IV ceftriaxone (2 g/day) for a total of 4 weeks, followed by oral amoxicillin (3×1 g/day) for another 2 months.

Learning Points

- This patient suffered from a complicated initial postoperative course due to a hematoma, which led to an initial arthrocentesis 8 weeks after surgery showing 6000 leukocytes per µl (24% neutrophils) without bacterial growth. This value was suspicious for PJI according to the established breakpoint of 4200 leukocytes per µl [25]. However, in this study only 7 of 55 patients were punctured within the first 6 weeks of surgery. In contrast, Yi *et al.* [101] investigated 73 patients with suspicion of PJI within the first 6 weeks of surgery, and found a cutoff value of 12,800 cells/µl. Therefore, in the described case, PJI was very unlikely at that time point.

- *S. dysgalactiae* causes about 3% of all episodes of PJI [33]. It is more frequent in patients with hematogenous PJI. In this patient, the duration of symptoms was short (<2 weeks), the implant was stable, and the tissue damage was rated minor during revision surgery. In patients with these conditions, the success rate for open debridement and retention of the implant followed by long-term antibiotic therapy is high (10/10), as reported by Sendi *et al.* [33].

Case 2: Two-Stage Exchange of Infected TKA with long interval without spacer due to difficult-to-treat microorganisms

A 71-year-old man received a TKA due to osteoarthritis in his right knee. After surgery, he suffered from wound healing disturbances, and was treated for 2 weeks with amoxicillin/clavulanic acid without further diagnostic workup. After a symptom-free period, he developed local pain and swelling, 7 months after implantation. An oral therapy with cefuroxime axetil was given for 6 weeks with clinical success. Five months later, local signs of arthritis reappeared. Therefore, arthrocentesis was performed and revealed 44,460 leukocytes per microliter (86% neutrophils), but no bacterial growth. Due to persisting symptoms, he was referred to our institution. At this time, culture of synovia showed growth of *S. aureus* small colony variants.

Two-stage exchange with external fixation instead of a spacer was performed. After 2 weeks of flucloxacillin (4 × 2 g i.v./day), followed by 4 weeks of oral levofloxacin (1 × 750 mg/day) plus rifampin (2 × 300 mg/day), antibiotics were stopped. Two weeks later, a new device was implanted. Cultures of intraoperative biopsies showed no growth. The patient was pain-free with a good function (120° flexion), and showed no signs of infection after a follow-up of 2.5 years.

Learning Points

- One year after surgery, the synovial cell counts and repartition were highly suspicious for PJI, even in the absence of bacterial growth. Trampuz *et al.* [27] found a sensitivity of 94% and a specificity of 88% for diagnosing prosthetic knee joint infection with a leukocyte count of greater than 1700/µl; a differential of greater than 65% neutrophils had a sensitivity of 97% and a specificity of 98%.
- *S. aureus* small colony variants are difficult to detect and can therefore be missed [102].
- *S. aureus* small colony variants should be considered as difficult-to-treat microorganisms [102] and should therefore be treated with a two-stage exchange without spacer with a long interval according to our algorithm [30].
- Two-stage exchange of TKA with intercurrent external fixation can give excellent functional results.

Case 3: Chronic infection of THA treated with two-stage exchange

A 69-year-old woman was treated with TKA for osteoarthritis. Two years later, she had a traumatic patella luxation, which was treated with reconstruction of the extensor mechanism. This intervention was followed by wound healing disturbances. Antibiotic therapy with amoxicillin/clavulanic acid (3 × 500/125 mg/day) was started. One week later, open debridement with installation of a vacuum-assisted closure (VAC) system was performed.

Intraoperative samples showed CNS susceptible to methicillin, and the oral therapy was continued. VAC exchange to an open system was planned 1 week later. Intraoperative biopsies showed *Enterobacter cloacae* and methicillin-resistant *S. epidermidis*. She was discharged with the VAC system, and regular VAC changes were performed on an outpatient basis. Treatment with amoxicillin/clavulanic acid was continued. Six months later, she was referred to our clinic with the VAC still in place (Figure 9.3) with a chronic wound discharge (200 ml/day).

After referral, antibiotics were stopped. Two weeks later, the TKA was explanted and the knee spanned with an external fixator without a spacer. Intraoperative samples showed polymicrobial growth of *S. aureus*, *S. epidermidis* (three different types), *Staphylococcus caprae*, and *E. cloacae*. She was treated for 6 weeks with vancomycin (2 × 1 g/day i.v.) plus imipenem (4 × 500 mg/day i.v.). Two weeks later, a rotating hinge knee was implanted. At reimplantation, six out of six biopsies showed no growth. She was still free of infection with an acceptable function with 90° flexion and mild pain at the latest follow-up 7 years after reimplantation.

Learning Points

- In order to improve the sensitivity of the intraoperative biopsy cultures, antibiotics should be stopped for at least 2 weeks [47].
- Open VAC systems are not able to clear infection, but represent an open wound, which is at high risk for polymicrobial superinfection and cannot be healed with the VAC system in situ.
- Diagnostic workup and treatment should be rapid and correct. If not, a more invasive procedure is generally required for the cure of infection. In the present case, infection after soft tissue reconstruction could have been cured with open debridement,

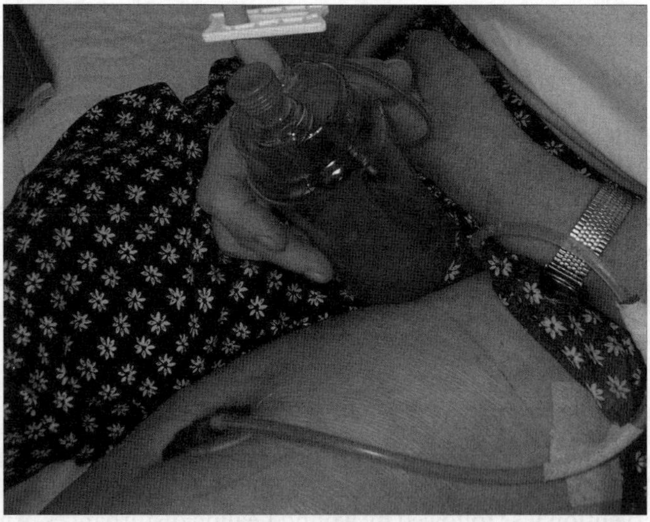

Figure 9.3. Vacuum-assisted closing (VAC) system. Clinical picture of a VAC system covering the open wound. After VAC removal the femoral component of the TKA was visible. (*See insert for color representation of the figure.*)

exchange of the tibial polyethylene inlay, and antibiotic treatment. Delayed and inadequate early treatment led to the need of the more invasive two-stage exchange and a hinged knee with functional inferior results.

Case 4: One-stage revision of a chronic THA infection after removal of ossifications

A 66-year-old man got a hip replacement for right-sided osteoarthritis. The initial course was uneventful, but he developed painful periarticular ossifications, which had to be removed 3 years later. Thereafter, he suffered from early wound infection due to CNS (PenR, OxaR, CiproR, FucS, RifS), considered as superficial surgical site infection. He was treated with debridement above the fascia and antibiotic therapy with fusidic acid (3 × 500 mg/day p.os) and rifampin (2 × 450 mg/day p.os) for 3 months. Nine months later, he still had a painful right hip and elevated CRP levels (23 mg/l). Arthrocentesis revealed growth of CNS with identical susceptibility pattern. The plain radiograph showed loosening of the prosthetic joint with signs of a chronic osteomyelitis surrounding the stem (Figure 9.4). He was treated with one-stage exchange. Sonication culture as well as cultures from five out of six intraoperative biopsies showed growth of the same CNS. He was treated with vancomycin (2 × 1 g/day i.v.) for 14 days, followed by an oral course of fusidic acid (3 × 500 mg/day p.os) and rifampin (2 × 450 mg/day p.os) for a total of 3 months. At the latest follow-up 4 years after surgery, he was free of infection with an excellent hip function and unrestricted walking time.

Learning Points

- Early infection of a THA needs open debridement of the joint with exchange of the mobile parts combined with long-term antibiotic therapy. Berbari *et al.* [14] showed an OR of 35.9 in case of early surgical site infection, which was considered as

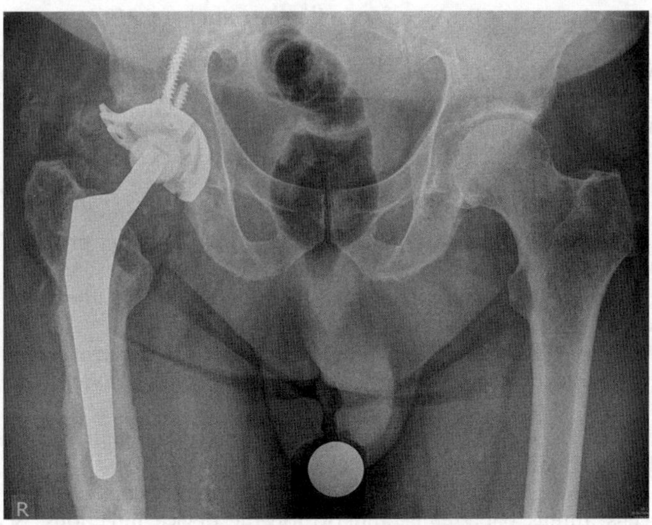

Figure 9.4. Plain radiograph of periarticular osteomyelitis.

superficial. Obviously, after joint replacement, the diagnosis of superficial wound infection is not reliable. Thus, in case of wound infection after joint replacement, superficial debridement is insufficient to cure possible PJI.
- One-stage exchange of THA can be successfully performed in case of infection with intact or only slightly damaged soft tissue, caused by a microorganism, which is susceptible to an antibiotic with excellent bioavailability and efficacy on biofilm bacteria [30, 47, 60a].

References

[1] Harris WH, Sledge CB. Total hip and total knee replacement (1). N Engl J Med 1990;323(11): 725–731.
[2] Harris WH, Sledge CB. Total hip and total knee replacement (2). N Engl J Med 1990;323(12): 801–807.
[3] Del Pozo JL, Patel R. Clinical practice. Infection associated with prosthetic joints. N Engl J Med 2009;361(8):787–794.
[4] Kurtz SM, Lau E, Schmier J, *et al.* Infection burden for hip and knee arthroplasty in the United States. J Arthroplasty 2008;23(7):984–991.
[5] Jamsen E, Virta LJ, Hakala M, *et al.* The decline in joint replacement surgery in rheumatoid arthritis is associated with a concomitant increase in the intensity of anti-rheumatic therapy: a nationwide register-based study from 1995 through 2010. Acta Orthop 2013;84(4):331–337.
[6] Zimmerli W, Waldvogel FA, Vaudaux P, *et al.* Pathogenesis of foreign body infection: description and characteristics of an animal model. J Infect Dis 1982;146(4):487–497.
[7] Zimmerli W, Lew PD, Waldvogel FA. Pathogenesis of foreign body infection. Evidence for a local granulocyte defect. J Clin Invest 1984;73(4):1191–1200.
[8] Zimmerli W, Sendi P. Pathogenesis of implant-associated infection: the role of the host. Semin Immunopathol 2011;33(3):295–306.
[9] Kurtz SM, Lau E, Watson H, *et al.* Economic burden of periprosthetic joint infection in the United States. J Arthroplasty 2012;27(8 Suppl):61–5 e1.
[10] Wymenga AB, van Horn JR, Theeuwes A, *et al.* Perioperative factors associated with septic arthritis after arthroplasty. Prospective multicenter study of 362 knee and 2,651 hip operations. Acta Orthop Scand 1992;63(6):665–671.
[11] Bozic KJ, Ward DT, Lau EC, *et al.* Risk factors for periprosthetic joint infection following primary total hip arthroplasty: a case control study. J Arthroplasty 2014;29(1):154–156. DOI: 10.1016/j.arth.2013.04.015 [Epub May 20, 2013].
[12] Workgroup of the American Association of Hip and Knee Surgeons Evidence Based Committee. Obesity and total joint arthroplasty: a literature based review. J Arthroplasty 2013 28(5):714–21.
[13] Bozic KJ, Lau E, Ong K, *et al.* Risk factors for early revision after primary TKA in medicare patients. Clin Orthop Relat Res 2014;472(2):449–454.
[14] Berbari EF, Hanssen AD, Duffy MC, *et al.* Risk factors for prosthetic joint infection: case-control study. Clin Infect Dis 1998;27(5):1247–1254.
[15] Sendi P, Banderet F, Graber P, *et al.* Clinical comparison between exogenous and haematogenous periprosthetic joint infections caused by *Staphylococcus aureus*. Clin Microbiol Infect 2011;17(7):1098–1100.
[16] Zimmerli W, Zak O, Vosbeck K. Experimental hematogenous infection of subcutaneously implanted foreign bodies. Scand J Infect Dis 1985;17(3):303–310.
[17] Lalani T, Chu VH, Grussemeyer CA, *et al.* Clinical outcomes and costs among patients with Staphylococcus aureus bacteremia and orthopedic device infections. Scand J Infect Dis 2008;40(11–12):973–977.

[18] Uckay I, Lubbeke A, Emonet S, *et al.* Low incidence of haematogenous seeding to total hip and knee prostheses in patients with remote infections. J Infect 2009;59(5):337–345.
[19] Harwin SF, Banerjee S, Issa K, *et al.* Tubercular prosthetic knee joint infection. Orthopedics. 2013;36(11):e1464–e1469.
[20] Wang SX, Yang CJ, Chen YC, *et al.* Septic arthritis caused by Mycobacterium fortuitum and Mycobacterium abscessus in a prosthetic knee joint: case report and review of literature. Intern Med 2011;50(19):2227–2232.
[21] Sneller M, Wellborne F, Barile MF, *et al.* Prosthetic joint infection with Mycoplasma hominis. J Infect Dis 1986;153(1):174–175.
[22] Fernandez-Cruz A, Marin M, Castelo L, *et al.* Legionella micdadei, a new cause of prosthetic joint infection. J Clin Microbiol 2011;49(9):3409–3410.
[23] Kuiper JW, van den Bekerom MP, van der Stappen J, *et al.* 2-stage revision recommended for treatment of fungal hip and knee prosthetic joint infections. Acta Orthop 2013;84(6):517–523.
[24] Giulieri SG, Graber P, Ochsner PE, *et al.* Management of infection associated with total hip arthroplasty according to a treatment algorithm. Infection 2004;32(4):222–228.
[25] Schinsky MF, Della Valle CJ, Sporer SM, *et al.* Perioperative testing for joint infection in patients undergoing revision total hip arthroplasty. J Bone Joint Surg Am 2008;90(9): 1869–1875.
[26] Laffer RR, Graber P, Ochsner PE, *et al.* Outcome of prosthetic knee-associated infection: evaluation of 40 consecutive episodes at a single centre. Clin Microbiol Infect 2006;12(5):433–439.
[27] Trampuz A, Hanssen AD, Osmon DR, *et al.* Synovial fluid leukocyte count and differential for the diagnosis of prosthetic knee infection. Am J Med 2004;117(8):556–562.
[28] Stefansdottir A, Johansson D, Knutson K, *et al.* Microbiology of the infected knee arthroplasty: report from the Swedish Knee Arthroplasty Register on 426 surgically revised cases. Scand J Infect Dis 2009;41(11–12):831–840.
[29] Peel TN, Cheng AC, Buising KL, *et al.* Microbiological aetiology, epidemiology, and clinical profile of prosthetic joint infections: are current antibiotic prophylaxis guidelines effective? Antimicrob Agents Chemother 2012;56(5):2386–2391.
[30] Zimmerli W, Trampuz A, Ochsner PE. Prosthetic-joint infections. N Engl J Med. 2004;351(16):1645–1654.
[31] Sendi P, Banderet F, Graber P, *et al.* Periprosthetic joint infection following *Staphylococcus aureus* bacteremia. J Infect 2011;63(1):17–22.
[32] Murdoch DR, Roberts SA, Fowler VG, Jr, *et al.* Infection of orthopedic prostheses after *Staphylococcus aureus* bacteremia. Clin Infect Dis 2001;32(4):647–649.
[33] Sendi P, Christensson B, Uckay I, *et al.* Group B streptococcus in prosthetic hip and knee joint-associated infections. J Hosp Infect 2011;79(1):64–69.
[34] Downes EM. Late infection after total hip replacement. J Bone Joint Surg Br 1977;59(1): 42–44.
[35] Simon L, Gauvin F, Amre DK, *et al.* Serum procalcitonin and C-reactive protein levels as markers of bacterial infection: a systematic review and meta-analysis. Clin Infect Dis 2004;39(2):206–217.
[36] Piper KE, Fernandez-Sampedro M, Steckelberg KE, *et al.* C-reactive protein, erythrocyte sedimentation rate and orthopedic implant infection. PLoS One 2010;5(2):e9358.
[37] Stumpe KD, Notzli HP, Zanetti M, *et al.* FDG PET for differentiation of infection and aseptic loosening in total hip replacements: comparison with conventional radiography and three-phase bone scintigraphy. Radiology 2004;231(2):333–341.
[38] Lee MJ, Kim S, Lee SA, *et al.* Overcoming artifacts from metallic orthopedic implants at high-field-strength MR imaging and multi-detector CT. Radiographics 2007;27(3):791–803.
[39] Shellock FG, Karacozoff AM. Detection of implants and other objects using a ferromagnetic detection system: implications for patient screening before MRI. AJR Am J Roentgenol 2013;201(4):720–725.

[40] Sutter R, Ulbrich EJ, Jellus V, *et al.* Reduction of metal artifacts in patients with total hip arthroplasty with slice-encoding metal artifact correction and view-angle tilting MR imaging. Radiology 2012;265(1):204–214.
[41] Gemmel F, Van den Wyngaert H, Love C, *et al.* Prosthetic joint infections: radionuclide state-of-the-art imaging. Eur J Nucl Med Mol Imaging 2012;39(5):892–909.
[42] Utz JA, Lull RJ, Galvin EG. Asymptomatic total hip prosthesis: natural history determined using Tc-99m MDP bone scans. Radiology 1986;161(2):509–512.
[43] Ivancevic V, Perka C, Hasart O, *et al.* Imaging of low-grade bone infection with a technetium-99m labelled monoclonal anti-NCA-90 Fab' fragment in patients with previous joint surgery. Eur J Nucl Med Mol Imaging 2002;29(4):547–551.
[44] Graute V, Feist M, Lehner S, *et al.* Detection of low-grade prosthetic joint infections using 99mTc-antigranulocyte SPECT/CT: initial clinical results. Eur J Nucl Med Mol Imaging 2010;37(9):1751–1759.
[45] Kwee TC, Kwee RM, Alavi A. FDG-PET for diagnosing prosthetic joint infection: systematic review and metaanalysis. Eur J Nucl Med Mol Imaging 2008;35(11):2122–2132.
[46] Sendi P, Zimmerli W. Antimicrobial treatment concepts for orthopaedic device-related infection. Clin Microbiol Infect 2012;18(12):1176–1184.
[47] Osmon DR, Berbari EF, Berendt AR, *et al.* Diagnosis and management of prosthetic joint infection: clinical practice guidelines by the Infectious Diseases Society of America. Clin Infect Dis 2013;56(1):e1–e25.
[48] Zimmerli W, Sendi P. Orthopedic implant-associated infections. In: Bennett JE, Dolin R, Blaser M, editors. Mandell, Douglas, and Bennett's Principles and Practice of Infectious Diseases, 8th edition. Philadelphia, PA: Elsevier; 2014.
[49] Zimmerli W, Widmer AF, Blatter M, *et al.* Role of rifampin for treatment of orthopedic implant-related staphylococcal infections: a randomized controlled trial. Foreign-Body Infection (FBI) Study Group. JAMA 1998;279(19):1537–1541.
[50] Puhto AP, Puhto T, Syrjala H. Short-course antibiotics for prosthetic joint infections treated with prosthesis retention. Clin Microbiol Infect 2012;18(11):1143–1148.
[51] Hsieh PH, Lee MS, Hsu KY, *et al.* Gram-negative prosthetic joint infections: risk factors and outcome of treatment. Clin Infect Dis 2009;49(7):1036–1043.
[52] Aboltins CA, Dowsey MM, Buising KL, *et al.* Gram-negative prosthetic joint infection treated with debridement, prosthesis retention and antibiotic regimens including a fluoroquinolone. Clin Microbiol Infect 2011;17(6):862–867.
[52a] Rodriguez-Pardo D, Pigrau C, Lora-Tamayo J, *et al.* Gram-negative prosthetic joint infection: outcome of a debridement, antibiotics, and implant retention approach. A large multicenter study. Clin Microbiol Infect 2014. DOI: 10.1111/1469-0691.12649 [Epub ahead of print].
[53] Wehrli W. Rifampin: mechanisms of action and resistance. Rev Infect Dis 1983; 983(Suppl 3):S407–S411.
[54] Achermann Y, Eigenmann K, Ledergerber B, *et al.* Factors associated with rifampin resistance in staphylococcal periprosthetic joint infections (PJI): a matched case-control study. Infection 2012;41(2):431–437.
[55] Zimmerli W, Ochsner PE. Management of infection associated with prosthetic joints. Infection 2003;31(2):99–108.
[56] Clauss M, Graf S, Gersbach S, *et al.* Material and biofilm load of K wires in toe surgery: titanium versus stainless steel. Clin Orthop Relat Res 2013;471(7):2312–2317.
[57] Clauss M, Tafin UF, Bizzini A, *et al.* Biofilm formation by staphylococci on fresh, fresh-frozen and processed human and bovine bone grafts. Eur Cell Mater 2013;25:159–166.
[58] Clauss M, Trampuz A, Borens O, *et al.* Biofilm formation on bone grafts and bone graft substitutes: comparison of different materials by a standard in vitro test and microcalorimetry. Acta Biomater 2010;6(9):3791–3797.
[59] Bahrs C, Schnabel M, Frank T, *et al.* Lavage of contaminated surfaces: an in vitro evaluation of the effectiveness of different systems. J Surg Res 2003;112(1):26–30.

[60] De Man FH, Sendi P, Zimmerli W, *et al.* Infectiological, functional, and radiographic outcome after revision for prosthetic hip infection according to a strict algorithm. Acta Orthop 2011;82(1):27–34.

[60a] Zeller V, Lhotellier L, Marmor S, *et al.* One-stage exchange arthroplasty for chronic periprosthetic hip infection: results of a large prospective cohort study. J Bone Joint Surg Am 2014;96(1):e1. DOI:10.2106/JBJS.L.01451.

[61] Buchholz HW, Elson RA, Engelbrecht E, *et al.* Management of deep infection of total hip replacement. J Bone Joint Surg Br 1981;63-B(3):342–353.

[62] Bae DK, Song SJ, Heo DB, *et al.* Long-term survival rate of implants and modes of failure after revision total knee arthroplasty by a single surgeon. J Arthroplasty 2013;28(7):1130–1134.

[63] Beckmann J, Luring C, Springorum R, *et al.* Fixation of revision TKA: a review of the literature. Knee Surg Sports Traumatol Arthrosc 2011;19(6):872–879.

[64] Edwards PK, Fehring TK, Hamilton WG, *et al.* Are cementless stems more durable than cemented stems in two-stage revisions of infected total knee arthroplasties? Clin Orthop Relat Res 2014;472(1):206–211.

[65] Kurtz SM, Ong KL, Lau E, *et al.* International survey of primary and revision total knee replacement. Int Orthop 2011;35(12):1783–1789.

[66] Richards CJ, Garbuz DS, Pugh L, *et al.* Revision total knee arthroplasty: clinical outcome comparison with and without the use of femoral head structural allograft. J Arthroplasty 2011;26(8):1299–1304.

[67] Schreurs BW, Arts JJ, Verdonschot N, *et al.* Femoral component revision with use of impaction bone-grafting and a cemented polished stem. J Bone Joint Surg Am 2005;87(11):2499–2507.

[68] Schreurs BW, Slooff TJ, Gardeniers JW, *et al.* Acetabular reconstruction with bone impaction grafting and a cemented cup: 20 years' experience. Clin Orthop Relat Res 2001;(393):202–215.

[69] Schreurs BW, Arts JJ, Verdonschot N, *et al.* Femoral component revision with use of impaction bone-grafting and a cemented polished stem. Surgical technique. J Bone Joint Surg Am 2006;88(Suppl 1 Pt 2):259–274.

[70] Winkler H, Stoiber A, Kaudela K, *et al.* One stage uncemented revision of infected total hip replacement using cancellous allograft bone impregnated with antibiotics. J Bone Joint Surg Br 2008;90(12):1580–1584.

[71] Westrich GH, Bornstein L, Brause BD, *et al.* Historical perspective on two-stage reimplantation for infection after total hip arthroplasty at Hospital for Special Surgery, New York City. Am J Orthop (Belle Mead NJ) 2011;40(11):E236–E240.

[72] Kamath AF, Austin D, Lee GC. Mating of a PROSTALAC spacer with an intramedullary nail for reconstruction of an infected interprosthetic femoral shaft fracture: a case report. J Orthop Surg (Hong Kong) 2012;20(2):263–268.

[73] Degen RM, Davey JR, Davey JR, *et al.* Does a prefabricated gentamicin-impregnated, load-bearing spacer control periprosthetic hip infection? Clin Orthop Relat Res 2012; 470(10):2724–2729.

[74] Fleck EE, Spangehl MJ, Rapuri VR, *et al.* An articulating antibiotic spacer controls infection and improves pain and function in a degenerative septic hip. Clin Orthop Relat Res 2011;469(11):3055–3064.

[75] Suda AJ, Heppert V. Vastus lateralis muscle flap for infected hips after resection arthroplasty. J Bone Joint Surg Br 2010;92(12):1654–1658.

[76] Dale H, Fenstad AM, Hallan G, *et al.* Increasing risk of prosthetic joint infection after total hip arthroplasty. Acta Orthop 2012;83(5):449–458.

[77] Dale H, Hallan G, Hallan G, *et al.* Increasing risk of revision due to deep infection after hip arthroplasty. Acta Orthop 2009;80(6):639–645.

[78] Dale H, Skramm I, Lower HL, *et al.* Infection after primary hip arthroplasty: a comparison of 3 Norwegian health registers. Acta Orthop 2011;82(6):646–654.

[79] Spaans AJ, van den Hout JA, Bolder SB. High complication rate in the early experience of minimally invasive total hip arthroplasty by the direct anterior approach. Acta Orthop 2012;83(4):342–346.

[80] Berend KR, Kavolus JJ, Morris MJ, *et al.* Primary and revision anterior supine total hip arthroplasty: an analysis of complications and reoperations. Instr Course Lect 2013;62: 251–263.
[81] Goytia RN, Jones LC, Hungerford MW. Learning curve for the anterior approach total hip arthroplasty. J Surg Orthop Adv 2012;21(2):78–83.
[82] Imamura M, Munro NA, Zhu S, *et al.* Single mini-incision total hip replacement for the management of arthritic disease of the hip: a systematic review and meta-analysis of randomized controlled trials. J Bone Joint Surg Am 2012;94(20):1897–1905.
[83] Pogliacomi F, Paraskevopoulos A, Costantino C, *et al.* Influence of surgical experience in the learning curve of a new approach in hip replacement: anterior mini-invasive vs. standard lateral. Hip Int 2012;22(5):555–561.
[84] Berend KR, Lombardi AV, Jr., Morris MJ, *et al.* Two-stage treatment of hip periprosthetic joint infection is associated with a high rate of infection control but high mortality. Clin Orthop Relat Res 2013;471(2):510–518.
[85] Bloomfield MR, Klika AK, Barsoum WK. Antibiotic-coated spacers for total hip arthroplasty infection. Orthopedics 2010;33(9):649.
[86] Diwanji SR, Kong IK, Park YH, *et al.* Two-stage reconstruction of infected hip joints. J Arthroplasty 2008;23(5):656–661.
[87] Macheras GA, Koutsostathis SD, Kateros K, *et al.* A two stage re-implantation protocol for the treatment of deep periprosthetic hip infection. Mid to long-term results. Hip Int 2012;22(Suppl 8):S54–S61.
[88] van Diemen MP, Colen S, Dalemans AA, *et al.* Two-stage revision of an infected total hip arthroplasty: a follow-up of 136 patients. Hip Int 2013;23(5):445–450.
[89] Parvizi J, Gehrke T. Executive summary. J Arthroplasty 2014 29(Suppl 2):5. DOI: 10.1016/j.arth.2013.09.023 [Epub October 17, 2013].
[90] Parvizi J, Gehrke T, Chen AF. Proceedings of the international consensus on periprosthetic joint infection. Bone Joint J 2013;95-B(11):1450–1452.
[91] Walenkamp G. Small PMMA beads improve gentamicin release. Acta Orthop Scand 1989;60(6):668–669.
[92] Walenkamp GH. Chronic osteomyelitis. Acta Orthop Scand 1997;68(5):497–506.
[93] Iarikov D, Demian H, Rubin D, *et al.* Choice and doses of antibacterial agents for cement spacers in treatment of prosthetic joint infections: review of published studies. Clin Infect Dis 2012;55(11):1474–1480.
[94] Walenkamp GH. Self-mixed antibiotic bone cement: western countries learn from developing countries. Acta Orthop 2009;80(5):505–507.
[95] Ammon P, Stockley I. Allograft bone in two-stage revision of the hip for infection. Is it safe? J Bone Joint Surg Br 2004;86(7):962–965.
[96] Wahl P, Livio F, Jacobi M, *et al.* Systemic exposure to tobramycin after local antibiotic treatment with calcium sulphate as carrier material. Arch Orthop Trauma Surg 2011;131(5): 657–662.
[97] Mont MA, Waldman B, Banerjee C, *et al.* Multiple irrigation, debridement, and retention of components in infected total knee arthroplasty. J Arthroplasty 1997;12(4):426–433.
[98] Waldman BJ, Hostin E, Mont MA, *et al.* Infected total knee arthroplasty treated by arthroscopic irrigation and debridement. J Arthroplasty 2000;15(4):430–436.
[99] Dixon P, Parish EN, Cross MJ. Arthroscopic debridement in the treatment of the infected total knee replacement. J Bone Joint Surg Br 2004;86(1):39–42.
[100] Liu CW, Kuo CL, Chuang SY, *et al.* Results of infected total knee arthroplasty treated with arthroscopic debridement and continuous antibiotic irrigation system. Indian J Orthop 2013;47(1):93–97.
[101] Yi PH, Cross MB, Moric M, *et al.* The 2013 Frank Stinchfield Award: diagnosis of infection in the early postoperative period after total hip arthroplasty. Clin Orthop Relat Res 2014;472(2):424–429.
[102] Sendi P, Rohrbach M, Graber P, *et al.* Staphylococcus aureus small colony variants in prosthetic joint infection. Clin Infect Dis 2006;43(8):961–967.

Chapter 10
Periprosthetic Joint Infection after Shoulder Arthroplasty

Parham Sendi, Beat K. Moor, and Matthias A. Zumstein

Introduction

The number of shoulder arthroplasties is rapidly increasing. In 2008, approximately 50,000 prosthetic shoulders were implanted in the United States. It is predicted that this number will increase by approximately 5000 implantations per year [1]. There are different types of shoulder arthroplasties. These include total shoulder arthroplasties with and without a stem, stemmed hemiarthroplasty, resurfacing hemiarthroplasty, and reverse total shoulder arthroplasty. Unless otherwise specified, the text in this chapter considers all types of shoulder arthroplasties.

The published incidence of periprosthetic shoulder joint infection (PSJI) varies from 0.9 to 1.8% for primary and from 3 to 4% for revision procedures [2–5]. However, the management for shoulder prosthesis has not been standardized. Most concepts stem from extrapolation of treatment strategies performed in other joints (see Chapter 8).

Risk Factors

Comorbidities

The risk factors represent the higher susceptibility for infection and/or wound complications due to the comorbidity or its treatment. These include rheumatoid arthritis, diabetes mellitus, the use of systemic corticosteroids, and/or other immunosuppressive medications [5–7].

Intra-articular Injections

Intra-articular corticosteroid injection is an obvious specific risk factor for PSJI [6]. However, to the best of our knowledge, this has been neither firmly analyzed nor quantified in a published study.

Bone and Joint Infections: From Microbiology to Diagnostics and Treatment, First Edition.
Edited by Werner Zimmerli.

Revision Surgery and Hematoma

As observed with other arthroplasties, the risk of infection increases with revision surgery. One retrospective study described the association of hematoma formation and PSJI [8]. Among 4147 shoulder arthroplasties (3643 patients), 12 (0.3%) had a revision surgery to evacuate the hematoma up to 31 days after the index surgery. In nine patients, cultures for microbiological analyses were taken. Six had positive cultures, and two underwent resection arthroplasty, 9 and 22 months, respectively, later [8]. The authors concluded that hematoma formation in patients undergoing operative debridement after shoulder arthroplasty is associated with positive culture results (6/12 or 6/9, respectively). In two of six patients with positive culture results, this ultimately manifested as PSJI.

Male gender

In the study by Singh *et al.* [3], male gender and younger age increased the risk for PSJI. Pottinger *et al.* [9] reported that male gender increased the likelihood (odds ratio [OR] 6.41; 95% confidence interval [CI] 3.10–14.42, $P<0.001$) of obtaining positive *Propionibacterium acnes* cultures in patients in whom revision arthroplasty was performed for stiffness, pain, or loosening. In the Norwegian Arthroplasty Register, the risk for revision due to infection in reversed total shoulder arthroplasties was also smaller in women than in men [10]. Interestingly, men have a higher burden of *P. acnes* colonization than women do [11].

Age and Trauma

The finding of Singh *et al.* [3] that older age was associated with a lower risk of infection is unclear (hazard ratio 0.78; 95% CI 0.61–0.98 per 10-year increase in age). The authors discussed that younger patients are more likely to have had previous trauma. In another study in which they focused only on hemiarthroplasties, they found that trauma as underlying diagnosis was associated with infection (hazard ratio 3.18; 95% CI 1.06–9.56), as compared to all other diagnoses [12].

Cement with or without Antibiotic Impregnation

Cement alone is not a risk factor for PSJI, according to a study that included 2588 total shoulder arthroplasties [3]. The authors compared 2485 arthoplasties that were implanted with cement (31 infections, 1.25%) and 103 without cement (1 infection, 0.97%). This observation was also confirmed in hemiarthroplasties (n = 875 with cement, 575 without cement, hazard ratio 0.88 (95% CI 0.30–2.56, P = 0.82) [12]. However, whether or not the cement is loaded with antibiotics appears to have an influence. In 501 consecutive primary reverse total shoulder arthroplasties, 265 shoulders (cement without antibiotics) were compared with 236 shoulders that were fixed with antibiotic-loaded cement (mostly aminoglycosides) [13]. The difference at 37 months follow-up was 3% versus 0% (n = 8) ($P<0.001$).

Type of Arthroplasty

It is unclear whether the type of arthroplasty has a significant influence on the risk for infection, since no direct comparison has been made. Reverse total shoulder arthroplasties have been postulated to have a higher infection incidence than anatomical arthroplasties [14, 15] because with reverse configuration, the dead space is large [15].

With hematoma/seroma, the large volume potentially enables a niche for the growth microorganisms. In a large systemic review on reverse total shoulder arthroplasty, 30 infection cases in 14 studies were reported. The calculated incidence was 3.8% [16]. This number is higher than the reported ranges for primary total arthroplasty in previous studies (0.9–1.8%), but comparable to the numbers of revision procedures (3–4%) [2–5]. Nevertheless, these comparisons are difficult to make, since the publication bias of infection cases can only be poorly corrected.

Microbiology

The most commonly identified microorganisms are illustrated in Table 10.1. Coagulase-negative staphylococci (CNS), *Staphylococcus aureus*, and *P. acnes* comprise approximately 70–80% of causative microorganisms in PSJI. The majority of PSJI series show only small absolute numbers. Therefore, the rate of a single pathogen may be under- or overreported. However, *Propionibacterium* spp. are more frequently found in PSJI than in other infected arthroplasties [17]. This organisms is slow-growing, and, hence, requires an extended incubation time (up 10–14 days) [22].

The relevance of CNS, and *Propionibacterium* spp. in particular, in a single tissue culture is difficult to interpret. Both organisms belong to the skin flora. *Propionibacterium* spp. commonly colonizes the skin of the shoulder and upper parts of the body [11] and interestingly the highest bacterial load is on the acromion and not in the axilla as one might expect. The responsible physician is faced with the challenge of whether the microbiological result should be regarded as a true pathogen or as a contaminant. One way to overcome this hurdle is to obtain several biopsies during revision surgery (see "Laboratory Investigation"). Also, it is important to note that the presence of a microorganism is not the only criterion for infection. In a retrospective study including 27 patients in whom 28 presumably aseptic revision procedures were performed, the relevance of routinely taken biopsies was investigated [23]. In 8 of 27 cases, an organism, mostly *Propionibacterium* spp., was isolated. However, only two of them developed a subsequent infection after 12 and 14 months, respectively. A similar result was reported by Grosso *et al.* [24]. The authors analyzed 17 patients with one-stage revision and unexpected positive culture results, mostly *P. acnes* and CNS. Only one patient developed an infection. Thus, the diagnosis of PSJI must not be based exclusively on microbiological results. In view of these results, we discourage taking biopsies routinely, if there is no suspicion of PSJI.

Table 10.1. Microorganisms causing periprosthetic shoulder joint infection [3, 7, 12, 17–21].

Coagulase-negative staphylococci	16–40%
Staphylococcus aureus (including MRSA)	21–31%
Propionibacterium spp.	14–40%
Streptococcus spp.	0–10%
Gram-negative bacteria	0–10%
Enterococcus spp.	0–10%
Polymicrobial	0–13%
Others	<5%
No isolates	0–10%

However, the threshold for suspicion of infection should be low in patients with a symptom-free interval of less than 2 years [25]. In addition, in PSJI due to *Propionibacterium* spp. the symptoms are often insidious and nonspecific [9, 26].

Pathogenesis

In clinical practice, it is relevant to differentiate between acute and chronic PSJI [27]. Thereby, three parameters should be evaluated: (1) the time interval from implantation to onset of symptoms, (2) the duration of symptoms, and (3) the type of symptoms (e.g., local or systemic). These parameters allow an appraisal regarding the pathogenesis of infection and the virulence of the pathogen. Such an assessment is important to make a decision regarding the treatment concept.

Exogenous or Hematogenous Infection Route

If the interval from index surgery to onset of symptoms and clinical findings is short, or if the patient has never had a symptom-free period after implantation, the infection is likely acquired exogenously. The longer the patient was symptom-free after index surgery, the more likely the infection is acquired hematogenously. Hematogenous infection can occur at any time after the index surgery, and the risk persists as long as the arthroplasty remains in situ [28] (see Chapter 8). Nevertheless, when the onset of symptoms is beyond 24–30 months after surgery, the infection route is almost always hematogenous.

Acute or Chronic Infection

The duration of symptoms is typically several months in chronic infections. If the duration of symptoms is ≤3 weeks irrespective of the index surgery, or the infection manifestation presents within 4 weeks after surgery, the infection is considered acute [27].

Virulent or Low-Virulent Pathogens

The type of symptoms is prominent at the local site of infection if virulent pathogens such as *S. aureus* or Gram-negative bacteria are involved. Often they are accompanied by systemic inflammation signs. In these cases, the infection manifestation is often acute. Virulent pathogens can cause both exogenous and hematogenous infections. In case of an exogenous route with a virulent pathogen, infection signs are often visible within a short interval after implantation (< 3 months) [29]. In contrast, when a low-virulent pathogen such as *Propionbacterium* spp. or CNS is the causative agent, the infection route is exogenous with only a few exceptions. In these cases, the patient complains of chronic symptoms. The symptoms are often nonspecific.

Clinical Features

Approximately 80% of PSJI are diagnosed several months after implantation, and the vast majority of them are acquired exogenously [2, 18, 19]. Hence, patients with PSJI present more often with chronic than acute symptoms.

Cardinal symptoms of chronic PSJI are pain, stiffness, and loss of or impaired function of the joint [9, 18, 20]. Redness and warmth are often but not necessarily additional symptoms [18]. Hence, symptoms for chronic PSJI are not specific, although the sum of clinical features should raise the suspicion of PSJI. Information from the patient history, such as wound healing problems, in the early postoperative period, can contribute to this suspicion.

Fever and chills are more often present in patients with acute symptoms and/or involvement of a virulent pathogen. In these cases, a distant source of infection must be actively looked for (e.g., skin and soft tissue infection).

Finally, the presence of sinus tract in the area of the prosthesis is indicative for chronic PSJI until proven otherwise.

Laboratory Investigation

Blood Tests

The most common parameters used in clinical practice are the C-reactive protein (CRP) level, the erythrocyte sedimentation rate (ESR), and the white blood cell (WBC) count. Yet, none of these is specific. In periprosthetic hip and knee infections, these parameters have a high sensitivity, only if normal values are defined as threshold (reviewed in [30]) (see Chapter 9). In contrast, in low-grade PSJI (e.g., with *Propionibacterium* spp.), CRP and/or WBC levels can be even normal [9, 18, 25, 26, 31, 32]. Thus, in PSJI, CRP and WBC are generally slightly elevated, but unreliable for the diagnosing infection.

Synovial Fluid Leukocyte Count

Till date, there are only few investigations that focus on the diagnosis of PSJI. Jerosch *et al.* [33] used a WBC count of 30,000/μl or more as evidence for PSJI. Piper *et al.* [17] compared the synovial fluid leukocyte count plus the proportion of neutrophils in patients with aseptic failures ($n = 18$) to those with PSJI ($n = 10$). The cutoff was set at >1700/μl, and >65% neutrophils. Patients with PSJI fulfilled these criteria more often (5/10 versus 1/18, $P = 0.01$, and 7/10 versus 4/18, $P = 0.02$, respectively). Yet, in PSJI the optimal cutoff values of synovial fluid leukocyte counts with clinically useful sensitivity and specificity still have to be defined.

Intraoperative Findings and the Number of Biopsies

In acute infections, pus and inflamed tissue are frequently evident. In chronic PSJI with low-virulent bacteria, macroscopic findings can be unspecific or even almost unremarkable. Yet, cloudy fluids and membrane formation should raise the suspicion of infection [9]. With respect to these difficulties in interpreting microbiological results, we recommend obtaining several biopsies, preferentially six. Thereby, a ratio of culture-positive to totally obtained biopsies can help to estimate the significance of bacterial growth. In addition, material should be sent for histopathological examinations. For each microbiological sample, we recommend obtaining a paired biopsy for histopathological examination (e.g., biopsy 1A for microbiology, 1B for histopathology). The results of these paired specimens can be correlated in terms of microbial growth and inflammation.

Microbiological Culture

The sensitivity of microbiological cultures is higher if bacteria have not been exposed to antibiotics. In patients with previous antimicrobial treatment, antibiotics should be stopped for at least 14 days prior to biopsy sampling [34]. In chronic infections and involvement of low-virulent bacteria, the risk of developing a sepsis syndrome is small. Therefore, stopping antibiotics does not generally jeopardize patients. Biopsies should be sent to a microbiological laboratory within a reasonable time (preferably < 1 h), because anaerobic bacteria require both special media and optimal conditions to grow. Also, an extended incubation time (10–14 days) should be asked for, because *Propionibacterium* spp., notably a frequent pathogen in PSJI, grow slowly [22].

Molecular Investigations

In cases of negative culture results and/or in patients who have been pretreated with antimicrobial agents, eubacterial polymerase chain reaction (PCR) analysis (amplifying a region of the bacterial 16S rRNA gene or 16S rDNA gene) is possible. Current data indicate that the addition of PCR to conventional culture mainly increases the sensitivity [35] (see Chapter 2).

Histopathological Investigations

This diagnostic tool is helpful in mainly two constellations. First, it is another criterion for the diagnosis of periprosthetic joint infection. This criterion must be considered in particular when culture results are negative [27, 36]. Second, if the sampling location can be corresponded to a microbiological result, this may facilitate interpretation of whether the single positive culture is a contaminant or the causative pathogen. Infection is usually accompanied by inflammation that can be detected within the tissue. However, it is important to define the threshold of neutrophils per a given high-power field. This definition varies among different experts. Suggestions range from ≥1 to ≥10 neutrophils per high-power field [36]. The specificity of a cutoff point increases with the number of neutrophils at the cost of a decline in sensitivity and vice versa [30]. Five or more neutrophils per high-power field in 40× magnification is commonly accepted as a positive indicator for infection [30].

Sonication of Removed Implants

This technique has been increasingly used in the past few years [17, 34]. It mainly improves the sensitivity of samples that have been obtained from antibiotic-treated patients. Trampuz *et al.* [34] showed on removed hip and knee implants that the sensitivity of the sonicated fluid was significantly higher than it was in periprosthetic biopsy samples (75% versus 45%, $P < 0.001$, in patients treated with antibiotics within 2 weeks before sampling). Piper *et al.* from the same group [17] investigated this technique in 33 patients with definite PSJI. The sensitivities of sonicated fluid and periprosthetic tissue culture were 66.7% (22/33) and 54.5% (18/33) ($P = 0.046$), respectively, and the specificities 98.0% (99/101) and 95.1% (96/101) ($P = 0.26$), respectively. Hence, 4 of 33 PSJI were detected only by culture of sonicated fluid. While the previous study suggested a cutoff level of ≥5 CFU per plate [34], Piper *et al.* [17] suggest a cutoff level of ≥20 CFU per plate.

Sonication of Removed Implants Combined with Molecular Methods

To further increase the sensitivity of the microbiological investigations, molecular diagnostics (PCR) on sonicated fluid can be applied [37, 38]. Portillo *et al.* [38] used a multiplex PCR on the sonicated fluid of 86 explanted prostheses (including two shoulder prostheses). They found a sensitivity of 96% and a specificity of 100%. It is important to note that currently commercially available multiplex PCR cannot detect *Propionibacterium* spp., a pathogen that is important in PSJI [37]. However, specific primers can be applied [17].

Imaging Procedures

Plain radiographs are not helpful in the early postoperative period. Although not firmly investigated, most experts additionally recommend computed tomography (CT) scans, magnetic resonance imaging (MRI), or ultrasound imaging to look for fluid collection [31]. In chronic infections, loosening of the implants (e.g., stem) is visible in anteroposterior and lateral views of radiographs. Humerus and glenoid loosening and signs of osteolysis are highly indicative for infection [2, 9, 18]. Comparisons of consecutive radiographs are important to detect subtle signs, such as medial calcar erosions or tuberositas resorption [14].

CT allows the detection of soft tissue infections (abscesses, sinus tracts), sequesters, and bone erosions. Often, this modality is applied to estimate the extent of an infection.

MRI offers a better resolution for soft tissue abnormalities. Considering the magnitude of the soft tissue and muscle mantle surrounding the shoulder joint, this imaging technique has gained importance. It is also helpful in preoperative planning with respect to the surgical technique.

Most data on the sensitivity, specificity, and accuracy of nuclear imaging for the diagnosis of infected arthroplasty stem from studies on hip and knee arthroplasties (reviewed in [30], see Chapter 9). Iyengar *et al.* [39] analyzed the results of 38 patients with clinical suspicion of periprosthetic joint infections. Four of 38 patients had a total shoulder arthroplasty. Imaging was done with 650 MBq of technetium-99m Sulesomab. When comparing the imaging result with the final clinical diagnosis, the overall sensitivity was 90.9% and the specificity was 81.5%. The sensitivity and specificity of shoulder prosthesis were 100% each, though the absolute numbers were small ($n = 2$ true positive and 2 true negative). Graute *et al.* [40] investigated 31 consecutive patients (including one shoulder arthroplasty) for the diagnosis of low-grade periprosthetic joint infections. Scintigraphy was performed with technetium-99m-labelled antigranulocyte antibodies. Nine patients had a final diagnosis of infection, indicating sensitivity, specificity, positive and negative predictive values of, respectively, 66, 60, 40, and 81%. With the addition of single-photon emission computed tomography plus conventional CT (SPECT/CT), values increased to 89, 73, 57, and 94%, respectively. In summary, the few available data on scintigraphy in patients with PSJI indicate similar diagnostic results than in patients with other arthroplasties. When performing a scintigraphy, it is important to evaluate the interdisciplinary optimal time point for imaging, since there may be signal overlaps between postoperative changes and infection [41].

Management

The management of PSJI requires—as in every case with orthopedic device–related infections—an interdisciplinary approach, including orthopedic surgeons, infectious diseases specialist, and microbiologists. Prior to the management, the responsible physicians should agree on a categorization of the infection. These include the following parameters as described earlier: (i) acute or chronic, (ii) interval from implantation to onset of symptoms, and (iii) potentially involved microorganisms (virulent, low-virulent). Ideally, the pathogen and its antimicrobial susceptibility patterns are known prior to the revision surgery. In addition, although more so in chronic than in acute infections, a thorough workup about the extent of the soft tissue and rotator cuff damage as well as the remaining bone stock is required. Finally, it is important to assess the function and strength of the shoulder joint (e.g., Constant–Murley score [CMS] [42]), and the quality of life and patient satisfaction corresponding to the shoulder joint (e.g., the disabilities of the arm, shoulder and hand (DASH) score [43], the Neer score [44]). The European Society for Surgery of the Shoulder and Elbow recommends for a precise assessment both a "patient-based outcome score" and a "clinical-based outcome score" [45]. For each patient, these evaluations are necessary to weigh the "best" versus the "feasible" treatment option in a risk–benefit assessment. For example, subsequent to a meticulous debridement, there is a lack of bone stock at the glenoid side [46] or an impaired rotator cuff; this has an impact on the treatment strategy. Moreover, such an evaluation can also be helpful to predict possible consequences of a failure after revision.

Surgical Interventions

Similar to the periprosthetic joint infections at other body sites, the treatment strategies for PSJI consists of debridement and implant retention (DAIR), direct prosthesis exchange (one-stage exchange), two-stage prosthesis exchange with or without a spacer, and resection arthroplasty. In selected cases, prolonged implantation of an articulating spacer has been suggested [21].

Debridement and Implant Retention

This treatment option is mainly considered in acute infections with a short duration of symptoms, including early postoperative infections. Also, implant components must not be loose. Since the majority of PSJIs are chronic, this treatment option is infrequently applied [18, 31, 47]. However, if DAIR is chosen, it is important to apply the following principles: (i) surgery must be performed rapidly, (ii) irrigation and debridement should be meticulous, and (iii) mobile elements must be exchanged. In infected hip and knee arthroplasties, DAIR should not be applied, if the soft tissue damage is moderate or severe (e.g., abscess, sinus tract), and if there is a difficult-to-treat pathogen (e.g., rifampin-resistant staphylococci, fungi) [29, 48] (see Chapter 9). In our view, it is reasonable to extrapolate these recommendations to PSJI also.

In the last decade, the outcome of this procedure has been reported to be good, provided that the correct patient population is selected. The proportion of patients with an infection-free interval ranges from 85 to 100%, with mean follow-ups ranging from 2.8 to 4 years [2, 18, 33, 47]. Reported failures with DAIR were attributed to inadequate patient selection (e.g., chronic infections) [19].

One-Stage Exchange

This procedure has the advantage of being a single operation with lower costs. Moreover, rehabilitation and functional improvement can be initiated earlier. Patients with chronic infections and no or only slightly damaged soft tissue are suitable for a one-stage exchange. The rationale to avoid this procedure in case of an abscess or a sinus tract is based on the notion that the extent of the soft tissue damage is often correlated with the magnitude of the bacterial load. Thus, adherence of a high bacterial load to the newly implanted prosthesis would increase the risk for failure. It is a prerequisite to identify the responsible pathogen and its antimicrobial susceptibility patterns prior to surgery. One-stage exchange is not recommended if the causative pathogen is difficult to treat (e.g., fungi) or if the pathogen is resistant to agents with activity against biofilm (e.g., rifampin-resistant staphylococci) [29].

If one-stage exchange is applied to patients who fulfill the earlier-mentioned criteria, the outcome is excellent. Coste *et al.* [2], Ince *et al.* [49], Cuff *et al.* [50], and Klatte *et al.* [20] treated 3, 9, 7, and 35 patients on the basis of this concept. The infection-free interval was 100, 100, 100, and 94%, respectively, with mean follow-ups ranging from 2.8 to 5.7 years. Again, it is important to note that a successful outcome includes both eradication of the pathogen and retention/improvement of the joint function. In the study by Klatte *et al.* [20], 26 of the 35 patients were available for functional follow-up investigations (CMS, pain 15 points possible, function 20 points, strength 25 points, range of motion 40 points, for a maximum 100 points [42]). The 26 implants consisted of 14 hemiprostheses (mean CMS 43.3), 5 hemiprostheses with bipolar head (mean CMS 56), and 7 reverse prostheses (mean 61). Although the best CMS results were achieved with reverse total shoulder arthroplasty, there was no statistical difference when comparing the mean CMS between the three prostheses types. However, the authors postulated that the infection and radical debridement caused rotator cuff deficiency, leading to lower CMS scores in the hemiarthroplasty group [20].

Two-Stage Exchange

This procedure involves, in the first stage, removal of all foreign material, debridement of the tissue, resection of infected and necrotic tissue, and (often) implantation of a spacer. In the second stage, the spacer is removed and the new prosthetic device is implanted. This concept is traditionally applied for infected arthroplasties and, is, hitherto, the most frequently applied treatment option for PSJI [18, 19, 47]. The time period after removal of the infected prosthesis allows treating the infection prior to implantation of a new arthroplasty. Thus, in most series, cure of infection is achieved in 100% of the cases [5, 18, 33, 47, 50–52]. On the other hand, an unsatisfactory functional outcome has been associated with a two-stage exchange [2, 47, 53]. However, whether or not two-stage exchange has a poorer functional outcome at the cost of infection eradication cannot be judged, unless a direct comparison with one-stage exchange is made in a randomized trial. Moreover, the type of arthroplasty (e.g., hemiarthroplasty versus reversed total shoulder arthroplasty) has a functional impact [20]. Consequently, it is important to stratify for the type of arthroplasty when comparing different surgical procedures. Cuff *et al.* [50] converted 17 infected hemiarthroplasties to reversed total shoulder arthroplasties, 7 of them in one stage and 10 in two stages. All outcome measures (i.e., ranges of motion) showed statistically significant improvements. There was no difference in outcome between the two surgical procedures.

Additional factors potentially influencing the functional outcome include the time interval between the stages. In PSJI, there are no recommendations on the optimal implant-free interval. When considering the outcome results from one-stage exchanges, it is reasonable to implant after a short interval. After a few weeks, the vast majority of pathogens do not cause persistent infections under correct antimicrobial treatment. On the basis of these arguments, we often perform the second stage (implantation) within 4–6 weeks after removal. However, in case of severely damaged soft tissue and/or frequently if the bone stock is deficient, implantation is delayed until optimal conditions are achieved (e.g., 3 months).

Spacers

Nowadays, spacers are loaded with antibiotics and allow joint articulation (reviewed in [54]). Gentamicin and vancomycin are commonly used compounds and are effective in the treatment of PSJI after implant removal [55–57]. In addition to conventional spacers, there are reports about hybrid spacers, made of a humeral nail and a custom-made polymethylmethacrylate (PMMA) spacer forming the humeral head. They can be used after removal of a reverse total shoulder joint arthroplasty [58]. Rarely, a spacer is left permanently [18], tough, if so, then a good function must be maintained [56, 58]. Also, fixed articulating spacers have been proposed, and they have shown a successful infection eradication rate and acceptable functional outcomes [21]. However, this treatment option should be offered to selected patients with multiple comorbidities. There are ambiguous results on functional outcome in patients with and without spacers. Verhelst *et al.* [59] compared 11 patients without a spacer and 10 with a spacer, after resection arthroplasty. Function was assessed with the visual analogue scale (VAS), the CMS, and the DASH score. Overall, 14 of 21 patients achieved acceptable or good results, with a mean improvement in all assessments. There was no significant difference between the two groups. Magnan *et al.* [57] reported significant functional improvement in seven patients, using a preformed antibiotic-loaded spacer. For assessment, the CMS, the Secec Elbow score (SES), and the American Shoulder and Elbow Society score (ASESS) were used. The CMS increased from 40.3 to 79.1, the mean SES from 34.4 to 77.3, and the ASESS ranged from 14.9 to 21.1. However, the role of the spacer is still unclear, because no comparison with patients without spacer was made.

We prefer to use a spacer in two-stage exchange. In our experience, this allows reimplantation of the new arthroplasty under good anatomical conditions. Considering that CNS and *Propionibacterium* spp. are the most frequently found bacteria in PSJI, we use vancomycin or vancomycin–gentamicin spacers. However, in case of difficult-to-treat microorganisms (e.g., small colony variants of bacteria), we do not implant a spacer.

Resection Arthroplasty

This procedure is associated with the poorest functional outcome. Consequently, it should be reserved for patients in whom pain relief and infection eradication is the primary goal (e.g., debilitated patients, short life expectancy). Previously, resection arthroplasty was also applied in case of severe infections. Reimplantation after a long period is possible, but the functional outcome is uncertain. Mileti *et al.* [52] presented four patients with a resection arthroplasty from a previously infected shoulder arthroplasty. The time interval from resection arthroplasty to reimplantation ranged from 7 months to 5.5 years. The outcome was satisfactory in two of the four patients.

Antimicrobial Treatment

In PSJI, antimicrobial concepts do not differ from those applied to other arthroplasties [60]. They have been described elsewhere (Chapter 8,Table 9.2). The concepts are briefly summarized. After microbiological sampling, antibiotic therapy is immediately started, initially by the intravenous (IV) route. After isolation of the pathogen, treatment should be streamlined from empirical to directed therapy (Table 9.2). If the postoperative course is uneventful, we switch from IV to oral formulation within 7–10 days. In our center, the total treatment duration is 3 months. In two constellations, shorter treatment duration is applied (i.e., 6 weeks): two-stage exchange without a spacer and resection arthroplasty. In these two cases, absence of a chronic osteomyelitis of the bone stock is a prerequisite. Long-term suppressive antimicrobial therapy is the exception, and only administered to patients with PSJI and short life expectancy.

PSJI due to Propionibacterium *spp.*

These pathogens are uniformly susceptible to penicillin. When switching from IV to the oral route, the question of bioavailability arises. Oral penicillin formulations have a poor bioavailability and—in comparison to many other antimicrobial agents—a low penetration rate into the bone [61] (see Chapter 3). Nevertheless, it is, possible that the achieved antibiotic levels in the bone and periprosthetic tissue are sufficient, thanks to the low penicillin minimum inhibitory concentration (MIC) of *Propionibacterium* spp. Due to the very low and unpredictable bioavailability of penicillin V [62], we prefer amoxicillin (3 × 1 g p.o. per day), or clindamycin (3 × 450–600 mg p.o. per day), or doxycycline (2 × 100 mg p.o. per day), which have good bone penetration [60]. Importantly, *Propionibacterium* spp. must be tested for its susceptibility to clindamycin or doxycycline. Alternatively, ceftriaxone (1 × 2 g IV per day) can be administered. In an outpatient setting, we apply this alternative treatment option for the first 2–4 weeks after DAIR. Thereafter, an oral formulation of the earlier-mentioned compound is administered to complete the treatment duration (i.e., 6 weeks or 3 months).

Rifampin Combination Therapy in PSJI due to Propionibacterium *spp.*

These pathogens are generally susceptible to rifampin, but emergence of resistance seems possible (patient 8 in [22]). The compound should not be administered alone. Also, we do not apply rifampin until the wound is dry [60]. Rifampin combinations have been used with clindamycin, amoxicillin, doxycycline, and daptomycin (reviewed in [60]). Yet, the benefit of adding rifampin remains unknown without a study comparing the combination versus monotherapy. In the case of two-stage exchange, we do not see a rationale for the use of rifampin.

Propionibacterium spp. is highly susceptible to antibiotics and commonly loses its pathogenicity as soon as the device is removed. In case of DAIR, administration of rifampin combination therapy is possible, but clinical data are lacking.

Instructive Cases

Case 1: Two-Stage Exchange from Internal Fixation to Reverse Total Shoulder Fracture-Arthroplasty

A 66-year-old previously healthy man presented with massive pain of his left shoulder. He reported a hard fall while skiing. Radiographs revealed a four-part fracture of the proximal humerus. An internal fixation with a PHILOS plate was performed (Figure 10.1a). The early postoperative course was uneventful, and the patient was discharged. Three months later, he claimed of persisting pain without other symptoms. Subsequent X-ray showed a secondary loss of reposition with progressive varus drift of the humeral head (Figure 10.1b). Revision surgery with implantation of a reverse fracture prosthesis was planned. Intraoperatively, necrotic tissue and cloudy fluid (but no pus) was detected. Because of this finding, the treatment concept was intraoperatively changed. All foreign material was removed, multiple biopsies were obtained, meticulous debridement was performed, and a vancomycin–gentamicin–loaded spacer was implanted (Figure 10.1c). Postoperatively, IV antimicrobial treatment was initiated (amoxicillin/clavulanate 4 × 2.2 g IV per day). *P. acnes* grew in five out of six obtained biopsy samples, 5 days after surgery. Antimicrobial treatment was streamlined to penicillin G (6 × 4 million IE IV per day). After a 3-week interval, the spacer was removed, and a reverse total shoulder fracture arthroplasty implanted (Figure 10.1d). Another week later, the patient was discharged. Antibiotics were switched to amoxicillin (3 × 1 g. p.o. per day), and continued until the 3-month course was completed. At the 2-year follow-up, the patient was infection-free with a satisfactory functional outcome.

Learning Points

- Male gender and trauma were typical risk factors for PSJI in this patient.
- In PSJI caused by *Propionibacterium* spp., symptoms are unspecific.
- Intraoperatively, necrotic tissue and cloudy fluid, even when there is no pus, should raise the suspicion of PSJI, irrespective of previous symptoms.

(a) (b) (c) (d)

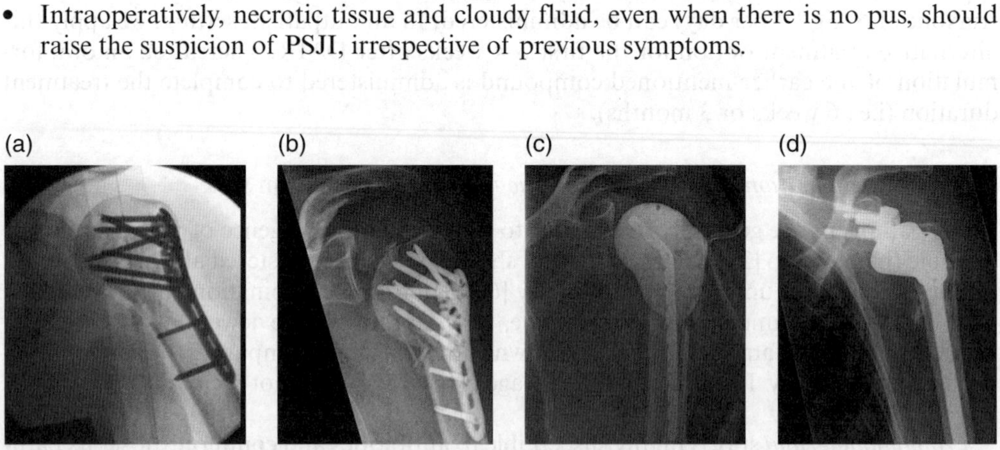

Figure 10.1. Anteroposterior view of the left shoulder with a four-part fracture of the proximal humerus. (a) Stabilization of the fracture with osteosynthesis, (b) dislocation of the plate, (c) implantation of a spacer, and (d) implantation of a reverse total shoulder arthroplasty.

Case 2: Chronic Polymicrobial PSJI Without Clinical Signs and Symptoms of Infection

A 62-year-old woman was referred to our center for an exchange of the right total shoulder arthroplasty. Her personal history included chronic obstructive pulmonary disease, insulin-dependent diabetes mellitus, arterial hypertension, and obesity. The indication for the primary arthroplasty was an avascular necrosis of the humerus head. Surgery was performed 8 months prior to presentation. After implantation, she had never been pain-free. She denied erythema, fever, or chills. On clinical examination, the range of motion of the shoulder joint was limited. The movements were painful. The blood test demonstrated the following results: CRP 21 mg/l (normal < 5 mg/l) and WBC 8.4 G/l (normal 4–10 G/l). An X-ray revealed a loose stem. A preoperative joint puncture was performed. The leukocyte count in the synovial fluid was 15,000 cells/μl, of which 67% were neutrophils. CNS grew in the aspirate sample. The organism was resistant to oxacillin and fusidic acid, but otherwise susceptible. The diagnosis of chronic PSJI was made. The pathogenesis was likely due to an exogenous route. Complete implant removal and meticulous debridement was performed, and seven biopsies were obtained. Large areas of necrotic tissue had to be removed. The rotator cuff was damaged, and the bone stock at the glenoid site was short. A spacer impregnated with vancomycin plus gentamicin was implanted. Postoperatively, antimicrobial treatment with vancomycin was initiated (2 × 15 mg IV per kg body weight per day). The removed arthroplasty was sonicated. CNS grew in all samples and in the sonicated fluid. In addition, *P. acnes* grew in four of seven biopsies and in the sonicated fluid. The postoperative course was uneventful. The wound healed quickly and adequately. After 2 weeks of IV antibiotics, the patient was discharged with oral antibiotics (clindamycin 3 × 600 mg p.o.) for another 10 weeks. Six months after implant removal, a reverse total shoulder joint arthroplasty was implanted. At 12-month follow-up, the patient was considered to be infection-free (probably cured). However, the functional outcome was only acceptable, but not satisfactory.

Learning Points

- In PSJI, empirical antimicrobial treatment should include the spectrum of *Propionibacterium* spp. and CNS until final culture results with susceptibility testing are available.
- Cases with delayed diagnosis and multiple revisions due to PSJI may end with an unsatisfactory functional outcome, even when the infection is cured.

References

[1] Kim SH, Wise BL, Zhang Y, Szabo RM. Increasing incidence of shoulder arthroplasty in the United States. J Bone Joint Surg Am 2011;93(24):2249–2254.

[2] Coste JS, Reig S, Trojani C, Berg M, Walch G, Boileau P. The management of infection in arthroplasty of the shoulder. J Bone Joint Surg Br 2004;86(1):65–69.

[3] Singh JA, Sperling JW, Schleck C, Harmsen WS, Cofield RH. Periprosthetic infections after total shoulder arthroplasty: a 33-year perspective. J Shoulder Elbow Surg 2012;21(11):1534–1541.

[4] van de Sande MA, Brand R, Rozing PM. Indications, complications, and results of shoulder arthroplasty. Scand J Rheumatol 2006;35(6):426–434.

[5] Sperling JW, Kozak TK, Hanssen AD, Cofield RH. Infection after shoulder arthroplasty. Clin Orthop Relat Res 2001;(382):206–216.

[6] Wirth MA, Rockwood CA, Jr. Complications of shoulder arthroplasty. Clin Orthop Relat Res 1994;(307):47–69.

[7] Zavala JA, Clark JC, Kissenberth MJ, Tolan SJ, Hawkins RJ. Management of deep infection after reverse total shoulder arthroplasty: a case series. J Shoulder Elbow Surg 2012;21(10): 1310–1315.

[8] Cheung EV, Sperling JW, Cofield RH. Infection associated with hematoma formation after shoulder arthroplasty. Clin Orthop Relat Res 2008;466(6):1363–1367.

[9] Pottinger P, Butler-Wu S, Neradilek MB, Merritt A, Bertelsen A, Jette JL, *et al.* Prognostic factors for bacterial cultures positive for Propionibacterium acnes and other organisms in a large series of revision shoulder arthroplasties performed for stiffness, pain, or loosening. J Bone Joint Surg Am 2012;94(22):2075–2083.

[10] Fevang BT, Lie SA, Havelin LI, Skredderstuen A, Furnes O. Risk factors for revision after shoulder arthroplasty: 1,825 shoulder arthroplasties from the Norwegian Arthroplasty Register. Acta Orthop 2009;80(1):83–91.

[11] Patel A, Calfee RP, Plante M, Fischer SA, Green A. Propionibacterium acnes colonization of the human shoulder. J Shoulder Elbow Surg 2009;18(6):897–902.

[12] Singh JA, Sperling JW, Schleck C, Harmsen W, Cofield RH. Periprosthetic infections after shoulder hemiarthroplasty. J Shoulder Elbow Surg 2012;21(10):1304–1309.

[13] Nowinski RJ, Gillespie RJ, Shishani Y, Cohen B, Walch G, Gobezie R. Antibiotic-loaded bone cement reduces deep infection rates for primary reverse total shoulder arthroplasty: a retrospective, cohort study of 501 shoulders. J Shoulder Elbow Surg 2012;21(3):324–328.

[14] Franceschini V, Chillemi C. Periprosthetic shoulder infection. Open Orthop J 2013;7: 243–249.

[15] Farshad M, Gerber C. Reverse total shoulder arthroplasty-from the most to the least common complication. Int Orthop 2010;34(8):1075–1082.

[16] Zumstein MA, Pinedo M, Old J, Boileau P. Problems, complications, reoperations, and revisions in reverse total shoulder arthroplasty: a systematic review. J Shoulder Elbow Surg 2011;20(1):146–157.

[17] Piper KE, Jacobson MJ, Cofield RH, Sperling JW, Sanchez-Sotelo J, Osmon DR, *et al.* Microbiologic diagnosis of prosthetic shoulder infection by use of implant sonication. J Clin Microbiol 2009;47(6):1878–1884.

[18] Romano CL, Borens O, Monti L, Meani E, Stuyck J. What treatment for periprosthetic shoulder infection? Results from a multicentre retrospective series. Int Orthop 2012;36(5): 1011–1017.

[19] Achermann Y, Sahin F, Schwyzer HK, Kolling C, Wust J, Vogt M. Characteristics and outcome of 16 periprosthetic shoulder joint infections. Infection 2013;41(3):613–620.

[20] Klatte TO, Junghans K, Al-Khateeb H, Rueger JM, Gehrke T, Kendoff D, Neumann J. Single-stage revision for peri-prosthetic shoulder infection: outcomes and results. Bone Joint J 2013;95-B(3):391–395.

[21] Stine IA, Lee B, Zalavras CG, Hatch G, 3rd, Itamura JM. Management of chronic shoulder infections utilizing a fixed articulating antibiotic-loaded spacer. J Shoulder Elbow Surg 2010;19(5):739–748.

[22] Zappe B, Graf S, Ochsner PE, Zimmerli W, Sendi P. Propionibacterium spp. in prosthetic joint infections: a diagnostic challenge. Arch Orthop Trauma Surg 2008;128(10):1039–1046.

[23] Kelly JD, 2nd, Hobgood ER. Positive culture rate in revision shoulder arthroplasty. Clin Orthop Relat Res 2009;467(9):2343–2348.

[24] Grosso MJ, Sabesan VJ, Ho JC, Ricchetti ET, Iannotti JP. Reinfection rates after 1-stage revision shoulder arthroplasty for patients with unexpected positive intraoperative cultures. J Shoulder Elbow Surg 2012;21(6):754–758.

[25] Zeller V, Ghorbani A, Strady C, Leonard P, Mamoudy P, Desplaces N. Propionibacterium acnes: an agent of prosthetic joint infection and colonization. J Infect 2007;55(2):119–124.
[26] Millett PJ, Yen YM, Price CS, Horan MP, van der Meijden OA, Elser F. Propionibacterium acnes infection as an occult cause of postoperative shoulder pain: a case series. Clin Orthop Relat Res 2011;469(10):2824–2830.
[27] Osmon DR, Berbari EF, Berendt AR, Lew D, Zimmerli W, Steckelberg JM, *et al.* Diagnosis and management of prosthetic joint infection: clinical practice guidelines by the Infectious Diseases Society of America. Clin Infect Dis 2013;56(1):e1–e25.
[28] Sendi P, Banderet F, Graber P, Zimmerli W. Periprosthetic joint infection following Staphylococcus aureus bacteremia. J Infect 2011;63(1):17–22.
[29] Zimmerli W, Trampuz A, Ochsner PE. Prosthetic-joint infections. N Engl J Med 2004;351(16): 1645–1654.
[30] Sendi P, Zimmerli W. Diagnosis of periprosthetic joint infections in clinical practice. Int J Artif Organs 2012;35(10):913–922.
[31] Dodson CC, Craig EV, Cordasco FA, Dines DM, Dines JS, Dicarlo E, *et al.* Propionibacterium acnes infection after shoulder arthroplasty: a diagnostic challenge. J Shoulder Elbow Surg 2010;19(2):303–307.
[32] Topolski MS, Chin PY, Sperling JW, Cofield RH. Revision shoulder arthroplasty with positive intraoperative cultures: the value of preoperative studies and intraoperative histology. J Shoulder Elbow Surg 2006;15(4):402–406.
[33] Jerosch J, Schneppenheim M. Management of infected shoulder replacement. Arch Orthop Trauma Surg 2003;123(5):209–214.
[34] Trampuz A, Piper KE, Jacobson MJ, Hanssen AD, Unni KK, Osmon DR, *et al.* Sonication of removed hip and knee prostheses for diagnosis of infection. N Engl J Med 2007;357(7): 654–663.
[35] Gallo J, Kolar M, Dendis M, Loveckova Y, Sauer P, Zapletalova J, Koukalova D. Culture and PCR analysis of joint fluid in the diagnosis of prosthetic joint infection. New Microbiol 2008;31(1):97–104.
[36] Trampuz A, Steckelberg JM, Osmon DR, Cockerill FR, Hanssen AD, Patel R. Advances in the laboratory diagnosis of prosthetic joint infection. Rev Med Microbiol. 2003;14(1):1–14.
[37] Achermann Y, Vogt M, Leunig M, Wust J, Trampuz A. Improved diagnosis of periprosthetic joint infection by multiplex PCR of sonication fluid from removed implants. J Clin Microbiol 2010;48(4):1208–1214.
[38] Portillo ME, Salvado M, Sorli L, Alier A, Martinez S, Trampuz A, *et al.* Multiplex PCR of sonication fluid accurately differentiates between prosthetic joint infection and aseptic failure. J Infect 2012;65(6):541–548.
[39] Iyengar KP, Vinjamuri S. Role of 99mTc Sulesomab in the diagnosis of prosthetic joint infections. Nucl Med Commun 2005;26(6):489–496.
[40] Graute V, Feist M, Lehner S, Haug A, Muller PE, Bartenstein P, Hacker M. Detection of low-grade prosthetic joint infections using 99mTc-antigranulocyte SPECT/CT: initial clinical results. Eur J Nucl Med Mol Imaging 2010;37(9):1751–1759.
[41] Buck FM, Jost B, Hodler J. Shoulder arthroplasty. Eur Radiol 2008;18(12):2937–2948.
[42] Constant CR, Murley AH. A clinical method of functional assessment of the shoulder. Clin Orthop Relat Res 1987;(214):160–164.
[43] Hudak PL, Amadio PC, Bombardier C. Development of an upper extremity outcome measure: the DASH (disabilities of the arm, shoulder and hand) [corrected]. The Upper Extremity Collaborative Group (UECG). Am J Ind Med 1996;29(6):602–608.
[44] Neer CS, 2nd, Watson KC, Stanton FJ. Recent experience in total shoulder replacement. J Bone Joint Surg Am 1982;64(3):319–337.
[45] The European Society for Surgery of the Shoulder and Elbow. http://www.secec.org/pages/r-d/assessments.php. Accessed on May 31, 2014.

[46] Antuna SA, Sperling JW, Cofield RH, Rowland CM. Glenoid revision surgery after total shoulder arthroplasty. J Shoulder Elbow Surg 2001;10(3):217–224.
[47] Weber P, Utzschneider S, Sadoghi P, Andress HJ, Jansson V, Muller PE. Management of the infected shoulder prosthesis: a retrospective analysis and review of the literature. Int Orthop 2011;35(3):365–373.
[48] Sendi P, Graber P, Zimmerli W. Risk factors associated with acute hip prosthetic joint infections and outcome of treatment with a rifampin-based regimen. Acta Orthop 2008;79(3):454; author reply 455.
[49] Ince A, Seemann K, Frommelt L, Katzer A, Loehr JF. One-stage exchange shoulder arthroplasty for peri-prosthetic infection. J Bone Joint Surg Br 2005;87(6):814–818.
[50] Cuff DJ, Virani NA, Levy J, Frankle MA, Derasari A, Hines B, *et al.* The treatment of deep shoulder infection and glenohumeral instability with debridement, reverse shoulder arthroplasty and postoperative antibiotics. J Bone Joint Surg Br 2008;90(3):336–342.
[51] Seitz WH, Jr., Damacen H. Staged exchange arthroplasty for shoulder sepsis. J Arthroplasty 2002;17(4 Suppl 1):36–40.
[52] Mileti J, Sperling JW, Cofield RH. Reimplantation of a shoulder arthroplasty after a previous infected arthroplasty. J Shoulder Elbow Surg 2004;13(5):528–531.
[53] Strickland JP, Sperling JW, Cofield RH. The results of two-stage re-implantation for infected shoulder replacement. J Bone Joint Surg Br 2008;90(4):460–465.
[54] Sendi P, Zimmerli W. Challenges in periprosthetic knee-joint infection. Int J Artif Organs 2011;34(9):947–956.
[55] Coffey MJ, Ely EE, Crosby LA. Treatment of glenohumeral sepsis with a commercially produced antibiotic-impregnated cement spacer. J Shoulder Elbow Surg 2010;19(6):868–873.
[56] Haddad S, Corona PS, Reverte MM, Amat C, Flores X. Antibiotic-impregnated cement spacer as a definitive treatment for post-arthroscopy shoulder destructive osteomyelitis: case report and review of literature. Strategies Trauma Limb Reconstr 2013;8(3):199–205.
[57] Magnan B, Bondi M, Vecchini E, Samaila E, Maluta T, Dall'oca C. A preformed antibiotic-loaded spacer for treatment for septic arthritis of the shoulder. Musculoskelet Surg 2013. DOI: 10.1007/s12306-013-0268-x.
[58] Brodt S, Hofmann GO, Tiemann AH. Preservation of the shoulder joint by the use of a hybrid-spacer after septic loosening of a reversed total shoulder joint arthroplasty: a case report. Strategies Trauma Limb Reconstr 2010;5(2):111–114.
[59] Verhelst L, Stuyck J, Bellemans J, Debeer P. Resection arthroplasty of the shoulder as a salvage procedure for deep shoulder infection: does the use of a cement spacer improve outcome? J Shoulder Elbow Surg 2011;20(8):1224–1233.
[60] Sendi P, Zimmerli W. Antimicrobial treatment concepts for orthopaedic device-related infection. Clin Microbiol Infect 2012;18(12):1176–1184.
[61] Landersdorfer CB, Bulitta JB, Kinzig M, Holzgrabe U, Sorgel F. Penetration of antibacterials into bone: pharmacokinetic, pharmacodynamic and bioanalytical considerations. Clin Pharmacokinet 2009;48(2):89–124.
[62] Sendi P, Zimmerli W. Comment on: prosthetic hip joint infection with a Streptococcus agalactiae isolate not susceptible to penicillin G and ceftriaxone. J Antimicrob Chemother 2012;67(4):1050–1051.

Chapter 11

Periprosthetic Joint Infection after Elbow Arthroplasty

Yvonne Achermann and Michael C. Glanzmann

Introduction

Total elbow arthroplasty (TEA) has become a popular reconstructive procedure among patients with elbow arthritis [1]. In the past, underlying joint disorders leading to implantation of a TEA were mainly different types of inflammatory arthritis [2]. Nowadays, more and more patients with posttraumatic osteoarthritis are becoming a new target group, because of new improved surgical techniques and implant designs [3].

Currently, the most commonly used implant designs are semiconstrained devices, such as Gschwend–Scheier–Bähler III (GSB III), Discovery, and Coonrad–Morrey prostheses. These implant types offer some sort of connection between the humeral and ulnar component. In contrast to fully constrained implants, semiconstrained prostheses allow some degree of varus–valgus movement. While semiconstrained designs provide biomechanical stability, unconstrained implants fully rely on natural ligaments and tendons. These designs are predominantly used in Scandinavian countries.

In 1996, a review report from Gschwend *et al.* [4] described a general high complication rate of up to 43% after elbow prosthesis implantation, including loosening, ulnar neuropathy, infection, dislocation, subluxation, uncoupling, intraoperative bone fractures, and failure of the implant. This is much higher in comparison to larger joints, such as knee and hip prostheses. Thanks to technical improvements during the last two decades, the frequency of complications after TEA has decreased. This confirms a newer review by Voloshin *et al.* [3] who reported a lower complication rate of 24%, mainly instability (dislocation and symptomatic subluxation), intraoperative fracture, infection, and aseptic loosening of the implant.

The overall rate of periprosthetic joint infection (PJI) is clearly higher after elbow replacement than after hip, knee, or shoulder arthroplasty [5–9]. The published infection rates are between 1.9 and 10.3% [4, 10–16]. The higher infection rates in elbow arthroplasty as compared to hip (<1%) and knee prostheses (<2%) may be due to several

Bone and Joint Infections: From Microbiology to Diagnostics and Treatment, First Edition.
Edited by Werner Zimmerli.

reasons. First, the main indication for hip or knee replacement is degenerative osteoarthritis. In contrast, in elbow arthroplasty, rheumatoid arthritis or posttraumatic osteoarthritis are the most common underlying diseases. Autoimmune rheumatic disorders are associated with a higher risk of infection due to a chronic inflammatory process, and the immunosuppressive disease-modifying treatment of the disease [17]. Second, the subcutaneous placement and lack of muscle coverage of the elbow prosthesis provides little protection against contiguous infection after bursitis or skin breakdown. Third, soft tissue is more vulnerable and prone to infection in patients with posttraumatic or inflammatory arthritis than in healthy individuals, due to traumatically damaged tissue, previous surgery, or skin atrophy by corticosteroids [3].

There is no published risk factor analysis for patients with elbow prosthesis, probably due to the small number of primary or revision surgeries. Assuming that evaluated risk factors in hip prosthesis are generally valid for all types of arthroplasty, obesity (hazard ratio [HR] 1.73), rheumatoid arthritis (HR 1.71), coagulopathy (HR 1.58), and presurgical anemia (HR 1.36) might be risk factors for elbow arthroplasty as well [18]. A group from the Mayo Clinic in Rochester, USA, recently developed a prognostic scoring system for the development of PJI for patients undergoing surgery for total hip or knee arthroplasty [19]. Multivariable modeling detected the following risk factors that should be considered before implantation of an arthroplasty: (i) body mass index, (ii) previous surgical procedure on the index joint, (iii) previous arthroplasty (revision surgery), (iv) immunosuppression, (v) higher ASA (American Society of Anaesthesiologists) score, and (vi) length of the surgical procedure.

Microbiology

Since the 1980s, there were several published studies reporting microbiological characteristics of elbow PJI, mainly using a retrospective study design. The most commonly isolated microorganisms were *Staphylococcus aureus* and coagulase-negative staphylococci (CNS) (Table 11.1) [2, 10, 11, 16, 20, 21]. *Propionibacterium acnes* as a skin commensal is only sporadically isolated in elbow PJI [21]. This might be due to the lower density of sweat glands in the elbow region, as compared to the shoulder, where *P. acnes* is the dominant pathogen found in prosthetic joint infection [5, 22–25].

Clinical Features

A differentiation between infective bursitis without deeper involvement of the joint and PJI is generally not possible. Therefore, in any patients with septic bursitis or surgical wound infection, PJI has to be considered. In a recently published study of Vergidis *et al.* [2], eight out of nine patients (89%) with elbow PJI presented with local inflammatory signs of infection (details not specified) and five out of nine (56%) with pain only. One patient suffered from fever. Cheung *et al.* [21] reported pain and swelling of the elbow tissue in the majority of their patients (18 of 29, 62%), followed by wound dehiscence with a draining sinus tract in 11 of 29 (38%) patients. In a study by Yamaguchi *et al.* [20], all patients had a diagnostic workup for PJI because of pain, loss of elbow function, erythema, or hyperthermic joint.

Table 11.1. Microbiological characteristics of elbow PJI.

	Wolfe *et al.* [13]	Yamaguchi *et al.* [20]	Gille *et al.* [16]	Cheung *et al.*[a] [21]	Vergidis *et al.* [2]	Achermann *et al.* [10]
Study design						
Cases	12	25	6	29	9	27
Infection rate	7.3%	3.3%	1.9%	Not available	Not available	7.5%
Study period	1974–1986	1981–1994	1978–1999	1976–2003	2007–2010	1994–2007
Study place	New York City, US	Rochester, US	Hamburg, Germany	Rochester, US	Rochester, US	Zurich, CH
Index surgery	164 primary EA	591 primary and 166 revision EA	305 primary EA	Primary EA	Primary and revision EA	262 primary and 96 revision EA
Microbiology						
S. aureus	10 (80%)	14 (56%)	6 (100%)	8 (28%)	2 (22.3%)	11 (41%)
CNS	1 (8.3%)	7 (28%)		13 (45%)	7 (77%)	9 (33%)
Streptococcus species	1 (8.3%)			1 (3.4%)		2 (7)
Gram-negative		3 (12%)		2 (7%)		1 (3.7%)
Anaerobes				1 (3.4%)[b]		0
Aerob diphteroids				1 (3.4%)		1 (3.7%)
Polymicrobial						2 (7%)[c]
No growth		1 (4%)		3 (10.3%)		1 (4%)

CNS, coagulase-negative staphylococci; EA, elbow arthroplasties.
[a]Only cases with sonicated prosthesis for diagnostic available.
[b]*Propionibacterium acnes*.
[c]Methicillin-resistant *Staphylococcus epidermidis* and *Enterobacter cloacae* in one episode; methicillin-resistant *S. epidermidis* and methicillin-susceptible *S. aureus* in one episode.

Diagnostic Procedures

Preoperative Workup

The main goal of all currently available diagnostic procedures is the confirmation of PJI with acute or chronic signs of inflammation, and the identification of the causative pathogen. As in any type of PJI, diagnosis relies on a combination of clinical signs with laboratory and microbiological tests. However, there is no published diagnostic algorithm for elbow PJI. Various diagnostic procedures, proposed in recent guidelines on the management of PJI, are presented later [26, 27]. The specific situation of elbow PJI is discussed.

Evaluation of any patient with a possible PJI should comprise an investigation with a thorough medical history, including the age of the prosthetic device (date of implantation), the type of prosthesis, previous surgical procedures, previous wound healing disorders or proven infection, comorbid conditions such as rheumatoid arthritis, and medical treatment, especially corticosteroids and disease-modifying immunosuppressive treatments. In addition, a careful physical examination is needed. Knowledge of the time of onset of symptoms after the last surgical procedure is crucial, since it influences the treatment strategy [26, 28].

Diagnosis of elbow PJI might be challenging in patients with inflammatory rheumatic diseases such as rheumatoid arthritis or psoriasis arthropathy, because of the patients' general compromised health condition, immune status, inflammatory signs, and symptoms from their underlying disease.

Acute infection is commonly obvious to the clinician's eyes and hands, showing acute inflammatory signs and symptoms such as pain, redness, swelling, and local warmth of the joint. Fever can be present in hematogenous or acute infection. The presence of a sinus tract is an explicit sign of infection, and is quite common in elbow PJI [21]. Other signs and symptoms are often nonspecific, such as pain and loss of function.

Erythrocyte sedimentation rate (ESR) and C-reactive protein (CRP) are valuable initial screening tests in the diagnostic algorithm, especially when an infection is clinically not evident. In a retrospective study with 20 episodes of elbow PJI (9 early, 1 delayed, and 10 late infections) published by Spormann *et al.* [11], the mean preoperative CRP level was 127 mg/l (range, 14–214 mg/l). In the study of Vergidis *et al.* [2], most patients presented with delayed PJI. Eighty-eight percent of the patients had radiographic signs of loosening after a median time to onset of symptoms of 13 months. They showed a median preoperative CRP of 56 mg/l (range, 10–269 mg/l).

In all patients with suspected acute elbow PJI, a diagnostic arthrocentesis should be performed prior to antimicrobial treatment and surgery, in order to find the causing pathogen. In patients with chronic joint pain, aspiration of joint fluid is also advised, if a delayed infection is suspected. Arthrocentesis with measurement of total white cell count, differential leukocyte count, and culture for aerobic and anaerobic organism is a cost-effective method to reliably confirm a PJI. White blood cell count of more than 1700 cells/μl and a differential of more than 65% of polymorphonuclear neutrophils are predictive for infection in knee PJI [29]. However, in this study, patients with inflammatory joint disease were excluded. Cutoffs vary between different joints, underlying diseases, and postoperative interval [30–34]. At this present time, there are no studies investigating this cutoff in elbow PJI. Aerobic and anaerobic cultures of the synovial fluid can help in identifying the causative microorganism, if the pathogen is not only in the biofilm but also in the planktonic phase of growth. Aspiration as a safe, quick, and inexpensive intervention is most commonly performed through a dorso-radial approach through the anconeus muscle (soft spot), where no major vessel or nerve traverses. With the elbow in approximately 135° of flexion, the needle should enter in the middle of a triangle formed by the lateral epicondyle, the radial head, and the tip of the olecranon (Figure 11.1).

In all patients, plain radiography should be performed (Figures 11.2 and 11.3). In the absence of obvious clinical signs of infection, a radiological exam offers important information such as periprosthetic fracture, implant failure, or loosening of implant, which can be a sign of delayed infection. In early infection, radiological signs of infection are often unspecific and difficult to detect. Periprosthetic endosteal scalloping and radiolucency in combination with other signs and symptoms give rise to suspicion of an infection. Data about the use of computed tomography or magnetic resonance imaging as a diagnostic test are inconclusive and are limited by hardware-induced artifacts [35]. Both methods are recommended for better visualization of anatomical changes (extent of soft tissue abscesses, sinus tract, or fractures) to guide surgical intervention. Nuclear imaging methods (labeled leukocyte imaging, bone marrow imaging, ^{18}F-fluorodeoxyglucose–positron

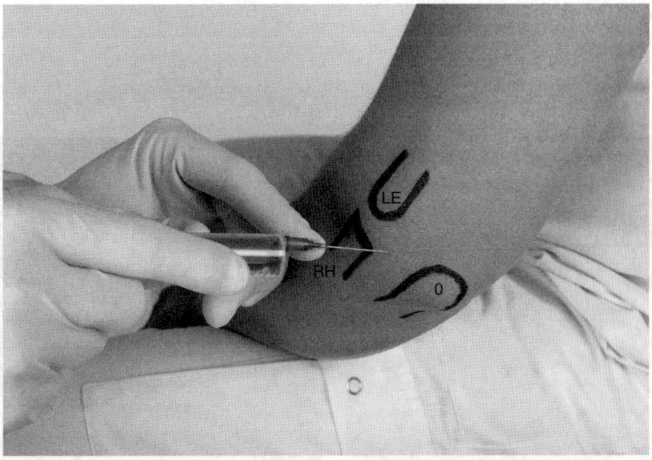

Figure 11.1. Arthrocentesis of the elbow through a dorso-radial approach. Landmarks are the olecranon (O), the lateral epicondylus (LE), and the radial head (RH). While the elbow is held in approximately 135° of flexion, a 22-gauge needle should enter in the middle of a triangle formed by the lateral epicondyle, the radial head, and the tip of the olecranon. (*See insert for color representation of the figure.*)

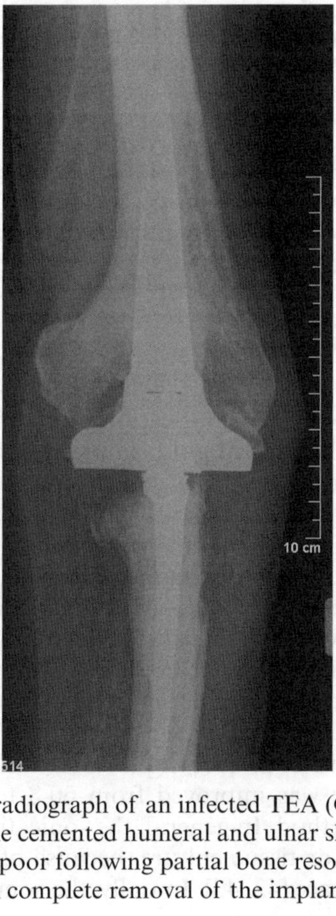

Figure 11.2. Anteroposterior radiograph of an infected TEA (GSB III, Zimmer). There is no observable radiolucency along the cemented humeral and ulnar shaft, but the bony containment between the humeral condyles is poor following partial bone resorption in this area. In this patient, a two-stage revision with complete removal of the implant was successfully performed.

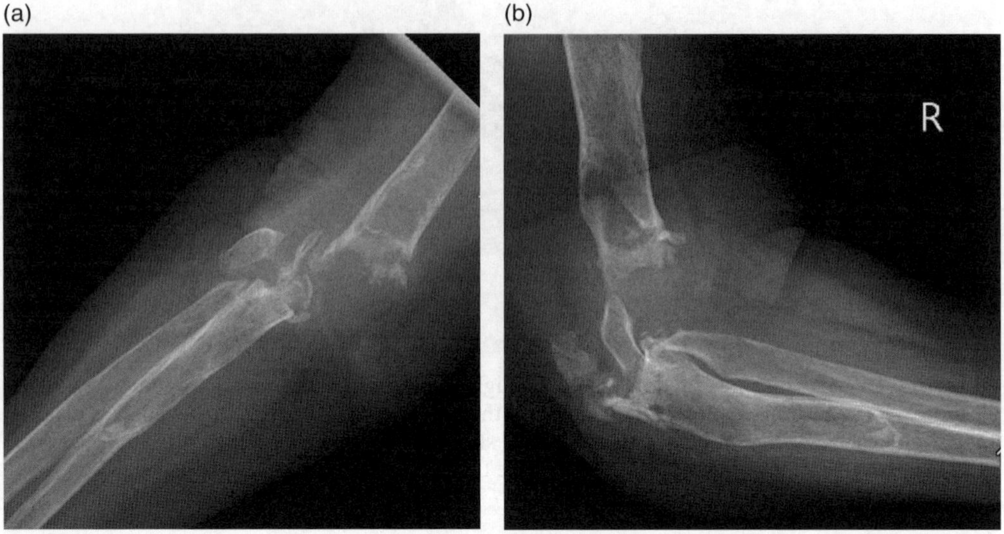

Figure 11.3. Anteroposterior (a) and lateral radiograph (b) of a right elbow after removal of an infected arthroplasty. There is significant loss of bone stock on the distal humerus. The olecranon shows severe osteolysis and is fractured into multiple thin fragments. After successful treatment of the infection, reimplantation of a long stem elbow prosthesis with structural bone allograft augmentation was performed.

emission tomography [FDG–PET] imaging, gallium imaging) are options in a patient with nonspecific symptoms, in whom a PJI is suspected, but not confirmed by arthrocentesis. Their role in the diagnostic algorithm in general is still debatable, and no consensus data on their use in elbow PJI is available at this time.

Peri- and Postoperative Workup

An antibiotic-free interval of 2 weeks prior to surgical treatment with sampling of tissue cultures increases the likelihood of identification of the pathogen with a sensitivity of 55–65% [2, 36–38], which is the current gold standard in diagnosing PJI. At least three but optimally six samples should be retrieved, in order to facilitate the differentiation between causing pathogen and contaminants [38]. The tissue samples should be incubated for 10–14 days, which increases the chance to identify slow-growing pathogens [39]. As a rule, swabs are not appropriate to diagnose PJI. The teeny amount of cell material decreases the sensitivity of the culture and makes performing polymerase chain reaction (PCR) in samples with negative culture results difficult [40, 41].

If the therapeutic procedure includes an exchange or removal of the implant, it is recommended to use a sonication bath in order to optimize diagnosis of implant-associated infection. The sonication method dislodges bacteria from the surface of an implant. It was first described by Trampuz *et al.* [37] in patients with hip and knee PJI. The sensitivity of a sonication culture was improved from 60.8 to 78.5%, as compared to the conventional tissue culture method. In general, the sonication method has been shown to be superior to synovial fluid or tissue culture, because of its ability to remove biofilm bacteria from the implant surface [42, 43]. Initially, the sonication method was developed for hip and knee PJI. It has also been successfully applied for other implant-associated infections such as shoulder and elbow PJI [2, 22], cardiovascular implants [44], breast implants [45, 46], ureteral stents [47], and spinal implants [48]. In the study by Vergidis *et al.* [2], sensitivity of sonicated versus conventional incubated tissue biopsies in elbow PJI improved from 55 to 89%.

Management

Depending on time of infection after the last surgical intervention, duration of symptoms, microbial susceptibility testing of the causing pathogen, and degree of the damaged infected tissue, surgical revision may include an extensive debridement with implant retention or alternatively an exchange or resection of the prosthesis [26, 28]. Exchange of an elbow prosthesis in a one- or two-stage intervention is a challenging surgical procedure because of several factors. First, identification and protection of the ulnar nerve can be difficult, because of scarring and septic alterations of the soft tissue. Second, removal of the prosthesis, the appropriate cement for fixing, and all necrotic tissue might endanger the intact distal humerus and proximal ulna bone for further reimplantation of the new prosthesis. Altogether, this is a risk for iatrogenic fracture (Figure 11.4). Third, preparation on the humerus might endanger the radial nerve with potentially disastrous functional consequences. In the case of severe bone deficiency after implant removal, necessary bone allografting may further complicate the procedure at time of reimplantation (Figure 11.5). Therefore, debridement with implant retention should be carried out whenever possible [11]. Table 11.2 summarizes the surgical procedures and success rates described in five recent publications [2, 10, 16, 20, 21].

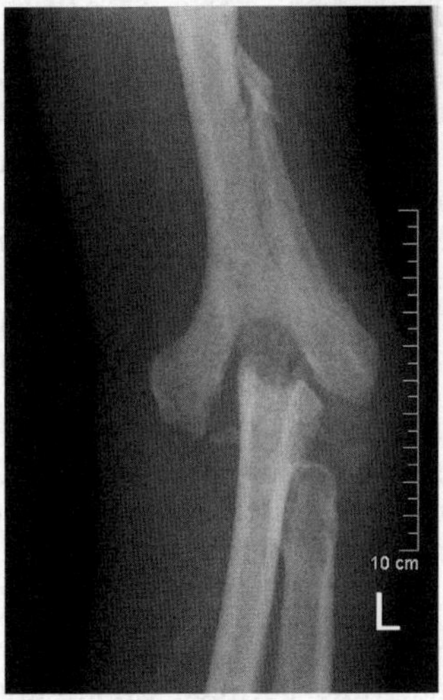

Figure 11.4. Anteroposterior radiograph of a left elbow after removal of an infected arthroplasty. While aiming for complete cement removal on the distal humerus, an iatrogenic fracture occurred. Due to the minimal displacement, stabilization of the fracture was not required. The fracture consolidated during the sine–sine period.

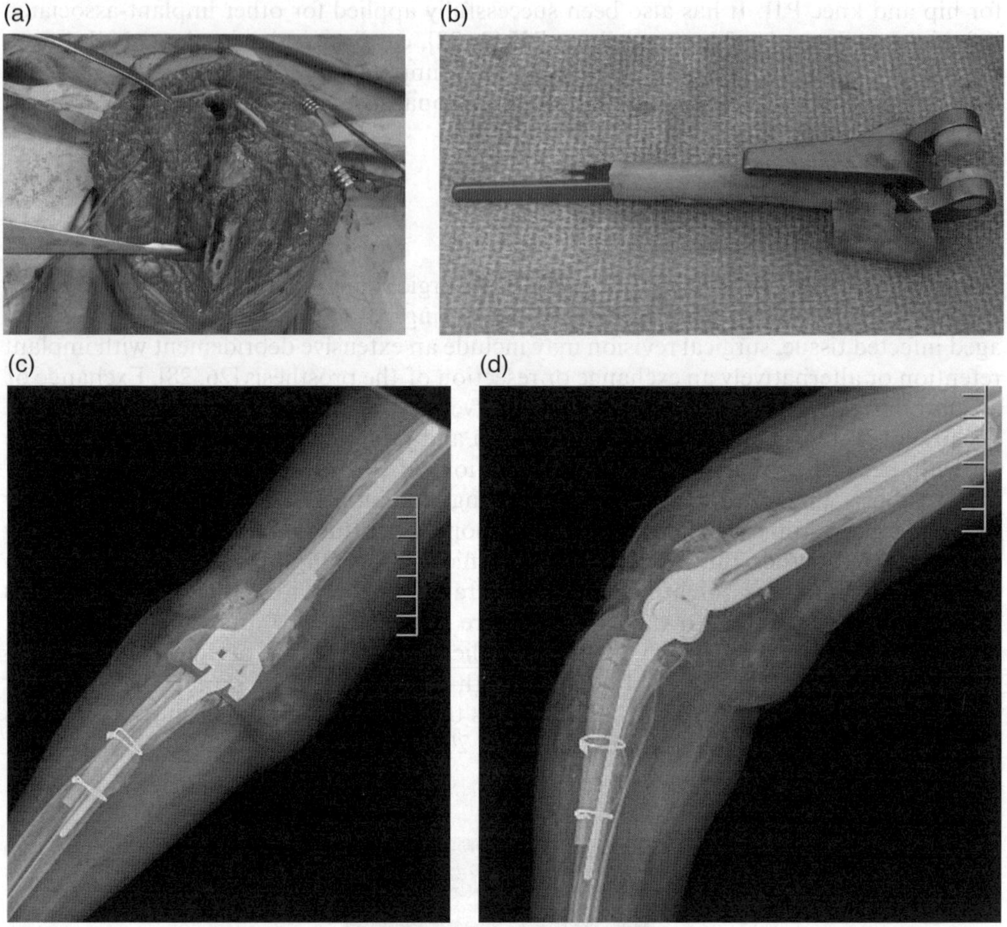

Figure 11.5. (a) Intraoperative view of an infected elbow arthroplasty operated through a posterior approach. Straps shown protect the ulnar and radial nerves while the retractors hold back the distal humerus and proximal ulna. Four months following removal of the infected prosthesis and meticulous debridement, severe bone deficiency challenged the reimplantation of a new prosthesis. Direct cementation of a new implant would lead to relevant shortening of the upper extremity and bear a high risk of fixation failure. (b) To bridge the humeral bone defect, a fresh frozen allograft was shaped to the patient's anatomy and prefixed to a long stem humeral component (Coonrad–Morrey prosthesis, Zimmer). (c) and (d) Elbow radiographs in two planes 12 months after reimplantation of a total elbow arthroplasty with structural bone graft augmentation. The films demonstrate a complete integration of the allografts and the cemented long stem prosthesis. Length of the elbow joint was properly restored. Each bar represents 1 cm.

An exchange of the prosthesis can be performed in a one- or two-stage exchange. In the two-stage exchange, the optimal time between explantation of the infected device and implantation of the new prosthesis has not been studied up to now. In hip and knee PJI, a short (2–4 weeks) or a long (8 weeks) interval has been studied with favored treatment success [6, 28]. In elbow PJI, there are no generally accepted recommendations. As a rule, orthopedic surgeons tend toward a prolonged device-free interval, because of the elbow complexity

Table. 11.2. Surgical procedures and success rate in elbow surgery.

	Yamaguchi *et al.* [20]	Gille *et al.* [16]	Cheung *et al.* [21]	Vergidis *et al.* [2]	Achermann *et al.* [10]
Study design					
Total examined cases	25	6	29	9	27
Infection rate	3.3%	1.9%	Not available	Not available	7.5%
Surgical treatment (success rate)					
Debridement and retention[a]	14 (50%)	0	0	1 (0%)	21 (62%)
One-stage exchange	1 (0%)	6 (83%)	0	2 (100%)	1 (100%)
Two-stage exchange	5 (80%)	0	29 (72.4%)	5 (100%)	2 (100%)
Resection	5 (100%)	0	0	1 (100%)	1 (100%)
Antibiotics only	0	0	0	0	2 (100%)

[a] Partial revision of the implant was classified as a debridement and retention procedure.

and difficulty in case of early reimplantation. According to several publications on elbow PJI, the mean interval varied from 7 weeks [20], to 7.6 weeks [2], and to 22 weeks [10].

Based on expert recommendations and European and American guidelines, patients with a PJI should receive 2–6 weeks of pathogen-specific intravenous (IV) antimicrobial therapy, followed by a prolonged oral antibiotic treatment for a total of 3 months for hip and 6 months for knee joint infections [26, 49]. For total elbow prosthesis infections, the same protocols as for periprosthetic hip joint infections should be followed, that is, a duration of 3 months. However, the quality and strength of this recommendation is only defined as C-III [26]. Recently published studies in hip PJI favor shorter treatment protocols [50–53]. However, these suggestions are not based on randomized controlled trials.

In case of a staphylococcal PJI, the IV pathogen-specific treatment should be combined with rifampin 300–450 mg orally twice daily, followed by rifampin plus a companion oral drug similar to those used with hip and knee prosthesis [54]. According to recent in vitro and animal studies, *P. acnes* device-associated infection, PJIs have a better outcome when rifampin is combined with vancomycin or daptomycin, as compared to these drugs alone [55]. However, in these animal experiments, standard drugs such as penicillin, ceftriaxone, and clindamycin could not be tested. Furthermore, only limited data exist about treatment outcomes in a clinical setting as of now [56–60].

A treatment algorithm was developed for the management of PJI in general on the basis of in vitro studies, animal models of foreign body infections [61, 62], and clinical studies [54, 63], in order to attain a high success rate with the least invasive surgical procedures. The cure rate in studies following this algorithm was 94.3% after knee arthroplasty [6], 83% [64] and 91% [65] after hip arthroplasty, and 100% in a population with different orthopedic devices [54]. In contrast, when the algorithm was not followed, cure rates were significantly lower, ranging from 57 to 60% [6, 66]. In elbow PJI there is only one study that analyzed the outcome in regard to this algorithm. It shows a general relapse-free survival of the elbow prosthesis of 65% after 2 years [10]. If the algorithm was not followed, a relapse-free survival of only 33% was documented, as compared to 100% when the management was in agreement with the algorithm. In all patients with relapse, debridement with retention of the prosthesis instead of an exchange of the prosthesis was performed. This study showed that a high chance of long-term success can be expected with proper selection of patients for debridement and retention. The crucial criteria were (i) no implant loosening; (ii) early or acute infections with symptom duration of less than 3 weeks; and (iii) susceptibility of the causative pathogen to a biofilm-active treatment regimen.

Instructive Cases

Case 1. Early Infection Treated with Debridement and Retention

A 71-year-old man was hospitalized in February 2006 with purulent wound secretion 2 weeks after primary implantation of an elbow prosthesis on the right side (GSB III, Zimmer). His medical history was remarkable for rheumatoid arthritis, treated with prednisone (6 mg daily) and methotrexate (20 mg weekly). Laboratory tests showed an elevated CRP of 160 mg/l (upper normal limit: 5 mg/l) and white blood cell count of 16,000/μl. Revision surgery with extensive debridement of necrotic scar and bone tissue, and a lavage was performed while retaining the prosthesis. After taking two intraoperative periprosthetic tissue biopsies, empirical IV antibiotic treatment with amoxicillin–clavulanate in combination with rifampin 300 mg twice daily oral was started. The next day, methicillin-susceptible *S. aureus* was isolated in both tissue culture biopsies. Treatment was streamlined to flucloxacillin (2 g intravenously q6 h) for 4 weeks, followed by ciprofloxacin 750 mg twice daily plus rifampin 300 mg twice daily for another 5 months according to the results of the susceptibility testing. There was no relapse of infection in the following 4 years.

Learning Points

- Exchanging an infected elbow prosthesis is a difficult surgical procedure associated with a high complication rate. Implant retention should therefore always be considered if the following conditions can be fulfilled: (i) early postoperative infection (within 3 months) or acute hematogenous infection at any time; (ii) short duration of signs and symptoms (i.e., <21 days); (iii) surrounding soft tissue without severe damage; and (iv) availability of an antimicrobial agent with activity against staphylococcal biofilms.
- In this case, the antimicrobial treatment was given for a total of 6 months. There is no controlled study dealing with the optimal duration of treatment in PJI. In this patient, therapy was given for a prolonged time, because the patient was treated with immunosuppressive drugs against rheumatoid arthritis.

Case 2. Delayed infection, Inappropriate Rifampin Therapy Leading to Emergence of Resistance

A 66-year-old male patient underwent spinal surgery 8 years earlier, following implantation of a total elbow prosthesis (GSB III, Zimmer) for posttraumatic osteoarthritis.

The elbow arthroplasty over the years following implantation was inconspicuous, until back surgery had to be performed. Four weeks following spinal surgery, the patient suffered from olecranon bursitis, initially treated by the patient himself before presenting to his elbow surgeon. According to the clinical presentation, the physician assumed acute PJI. Debridement with retention was performed, because the elbow implant was considered to be radiologically stable and the patient denied the explantation of the prosthesis. Five out of seven intraoperative tissue samples tested positive for *Staphylococcus capitis* (methicillin- and rifampin-susceptible), a species from the CNS group. An antimicrobial treatment of total 6 months was given, starting with vancomycin intravenously plus rifampin (orally), followed by an oral combination of ciprofloxacin plus rifampin. Due to a misunderstanding between the physician and the patient after the IV antimicrobial treatment, rifampin was given as a monotherapy for 5 days. One month after stopping antimicrobial treatment, a relapse of infection with isolation of *S. capitis* (6 of 10 biopsies) and *Micrococcus* species (4 of 10 biopsies) was diagnosed. The *S. capitis* at that time was multiresistant including resistance to rifampin. After a two-stage exchange of the prosthesis with a 3-month interval without prosthesis, a new elbow prosthesis (type Discovery, Biomet) was successfully implanted with no further suspicion of infection. Temporary symptoms of ulnar nerve irritation resolved without further surgical intervention until final follow-up 3 years after reimplantation of the elbow prosthesis.

Learning Points

- Isolation of CNS should raise suspicion of delayed infection, requiring explantation of the prosthesis for cure.
- Although rifampin resistance is rare, resistance may quickly emerge due to a single-step point mutation [67, 68]. Emergence of rifampin resistance is associated with several risk factors, including (i) several revision interventions interventions, (ii) previous PJI treated with inadequate surgical debridement and/or less than 2 weeks of IV antimicrobial treatment, or (iii) inadequate rifampin treatment (e.g., monotherapy) [69].
- Rifampin-resistant staphylococci are difficult-to-treat organisms, and patients with confirmed infections are probably best treated by a two-stage revision with a minimum implant-free interval of 8 weeks [70].

Case 3. Delayed Diagnosis of Late Infections Treated with Two-Stage Exchange

A 60-year-old female with rheumatoid arthritis got a semiconstrained and cemented prosthesis (GSB III, Zimmer) in the right elbow, 8 years before the current presentation. Before consultation, she suffered from a superficial skin lesion over the prosthetic joint, which was inappropriately mechanically treated by herself. In addition, she was treated with a course of oral antibiotics for several weeks by her general practitioner. She was not treated with disease-modifying medication for her rheumatoid arthritis. Due to persistent signs of local infection, she was referred to an orthopedic clinic for further management. The preoperative plain films were unremarkable, and implant stability was considered to be intact. Her CRP was 10 times above the upper normal limit. Because the duration of symptoms lasted more than 3 weeks, it was decided to perform a two-stage exchange of the implant. The first-stage intervention included removal of the stable implant and most of the cement followed by an aggressive debridement and synovectomy. A total of eight soft and bone tissue samples for microbiological culture were obtained. Furthermore, the removed prosthesis and a large cement remnant were sonicated and the collected fluid was cultured for additional pathogen identification as previously described [36, 37]. Immediately after collection of the specimens, she was started on piperacillin–tazobactam (4.5 g i.v. three times a day) until the microbiological diagnosis was available. All six tissue biopsies of soft tissue ($n = 4$) and bone ($n = 2$) as well as the sonicated fluid from the implant showed growth of *S. aureus*. The strain was susceptible to oxacillin (minimum inhibitory concentration [MIC] 0.125 mg/l), levofloxacin (MIC 0.25), clindamycin (MIC 0.0914), sulfamethoxazole/trimethoprim (MIC 0.64), and rifampin (<0.06). Piperacillin–tazobactam was switched to a pathogen-specific IV treatment with flucloxacillin in combination with rifampin. After an IV treatment of 14 days, an oral course with levofloxacin and rifampin for additional 4 weeks was followed. Wound healing was uneventful and the sine–sine situation without an elbow joint was well-tolerated with additional help from a brace. Reimplantation of a TEA was performed after a 6-week antibiotic-free interval following antimicrobial therapy, that is, 12 weeks after explantation. At reimplantation, a preemptive IV antibiotic regimen with amoxicillin–clavulanate was started after taking several tissue biopsies, and a residual cement body was sent for sonication. After exclusion of an infection, the preemptive antibiotic regimen was stopped when cultures remained negative. The clinical and laboratory evolution was favorable with a nearly complete restoration of elbow function without signs of persistent infection.

Learning Points

- If a biofilm infection has already been established (indicated here with a duration of symptoms >3 weeks and/or possible signs of an osteomyelitis on X-ray), an exchange of the prosthesis should be performed for a successful outcome.
- In this patient, the interval time without prosthesis was about 3 months. However, there is no data about the optimal time interval between a two-stage exchange especially not in elbow surgery. In hip and knee PJI, there are data describing a favorable outcome with a short (2–4 weeks) or a long (6 weeks) interval [6, 66]. In elbow PJI, further studies are needed taking into account the elbow complexity.
- When using a hinged brace, it is surprising how well patients tolerate the time without the elbow prosthesis. In the authors' experience, it is not uncommon that patients report to be pain-free and show a remarkable function of the upper extremity (see Case 4).

Case 4. Late Polymicrobial Infection Treated with Removal without Replacement

In 2000, a 39-year-old woman underwent total elbow replacement surgery (GSB III, Zimmer) on the right side after a complex traumatic fracture. Four years later, she suffered from traumatic periprosthetic fracture, which had to be stabilized by open reduction and internal fixation of the ulna. Four years following this second intervention, she presented with pain and a sinus tract above the elbow, present for several weeks. Conventional radiological imaging was unremarkable. The prosthesis was removed, and polymicrobial infection was diagnosed by isolation of *S. aureus* and CNS (sonication cultures >1000 CFU/ml *S. aureus* and >1000 CFU/ml CNS, conventional tissue cultures with six of six positive biopsies with *S. aureus*). Histopathology of the resected bone showed chronic osteomyelitis. After treatment with vancomycin and rifampin for 14 days, an oral regimen using levofloxacin and rifampin was started. The patient was satisfied without having an elbow joint, and she autonomously stopped the antibiotics after 6 weeks of treatment. The sine–sine situation was very well-tolerated by the active patient and made reimplantation of a prosthesis both subjectively and objectively unneeded. Until final follow-up 3 years after implant explantation, the patient was still free of signs of infection.

Learning Point

- Elbow revision surgery is demanding and potential complications including neurovascular injury, insufficient bone stock endangering secure implant fixation, and extensor mechanism deficiency are quite frequent. In case of severe bone deficiency, fragile soft tissue coverage, or recurrent signs of infection, resection arthroplasty without reimplantation of a new TEA is a feasible treatment option without major loss of quality of life if the condyles are preserved for static and dynamic stability.

References

[1] Mackay DC, Hudson B, Williams JR. Which primary shoulder and elbow replacement? A review of the results of prostheses available in the UK. Ann R Coll Surg Engl 2001;83(4): 258–265.

[2] Vergidis P, Greenwood-Quaintance KE, Sanchez-Sotelo J, *et al.* Implant sonication for the diagnosis of prosthetic elbow infection. J Shoulder Elbow Surg 2011;20(8):1275–1281.

[3] Voloshin I, Schippert DW, Kakar S, *et al.* Complications of total elbow replacement: a systematic review. J Shoulder Elbow Surg 2011;20(1):158–168.

[4] Gschwend N, Simmen BR, Matejovsky Z. Late complications in elbow arthroplasty. J Shoulder Elbow Surg 1996;5(2 Pt 1):86–96.

[5] Achermann Y, Sahin F, Schwyzer H, *et al.* Characteristics and outcome of 16 periprosthetic shoulder joint infections. Infection 2012;41(3);613–620.

[6] Laffer RR, Graber P, Ochsner PE, Zimmerli W. Outcome of prosthetic knee-associated infection: evaluation of 40 consecutive episodes at a single centre. Clin Microbiol Infect 2006;12(5):433–439.

[7] Tsaras G, Osmon DR, Mabry T, *et al.* Incidence, secular trends, and outcomes of prosthetic joint infection: a population-based study, olmsted county, Minnesota, 1969–2007. Infect Control Hosp Epidemiol 2012;33(12):1207–1212.

[8] Sperling JW, Kozak TK, Hanssen AD, Cofield RH. Infection after shoulder arthroplasty. Clin Orthop Relat Res 2001;(382):206–216.

[9] Lidgren L, Knutson K, Stefansdottir A. Infection and arthritis. Infection of prosthetic joints. Best Pract Res Clin Rheumatol 2003;17(2):209–218.

[10] Achermann Y, Vogt M, Spormann C, *et al.* Characteristics and outcome of 27 elbow periprosthetic joint infections: results from a 14-year cohort study of 358 elbow prostheses. Clin Microbiol Infect 2011;17(3):432–438.

[11] Spormann C, Achermann Y, Simmen BR, *et al.* Treatment strategies for periprosthetic infections after primary elbow. J Shoulder Elbow Surg 2012;21(8):992–1000.

[12] Morrey BF, Bryan RS. Infection after total elbow arthroplasty. J Bone Joint Surg Am 1983;65 3:330–338.

[13] Wolfe SW, Figgie MP, Inglis AE, *et al.* Management of infection about total elbow prostheses. J Bone Joint Surg Am 1990;72(2):198–212.

[14] Schmidt K, Hilker A, Miehlke RK. Differences in elbow replacement in rheumatoid arthritis. Orthopade 2007;36(8):714–722.

[15] Yamaguchi K, Adams RA, Morrey BF. Semiconstrained total elbow arthroplasty in the context of treated previous infection. J Shoulder Elbow Surg 1999;8(5):461–465.
[16] Gille J, Ince A, Gonzalez O, *et al.* Single-stage revision of peri-prosthetic infection following total elbow replacement. J Bone Joint Surg Br 2006;88(10):1341–1346.
[17] Berbari EF, Osmon DR, Duffy MC, *et al.* Outcome of prosthetic joint infection in patients with rheumatoid arthritis: the impact of medical and surgical therapy in 200 episodes. Clin Infect Dis 2006;42(2):216–223.
[18] Bozic KJ, Lau E, Kurtz S, *et al.* Patient-related risk factors for periprosthetic joint infection and postoperative mortality following total hip arthroplasty in Medicare patients. J Bone Joint Surg Am 2012;94(9):794–800.
[19] Berbari EF, Osmon DR, Lahr B, *et al.* The Mayo prosthetic joint infection risk score: implication for surgical site infection reporting and risk stratification. Infect Control Hosp Epidemiol 2012;33(8):774–781.
[20] Yamaguchi K, Adams RA, Morrey BF. Infection after total elbow arthroplasty. J Bone Joint Surg Am 1998;80(4):481–491.
[21] Cheung EV, Adams RA, Morrey BF. Reimplantation of a total elbow prosthesis following resection arthroplasty for infection. J Bone Joint Surg Am 2008;90(3):589–594.
[22] Piper KE, Jacobson MJ, Cofield RH, *et al.* Microbiologic diagnosis of prosthetic shoulder infection using implant sonication. J Clin Microbiol 2009:47(6):1878–1884.
[23] Topolski MS, Chin PY, Sperling JW, Cofield RH. Revision shoulder arthroplasty with positive intraoperative cultures: the value of preoperative studies and intraoperative histology. J Shoulder Elbow Surg 2006;15(4):402–406.
[24] Patel A, Calfee RP, Plante M, *et al. Propionibacterium acnes* colonization of the human shoulder. J Shoulder Elbow Surg 2009;18(6):897–902.
[25] Grice EA, Kong HH, Conlan S, *et al.* Topographical and temporal diversity of the human skin microbiome. Science 2009;324(5931):1190–1192.
[26] Osmon DR, Berbari EF, Berendt AR, *et al.* Diagnosis and management of prosthetic joint infection: clinical practice. Clin Infect Dis 2013;56(1):e1–e25.
[27] Parvizi J, Della Valle CJ. AAOS Clinical Practice Guideline: diagnosis and treatment of periprosthetic joint infections of the hip and knee. J Am Acad Orthop Surg 2010;18(12):771–772.
[28] Zimmerli W, Trampuz A, Ochsner PE. Prosthetic-joint infections. N Engl J Med 2004;351(16):1645–1654.
[29] Trampuz A, Hanssen AD, Osmon DR, *et al.* Synovial fluid leukocyte count and differential for the diagnosis of prosthetic knee infection. Am J Med 2004;117(8):556–562.
[30] Schinsky MF, Della Valle CJ, Sporer SM, Paprosky WG. Perioperative testing for joint infection in patients undergoing revision total hip arthroplasty. J Bone Joint Surg Am 2008;90(9): 1869–1875.
[31] Ghanem E, Parvizi J, Burnett RS, *et al.* Cell count and differential of aspirated fluid in the diagnosis of infection at the site of total knee arthroplasty. J Bone Joint Surg Am 2008;90(8): 1637–1643.
[32] Dinneen A, Guyot A, Clements J, Bradley N. Synovial fluid white cell and differential count in the diagnosis or exclusion of prosthetic joint infection. Bone Joint J 2013;95-B(4):554–557.
[33] Cipriano CA, Brown NM, Michael AM, *et al.* Serum and synovial fluid analysis for diagnosing chronic periprosthetic infection in patients with inflammatory arthritis. J Bone Joint Surg Am 2012;94(7) 594–600.
[34] Bedair H, Ting N, Jacovides C, *et al.* The Mark Coventry Award: diagnosis of early postoperative TKA infection using synovial fluid analysis. Clin Orthop Relat Res 2011;469(1): 34–40.
[35] Gemmel F, Van den Wyngaert H, Love C, *et al.* Prosthetic joint infections: radionuclide state-of-the-art imaging. Eur J Nucl Med Mol Imaging 2012;39(5):892–909.
[36] Achermann Y, Vogt M, Leunig M, *et al.* Improved diagnosis of periprosthetic joint infection by multiplex PCR of sonication fluid from removed implants. J Clin Microbiol 2010;48(4): 1208–1214.

[37] Trampuz A, Piper KE, Jacobson MJ, *et al.* Sonication of removed hip and knee prostheses for diagnosis of infection. N Engl J Med 2007;357(7):654–663.
[38] Atkins BL, Athanasou N, Deeks JJ, *et al.* Prospective evaluation of criteria for microbiological diagnosis of prosthetic-joint infection at revision arthroplasty. The OSIRIS Collaborative Study Group. J Clin Microbiol 1998;36(10):2932–2939.
[39] Schafer P, Fink B, Sandow D, *et al.* Prolonged bacterial culture to identify late periprosthetic joint infection: a promising strategy. Clin Infect Dis 2008;47(11):1403–1409.
[40] Font-Vizcarra L, Garcia S, Martinez-Pastor JC, *et al.* Blood culture flasks for culturing synovial fluid in prosthetic joint infections. Clin Orthop Relat Res 2010;468(8):2238–2243.
[41] Aggarwal VK, Higuera C, Deirmengian G, *et al.* Swab cultures are not as effective as tissue cultures for diagnosis of periprosthetic joint infection. Clin Orthop Relat Res 2013;471(10):3196–3203.
[42] Portillo ME, Salvado M, Trampuz A, *et al.* Sonication versus vortexing of implants for diagnosis of prosthetic joint infection. J Clin Microbiol 2013;51(2):591–594.
[43] Monsen T, Lövgren E, Widerström M, Wallinder L. In vitro effect of ultrasound on bacteria and suggested protocol for sonication and diagnosis of prosthetic infections. J Clin Microbiol 2009;47(8):2496–2501.
[44] Rohacek M, Weisser M, Kobza R, *et al.* Bacterial colonization and infection of electrophysiological cardiac devices. Circulation 2010;121(15):1691–1697.
[45] Rieger UM, Pierer G, Luscher NJ, Trampuz A. Sonication of removed breast implants for improved detection of subclinical infection. Aesthetic Plast Surg 2009;33(3):404–408.
[46] Rieger UM, Mesina J, Kalbermatten DF, *et al.* Bacterial biofilms and capsular contracture in patients with breast implants. Br J Surg 2013;100(6):768–774.
[47] Bonkat G, Rieken M, Rentsch CA, *et al.* Improved detection of microbial ureteral stent colonisation by sonication. World J Urol 2011;29(1):133–138.
[48] Sampedro MF, Huddleston PM, Piper KE, *et al.* A biofilm approach to detect bacteria on removed spinal implants. Spine 2010;35(12):1218–1224
[49] Esposito S, Leone S, Bassetti M, *et al.* Italian guidelines for the diagnosis and infectious disease management of osteomyelitis and prosthetic joint infections in adults. Infection 2009;37(6):478–496.
[50] Bernard L, Legout L, Zurcher-Pfund L, *et al.* Six weeks of antibiotic treatment is sufficient following surgery for septic arthroplasty. J Infect 2010;61(2):125–132.
[51] Farhad R, Roger PM, Albert C, *et al.* Six weeks antibiotic therapy for all bone infections: results of a cohort study. Eur J Clin Microbiol Infect Dis 2010;29(2):217–222.
[52] Hsieh PH, Huang KC, Lee PC, Lee MS. Two-stage revision of infected hip arthroplasty using an antibiotic-loaded spacer: retrospective comparison between short-term and prolonged antibiotic therapy. J Antimicrob Chemother 2009;64(2):392–397.
[53] Puhto AP, Puhto T, Syrjala H. Short-course antibiotics for prosthetic joint infections treated with prosthesis retention. Clin Microbiol Infect 2012;18(11):1143–1148.
[54] Zimmerli W, Widmer AF, Blatter M, *et al.* Role of rifampin for treatment of orthopedic implant-related staphylococcal infections: a randomized controlled trial. Foreign-Body Infection (FBI) Study Group. JAMA 1998;279(19):1537–1541.
[55] Furustrand TU, Corvec S, Betrisey B, *et al.* Role of rifampin against *Propionibacterium acnes* biofilm in vitro and in an experimental foreign-body infection model. Antimicrob Agents Chemother 2012;56(4):1885–1891.
[56] Jakab E, Zbinden R, Gubler J, *et al.* Severe infections caused by *Propionibacterium acnes*: an underestimated pathogen in late postoperative infections. Yale J Biol Med 1996;69(6):477–482.
[57] Zeller V, Ghorbani A, Strady C, *et al.* *Propionibacterium acnes*: an agent of prosthetic joint infection and colonization. J Infect 2007;55 2:119–124.
[58] Soderquist B, Holmberg A, Unemo M. *Propionibacterium acnes* as an etiological agent of arthroplastic and osteosynthetic infections—two cases with specific clinical presentation including formation of draining fistulae. Anaerobe 2010;16(3):304–306.

[59] Lutz MF, Berthelot P, Fresard A, *et al.* Arthroplastic and osteosynthetic infections due to *Propionibacterium acnes*: a retrospective study of 52 cases, 1995–2002. Eur J Clin Microbiol Infect Dis 2005;24(11):739–744.
[60] Levy PY, Fenollar F, Stein A, *et al. Propionibacterium acnes* postoperative shoulder arthritis: an emerging clinical. Clin Infect Dis 2008;46(12):1884–1886.
[61] Zimmerli W, Frei R, Widmer AF, Rajacic Z. Microbiological tests to predict treatment outcome in experimental device-related infections due to *Staphylococcus aureus*. J Antimicrob Chemother 1994;33(5):959–967.
[62] Schwank S, Rajacic Z, Zimmerli W, Blaser J. Impact of bacterial biofilm formation on in vitro and in vivo activities of antibiotics. Antimicrob Agents Chemother 1998;42(4):895–898.
[63] Widmer AF, Gaechter A, Ochsner PE, Zimmerli W. Antimicrobial treatment of orthopedic implant-related infections with rifampin combinations. Clin Infect Dis 1992;14(6):1251–1253.
[64] Giulieri SG, Graber P, Ochsner PE, Zimmerli W. Management of infection associated with total hip arthroplasty according to a treatment algorithm. Infection 2004;32(4):222–228.
[65] Tsukayama DT, Estrada R, Gustilo RB. Infection after total hip arthroplasty. A study of the treatment of one hundred and six infections. J Bone Joint Surg Am 1996;78(4):512–523.
[66] Betsch BY, Eggli S, Siebenrock KA, *et al.* Treatment of joint prosthesis infection in accordance with current recommendations improves outcome. Clin Infect Dis 2008;46(8):1221–1226.
[67] Wehrli W. Rifampin: mechanisms of action and resistance. Rev Infect Dis 1983;5(Suppl 3): S407–S411.
[68] Villar M, Marimon JM, Garcia-Arenzana JM, *et al.* Epidemiological and molecular aspects of rifampicin-resistant *Staphylococcus aureus* isolated from wounds, blood and respiratory samples. J Antimicrob Chemother 2011;66(5):997–1000.
[69] Achermann Y, Eigenmann K, Ledergerber B, *et al.* Factors associated with rifampin resistance in staphylococcal periprosthetic. Infection 2012;41(2):431–437.
[70] Eigenmann K, Achermann Y, Clauss M, *et al.* Outcome of periprosthetic joint infections caused by rifampin resistant staphylococci: 21st ECCMID. Milan, Italy, 7–10 May, 2011.

Chapter 12
Periprosthetic Joint Infection after Ankle Arthroplasty

Parham Sendi, Bernhard Kessler, and Markus Knupp

Introduction

Arthrodesis has been the standard treatment for end-stage ankle osteoarthritis for many years. Thanks to the improvements in surgical technique and implant design, total ankle arthroplasty (TAA) is now a promising treatment option. In contrast to ankle arthrodesis, TAA allows a range of motion and therefore provides a more physiological gait [1]. It also reduces the risk of overload to the adjacent subtalar joint. As with every arthroplasty, there are risks of complications after implantation. Aseptic loosening and infections are considered to be high-grade complications, because in the majority of these cases, the implant fails [2]. The reported rates of periprosthetic ankle joint infection (PAJI) range from 2 to 8.6% [3–10]. This is clearly higher than the infection rates after total hip and knee arthroplasties [11, 12]. It is important to note that both the patient population and the soft tissue conditions surrounding the joint are different from those observed in patients with hip or knee osteoarthritis. Furthermore, knowledge about PAJI is limited because of the relatively short development history of TAA.

Approximately 1% of the adult population is affected by ankle osteoarthritis [13]. While knee and hip osteoarthritis is mainly primary in origin, this is different in ankle osteoarthritis. Most cases are posttraumatic, as confirmed in an epidemiological study [14]. Among 390 patients (406 ankles) with painful end-stage ankle osteoarthritis, the origin was posttraumatic in 78% (number of ankles = 318), secondary in 13% (n = 52), and primary in only 9% (n = 36) of cases. Secondary arthritis consisted of patients with rheumatoid arthritis (n = 22 ankles), hemochromatosis, hemophilia, clubfoot, and avascular talus necrosis. The high proportion of posttraumatic osteoarthritis is associated with two findings that are relevant for risk factor analysis, namely, younger age and soft tissue mantle surrounding the ankle joint.

Patients with ankle osteoarthritis are 12–15 years younger than those with knee or hip osteoarthritis [15]. On an epidemiological level, older age is associated with the number

Bone and Joint Infections: From Microbiology to Diagnostics and Treatment, First Edition.
Edited by Werner Zimmerli.

and types of comorbidities. Hence, risk factors commonly associated with infection (e.g., diabetes mellitus, corticosteroid use) are less frequent in patients with ankle osteoarthritis [10].

The younger age of patients and the relatively new TAA procedure, in comparison with other types of joint replacement, make the prediction of long-term outcome results difficult. In a prospective study, Esparragoza *et al.* [16] compared outcome results of patients undergoing arthrodesis ($n=16$) with those receiving TAA ($n=14$). They used the American Orthopedic Foot and Ankle Society (AOFAS) scale for investigating clinical function and the short form-36 questionnaire for quality of life. Time points for comparison of results were prior to surgery and after 2 years (mean 25.2 months) of follow-up. In both groups, function and quality of life improved, though in all cases the improvement was statistically greater in patients who underwent TAA. These results are promising, although a follow-up of 2 years is short in relation to the life expectancy of these patients and the (unknown) duration of implant stability.

The posttraumatic etiology implies that a majority of these patients had previous surgeries with subsequent scars, impaired soft tissue conditions, previous infections, skin flaps, or reduced bone stock. Conceivably, there is a predisposition for delayed wound healing, which again can increase the risk of infection.

For implantation, the manufacturers of third-generation ankle prostheses provide reliable instrumentation to perform precise bone cuts and to prepare the resection surfaces to accommodate the prosthesis' components. Most surgeons use an anterior approach to the ankle. This involves splitting of the extensor retinaculum and exposure of the anterior tibial tendon. The prosthesis is then implanted according to the preoperative plan and the skin closed in layers over a drain. Postoperatively, the joint is protected in an orthesis or a cast for the first 6 weeks. This allows partial or full weight bearing [17].

The success of the prosthesis is highly associated with the bony support of the components, ligament balancing, and alignment of the foot. In order to prevent early mechanical failure, the foot must be in a plantigrade position at the end of the implantation and the ligaments well balanced. This requirement explains the numerous additional procedures, such as fusions of the adjacent joints or ligament reconstructions, at the time of replacement surgery.

History of Total Ankle Arthroplasty

Total ankle replacements have been available since the early 1970s [18]. The concept of preserving mobility in the main joint of the hindfoot seemed promising, as ankle fusion may lead to gait alterations and adjacent joint arthritis [19]. The first prosthesis consisted of two components and relied on cement fixation to the bone. Although the implants were constrained, they provided only limited intrinsic stability (ball and socket type of joint), and the implantation involved an enormous amount of bone resection. Both the fixation method and the excessive load on the surrounding soft tissues led to unacceptably high rates of early implant failure. The concept of ankle replacement was therefore questioned in the 1980s, and the number of total ankle joint replacements decreased [20].

The second generation of total ankle replacements was first used in 1984. For the new design, attention was paid to reproducing normal ankle anatomy, joint kinematics, and ligament stability [6]. These new prostheses had either two or three components and were less constrained, allowing sliding and rotational motions. The concept of cement fixation was replaced by biological fixation methods with porous surfaces. This step allowed bony ingrowth of the implant.

For the modern TAA design (third generation) the shapes of the components were further refined to mimic the anatomy of the ankle joint. The coating of the nonarticulating surfaces was improved, facilitating bony ingrowth and reducing the risk for aseptic loosening [21]. Finally, ultra-high-molecular-weight polyethylene was introduced. This reduces wear and makes the implants more durable.

Risk Factors

Several comorbid conditions have been described as risk factors for periprosthetic hip and knee joint-associated infections [22, 23] (see Chapter 9). They are commonly categorized as related to patient characteristics, surgery, and postoperative course. In the following section, we evaluate whether the risk factors for periprosthetic hip and knee joint infections apply to PAJI as well.

Patient-Related Risk Factors

In prosthetic hip- and knee joint–associated infections, patient-related risk factors include tobacco abuse, obesity, rheumatoid arthritis, neoplasia, immunosuppression, and diabetes mellitus. However, specific patient-related risk factors for PAJI are rarely evaluated. In a recent case control study by Kessler *et al.* [10], none of the conditions occurred significantly more often than they did in the control groups. This may be explained by two observations. First, the absolute numbers of cases and controls were relatively small (26 cases, 2 control groups, with 52 patients each). Among all included patients ($n = 120$), only three were smokers, six used corticosteroids, and eight had diabetes mellitus. Second, and as mentioned earlier, the population of patients with TAA is often younger than that of patients with hip and knee arthroplasties. Therefore, the proportion of comorbid conditions is smaller. Nevertheless, it is reasonable to assume that certain comorbid conditions are associated with delayed wound healing and hence favor exogenous infections. Similarly, certain comorbid conditions are associated with a higher susceptibility to *Staphylococcus aureus* bacteremia (e.g., diabetes mellitus, hemodialysis [24]) and hence increase the risk for hematogenous infections [25]. Van der Heide *et al.* [5] reported an incidence of PAJI of 8.6% in patients with rheumatoid arthritis but did not make a comparison with other types of arthritis. Nonetheless, it is likely that patients with rheumatoid arthritis are at increased risk for infection.

Surgery-Related Risk Factors

Patients with end-stage ankle osteoarthritis due to posttraumatic etiology often have had previous operations with subsequent scars and impaired soft tissue mantle. Kessler *et al.* [10] evaluated whether such conditions increase the risk for infection. When comparing PAJI cases with two control groups, they found that prior surgery at the site of infection had an odds ratio (OR) of 4.56 and 4.78. This finding is in agreement with results from previously published studies on orthopedic implants [26], showing that more than one operation prior to the implantation of a prosthesis is a recognized risk factor. Another risk for the development of PAJI was a prolonged time of the surgical intervention. The mean duration of the index surgery was significantly longer in the case group than it was in either control group (119 versus 84 and 93 min, $P \leq 0.02$) [10]. The operation time is often associated with the complexity of the intervention. For example, TAAs without arthrodesis of subtalar joints have a shorter operation time and are less prone to infection (unpublished observation). Taken together, these results show that a prolonged operative time is a risk factor for infection. This has also been shown in patients with knee and hip arthroplasties [27, 28].

Risk Factors Related to the Postoperative Course

In the postoperative course, wound dehiscence persisting ≥14 days and the presence of secondary wound secretion are associated with PAJI [10]. In addition, a wet wound after initially adequate wound healing in the post-operative course should raise the suspicion of PAJI. Delayed wound healing after TAA occurs in up to 15% of cases [2, 20]. It is conceivable that the longer the wound dehiscence persists, the higher the risk for PAJI. However, there is no precise cut-off point for the duration of dehiscence at which the risk for infection significantly increases. Retrospectively, the data from Kessler *et al.* [10] suggested that the risk increases after 8–10 days. However, in that study, the cutoff was set at 14 days prior to data collection. In hip arthroplasty, prolonged wound drainage is a risk factor for infection as well [29]. Prolonged wound drainage was a significant predictor of wound infection (OR = 1.42; 95% confidence interval [CI] 1.18–1.71, $P < 0.001$), and each day of prolonged drainage was associated with a 42% increase in this risk [29].

Pitfalls in the Postoperative Course

If wound healing disturbances are treated with antibiotics without obtaining specimens for microbiological culture, both the indication and the appropriateness of antimicrobial treatment are unclear. Furthermore, in the case of PAJI, such a procedure delays a correct diagnosis and may worsen the outcome (i.e., several operations). In our institution, one or more courses of antimicrobial treatment in the postoperative period were associated with PAJI (unpublished observation). Hence, antibiotic therapy for wound complications without proper diagnostics should be discouraged. On the other hand, physicians should consider PAJI in patients who present with local signs of inflammation and a history of antimicrobial treatment.

Microbiology

Staphylococci, including both *S. aureus* and coagulase-negative staphylococci, are the most common pathogens, causing approximately 70% of episodes. In comparison to infection in other arthroplasties, the infection is more often polymicrobial (~20%). The proportion of isolated microorganisms that were found in two studies (total $n = 46$) are represented in Table 12.1 [30, 31].

Clinical Features

In implant-associated infections, clinical features are often associated with the pathogenesis of the infections. The route of infection is commonly classified as exogenous and hematogenous [32]. Approximately 80% of PAJI cases are acquired exogenously and 20%

Table 12.1. Microbiological findings in 46 episodes with periprosthetic ankle joint infections [30, 31].

Microorganism	Percent
Single microorganism	80.4
Multiple microorganisms	19.6
S. aureus (including MRSA)	34.8
Coagulase-negative staphylococci	32.6
Gram-negative bacteria (*Enterobacter* sp., *Klebsiella* sp., *Pseudomonas* sp.)	15.2
Enterococcus spp.	8.7
Propionibacterium acnes	4.3
Streptococcus milleri	4.3
Others (*Achromobacter* sp., anaerobes)	6.5

hematogenously [10, 31]. Hematogenous infections are often caused by virulent pathogens, while exogenous infections can be caused by both virulent and low-virulent organisms. In hip and knee arthroplasties, exogenous infections with virulent pathogens (e.g., *S. aureus*) have an acute onset of symptoms and present more often with damaged periprosthetic soft tissue. Hematogenous cases more often have systemic signs of infection, such as fever, chills, and sepsis syndrome [32]. In PAJI, however, the clinical presentation differs to a certain degree from that in periprosthetic hip and knee joint infections (see Chapter 9). In a recent study, we analyzed the clinical features of 34 patients with PAJI [31]. Three of the six patients with hematogenous PAJI presented with a chronic infection. These patients had symptoms between 61 and 112 days. Of 28 exogenous episodes, 16 (57.1%) were caused either by a pathogen with virulent properties or by organisms not belonging to the typical skin flora (i.e., *S. aureus*, *Enterococcus* spp., *Streptococcus* spp., *Klebsiella* spp., *Pseudomonas* spp.). Yet, in 4 (25%) of these 16 episodes, there was a chronic infection. Hence, a subset of PAJI patients (7 of 34; 20.6%) presents with chronic symptoms, although one would theoretically expect a shorter duration of symptoms.

In the ankle joint, low-grade infections are very challenging from a clinical perspective. On the one hand, the surrounding soft tissue can be inflamed irrespective of the presence of an infection. Anatomically, the foot and ankle are more strongly affected by systemic neurological and vascular disorders than the proximal joints. On the other hand, clinical findings can be unremarkable, as typically seen in low-grade hip prosthesis infection. Nevertheless, a key feature on clinical examination is the inflamed and/or damaged soft tissue (Figure 12.1). Erythema, edema, and wound discharge are typical clinical signs. In our recent study [31], the soft tissue surrounding the implant was severely damaged at the time of presentation with the infection in 61.8% (21) of the episodes. Further frequent clinical features are pain and immobility of the joint. Many of these patients have chronic symptoms without ever developing a sepsis syndrome. On radiological examination, the implant was considered stable in 22 (64.7%) patients.

Laboratory Investigation

In PAJI, the clinical suspicion of infection is often high. Clinical features are an important element in the diagnosis of PAJI. Therefore, diagnostic procedures mainly focus on the extent of infection and isolation of the pathogen.

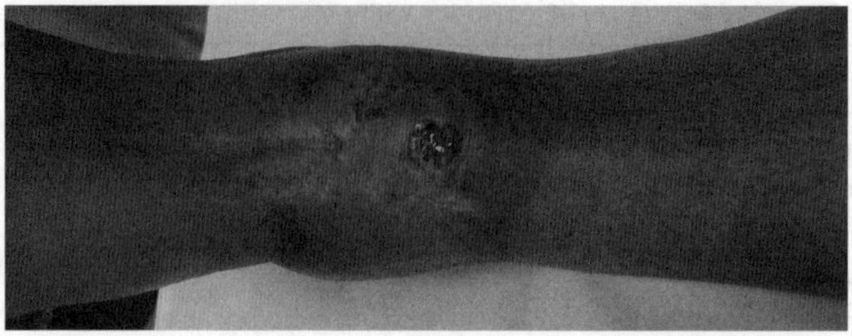

Figure 12.1. A 61-year-old patient with chronic periprosthetic ankle joint infection. Two months after implantation, he presented with a swollen, tender ankle joint and a sinus tract. The implant was removed, five samples were obtained, and a spacer was implanted. *S. aureus* grew in all biopsy samples and in sonicated fluid culture from the implant. (*See insert for color representation of the figure.*)

Blood Tests

To the best of our knowledge, no study has investigated the laboratory parameters of C-reactive protein (CRP), erythrocyte sedimentation rate (ESR), white blood cell (WBC) count, or procalcitonin specifically for PAJI. Thus, interpretation of these values has to be extrapolated from data for hip and knee arthroplasties (reviewed in [33]) (see Chapter 9). In brief, the majority of studies use cutoff values of 30 mm/h for ESR and 10 mg/l for CRP. Results below these values (either alone or in combination) have high sensitivity to exclude periprosthetic joint infection (91–97%). In our experience, the laboratory values in PAJI are less sensitive than those in periprosthetic hip or knee infection. A WBC count in blood is not helpful, because it shows poor sensitivity and low predictive values, in particular in chronic periprosthetic joint infection.

Synovial Fluid Leukocyte Count

While this diagnostic tool has proven helpful in periprosthetic hip and knee joint infections [33, 34] (see Chapters 8 and 9), it has limited value in PAJI. The joint volume of the ankle is limited. This is true in particular after previous surgery because of the scar tissue. Therefore synovial fluid can rarely be aspirated from the ankle joint (*punctio sicca*). Even if synovial fluid can be aspirated, the WBC and neutrophil cutoff values for optimal sensitivity, specificity, and accuracy are unknown in PAJI. Considering that values found in periprosthetic knee joint infections [34] differ from those in hip infections [35], and likely also from shoulder infections [36], it is evident that these values cannot be uncritically extrapolated to ankle joint infections.

Intraoperative Samples

Considering that joint puncture and blood tests are less helpful for the diagnosis of infection in TAA than they are in other types of joint replacement, intraoperative samples are one of most important diagnostic means in PAJI. Swabs must not be used for microbiological sampling. Their sensitivity is low, the cell material collected is small, and, importantly, with the majority of swabs used in routine laboratory, polymerase chain reaction analysis as an adjunct is not possible for technical reasons. In addition, biopsy samples should not be from the tissue in the proximity of a sinus tract. A fistula is colonized with the bacterial flora of the surrounding skin. Therefore, synovial fluid and periprosthetic tissue should be obtained. We recommend obtaining at least three biopsy samples, optimally six samples or more.

The microbiological and histopathological processing and the diagnostic criteria of the biopsies are described elsewhere (Chapter 10).

In our center, removed implants are sonicated. The principles and practice of this method are described elsewhere (Chapter 2).

Imaging Procedures

Overall, plain radiographs have a low sensitivity and specificity for the diagnosis of infection. Radiolucency, osteolysis, and migration are signs of not only infection, but also aseptic loosening, as shown in patients with hip arthroplasties [37]. In PAJI, osteolysis is typically a late sign of infection. In addition, the small size of ankle implants limits the value of plain radiographs for assessing the bone–implant interface. Serial radiographs that show prosthetic migration or progressing osteolysis are suggestive of implant-associated infections. However, this finding is less frequently seen in PAJI than in other reported joints. In our series of PAJI, an implant was considered stable in 22 of 34 (64.7%) patients [31].

Ultrasonography plays only a minimal role in detecting PAJI, since significant joint effusion is rarely found in PAJI.

Computed tomography (CT) allows detection of bone cysts, prosthetic loosening/migration, and bone erosion. Although in other types of joint replacement CT is helpful in detecting a sinus tract or an abscess, this is rarely true in PAJI. Since the ankle joint is surrounded only by a thin soft tissue mantle, the extent of infection is often clinically evident. Nevertheless, CT is often routinely performed in addition to plain radiography.

Magnetic resonance imaging (MRI) is rarely used for the detection of PJI at our center. Artifacts from the prosthetic device hinder reliable assessment of the periarticular soft tissue. Therefore, it has no advantage over CT.

Single-photon emission computed tomography (SPECT/CT) is performed with an integrated hybrid machine that allows simultaneous imaging with a radionuclide and CT. It is mainly performed with technetium-99m-labeled methylene diphosphonate (^{99m}Tc-MDP), labeled leukocytes, or labeled antigranulocyte monoclonal antibodies. By using this method, Graute *et al.* [38] reported sensitivity, specificity, and accuracy of 89, 73, and 77%, respectively, in patients with suspected low-grade periprosthetic joint infections. Nevertheless, considering the costs, this method cannot be recommended for routine clinical practice. Moreover, it has not been specifically investigated for PAJI. In selected cases, however, it can be applied (see Instructive Case 2) [39].

Management

Surgical Interventions

As with every type of periprosthetic joint infection, the principles of surgical treatment include debridement and implant retention (DAIR), one-stage exchange, two-stage exchange, and removal of the implant without replacement. However, there are significant differences in the surgical management of infection in the different types of prosthetic joints (see Chapters 9–12). In PAJI, the assessment of the surrounding soft tissue mantle is especially crucial. The number of surgical interventions, and in particular device exchanges, is limited. Because of the thin soft tissue conditions and the limited bone stock, fewer exchanges are possible than is the case in periprosthetic hip joint infection. From our experience, two or more exchanges of ankle joint prostheses are rarely seen. In contrast, in patients with hip arthroplasty, it is not uncommon to get two or more exchanges during their lifetime. However, no comparison can be made to directly address this question. Moreover, as mentioned earlier, the development and experience on ankle joint arthroplasties is considerably beyond that in hip or knee joint arthroplasty. This is even more evident for PAJI. As a rule, the first surgical approach to PAJI must be chosen wisely because failure can end in amputation.

Debridement and Implant Retention

This is the most common procedure in PAJI [31]. Among 34 episodes of PAJI, 61.8% (21 of 34) were treated with surgery and implant retention. In selected cases, only partial exchange of the implant (e.g., only the tibia or talus component) is possible, although we do not recommend this procedure in DAIR. Previously, Zimmerli *et al.* [40] developed an algorithm for DAIR in hip and knee arthroplasties. Criteria for this concept include an acute infection, only slightly damaged soft tissue, a stable implant, and a causative pathogen susceptible to an agent with activity against biofilm microorganisms. This algorithm has been validated in different centers [41, 42] and applied to other types of prosthetic joints [43]. The analysis of DAIR within our cohort of PAJI revealed, however, that the application to the ankle joint must be adapted. Only 4 of 21 patients with PAJI strictly fulfilled all criteria for DAIR, and these 4 patients were cured. However, of the remaining 17 patients, who did not fulfill one or more criteria, the cure rate was also high. Fourteen (82.4%) had relapse-free survival, and 11 (64.7%) had infection-free survival of ≥2 years [31]. Among the 14 patients, 3 had a second infection with another organism within the 2-year follow-up. Hence, in PAJIs that are treated with DAIR despite not fulfilling the criteria of the algorithm [40], a subset of patients is still cured. This led to the hypothesis that in PAJI, in contrast to hip, knee, and elbow arthroplasties [41–43], the algorithm for DAIR should be adapted accordingly.

Further analysis indicated that the severity grading of soft tissue damage as a criterion should at least be adapted. Eleven of 17 patients not fulfilling the criteria for DAIR, according to Zimmerli *et al.* [40], had soft tissue damage grading as the only criterion that conflicted with the recommendation. Ten of these patients (10 of 11, 90.9%) had a relapse-free survival, and 7 of 11 (63.6%) had an infection-free survival of ≥2 years. Four patients had a second infection with another microorganism within the 2-year follow-up [31]. Taken together, these results again indicate how difficult the clinical judgment is regarding the surrounding tissue of the ankle joint. They also indicate how important the soft tissue condition is for the outcome. Nevertheless, until criteria are better defined for DAIR in PAJI, we suggest following those recommended in the algorithm [40]. In our series [31], all failures due to infection (7 of 21 patients) were in the DAIR group (4 relapses, 3 reinfections). These failures had severe consequences (see "Removal of Implant"). In four of the seven patients, the definitive treatment was either arthrodesis or amputation.

One-Stage Exchange

In our experience, this procedure plays only a minor role in the surgical management of PAJI. Among our series of 34 patients [31], 10 were treated with an exchange of the device, and only one of them with a one-stage exchange (1 of 34 = 2.9%, 1 of 10 = 10%). The rationale behind the low numbers is the potential for a poor outcome. The soft tissue conditions in TAA are poorer and more vulnerable than those found in total hip or knee arthroplasty. For this reason and the limited joint mobility in PAJI, a lower number of surgical interventions—as compared with other arthroplasties—is possible. As a consequence, early failure of the revision arthroplasty may end in an amputation. Early failure can occur in cases of delayed or absent wound healing. Furthermore, in the presence of difficult-to-treat microorganisms (e.g., rifampin-resistant staphylococci, small-colony variants of microorganisms, fungi, enterococci) or a high bacterial load (e.g., sinus tract), a one-stage exchange can fail. Therefore, and if DAIR is not an option, we prefer removing the implant and obtaining multiple biopsy samples. After isolation of the pathogen and its susceptibility patterns, as well as healing of the soft tissue conditions, reimplantation of a new prosthesis can be planned.

Two-Stage Exchange

On the basis of our PAJI series [10, 31], this was the treatment of choice if the implant could not be retained. However, it is important to assess the residual bone capital [44]. Although it is evident that the implant must be removed and debridement should be meticulous, the bone capital should be preserved to the maximum extent possible while still being infection-free. In the same context, it is important to remove metal particles and ossifications. In general, a gentamicin-loaded spacer is implanted. Postoperatively, a CT scan should be considered to assess the remaining bone stock. Reimplantation is targeted within 2–6 weeks. In cases with a difficult-to-treat organism, a change of treatment concept should be evaluated (e.g., arthrodesis without internal foreign body material). In our series [31], 9 of 34 patients were treated with a two-stage exchange, and all of them had infection-free survival of ≥2 years.

Removal of Implant without Replacement

This procedure results in a loss of joint mobility. Therefore, it should be reserved for patients with no other treatment option or who have already lost joint function. However, when evaluating TAA exchange versus conversion to arthrodesis in PAJI treatment, it should be noted that long-term functional and quality-of-life outcomes are missing. Reoperation of failed TAA, irrespective of the cause of failure, can have immobilizing consequences. Kotnis *et al.* [45] reported the management of 16 patients with failed ankle replacement. The definite treatment was arthrodesis in 9 of 16 (56.3%) cases. Two of 16 TAA cases failed because of infection. In one case, the definite treatment was amputation, whereas in the other, it was fusion with an Ilizarov frame [45]. Thus, amputation is the worst case scenario to consider. In our series, there were two infection relapses and two reinfections with another organism. In 3 of the 4 cases, definite treatment was arthrodesis, and in 1 it was amputation. Therefore, removal of the implant always requires evaluation of the impact of functional failure on the patient's quality of life. In a few selected cases, it might be wiser to choose arthrodesis instead of performing multiple revisions, which may finally end in amputation. Taken together, in evaluating treatment concepts for PAJI, permanent implant removal should be the exception.

Antimicrobial Treatment

In PAJI, antimicrobial concepts do not differ from those applied to other arthroplasties [46]. They have been described elsewhere (Chapter 8, Table 9.2, and Chapter 10). In brief, after microbiological sampling, antibiotic therapy is immediately started, initially by the intravenous (IV) route. After the isolation of the pathogen, treatment should be streamlined from empirical to directed therapy (Table 9.2). If the postoperative course is uneventful, the patient can be switched from an IV to an oral regimen within 14 days. In our center, total treatment duration is 3 months.

Instructive Cases

Case 1. Acute hematogenous PAJI

A 52-year-old man presented with a 3-day history of fever, chills, and immobilizing pain in his right ankle. His personal history included tobacco use and a posttraumatic end-stage ankle osteoarthritis. Fourteen months prior to presentation, a TAA was implanted. The postoperative course was uneventful. At the 12-month follow-up investigations, function was satisfactory and there were no signs of infection. On clinical examination, his right ankle was erythematous, swollen, and tender (Figure 12.2). Blood tests revealed the following values: WBC counts 4.7 g/l (normal 4–10 g/l), CRP 248 mg/l (normal <5 mg/l). A plain radiograph showed a stable implant. Blood cultures were drawn. On the basis of (i) the time interval between implantation and onset of symptoms, (ii) the duration of symptoms, and (iii) clinical features, an acute hematogenous PAJI was postulated. The patient was rapidly transferred to the operating theater. After drainage of pus, debridement

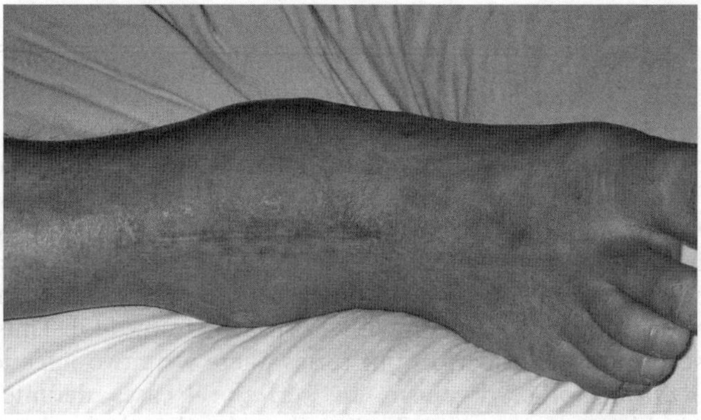

Figure 12.2. A 52-year-old patient with acute hematogenous PAJI. Fourteen months after implantation, the right ankle was suddenly erythematous, swollen, and tender. (*See insert for color representation of the figure.*)

was performed, five biopsy samples were obtained, and the inlay was exchanged. Postoperatively, amoxicillin/clavulanate (4 × 2.2 g i.v. per day) was administered. *S. aureus* grew all biopsy samples and blood cultures. The postoperative course was without any further complications. Ten days after surgery, the patient was discharged and antimicrobial treatment was switched to levofloxacin (2 × 500 mg p.o. per day) plus rifampin (2 × 450 mg p.o. per day). After a 3-month course, the antibiotics were stopped. Two years later, the patient remained infection-free.

Learning Points

- Acute onset of symptoms after an uneventful postoperative period points toward a hematogenous origin of infection.
- When the duration of symptoms is short and the implant stable, DAIR can be successfully performed in PAJI.

Case 2. Chronic exogenous PAJI

A 62-year-old woman was referred to our center because of chronic pain in her left ankle joint. Her personal history included controlled hyperthyroidism, arterial hypertension, and obesity. Twenty months prior to presentation, she received a TAA. The etiology of end-stage ankle osteoarthritis was osteochondrosis dissecans. After surgery, she was never pain-free and was therefore treated with repeated intra-articular steroid injections. On clinical examination, the left ankle was swollen and tender. Warmth or redness was denied. Blood tests were unremarkable: WBC counts 7.4 g/l (normal 4–10 g/l), CRP 6 mg/l (normal <5 mg/l). The plain radiograph showed a loose implant (Figure 12.3a). SPECT/CT was highly suggestive for infection (Figure 12.3b). Pathogenetically, chronic exogenous PAJI was postulated, based on (i) continuous pain as bridging symptoms since implantation, (ii) duration of symptoms, (iii) clinical features, and (iv) imaging

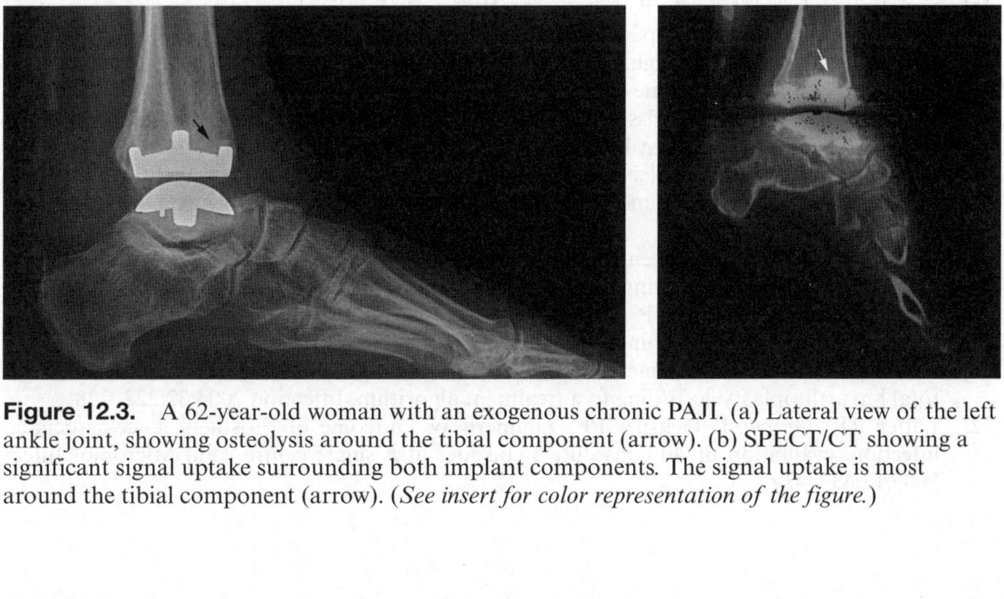

Figure 12.3. A 62-year-old woman with an exogenous chronic PAJI. (a) Lateral view of the left ankle joint, showing osteolysis around the tibial component (arrow). (b) SPECT/CT showing a significant signal uptake surrounding both implant components. The signal uptake is most around the tibial component (arrow). (*See insert for color representation of the figure.*)

results. It was decided that the most appropriate surgical approach would be a two-stage exchange. All foreign body material was removed, the joint was debrided, and six biopsy samples were obtained for microbiological and histopathological examination. A gentamicin-loaded spacer was inserted and the implants sent for sonication. Postoperatively, antimicrobial treatment with vancomycin (2 × 15 mg per kg body weight, i.v. per day) was administered. *Staphylococcus epidermidis* grew in all biopsy samples and in sonicated fluid. The microorganism was resistant to oxacillin, clindamycin, and erythromycin. Histopathological examination was consistent with chronic inflammation. Antimicrobial treatment was continued, and 14 days later, a new arthroplasty implanted. The postoperative course was uneventful, and the patient was discharged 25 days after referral. Oral antimicrobial treatment was switched to levofloxacin (2 × 500 mg p.o. per day) plus rifampin (2 × 450 mg p.o. per day), and continued until the 3-month course of therapy was completed. At follow-up, 24 months after reimplantation, the patient remained infection-free.

Learning Points

- In chronic PAJI, blood tests are not helpful for diagnosis.
- In a patient who is never pain-free after surgery, in particular when imaging shows a loose implant, a PAJI must be postulated until proven otherwise.

References

[1] Hahn ME, Wright ES, Segal AD, *et al.* Comparative gait analysis of ankle arthrodesis and arthroplasty: initial findings of a prospective study. Foot Ankle Int 2012;33:282–289.

[2] Glazebrook MA, Arsenault K, Dunbar M. Evidence-based classification of complications in total ankle arthroplasty. Foot Ankle Int 2009;30:945–949.

[3] Kitaoka HB, Patzer GL. Clinical results of the Mayo total ankle arthroplasty. J Bone Joint Surg Am 1996;78:1658–1664.

[4] Henricson A, Knutson K, Lindahl J, Rydholm U. The AES total ankle replacement: a mid-term analysis of 93 cases. Foot Ankle Surg 2010;16:61–64.

[5] van der Heide HJ, Schutte B, Louwerens JW, *et al.* Total ankle prostheses in rheumatoid arthropathy: outcome in 52 patients followed for 1–9 years. Acta Orthop 2009;80:440–444.

[6] Knecht SI, Estin M, Callaghan JJ, *et al.* The Agility total ankle arthroplasty. Seven to sixteen-year follow-up. J Bone Joint Surg Am 2004;86-A:1161–1171.

[7] Henricson A, Skoog A, Carlsson A. The Swedish Ankle Arthroplasty Register: an analysis of 531 arthroplasties between 1993 and 2005. Acta Orthop 2007;78:569–574.

[8] Hurowitz EJ, Gould JS, Fleisig GS, Fowler R. Outcome analysis of agility total ankle replacement with prior adjunctive procedures: two to six year followup. Foot Ankle Int 2007;28:308–312.

[9] Doets HC, Brand R, Nelissen RG. Total ankle arthroplasty in inflammatory joint disease with use of two mobile-bearing designs. J Bone Joint Surg Am 2006;88:1272–1284.

[10] Kessler B, Sendi P, Graber P, *et al.* Risk factors for periprosthetic ankle joint infection: a case-control study. J Bone Joint Surg Am 2012;94:1871–1876.

[11] Giulieri SG, Graber P, Ochsner PE, Zimmerli W. Management of infection associated with total hip arthroplasty according to a treatment algorithm. Infection 2004;32:222–228.

[12] Laffer RR, Graber P, Ochsner PE, Zimmerli W. Outcome of prosthetic knee-associated infection: evaluation of 40 consecutive episodes at a single centre. Clin Microbiol Infect 2006;12:433–439.

[13] Peyron J. The epidemiology of osteoarthritis. In: Moskowitz R, Howell D, Goldberg V, Mankin H, editors. *Osteoarthritis: Diagnosis and Treatment*. Philadelphia: WB Saunders; 1984, pp. 9–27.

[14] Valderrabano V, Horisberger M, Russell I, *et al.* Etiology of ankle osteoarthritis. Clin Orthop Relat Res 2009;467:1800–1806.

[15] Haddad SL, Coetzee JC, Estok R, *et al.* Intermediate and long-term outcomes of total ankle arthroplasty and ankle arthrodesis. A systematic review of the literature. J Bone Joint Surg Am 2007;89:1899–1905.

[16] Esparragoza L, Vidal C, Vaquero J. Comparative study of the quality of life between arthrodesis and total arthroplasty substitution of the ankle. J Foot Ankle Surg 2011;50:383–387.

[17] Barg A, Zwicky L, Knupp M, *et al.* HINTEGRA total ankle replacement: survivorship analysis in 684 patients. J Bone Joint Surg Am 2012;95:1175–1183.

[18] Mazur JM, Schwartz E, Simon SR. Ankle arthrodesis. Long-term follow-up with gait analysis. J Bone Joint Surg Am 1979;61:964–975.

[19] Fuchs S, Sandmann C, Skwara A, Chylarecki C. Quality of life 20 years after arthrodesis of the ankle. A study of adjacent joints. J Bone Joint Surg Br 2003;85:994–998.

[20] Gougoulias NE, Khanna A, Maffulli N. History and evolution in total ankle arthroplasty. Br Med Bull 2009;89:111–151.

[21] DiDomenico LA, Anania MC. Total ankle replacements: an overview. Clin Podiatr Med Surg 2011;28:727–744.

[22] Del Pozo JL, Patel R. Clinical practice. Infection associated with prosthetic joints. N Engl J Med 2009;361:787–794.

[23] Pulido L, Ghanem E, Joshi A, *et al.* Periprosthetic joint infection: the incidence, timing, and predisposing factors. Clin Orthop Relat Res 2008;466:1710–1715.

[24] Laupland KB, Ross T, Gregson DB. *Staphylococcus aureus* bloodstream infections: risk factors, outcomes, and the influence of methicillin resistance in Calgary, Canada, 2000–2006. J Infect Dis 2008;198:336–343.

[25] Sendi P, Banderet F, Graber P, Zimmerli W. Periprosthetic joint infection following *Staphylococcus aureus* bacteremia. J Infect 2011;63:17–22.

[26] de Boer AS, Mintjes-de Groot AJ, Severijnen AJ, *et al.* Risk assessment for surgical-site infections in orthopedic patients. Infect Control Hosp Epidemiol 1999;20:402–407.

[27] Peersman G, Laskin R, Davis J, *et al.* Prolonged operative time correlates with increased infection rate after total knee arthroplasty. HSS J 2006;2:70–72.

[28] Ridgeway S, Wilson J, Charlet A, *et al.* Infection of the surgical site after arthroplasty of the hip. J Bone Joint Surg Br 2005;87:844–850.

[29] Patel VP, Walsh M, Sehgal B, *et al.* Factors associated with prolonged wound drainage after primary total hip and knee arthroplasty. J Bone Joint Surg Am 2007;89:33–38.

[30] Hintermann B, Schneiderbauer MM, Trampuz A, Widmer AF. Infection rates after primary ankle replacement: a cohort study of 386 consecutive ankel replacements. *Annual Meeting AAOS*. San Diego, February 14–18, 2007, p. #74.

[31] Kessler B, Knupp M, Graber P, *et al.* Treatment and outcome of periprosthetic ankle joint infection: a single cohort center experience of 34 cases. 2014 Bone Joint J, 96B:772–777.

[32] Sendi P, Banderet F, Graber P, Zimmerli W. Clinical comparison between exogenous and haematogenous periprosthetic joint infections caused by *Staphylococcus aureus*. Clin Microbiol Infect 2011;17:1098–1100.

[33] Sendi P, Zimmerli W. Diagnosis of periprosthetic joint infections in clinical practice. Int J Artif Organs 2012;35:913–922.

[34] Trampuz A, Hanssen AD, Osmon DR, *et al.* Synovial fluid leukocyte count and differential for the diagnosis of prosthetic knee infection. Am J Med 2004;117:556–562.

[35] Schinsky MF, Della Valle CJ, Sporer SM, Paprosky WG. Perioperative testing for joint infection in patients undergoing revision total hip arthroplasty. J Bone Joint Surg Am 2008;90: 1869–1875.

[36] Piper KE, Jacobson MJ, Cofield RH, *et al.* Microbiologic diagnosis of prosthetic shoulder infection by use of implant sonication. J Clin Microbiol 2009;47:1878–1884.
[37] Stumpe KD, Notzli HP, Zanetti M, *et al.* FDG PET for differentiation of infection and aseptic loosening in total hip replacements: comparison with conventional radiography and three-phase bone scintigraphy. Radiology 2004;231:333–341.
[38] Graute V, Feist M, Lehner S, *et al.* Detection of low-grade prosthetic joint infections using 99mTc-antigranulocyte SPECT/CT: initial clinical results. Eur J Nucl Med Mol Imaging 2010;37:1751–1759.
[39] Stumpe KD, Strobel K. Osteomyelitis and arthritis. Semin Nucl Med 2009;39:27–35.
[40] Zimmerli W, Trampuz A, Ochsner PE. Prosthetic-joint infections. N Engl J Med 2004; 351:1645–1654.
[41] Betsch BY, Eggli S, Siebenrock KA, *et al.* Treatment of joint prosthesis infection in accordance with current recommendations improves outcome. Clin Infect Dis 2008;46:1221–1226.
[42] Sendi P, Christensson B, Uckay I, *et al.* Group B streptococcus in prosthetic hip and knee joint-associated infections. J Hosp Infect 2011;79:64–69.
[43] Achermann Y, Vogt M, Spormann C, *et al.* Characteristics and outcome of 27 elbow periprosthetic joint infections: results from a 14-year cohort study of 358 elbow prostheses. Clin Microbiol Infect 2011;17:432–438.
[44] Besse JL, Colombier JA, Asencio J, *et al.* Total ankle arthroplasty in France. Orthop Traumatol Surg Res 2010;96:291–303.
[45] Kotnis R, Pasapula C, Anwar F, *et al.* The management of failed ankle replacement. J Bone Joint Surg Br 2006;88:1039–1047.
[46] Sendi P, Zimmerli W. Antimicrobial treatment concepts for orthopaedic device-related infection. Clin Microbiol Infect 2012;18:1176–1184.

Chapter 13
Osteomyelitis: Classification

Werner Zimmerli

There is no comprehensive universally accepted system for classification of osteomyelitis. This is mainly due to the multifaceted presentation of bone infection. Different specialists are confronted with various facets of osteomyelitis and therefore prefer different classification systems. For a meaningful appraisal of antimicrobial treatment results, well-defined categories of diseases are required [1]. For this purpose, classification according to pathogenesis, as first described by Waldvogel *et al.*, is the best suited system [2, 3]. In contrast, the staging system of Cierny–Mader has been developed for orthopedic surgeons dealing with osteomyelitis [4]. This classification system is a guideline for surgical management. It allows sophisticated stratification of long bone osteomyelitis, based on the affected portion of the bone, the local wound situation, as well as the comorbidity of the host.

Classification According to Pathogenesis

Regarding pathogenesis, there are four different groups of etiologies: (i) hematogenous osteomyelitis, (ii) spread from a contiguous source following surgery, (iii) secondary osteomyelitis in patients with vascular insufficiency or concomitant neuropathy, and (iv) osteomyelitis in a heterogenous group of special hosts with an increased risk for bone infection caused by different mechanisms. This latter group has not been included as separate class by Waldvogel *et al.* [2]. It can be labeled as osteomyelitis in the special host, including intravenous drug users (IVDU), individuals with sickle cell anemia, and patients with Gaucher's disease [5–10].

Hematogenous Osteomyelitis

This type of osteomyelitis is most frequent in prepubertal children, where microorganisms predominantly seed in the metaphysis of long bones, particularly in the femur and the tibia [2, 3, 11, 12]. In adults, hematogenous osteomyelitis is much rarer and typically

Bone and Joint Infections: From Microbiology to Diagnostics and Treatment, First Edition.
Edited by Werner Zimmerli.

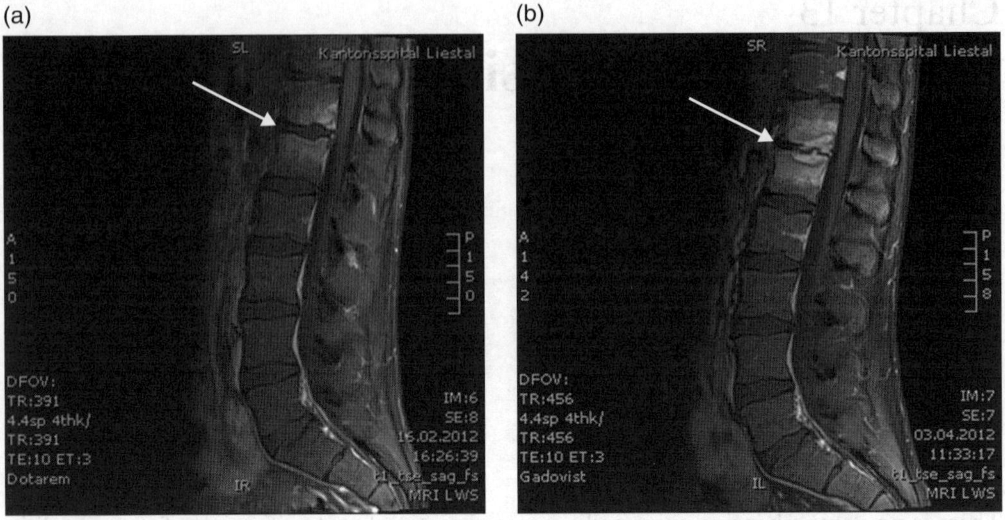

Figure 13.1. A 20-year-old man with hematogenous vertebral osteomyelitis due to *Salmonella enterica* subsp enterica Tennessee. (a) Magnetic resonance imaging (MRI) T1 with gadolinium 5 weeks after episode with diarrhea. No disk enhancement is visible (arrow). (b) MRI T1 with gadolinium 11 weeks after episode with diarrhea and after 6 weeks of adequate antimicrobial therapy. Inflammation and destruction of intervertebral disk is visible (arrow). (*See insert for color representation of the figure.*)

involves the vertebral column. Vertebral osteomyelitis rarely occurs in children and young adults. Its prevalence increases above age 60 [13–15]. It involves two adjacent vertebrae. In children, it starts in the vascularized intravertebral disk, before it spreads to the lower and upper plane of neighboring vertebra [16]. In contrast, in adults, microorganisms seed through bifurcated segmented arteries in adjacent vertebrae before the avascular disk is involved (Figure 13.1). [3, 17] In about half of the patients, a primary focus can be detected. The most frequent primary foci of infection are the urinary tract, skin/soft tissue, intravascular catheters, and endocarditis [18].

Osteomyelitis Secondary to a Contiguous Focus

Spreading from a contiguous source follows either bone trauma or surgical interventions. This type of osteomyelitis has a biphasic age distribution. In younger persons, it is generally a complication of open bone fracture or bone surgery [19–21]. In contrast, in the elderly it is a sequel of decubitus ulcers, chronic soft tissue infections, or dental abscesses [22–25]. In addition, in each age group, it may occur as a complication of surgical site infection after cardiovascular intervention involving the sternum (sternal osteomyelitis), internal bone fixation, or joint arthroplasty [19, 26, 27].

Osteomyelitis Secondary to Vascular Insufficiency or to Peripheral Neuropathy

This type of bone infection generally starts with chronic and progressively deep skin and soft tissue infection of the foot. The most frequent underlying condition is diabetes [25, 28]. In metabolically poorly controlled diabetes, there is a combination of skin, soft tissue and bone ischemia as well as motor, sensory, and autonomic neuropathy [28–31]. This type of osteomyelitis is also called diabetic foot syndrome.

Osteomyelitis in the Special Host

Hematogenous seeding is the principal route of infection in this type of bone infection. Distinct pathogenic mechanisms are responsible for the increased risk of these individuals for bone infection. IVDU with and without HIV infection have an increased risk for vertebral osteomyelitis and osteomyelitis on unusual sites such as the sternum, pubic symphysis, sacroiliac joint, and sternoclavicular joint [32–35]. IVDU have more frequent episodes of bacteremia for several reasons. First, they are more often colonized by *Staphylococcus aureus*, increasing the risk for sepsis. Second, some use nonsterile paraphernalia. If lemon juice is used as solvent of heroin, there is an increased risk for candidemia [5, 6, 36]. Licking needles or the injection site may result in bacteremia by oral flora microorganisms [37]. In addition, microparticles in lengthened drugs may cause bone microinfarction, which predisposes to osteomyelitis.

There are different bone problems in patients with sickle cell disease; one of them is osteomyelitis. The common pathogenic mechanism is microvascular occlusion. Clinically, it may be difficult to differentiate between osteomyelitis and other reasons of bone pain caused by stress fracture, dental complication, vertebral collapse, or bone marrow necrosis [38]. In a study from Nigeria, osteomyelitis caused only one-third of the bone disorders of patients with sickle cell disease presenting with musculoskeletal signs and symptoms [8].

Gaucher's disease is the most common lysosomal storage diseases. It is a sphingolipidosis characterized by abnormal accumulation of glucocerebroside in monocytes and macrophages. Osteomyelitis is a relatively common complication of Gaucher's disease [10, 39]. It has to be differentiated from so-called Gaucher crisis which is presumably due to sudden infarction of a segment of bone. Osteomyelitis may be due to hematogenous seeding in the necrotic bone, or alternatively it may be caused by exogenous inoculation by biopsy of the infected bone [10].

Classification According to the Duration of Infection

There is no clearly defined time limit between acute and chronic osteomyelitis [2]. Since symptoms may start gradually, the length of the clinical course is sometimes not even known. Nevertheless, the duration of infection is crucial, because the management differs according to the chronicity of osteomyelitis. Whereas acute osteomyelitis can be generally treated with antibiotics alone, chronic osteomyelitis always requires combined surgical and antibiotic treatment. Acute hematogenous or contiguous osteomyelitis evolves over a short period of a few days or weeks. In contrast, subacute or chronic osteomyelitis lasts for weeks or months before treatment is started. Typical examples of a subacute course are vertebral osteomyelitis due to tuberculosis or brucellosis on the one hand, or delayed implant-associated infections, mainly caused by low-virulence microorganisms (coagulase-negative staphylococci, *Propionibacterium acnes*) on the other hand [15, 27]. Chronic osteomyelitis is observed in case of insufficient therapy leading to persistence or recurrence. This is mainly observed after sternal, mandibular, or foot osteomyelitis but is less frequent after vertebral osteomyelitis.

Classification According to the Localization

Classification according to the localization comprises (i) long bones, (ii) the vertebral column, (iii) the periarticular bone, and (iv) unusual sites such as clavicle, sternum, os pubis, mandibule, and multifocal sites. Long bones are mainly involved after hematogenous seeding in children or by contiguous spread after trauma or surgery [2, 3, 40]. The risk for vertebral osteomyelitis in adults increases with age [2, 14, 15]. Periarticular osteomyelitis complicates septic arthritis, which has not been adequately treated. It is especially frequent in periprosthetic joint infection [27].

If unusual sites of osteomyelitis are involved, special host conditions (comorbidity, surgical trauma, etc.) or noninfectious bone disorders have to be considered. Osteomyelitis of the clavicle is generally a manifestation of chronic recurrent multifocal osteomyelitis, which is an inflammatory noninfectious disorder of unknown origin. It may also be a complication of neck surgery or subclavian vein catheterization. Occasionally, it follows sternoclavicular arthritis, which is mainly seen in patients with IV drug use. Sternal osteomyelitis generally occurs after thorax surgery; hematogenous infection is rare and mainly occurs in IV drug addicts [32] (see also Chapter 22). Pubic osteomyelitis rarely occurs as acute hematogenous osteomyelitis in children and IVDU [33]. More frequently, it is exogenously acquired after gyneco-urological surgery, especially when combined with irradiation [41–43]. It has also been observed during pregnancy [44]. Osteitis pubis in athletes is generally not an infection [45]. However, because osteitis pubis and osteomyelitis pubis can both occur in athletes, an appropriate diagnostic workup is needed in all cases [46]. Mandibular osteomyelitis has a broad differential diagnosis including odontogenic osteomyelitis, bisphosphonate-related osteonecrosis, and autoimmune osteitis in the context of chronic recurrent multifocal osteomyelitis [47].

In general, multifocal osteomyelitis is a noninfectious inflammatory bone disease of children and adolescents [48]. It mainly involves long bone metaphyses, the pelvis, the spine, the clavicle, and the mandible [49]. It is typically but not always part of the SAPHO syndrome, which includes synovitis, acne, pustulosis, hyperostosis, and osteitis [50]. In case of acute presentation, septic multifocal hematogenous osteomyelitis has to be considered in children [51].

Classification According to the Presence of an Implant

Osteomyelitis involving a foreign device occurs as complication of joint replacement (periprosthetic joint infection), after internal fixation (plate-associated or intramedullary nail-associated osteomyelitis), or after external fixation (pin-track infection) [19, 52]. All types of device-associated osteomyelitis must be surgically treated [19, 27, 53]. Even acute implant-associated infection needs prolonged antimicrobial therapy. Therefore, this type of classification is of practical importance.

Classification According to Anatomy and Comorbidity

The staging system described by Cierny–Mader allows stratification of long bone osteomyelitis in view of an optimal surgical management [4, 54, 55]. It considers system and local factors affecting immune surveillance, metabolism, and local vascularity (Tables 13.1 and 13.2). This type of classification is especially useful for the management of chronic posttraumatic osteomyelitis, for which it provides guidelines for surgical management.

Table 13.1. Staging system according to Cierny and Mader [4, 54, 55].

Anatomical type	Stage 1: Medullary osteomyelitis
	Stage 2: Superficial osteomyelitis
	Stage 3: Localized osteomyelitis
	Stage 4: Diffuse osteomyelitis
Physiological class	A Host: Normal host
	B Host:
	Systemically compromised
	Locally compromised
	Systemically and locally compromised
	C Host: Treatment worse than the disease

Table 13.2. Staging system according to Cierny and Mader: systemic and local factors in class B hosts that affect immune surveillance, metabolism, and local vascularity.

Systemic comorbidity	Local detriment
Malnutrition	Chronic lymphedema
Renal or hepatic failure	Venous stasis
Diabetes mellitus	Major vessel compromise including arteritis
Chronic hypoxia	Extensive scarring
Immune disease	Radiation fibrosis
Malignancy	Small vessel disease
Extremes of age	Neuropathy
Immunosuppression/deficiency	Tobacco abuse

References

[1] Lazzarini L, Lipsky BA, Mader JT. Antibiotic treatment of osteomyelitis: what have we learned from 30 years of clinical trials? Int J Infect Dis 2005;9(3):127–138.

[2] Waldvogel FA, Medoff G, Swartz MN. Osteomyelitis: a review of clinical features, therapeutic considerations and unusual aspects. N Engl J Med 1970;282(4):198–206.

[3] Lew DP, Waldvogel FA. Osteomyelitis. Lancet 2004;364(9431):369–379.

[4] Cierny G 3rd, Mader JT, Penninck JJ. A clinical staging system for adult osteomyelitis. Clin Orthop Relat Res 2003;414:7–23.

[5] Sapico FL, Montgomerie JZ. Vertebral osteomyelitis in intravenous drug abusers: report of three cases and review of the literature. Rev Infect Dis 1980;2(2):196–206.

[6] Scheidegger C, Zimmerli W. Infectious complications in drug addicts: seven-year review of 269 hospitalized narcotics abusers in Switzerland. Rev Infect Dis 1989;11(3):486–493.

[7] Chandrasekar PH, Narula AP. Bone and joint infections in intravenous drug abusers. Rev Infect Dis 1986;8(6):904–911.

[8] Balogun RA, Obalum DC, Giwa SO, *et al*. Spectrum of musculo-skeletal disorders in sickle cell disease in Lagos, Nigeria. J Orthop Surg Res 2010;5:2.

[9] Martinoli C, Bacigalupo L, Forni GL, *et al*. Musculoskeletal manifestations of chronic anemias. Semin Musculoskelet Radiol 2011;15(3):269–280.

[10] Bell RS, Mankin HJ, Doppelt SH. Osteomyelitis in Gaucher disease. J Bone Joint Surg Am 1986;68(9):1380–1388.

[11] Dartnell J, Ramachandran M, Katchburian M. Haematogenous acute and subacute paediatric osteomyelitis: a systematic review of the literature. J Bone Joint Surg Br 2012;94(5): 584–595.

[12] Paakkonen M, Peltola H. Bone and joint infections. Pediatr Clin North Am 2013;60(2): 425–436.

[13] Waldvogel FA, Vasey H. Osteomyelitis: the past decade. N Engl J Med 1980;303(7): 360–370.

[14] Grammatico L, Baron S, Rusch E, *et al.* Epidemiology of vertebral osteomyelitis (VO) in France: analysis of hospital-discharge data 2002–2003. Epidemiol Infect 2008;136:653–660.

[15] Zimmerli W. Clinical practice. Vertebral osteomyelitis. N Engl J Med 2010;362:1022–1029.

[16] Waldvogel FA, Medoff G, Swartz MN. Osteomyelitis: a review of clinical features, therapeutic considerations and unusual aspects. 3. Osteomyelitis associated with vascular insufficiency. N Engl J Med 1970;282(6):316–322.

[17] Urban JP, Smith S, Fairbank JC. Nutrition of the intervertebral disc. Spine (Phila Pa 1976) 2004;29(23):2700–2709.

[18] Mylona E, Samarkos M, Kakalou E, *et al.* Pyogenic vertebral osteomyelitis: a systematic review of clinical characteristics. Semin Arthritis Rheum 2009;39:10–17.

[19] Trampuz A, Zimmerli W. Diagnosis and treatment of infections associated with fracture-fixation devices. Injury 2006;37(Suppl 2):S59–S66.

[20] Arens S, Hansis M, Schlegel U, *et al.* Infection after open reduction and internal fixation with dynamic compression plates—clinical and experimental data. Injury 1996;27(Suppl 3): SC27–SC33.

[21] Gustilo RB, Gruninger RP, Davis T. Classification of type III (severe) open fractures relative to treatment and results. Orthopedics 1987;10(12):1781–1788.

[22] Larson DL, Gilstrap J, Simonelic K, Carrera GF. Is there a simple, definitive, and cost-effective way to diagnose osteomyelitis in the pressure ulcer patient? Plast Reconstr Surg 2011; 127(2):670–676.

[23] Marriott R, Rubayi S. Successful truncated osteomyelitis treatment for chronic osteomyelitis secondary to pressure ulcers in spinal cord injury patients. Ann Plast Surg 2008;61(4): 425–429.

[24] Rokadiya S, Malden NJ. An implant periapical lesion leading to acute osteomyelitis with isolation of *Staphylococcus aureus*. Br Dent J 2008;205(9):489–491.

[25] Widatalla AH, Mahadi SE, Shawer MA, *et al.* Diabetic foot infections with osteomyelitis: efficacy of combined surgical and medical treatment. Diabet Foot Ankle 2012;3. DOI:10.3402/dfa.v3i0.18809.

[26] Tocco MP, Ballardini M, Masala M, Perozzi A. Post-sternotomy chronic osteomyelitis: is sternal resection always necessary? Eur J Cardiothorac Surg 2013;43(4):715–721.

[27] Zimmerli W, Trampuz A, Ochsner PE. Prosthetic-joint infections. N Engl J Med 2004; 351(16):1645–1654.

[28] Lipsky BA, Berendt AR, Cornia PB, *et al.*, Executive summary: 2012 Infectious Diseases Society of America clinical practice guideline for the diagnosis and treatment of diabetic foot infections. Clin Infect Dis 2012;54(12):1679–1684.

[29] Reiber GE, Pecoraro RE, Koepsell TD. Risk factors for amputation in patients with diabetes mellitus. A case-control study. Ann Intern Med 1992;117(2):97–105.

[30] Lavery LA, Barnes SA, Keith MS, *et al.* Prediction of healing for postoperative diabetic foot wounds based on early wound area progression. Diabetes Care 2008;31:26–29.

[31] Butalia S, Palda VA, Sargeant RJ, *et al.* Does this patient with diabetes have osteomyelitis of the lower extremity? JAMA 2008;299(7):806–813.

[32] Chen YL, Tsai SH, Hsu KC, *et al.* Primary sternal osteomyelitis due to Peptostreptococcus anaerobius. Infection 2013;40(2):195–197.

[33] Ross JJ, Hu LT. Septic arthritis of the pubic symphysis: review of 100 cases. Medicine (Baltimore) 2003;82(5):340–345.

[34] Ferraro K, Cohen MA. Acute septic sacroiliitis in an injection drug user. Am J Emerg Med 2004;22(1):60–61.
[35] Ross JJ, Shamsuddin H. Sternoclavicular septic arthritis: review of 180 cases. Medicine (Baltimore) 2004;83(3):139–148.
[36] Miller DJ, Mejicano GC. Vertebral osteomyelitis due to Candida species: case report and literature review. Clin Infect Dis 2001;33(4):523–530.
[37] Deutscher M, Perlman D. Why some injection drug users lick their needles: a preliminary survey. Int J Drug Policy 2008;19:342–345.
[38] Almeida A, Roberts I. Bone involvement in sickle cell disease. Br J Haematol 2005;129(4):482–490.
[39] Margalit M, Ash N, Zimran A, Halkin H. Enzyme replacement therapy in the management of longstanding skeletal and soft tissue salmonella infection in a patient with Gaucher's disease. Postgrad Med J 2002;78(923):564–565.
[40] Lazzarini L, Mader JT, Calhoun JH. Osteomyelitis in long bones. J Bone Joint Surg Am 2004;86-A(10):2305–2318.
[41] Lentz SS. Osteitis pubis: a review. Obstet Gynecol Surv 1995;50(4):310–315.
[42] Almeras C, Madi F, Desplaces N, Mamoudy P. Postoperative osteitis of the pubis: diagnosis, treatment and results. Prog Urol 2002;12(2):253–259.
[43] Garcia-Porrua C, Picallo JA, Gonzalez-Gay MA. Osteitis pubis after Marshall-Marchetti-Krantz urethropexy. Joint Bone Spine 2003;70(1):61–63.
[44] Gamble K, Dardarian TS, Finstein J, *et al.* Osteomyelitis of the pubic symphysis in pregnancy. Obstet Gynecol 2006;107(2 Pt 2):477–481.
[45] Beatty T. Osteitis pubis in athletes. Curr Sports Med Rep 2012;11(2):96–98.
[46] Pauli S, Willemsen P, Declerck K, *et al.* Osteomyelitis pubis versus osteitis pubis: a case presentation and review of the literature. Br J Sports Med 2002;36(1):71–73.
[47] Tolstunov L. Influence of immediate post-extraction socket irrigation on development of alveolar osteitis after mandibular third molar removal: a prospective split-mouth study, preliminary report. Br Dent J 2012;213(12):597–601.
[48] Roderick MR, Ramanan AV. Chronic recurrent multifocal osteomyelitis. Adv Exp Med Biol 2013;764:99–107.
[49] Falip C, Alison M, Boutry N, *et al.* Chronic recurrent multifocal osteomyelitis (CRMO): a longitudinal case series review. Pediatr Radiol 2013;43(3):355–375.
[50] Nguyen MT, Borchers A, Selmi C, *et al.* The SAPHO syndrome. Semin Arthritis Rheum 2012;42(3):254–265.
[51] Sreenivas T, Nataraj AR, Menon J, Patro DK. Acute multifocal haematogenous osteomyelitis in children. J Child Orthop 2012;5(3):231–235.
[52] Saithna A. The influence of hydroxyapatite coating of external fixator pins on pin loosening and pin track infection: a systematic review. Injury 2010;41(2):128–132.
[53] Hamel A, Caillon J, Jacqueline C, *et al.* Internal device decreases antibiotic's efficacy on experimental osteomyelitis. J Child Orthop 2008;2(3):239–243.
[54] Levine SE, Esterhai JL Jr, Heppenstall RB, *et al.* Diagnoses and staging. Osteomyelitis and prosthetic joint infections. Clin Orthop Relat Res 1993;295:77–86.
[55] Mader JT, Shirtliff M, Calhoun JH. Staging and staging application in osteomyelitis. Clin Infect Dis 1997;25(6):1303–1309.

Chapter 14
Osteomyelitis in Children

Priya Sukhtankar, Saul N. Faust, and Nicholas M.P. Clarke

Introduction

Osteomyelitis (OM) and septic arthritis are relatively uncommon infections in infants and children. Clinically, these patients present with a range of symptoms depending on age, from nonspecific illness and pseudoparalysis in infancy to more classical localizing pain and fever in older children and teenagers. OM and septic arthritis may present with pyrexia of unknown origin (PUO) and may be difficult to diagnose.

This chapter will discuss the pathophysiology of infection, pathogens implicated, diagnosis, and medical and surgical treatment options. Diagnosis and treatment depend on the age of the child, comorbidities, and causative organism.

Epidemiology

The infectious epidemiology of osteoarticular infection is changing in an era of resistant organisms, as well as congenital and acquired immunodeficiency. Improvements in microbiological diagnosis now mean that more organisms are identified. A study looking at children in the United States between 2002 and 2004 showed a 2.8-fold increase in the incidence of osteoarticular infections compared with data from 1982 to 1984 [1].

An epidemiological study carried out in Norway in 2008 showed an incidence of 13 per 100,000 children per year [2]. A study looking at all cases of osteoarticular infections presenting to a teaching hospital in Cambodia between 2007 and 2011 showed a similar incidence of 13.8 per 100,000 hospital attendances [3]. In both of these studies, conducted before the routine availability of molecular diagnostic tests, methicillin-sensitive *Staphylococcus aureus* (MSSA) was the most commonly isolated organism. Tuberculous osteoarticular infection is important to consider in at-risk groups, and primarily affects older children, usually from countries with endemic tuberculosis (TB). Tuberculous OM

Bone and Joint Infections: From Microbiology to Diagnostics and Treatment, First Edition.
Edited by Werner Zimmerli.

often affects the spine and is an important differential diagnosis in children presenting with back pain [4].

The incidence of OM and septic arthritis in children in the United Kingdom is currently being ascertained by a national study. Unpublished data from the United Kingdom in 2009 (Southampton, Newcastle, and South London) show an admission rate of between 0.048 and 0.07 per 1000 child-years in all children from birth to 18 years [5]. Children under 5 years are most commonly affected, and from 1 to 4 years of age boys are twice as likely to be affected as girls. There is no gender predilection in the first year of life [6].

Children with primary immunodeficiency, sickle cell disease, and human immunodeficiency virus (HIV), as well as preterm neonates, are at increased risk of osteoarticular infection, caused by a different range of organisms compared to children without comorbidities.

The most common causative organisms in different age groups will be discussed (see microbiology). Certain organisms such as *Haemophilus influenzae* type b have declined. Others such as *S. aureus* containing the Panton–Valentine Leucocidin (PVL) virulence gene and methicillin-resistant *S. aureus* (MRSA) have increased reports in the literature, although worldwide there is considerable geographical variation.

Pathophysiology

The majority of childhood OM is acute and hematogenous in origin. A suspected port of entry or primary focus may be identified in approximately 44% of cases, most frequently from ear, nose, and throat infections [7]. The child develops a bacteremia and the organism enters the bone through the nutrient artery. The metaphyseal region is most commonly affected in children as this has tight capillary loops, allowing deposition of bacteria and formation of an infective nidus, where virulence factors facilitate adhesion and cause bone lysis. Localized necrosis occurs and inflammatory exudate causes destruction of the bony cortex, allowing infection to spread. The Haversian and Volkmann canals become compressed and disrupted, and, if untreated, the infection and necrosis spread to the periosteum. This causes capillary compression and ischemia, further necrosis, and eventually a sequestrum. Animal models inoculating rabbits with MRSA suggest that a critical level of bacteremia is necessary for OM to become established [8], although in humans host immunity is likely to play a significant role in individual susceptibility.

A subset of cases will have subacute OM, without preceding illness or bacteremia. These will usually have direct inoculation of the bone or joint by a puncture wound, injury, or surgery. Infection is more insidious, and a Brodie abscess may develop within the bone, without breaching the periosteum.

Septic arthritis affects synovial joints and involves the synovial membrane and joint space. In children, the joint capsule extends to the metaphysis and therefore septic arthritis may develop as a contiguous infection from adjacent OM, or from direct inoculation of the joint. Conversely, OM may develop as bacteria and breach the periosteum from an adjacent septic arthritis. Contiguous infection is more likely in young infants, because there are transphyseal vessels crossing the growth plate, allowing seeding of bacteria to the joint. In older children, the growth plate may conversely act as a barrier to spread of infection in the joint. This barrier disappears once the growth plate cartilage ossifies.

Diskitis is specifically infection/inflammation of an intervertebral disk or vertebral end-plate and is rare in children. It is difficult to diagnose, particularly in nonverbal toddlers [9]. Unlike in adults, the intervertebral disks of children under the age of 8 years are vascularized, as these are growth regions. This allows hematogenous spread of bacteria to the intervertebral disk, which may occasionally be identified on biopsy (which is frequently done to exclude malignancy), but is usually not identified. There is usually a clinical response to antimicrobial therapy [10].

Clinical Presentation, Diagnosis, and Microbiology

Neonates

Babies are likely to present with nonspecific symptoms or signs such as poor feeding, lethargy, and fever or pseudoparalysis of a limb. Parents sometimes report their child as being in pain on changing the diaper or dressing. Differential diagnoses when the upper limb is involved are Erb's palsy or clavicle fracture related to birth trauma. Infants may also be irritable and have swelling or erythema. Risk factors as for any neonatal sepsis include prolonged rupture of membranes, maternal pyrexia, and known maternal carriage of Group-B *Streptococcus* [11]. It is the author's experience, supported by several case reports [12–14], that neonatal osteoarticular infections often present late due to lack of awareness or suspicion by the doctor or healthcare professional who the parents first consult.

The neonatal group is divided into those admitted for prolonged management in the neonatal intensive care (NICU) and those presenting between discharge from hospital up to 3 months of age. Both groups are at risk of overwhelming sepsis. The commonest organisms causing osteoarticular infection in both these groups are Group-B *Streptococcus* and *S. aureus*.

Babies in NICU are at increased risk of bacteremia, and therefore acute hematogenous osteoarticular infection, owing to frequent invasive procedures, ventilation, central venous access, parenteral nutrition, and immature immunity [15]. This increases the risk for coagulase-negative staphylococci, group-B streptococci, *S. aureus*, and *Candida* spp. Premature babies in particular are at risk of bacterial translocation into the blood from the bowel, especially if there is necrotizing enterocolitis. They are therefore more likely to be infected with enterococci or Gram-negative organisms such as *Escherichia coli* and *Klebsiella pneumoniae*.

Usually the long bones and large joints, especially the hip, are affected. Osteoarticular infections are more likely to be multifocal in this age group and, if suspected, all potential sites should be assessed. Newborns with cephalohematoma are at risk of this becoming infected with secondary infection of the skull and conferring a risk of meningitis [16]. Calcaneal OM following heel prick test for Guthrie screening has also been reported [17].

All babies aged less than 3 months presenting with pyrexia greater than 38°C or unwell with risk factors for sepsis should have a "septic screen" consisting of blood culture, lumbar puncture, urine culture, and consideration of chest x-ray. Usually infants are then treated empirically with broad-spectrum intravenous (IV) antibiotics pending culture results, and/or resolution of symptoms.

A proportion of newborns present without overt sepsis, but with localized swelling and/or pseudoparalysis. This is most common in term babies who have been discharged from hospital [18]. There may be an overlying soft tissue abscess or erythema.

White cell count (WCC), erythrocyte sedimentation rate (ESR), and C-reactive protein (CRP) are poor diagnostic indicators in the neonatal age group, and normal values may not exclude serious infection.

Plain x-ray is usually the first investigation as this is rapid, accessible, and excludes fractures and dislocations. X-rays are relatively specific but lack sensitivity. Up to 90% are normal in confirmed osteoarticular infection [7]. For bone and joint infections in all age groups, magnetic resonance imaging (MRI) scan is the imaging modality of choice. In neonates, it may be possible to do this without general anesthesia ("feed and wrap").

An ultrasound scan is particularly useful in premature neonates who cannot be moved from the incubator, and may reveal bone abscess or joint effusion. Effusions, abscesses, and joint involvement are easily seen, and ultrasound guidance is also useful for aspiration of these to confirm diagnosis and identify the causative organism [19, 20].

Children: 3 Months to 5 Years

After 3 months of age, it is likely that infection has been acquired from the infants' environment rather than perinatally. The organisms most commonly implicated are *S. aureus*, *Streptococcus pyogenes* (Group-A streptococci), and *Streptococcus pneumoniae. Kingella kingae* has been recognized as a significant pathogen in this age group in the last decade, as a result of improved diagnostic tools [21]. *H. influenzae* type b is also seen at this age, but has considerably declined since the introduction of routine HiB conjugate immunization. "Nontypeable" *Haemophilus* is still seen, particularly in skull and facial osteoarticular infection [22].

Children present with any combination of pseudoparalysis, pain, lethargy, refusal to weight-bear, and occasionally PUO. There may be deformity, erythema, pain, swelling, or reduced range of movement. Differential diagnosis of the limping child or a child with a painful limb includes transient synovitis or "irritable hip," trauma (where there is a need to also consider nonaccidental injury), leukemia, and autoimmune arthritis, such as juvenile idiopathic arthritis. The latter remains the most common cause of monoarthritis in children. However, if a child presents with monoarthritis, the most important diagnosis to exclude is septic arthritis, even in children with a known diagnosis of juvenile idiopathic arthritis [23].

The CRP and ESR are important diagnostic tools in all children over 3 months. A Finnish study showed that CRP was greater than 20 mg/l and ESR was greater than 50 mm/h in 95% of culture-positive patients between 3 months and 16 years. If either CRP or ESR was used, a sensitivity of 98% was seen in children presenting with symptoms consistent with osteoarticular infection [24]. Therefore, if either CRP or ESR is elevated, osteoarticular infection is a likely diagnosis. However, a normal CRP does not exclude culture-negative osteoarticular infection, as will be discussed. Serum procalcitonin is sensitive and has been reported to be more specific, but not more sensitive to bone and joint infections than CRP or ESR, particularly if a lower cutoff value of 0.4 ng/ml is used [25, 26]. However, this test is not routinely available in many centers.

Ambulant children may also present with subacute OM, with more insidious symptoms of pain and reduced movement, without fever. These children may have normal CRP and ESR. Differential diagnosis is Ewing's sarcoma or other bony tumor, and imaging and biopsy may be indicated [27].

Blood cultures of adequate volume should always be taken if the child is febrile and if osteoarticular infection is suspected. However, while published studies of selected case series show positive blood cultures in 30–70% of all children with osteoarticular infection, unpublished data from the United Kingdom suggests that the actual number of positive blood cultures is less than 10%. Negative blood culture is therefore not

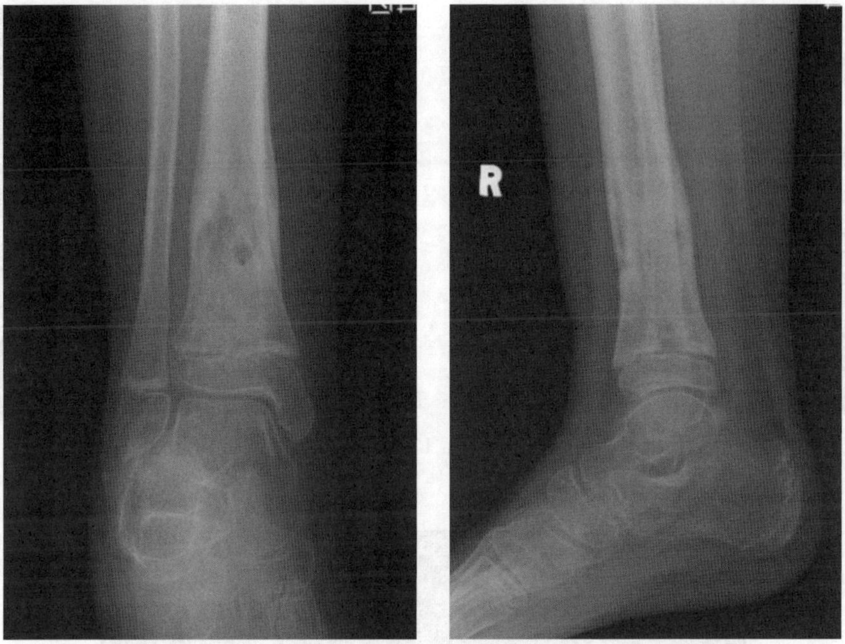

Figure 14.1. Radiographs showing periosteal elevation and lytic change in the tibia of an 11-year-old boy with osteomyelitis.

reassuring [28]. A national publically funded service evaluation study is currently under way in the United Kingdom that will establish the true "real-life" infecting organism rate from blood and tissue cultures (www.dinosaur-study.org.uk/).

Plain x-rays are commonly done to exclude fracture and conditions such as Perthes' disease and slipped femoral epiphysis when the hip is involved. If infection has been present for 10 days or more, periosteal elevation or lytic changes may be seen on plain x-ray (Figure 14.1).

Computed tomography (CT) may be useful in young children, although it is discouraged because of the ionizing radiation dose. However, it does show more subtle changes and takes less time than MRI so that the child does not need to be sedated or anesthetized. Additionally, CT may be useful for guided biopsy, particularly where the spine is involved (Figure 14.2).

MRI is the preferred imaging test for diagnosing osteoarticular infection. It is sensitive and specific, and is particularly useful in assessing bones, joints, and soft tissue simultaneously without ionizing radiation [29]. Acute OM typically shows on MRI as hypointensity of the bone marrow on T1-weighted images without gadolinium enhancement. If gadolinium is used, abnormal enhancement is seen on fat-saturated T1-weighted sequencing. On T2-weighted and "Short tau-inversion recovery" (STIR) sequences, there is hyperintensity of the marrow. STIR sequences are fluid-sensitive and may be useful in localizing the infection to a particular site in challenging cases (Figure 14.3). MRI is also useful in differentiating between subacute OM or Brodie abscess and bone tumors [30].

In young children, the epiphysis and growth plate are more likely to be involved, and there may be disruption of normal bone growth causing deformity. MRI with

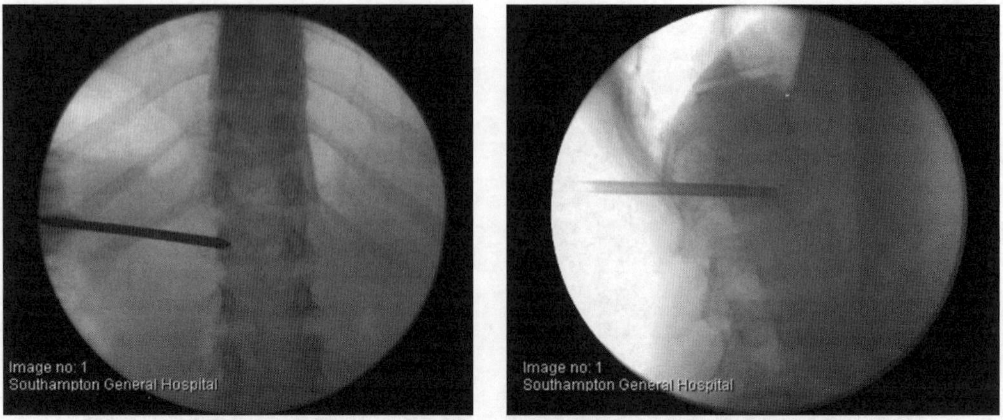

Figure 14.2. Guided biopsy of the spine of a 17-year-old boy with final diagnosis of tuberculosis osteomyelitis.

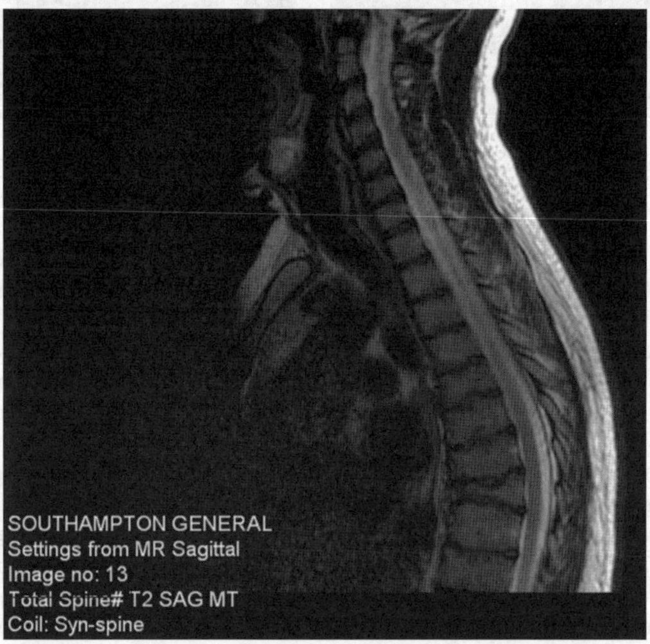

Figure 14.3. T2-weighted magnetic resonance image showing the spine of a 17-year-old Nepalese boy with tuberculosis osteomyelitis and collapse of T9 vertebra.

gadolinium shows hypoenhancing foci in the growth cartilage, if there is involvement. This is not seen on unenhanced images; therefore, gadolinium enhancement is recommended in this group [31].

Figure 14.4 shows a technetium bone scan of a 17-year-old Nepalese boy with tuberculous osteoarthritis. This technique is now used less frequently in UK practice, because of the associated radiation burden. However, it is useful in multifocal disease.

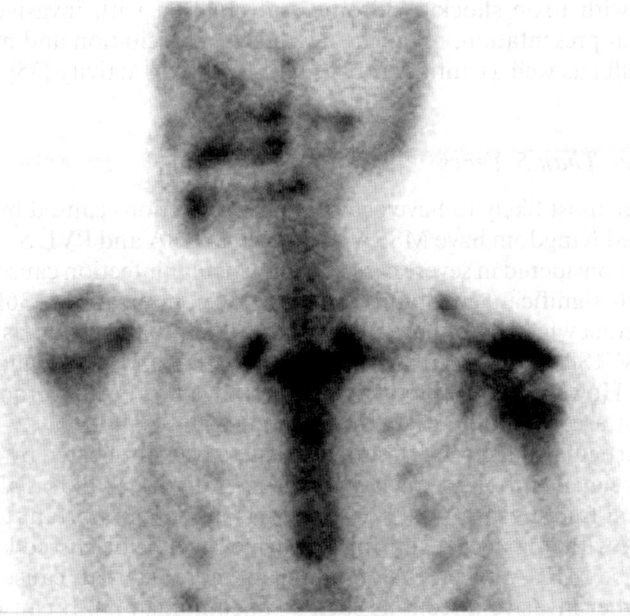

Figure 14.4. Technetium bone scan showing high signal intensity in the spine and left shoulder of a Nepalese boy with tuberculosis osteomyelitis of the spine, and tuberculosis arthritis and osteomyelitis of the shoulder.

Children under 5 years are more likely to have osteoarthritis caused by *K. kingae* and more recent studies using polymerase chain reaction (PCR) show that this Gram-negative organism is more common in this age group than *S. aureus* [7, 21]. It is fastidious and therefore does not grow using conventional culture techniques on solid culture media, although it may be grown more successfully from broth. Invasive infection probably originates in young children following *K. kingae* carriage in the pharynx [32]. PCR is now more widely used to detect *K. kingae* (and other organisms) and has consistently shown this to be the causative organism in previously culture-negative osteoarticular infection.

Children with *K. kingae* osteoarticular infection are more likely to be afebrile, with normal CRP and WCC than those with infection caused by *S. aureus* and other Gram-positive organisms. This may lead to diagnostic delay [21]. However, a minority of children becomes seriously ill with *K. kingae* osteoarticular infection [33].

If *S. aureus* is identified as the causative organism, MRSA, community-associated MRSA (CA-MRSA), and PVL strains should be considered, especially in severe disease. This will be discussed further, as, although less common than MSSA infection in European pediatric osteoarticular infection, these strains more commonly affect children greater than 5 years.

Children who have had recent Varicella infection are at risk of developing invasive group-A *Streptococcus* (GAS) disease including osteoarticular infection and may present shocked [34]. GAS also colonizes the oropharynx, and causes tonsillitis and pharyngitis in young children. GAS osteoarticular infections are almost always hematogenously spread.

It is associated with toxic shock syndrome and children with invasive GAS infection may be shocked at presentation, requiring aggressive resuscitation and management with IV immunoglobulin as well as antimicrobials with antitoxin activity [35].

Children Greater Than 5 Years

Older children are most likely to have osteoarticular infections caused by *S. aureus*. Most cases in the United Kingdom have MSSA. However, MRSA and PVL *S. aureus* (PVL-SA) should always be considered in severe cases. Osteoarticular infection caused by CA-MRSA is associated with significant morbidity and risk of complications [36]. Attempts have been made to predict which cases have CA-MRSA rather than MSSA, as there is evidence suggesting that MRSA causes a greater elevation of CRP and WCC, and more persistent pyrexia [37, 38]. However, these predictive values have not been reproducible and CRP and WCC are not used to differentiate the infecting agent [39].

MRSA osteoarticular infection is far more likely to be complicated by deep vein thrombosis and its associated embolic sequelae than MSSA disease [40]. It is also more likely to cause pathological fractures [41, 42]. For these reasons, it is important to treat suspected MRSA promptly with appropriate antibiotics, which will be discussed later.

PVL-SA may be MRSA or MSSA and appears more likely to cause musculoskeletal infection than other invasive infections such as pneumonia [43]. It causes recurrent skin and soft tissue infections and should be suspected in all cases that have a past medical history of these. PVL is a necrotizing exotoxin, which causes leucocyte death and severe disease with shock [44]. It is also a virulence factor that is associated with infections causing bone destruction [45]. Animal experiments support clinical observation that PVL positive strains cause more severe disease with extensive bone destruction [46].

PVL-SA is reported with increasing frequency. PVL-producing strains are identified either by using PCR to detect the PVL-producing gene or by conducting an immunosorbent assay to detect the PVL toxin. The toxin itself is known to cause leucocyte cell death; however, in osteoarticular infection, both normal and elevated WCC are seen. PVL-producing strains have been shown to cause more severe pneumonia and osteoarticular infection than non-PVL-producing *S. aureus*, although this is not seen in skin and soft tissue infections. Additionally, children with PVL-positive infection are more likely to present with shock and to have a longer duration of fever and raised inflammatory markers. They are also more likely to require intensive care [47].

Older children and adolescents are the group most likely to present with subacute osteoarticular infection, partly because of increased likelihood of puncture wounds and decreased incidence of bacteremia. It is important to consider *Pseudomonas aeruginosa* in this group, particularly where there is a history of puncture wound of the foot.

Rarer organisms include nontuberculous mycobacteria, *Pantoea agglomerans* following palm thorn injury, *Fusobacterium necrophorum* associated with chronic sinusitis, *Bartonella henselae* associated with cat scratch, Brucellosis, and *Candida albicans*.

Special Groups

There are particular groups of children who may be more susceptible to certain organisms, or more likely to develop bone and joint infections.

Sickle Cell Disease

Children with sickle cell disease (SCD) are more likely to have osteoarticular infection caused by *Salmonella* species [48], although it is likely that *S. aureus* has a similar incidence in this group. Importantly, sickle cell crisis may be difficult to distinguish from osteoarticular infection. Children with SCD are also at risk of avascular necrosis of the femoral head, which should also be excluded as a diagnosis. The most commonly affected sites for osteoarticular infections are the wrists and hands.

Primary Immunodeficiency

Children with a known primary immunodeficiency are more susceptible to osteoarticular infection, as other invasive infections. Osteoarticular infection is more likely to be multifocal in this group, and may be caused by a wider range of organisms, including those that are not usually pathogenic. Frequently, these patients have had prolonged hospital admissions or multiple courses of antibiotic treatment, and, therefore, resistant organisms should always be considered. For example, chronic granulomatous disease (CGD) is an X-linked primary immune deficiency in which children are at increased risk of infection with catalase-positive organisms, including *S. aureus*, *Nocardia* spp., *Serratia marcescens*, and *Aspergillus* spp. Fungal OM should always be considered in this group. The sites most commonly affected are the ribs and sternum, small bones, and spine.

Postoperative Infection

Children who have had orthopedic surgery, especially those who have had metal-work in situ, such as scoliosis repairs, or arthroplasty are at particular risk of osteoarticular infection. This frequently includes children with significant comorbidities such as cerebral palsy, who may not be able to communicate pain and therefore have delay in diagnosis.

Implant-associated infections following orthopedic surgery are most commonly caused by *S. aureus*, but organisms such as *S. pneumoniae* and GAS should also be considered. *Propionibacterium acnes* is now recognized as a cause of osteoarticular infection in these children. It typically has delayed presentation as infection is insidious [49, 50].

Tuberculous Bone Infection

This is usually seen in the adolescent age group. Half of all children with probable *Mycobacterium tuberculosis* (TB) infection do not have an active pulmonary focus. In children, pulmonary TB is frequently silent without signs or symptoms. The initial infection may be missed, and subsequent bone and joint infection may not become apparent until much later when localized symptoms arise.

Pathogenesis of TB bone infection is hematogenous or occasionally by direct spread from an adjacent lymph node. Tuberculous OM is characterized by necrotizing (caseating) granulomatous inflammation. Diagnosis includes Mantoux test, interferon gamma release assay (IGRA), imaging, and biopsy of suspected lesions as these may be difficult to distinguish from neoplasia on imaging.

At least half of all cases with tuberculous bone infection have spinal involvement, typically the lower thoracic and lumbar spine (see also Chapter 16) [4, 51]. The remainder have infection of long bones and synovial joints. The most commonly affected joints after the spine are the hips, knees, ankles, and elbows [52].

Pott's disease is vertebral OM with adjacent paravertebral abscess, later associated with extensive bone destruction and vertebral collapse (Figure 14.3). Children with Pott's disease are at risk of spinal cord compression and should be assessed for this. They may require internal or orthotic stabilization of the spine. Gibbus deformity occurs when there are multiple levels of collapse late in the disease [4, 51].

Tuberculous dactylitis is largely a disease of young children (< 10 years) and may remain unrecognized for long periods of time. It usually affects the hands, but can also affect the feet [53].

Treatment

Antibiotic Treatment

Neonates

Antibiotic treatment of acute hematogenous OM is usually started empirically and aims to treat all common organisms causing the infection in each age group.

In the neonatal period, treatment should include a broad-spectrum antibiotic covering Gram-positive and Gram-negative microorganisms. However, since the majority of cases are caused by group-B streptococci, treatment should be rapidly streamlined, once this is confirmed. Ideally, there should also be adequate penetration of the blood–brain barrier, as there is high risk of developing meningitis following bacteremia in this group. Third-generation cephalosporins are often used, or a combination of beta-lactam and gentamicin.

Additionally, in premature neonates and babies with significant perinatal morbidity admitted to the NICU, antibiotic cover for resistant organisms such as MRSA (usually vancomycin) and organisms producing extended-spectrum beta-lactamase (ESBL) should be considered. Antibiotics are widely used in neonatal units, and there is high risk of infections being caused by these, particularly with prolonged stay. Antimicrobial treatment of late-onset neonatal infections causing bone and joint infections must be considered against local bacterial antibiotic resistance profiles.

Babies who have central venous access and total parenteral nutrition (TPN) are at risk of fungal sepsis, and antifungal agents (usually fluconazole, liposomal amphotericin B, or caspofungin) should be considered if there is no response to 24–48 h of antibiotic treatment.

IV antibiotic therapy is advised for a minimum of 2 weeks and up to 6 weeks in neonates. The role of oral therapy is not well defined. Absorption of oral antibiotics is very variable in the neonatal age group; however, there are now reports of successful IV to oral switch in babies with few comorbidities and uncomplicated osteoarticular infection [54].

Children Aged 3 Months to 5 Years

Children aged 3 months to 5 years should receive initial IV antibiotic therapy that covers both *S. aureus* and *K. kingae*. *K. kingae* is intrinsically resistant to glycopeptide antibiotics and also has poor susceptibility to clindamycin [55]. Isolates of *K. kingae* from the oropharynx increasingly produce beta-lactamases [56, 57]. The IV antibiotic of choice should therefore be second- or third-generation cephalosporin. If IV antibiotic can be given out of hospital, a once-daily antibiotic such as ceftriaxone (80 mg/kg/dose) can be an appropriate choice.

It is not well established how long children with OM should be treated with IV antibiotics. Historically, long courses of up to 6 weeks have been advocated. However, more recent randomized controlled trials suggest that a short course of IV antibiotics (2–7 days), followed by 1–2 weeks of oral therapy, is sufficient [58]. However, there is still much debate regarding the applicability to other populations due to case ascertainment, length of time to conduct the trial, and differences in reported treatments [59, 60]. Retrospective data from Finland suggests that oral switch may be safe with a reduction in complications due to prolonged IV therapy [58].

Oral switch is particularly important in young children, as adherence is dependent on palatability of the chosen drug. At this age, children tend to take antibiotics in liquid form. Certain antibiotics such as flucloxacillin and clindamycin do not have a pleasant taste and are not well tolerated. Erythromycin is often used in penicillin-allergic children. However, because of gastrointestinal side-effects, the oral course is frequently not completed. For flucloxacillin, clindamycin, and erythromycin, the need to dose 6-hourly precludes most parents successfully dosing their children with adequate antimicrobial therapy. Oral cephalosporins and amoxicillin/clavulanic acid tend to taste pleasant, can be dosed three times per day, and do not usually have significant adverse effects. Genuinely penicillin-allergic children rarely have cross-allergy to cephalosporins. Nevertheless, clindamycin has to be considered in patients with immediate hypersensitivity if no macrolide resistance is suspected.

Children Greater Than 5 Years

In areas where there is low incidence of MRSA, such as Northern Europe and Oceania, older children should be treated initially with IV high-dose flucloxacillin (50 mg/kg four times per day), unless there is a history of multiple hospital admissions, or other risk factors such as recent travel to an area where MRSA is prevalent. If there is no clinical response after 24–48 h treatment, the change to an alternative antibiotic, covering resistant organisms, should be considered. In addition, evidence of an abscess has to be investigated. Ideally, if MRSA is suspected, it is advantageous to treat early in view of higher rates of serious complications.

If MRSA is suspected, glycopeptides are usually first line, often clindamycin. Linezolid and daptomycin are also increasingly being used for complex cases with specialist microbiological advice. Patients with MRSA OM should be carefully monitored for abscess formation and deep vein thrombosis. IV treatment may be prolonged.

Oral switch for MSSA infection may be to flucloxacillin tablets or to amoxicillin/clavulanic acid or cephalexin suspension, where tablets cannot be taken. If flucloxacillin is prescribed, the 6-hourly dose regimen must be strictly adhered to. In uncomplicated MRSA infection, clindamycin (also 6-hourly dosing) and linezolid are suitable agents, although linezolid should be reserved for resistant organisms or recurrent difficult-to-treat infections. Clindamycin suspension has an unpleasant taste and may not be well tolerated; therefore, tablets are advised where possible. Of note, *Clostridium difficile* enterocolitis rarely occurs in children treated with clindamycin.

In subacute cases where *P. aeruginosa* infection is suspected, antipseudomonal agents such as ceftazidime, tobramycin, or ciprofloxacin should be included in therapy [61].

For children with sickle cell disease, second- and third-generation cephalosporins may be used where MRSA is not suspected, as these will cover *Salmonella* species and *S. aureus*. Where *Salmonella* is identified as the causative organism, ciprofloxacin may also be used, although sensitivity should be confirmed.

In immunocompromised patients, prolonged courses of IV antibiotics are indicated, and insertion of a percutaneous IV central catheter (PICC) is usually required. These patients often require broad-spectrum Gram-positive and Gram-negative cover, at high doses. Usually, more than one antibiotic is used. Sequential imaging of these cases may be indicated to ensure that the disease resolves and does not progress. In cases of fungal OM, IV antifungal agents are used, such as fluconazole, voriconazole, liposomal amphotericin B, and caspofungin.

Tuberculous OM is treated initially with a four-drug therapy (rifampicin, isoniazid, pyrazinamide, and ethambutol) for 2 months, followed by a two-drug therapy (rifampicin and isoniazid) for 4–10 months. Diagnosis usually includes biopsy and drainage of any collection.

In all children with osteoarticular infection, the treatment is multidisciplinary and should include input from pediatric, orthopedic, radiology, and microbiology departments. Where the spine is involved or there is significant disability, orthotics and physiotherapy may also be required.

Surgical Treatment

Surgical management of OM is infrequently needed. If a collection in the periosteum, soft tissue, or a cyst is visible in an imaging procedure, surgery may be indicated for diagnostic (identification of a microorganism) and therapeutic purposes. Bone biopsy or drill decompression may be done if there is no response to antibiotic therapy in the first 72 h, although this may also suggest resistant organisms or fungal infection. Figure 14.5 shows guided decompression of OM of the tibia.

In cases of postoperative complication such as an infected implant, all metal work should be removed and any necrotic tissue debrided.

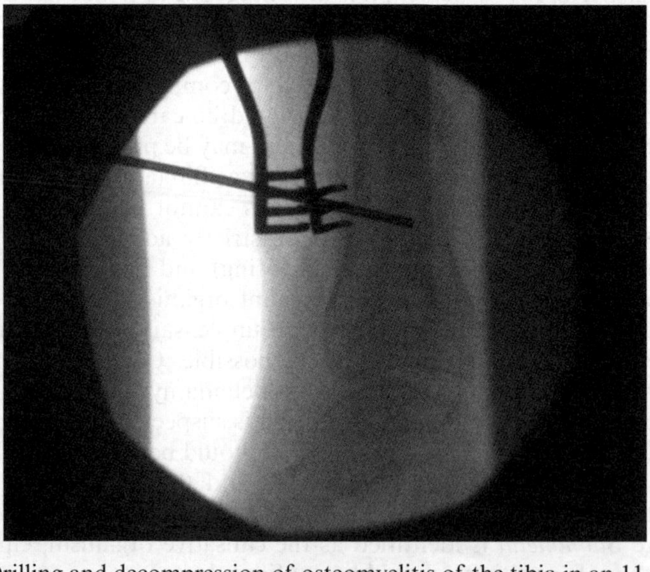

Figure 14.5. Drilling and decompression of osteomyelitis of the tibia in an 11-year-old boy.

If there is subacute or chronic OM with lytic or cystic changes of the bone, sinus, or sequestrum, these may also be debrided and drained. This may be done with ultrasound or CT guidance and may be very useful for identifying an organism and excluding malignancy.

Complications

Complications of bone and joint infections may be severe and are generally preventable by timely treatment.

Where there is a delay in presentation of OM and pathological fractures, formation of a sinus or sequestrum and persistent abscesses are more common. Deep venous thrombosis and thromboembolic events are more common where MRSA is the causative organism. These are therefore increasing as MRSA becomes more prevalent [36].

Key Points

- OM is an uncommon but important diagnosis in children.
- OM may present with nonspecific signs and symptoms, particularly in neonates and infants.
- Diagnosis and treatment of OM is multidisciplinary, and duration of antibiotic treatment is currently being researched.

References

[1] Gafur OA, Copley LA, Hollmig ST, *et al.* The impact of the current epidemiology of pediatric musculoskeletal infection on evaluation and treatment guidelines. J Pediatr Orthop 2008; 28(7): 777–785.

[2] Riise OR, Kirkhus E, Handeland KS, *et al.* Childhood osteomyelitis-incidence and differentiation from other acute onset musculoskeletal features in a population-based study. BMC Pediatr 2008; 8:45.

[3] Stoesser N, Pocock J, Moore CE, *et al.* The epidemiology of pediatric bone and joint infections in Cambodia, 2007–11. J Trop Pediatr 2013;59(1):36–42.

[4] Eisen S, Honywood L, Shingadia D, Novelli V. Spinal tuberculosis in children. Arch Dis Child 2012;97(8):724–729.

[5] Faust S.N., Clark J, Pallett A, Clarke NMP Managing bone and joint infection in children. Arch Dis Child 2012;97(6):545–553.

[6] Grammatico-Guillon L, Maakaroun Vermesse Z, Baron S, *et al.* Paediatric bone and joint infections are more common in boys and toddlers: a national epidemiology study. Acta Paediatr 2013;102(3):e120–e125.

[7] Ferroni A, Al Khoury H, Dana C, *et al.* Prospective survey of acute osteoarticular infections in a French paediatric orthopedic surgery unit. Clin Microbiol Infect 2013;19(9):822–828.

[8] Gaudin A, Amador Del Valle G, Hamel A, *et al.* A new experimental model of acute osteomyelitis due to methicillin-resistant *Staphylococcus aureus* in rabbit. Lett Appl Microbiol 2011; 52(3):253–257.

[9] Spencer SJ, Wilson NI. Childhood discitis in a regional children's hospital. J Pediatr Orthop B 2012; 21(3):264–268.

[10] Brown R., Hussain M, McHugh K, *et al.* Discitis in young children. J Bone Joint Surg Br 2001;83(1):106–111.

[11] Liao SL, Lai SH, Lin TY, *et al.* Premature rupture of the membranes: a cause for neonatal osteomyelitis? Am J Perinatol 2005; 22(2):63–66.

[12] Waseem M, Devas G, Laureta E. A neonate with asymmetric arm movements. Pediatr Emerg Care 2009; 25(2):98–99.

[13] Dessi A, Crisafulli M, Accossu S, *et al.* Osteo-articular infections in newborns: diagnosis and treatment. J Chemother 2008; 20(5):542–550.

[14] McPherson DM. Osteomyelitis in the neonate. Neonatal Netw 2002; 21(1):9–22.

[15] Williamson JB, Galasko CS, Robinson MJ. Outcome after acute osteomyelitis in preterm infants. Arch Dis Child 1990;65(10 Spec No):1060–1062.

[16] Brook I. Infected neonatal cephalohematomas caused by anaerobic bacteria. J Perinat Med 2005;33(3):255–258.

[17] Yüksel S, Yüksel G, Oncel S, Divani E. Osteomyelitis of the calcaneus in the newborn: an ongoing complication of Guthrie test. Eur J Pediatr 2007;166(5):503–504.

[18] Knudsen CJ, Hoffman EB. Neonatal osteomyelitis. J Bone Joint Surg Br 1990;72(5): 846–851.

[19] Azam Q, Ahmad I, Abbas M, *et al.* Ultrasound and colour Doppler sonography in acute osteomyelitis in children. Acta Orthop Belg 2005;71(5):590–596.

[20] Riebe TW, Nasir R, Nazarenko O. The value of sonography in the detection of osteomyelitis. Pediatr Radiol 1996; 26(4):291–297.

[21] Ceroni D, Cherkaoui A, Ferey S, *et al.* Kingella kingae osteoarticular infections in young children: clinical features and contribution of a new specific real-time PCR assay to the diagnosis. J Pediatr Orthop 2010; 30(3): 301–304.

[22] Howard AW, Viskontas D, Sabbagh C. Reduction in osteomyelitis and septic arthritis related to *Haemophilus influenzae* type B vaccination. J Pediatr Orthop 1999; 19(6):705–709.

[23] Toussi SS, Pan N, Walters HM, *et al.* Infections in children and adolescents with juvenile idiopathic arthritis and inflammatory bowel disease treated with tumor necrosis factor-alpha inhibitors: systematic review of the literature. Clin Infect Dis 2013;57(9):1318–1330.

[24] Paakkonen M, Kallio MJ, Kallio PE, Peltola H. C-reactive protein versus erythrocyte sedimentation rate, white blood cell count and alkaline phosphatase in diagnosing bacteraemia in bone and joint infections. J Paediatr Child Health 2013;49(3):E189–E192.

[25] Maharajan K, Patro DK, Menon J, *et al.* Serum Procalcitonin is a sensitive and specific marker in the diagnosis of septic arthritis and acute osteomyelitis. J Orthop Surg Res 2013;8(1):19.

[26] Shen CJ, Wu MS, Lin KH, *et al.* The use of procalcitonin in the diagnosis of bone and joint infection: a systemic review and meta-analysis. Eur J Clin Microbiol Infect Dis 2013;32(6): 807–814.

[27] Dhanoa A, Singh VA. Subacute osteomyelitis masquerading as primary bone sarcoma: report of six cases. Surg Infect (Larchmt) 2010;11(5):475–478.

[28] Sreenivas T, Nataraj AR, Menon J, *et al.* Acute multifocal haematogenous osteomyelitis in children. J Child Orthop 2011;5(3):231–235.

[29] Montgomery CO, Siegel E, Blasier RD, Suva LJ. Concurrent septic arthritis and osteomyelitis in children. J Pediatr Orthop 2013;33(4):464–467.

[30] Guillerman RP. Osteomyelitis and beyond. Pediatr Radiol 2013;43(Suppl 1):S193–S203.

[31] Browne LP, Guillerman RP, Orth RC, *et al.* Community-acquired staphylococcal musculoskeletal infection in infants and young children: necessity of contrast-enhanced MRI for the diagnosis of growth cartilage involvement. AJR Am J Roentgenol 2012;198(1):194–199.

[32] Ceroni D, Llana RA, Kherad O, *et al.* Comparing the oropharyngeal colonization density of kingella kingae between asymptomatic carriers and children with invasive osteoarticular infections. Pediatr Infect Dis J 2013;32(4):412–414.

[33] Mallet C, Ceroni D, Litzelmann E, *et al.* Unusually severe cases of Kingella kingae Osteoarticular infections in children. Pediatr Infect Dis J 2013.

[34] Bozzola E, Krzystofiak A, Lancella L, *et al.* A severe case of paediatric group A streptococcal osteomyelitis in varicella. Infection 2012;40(3):343–345.

[35] Steer AC, Lamagni T, Curtis N, *et al.* Invasive group a streptococcal disease: epidemiology, pathogenesis and management. Drugs 2012;72(9):1213–1227.
[36] Schaub RL, Rodkey ML. Deep vein thrombosis and septic pulmonary emboli with MRSA osteomyelitis in a pediatric patient. Pediatr Emerg Care 2012;28(9):911–912.
[37] Ju KL, Zurakowski D, Kocher MS. Differentiating between methicillin-resistant and methicillin-sensitive *Staphylococcus aureus* osteomyelitis in children: an evidence-based clinical prediction algorithm. J Bone Joint Surg Am 2011;93(18):1693–1701.
[38] Kini AR, Shetty V, Kumar AM, *et al.* Community-associated, methicillin-susceptible, and methicillin-resistant *Staphylococcus aureus* bone and joint infections in children: experience from India. J Pediatr Orthop B 2013;22(2):158–166.
[39] Wade Shrader M, Nowlin M, Segal LS. Independent analysis of a clinical predictive algorithm to identify methicillin-resistant *Staphylococcus aureus* Osteomyelitis in children. J Pediatr Orthop 2013;33(7):759–762.
[40] Mantadakis E, Plessa E, Vouloumanou EK, *et al.* Deep venous thrombosis in children with musculoskeletal infections: the clinical evidence. Int J Infect Dis 2012;16(4):e236–e243.
[41] Belthur MV, Birchansky SB, Verdugo AA, *et al.* Pathologic fractures in children with acute *Staphylococcus aureus* osteomyelitis. J Bone Joint Surg Am 2012;94(1):34–42.
[42] Gelfand MS, Cleveland KO, Heck RK, Goswami R. Pathological fracture in acute osteomyelitis of long bones secondary to community-acquired methicillin-resistant *Staphylococcus aureus*: two cases and review of the literature. Am J Med Sci 2006; 332(6):357–360.
[43] Cunnington A, Brick T, Cooper M, *et al.* Severe invasive Panton-Valentine Leucocidin positive *Staphylococcus aureus* infections in children in London, UK. J Infect 2009; 59(1):28–36.
[44] Dohin B, Gillet Y, Kohler R, *et al.* Pediatric bone and joint infections caused by Panton-Valentine leukocidin-positive *Staphylococcus aureus*. Pediatr Infect Dis J 2007;26(11): 1042–1048.
[45] Jin T, Zhu YL, Li J, *et al.* Staphylococcal protein A, Panton-Valentine Leukocidin and coagulase aggravate the bone loss and bone destruction in Osteomyelitis. Cell Physiol Biochem 2013;32(2):322–333.
[46] Cremieux AC, Dumitrescu O, Lina G, *et al.* Panton-valentine leukocidin enhances the severity of community-associated methicillin-resistant *Staphylococcus aureus* rabbit osteomyelitis. PLoS One 2009;4(9):e7204.
[47] Ritz N, Curtis N. The role of Panton-Valentine leukocidin in *Staphylococcus aureus* musculoskeletal infections in children. Pediatr Infect Dis J 2012;31(5):514–518.
[48] Hann G, Roberts-Harewood M, Yardumian DA, Wilkey O. Salmonella bacteraemia in children with sickle cell disease. Arch Dis Child 2012;97(Suppl 1):A1–A186 A89.
[49] Jakab E, Zbinden R, Gubler J, *et al.* Severe infections caused by Propionibacterium acnes: an underestimated pathogen in late postoperative infections. Yale J Biol Med 1996; 69(6): 477–482.
[50] Haidar R, Najjar M, Der Boghossian A, Tabbarah Z. Propionibacterium acnes causing delayed postoperative spine infection: review. Scand J Infect Dis 2010; 42(6–7):405–411.
[51] Sahu R, Kumar R, Srivasta A, *et al.* Tuberculosis of the thoracic spine, an analysis of 22 surgically treated children. Childs Nerv Syst 2012;28 (9):1659.
[52] Talbot JC, Bismi Q, Saralaya D, Newton DA, Frizzel RM, Shaw DL. Musculoskeletal tuberculosis in Bradford—a 6-year review. Ann R Coll Surg Engl 2007;89(4):405–409.
[53] Ritz N, Connell TG, Tebruegge M, *et al.* Tuberculous dactylitis—an easily missed diagnosis. Eur J Clin Microbiol Infect Dis 2011;30(11):1303–1310.
[54] Ecury-Goosssen GM, Huysman MA, Verhallen-Dantuma JCTM, Man P. Sequential intravenous-oral antibiotic therapy for neonatal osteomyelitis. Pediatr Infect Dis J 2009;28(1): 72–73.
[55] Yagupsky P. Antibiotic susceptibility of Kingella kingae isolates from children with skeletal system infections. Pediatr Infect Dis J 2012;31(2):212.

[56] Banerjee A, Kaplan JB, Soherwardy A. Characterization of TEM-1 beta-lactamase producing Kingella kingae clinical isolates. Antimicrob Agents Chemother 2013; 57(9): 4300–4306.
[57] Yagupsky P, Slonim A, Amit U, *et al.* Beta-lactamase production by Kingella kingae in Israel is clonal and common in carriage organisms but rare among invasive strains. Eur J Clin Microbiol Infect Dis 2013;32(8):1049–1053.
[58] Paakkonen M, Kallio MJT, Kallio PE, Peltola H. Shortened hospital stay for childhood bone and joint infections: Analysis of 265 prospectively collected culture-positive cases in 1983–2005. Scand J Infect Dis 2012; 44(9):683–688.
[59] Zaoutis T, Russell Localio A, Leckerman K *et al.* Prolonged intravenous therapy versus early transition to oral antimicrobial therapy for acute osteomyelitis in children. Pediatrics 2009; 123(2):636–642.
[60] Inaba AS, Zukin DD, Perro M. An update on the evaluation and management of plantar puncture wounds and Pseudomonas osteomyelitis. Pediatr Emerg Care 1992; 8(1):38–44.
[61] Baghdadi T, Saberi S, Sobhani Eraghi, A, *et al.* Late sequelae of hip septic arthritis in children. Acta Med Iran 2012; 50(7):463–467.

Chapter 15

Acute Osteomyelitis in Adults

Werner Zimmerli

Introduction

Acute osteomyelitis evolves over a period of a few days to weeks. There is no established time limit between acute, subacute, and chronic osteomyelitis. The differentiation is based on histological characteristics rather than on a defined time interval between start of infection and presentation [1]. Subacute osteomyelitis is generally caused by *Brucella* spp. or *Mycobacterium tuberculosis* (see Chapter 16), or alternatively by low-virulence microorganisms (e.g., coagulase-negative staphylococci or *Propionibacterium acnes*) in the presence of an implant (see Chapter 21). In most cases, it has a smoldering course, which may delay the diagnosis for several weeks or even months. Chronic osteomyelitis is characterized by dead bone. It is the result of an ineffectively treated acute osteomyelitis, during the first couple of weeks after infection. It may persist up to several decades [2, 3]. In case of late recurrence with acute presentation, it has to be considered as chronic osteomyelitis and treated as such.

Hematogenous osteomyelitis involves long bones in children, but predominantly the vertebral column in adults. According to a large series of hematogenous osteomyelitis reported by Jensen *et al.* [4], the median age of patients with vertebral osteomyelitis is 65 years, as compared to 16 years in patients with other types of osteomyelitis. This chapter is focused on acute vertebral osteomyelitis, also termed spinal osteomyelitis, spondylodiskitis, septic diskitis, or disk-space infection [5]. Acute osteomyelitis in long bones is presented in Chapter 14 (osteomyelitis in children) and in Chapter 20 (implant-associated osteomyelitis of long bones).

Pathogenesis

Acute osteomyelitis occurs either by the exogenous or the hematogenous route. Exogenous infection is directly inoculated as the result of a penetrating trauma or contamination during bone surgery. In the majority of cases, acute exogenous osteomyelitis is related to

Bone and Joint Infections: From Microbiology to Diagnostics and Treatment, First Edition.
Edited by Werner Zimmerli.

Table 15.1. Microorganisms in acute vertebral osteomyelitis.

	McHenry *et al.* [10] 1972–1982	Nolla *et al.* [11] 1980–1999	Hadjipavlon *et al.* [12] 1986–1996	Total number	Total percent[a]
Number of patients	253	64	101	418	
Number of episodes with data on microbiology	255	64	98	417	
Staphylococcus aureus	123 (56.7%)[a]	23 (35.9%)	44 (59.5%)	190	50.7
Coagulase-negative staphylococci	17 (6.7%)	2 (3.1%)	20 (27%)	39	12.3
Streptococcus spp.	24 (11.1%)	12 (18.8%)	20 (27%)	56	19
Enterococcus faecalis	0	1 (1.6%)	4 (5.4%)	5	2.3
Escherichia coli	30 (13.8%)	15 (23.4%)	3 (4.1%)	48	13.8
Proteus spp.	5 (2.3%)	3 (4.7%)	1 (1.4%)	9	2.8
Klebsiella spp.	5 (2.3%)	0	0	5	0.8
Pseudomonas aeruginosa	13 (6.0%)	3 (4.7%)	4 (5.4%)	20	5.4
Fungi	2 (0.9%)	0	0	2	0.3
Others	10 (4.6%)	5 (7.8%)	10 (13.5%)	25	8.6
No growth	38 (15.0%)	0	24 (23.8%)	62	

[a] All percentages are calculated per total cases with positive microbiology. The sum exceeds 100% in series with polymicrobial episodes.

an internal fixation device (see Chapter 21) [6]. In hematogenous osteomyelitis, microorganisms reach the bone via the bloodstream from a distant focus of infection. The presentation of this type of bone infection differs in children and adults. In prepubertal children, microorganisms generally seed to the metaphysis of long bones, mainly the femur, tibia, and humerus (see Chapter 14) [7]. In adults, hematogenous osteomyelitis predominantly involves the vertebral bodies. It generally originates at the subchondral part of the vertebral body before involving the disk space. Hematogenous long-bone osteomyelitis is extremely rare in adults.

Acute osteomyelitis caused by nonspecific microorganisms is characterized by pus formation. Therefore, it is also called pyogenic infection as opposed to specific osteomyelitis caused by *Brucella* spp. or *M. tuberculosis*. However, this nomenclature is somewhat misleading, because conventional bacteria such as viridans streptococci may cause nonpyogenic infection, whereas gravitation abscesses are a typical complication of tuberculous vertebral osteomyelitis [8, 9]. Thus, classifying in acute, subacute, and chronic infection may be more appropriate than classifying in pyogenic and nonpyogenic osteomyelitis (see Chapter 13).

Histologic characteristics of acute osteomyelitis are suppurative inflammation, followed by abnormal bone remodeling, resulting in uncontrolled bone resorption. *Staphylococcus aureus* is by far the most frequent microorganism causing hematogenous osteomyelitis (Table 15.1) [10–12]. Therefore, the pathogenetic mechanisms of staphylococcal osteomyelitis are particularly well studied [13–15]. *S. aureus* targets bone by expressing adhesins for bone matrix components such as fibronectin, collagen, and sialoprotein [16–18]. Fibronectin-binding proteins are expressed on most clinically relevant strains of *S. aureus*, and are crucial for adherence not only to implants but also to osteoblasts [16, 19]. Bacterial adhesion is followed by internalization in osteoblasts, which explains the long-term persistence of the microorganism in case of insufficient therapy of acute osteomyelitis [16, 20]. *S. aureus* is bound via surface-expressed fibronectin-binding proteins to integrins clustering on osteoblasts [16]. In addition, *S. aureus* can also bind to osteoblasts via protein A. This process results in inhibition of osteoblast proliferation and mineralization, apoptosis, and activation of osteoclasts [14]. It has been shown that *S. aureus* protein A binds to tumor necrosis factor receptor-1 on osteoblasts, resulting in the release of interleukin 6 (IL-6). This proinflammatory cytokine activates osteoclasts, and thereby bone destruction [15].

Histologically, bone edema, vascular congestion, and small-vessel thrombosis are the hallmarks of acute osteomyelitis. If this stage is not adequately managed with antimicrobial therapy, medullary and periosteal blood supplies are impaired, leading to bone necrosis. Fragments of dead cortical bone, which are completely detached from living bone, are called sequestra. Such sequestrated pieces of dead bone are colonized by an amorphous biofilm of bacteria, which resist to antimicrobial therapy [21]. This biofilm is formed by exopolysaccharides (glycocalyx) of the infecting microorganisms. Thus, sequestra behave similar to implanted foreign devices [22, 23]. Consequently, if acute osteomyelitis progresses to the chronic stage, antimicrobial therapy has to be combined with debridement surgery.

Epidemiology

In theory, the incidence of hematogenous osteomyelitis should have decreased in the antibiotic era by the prevention of microbial seeding in the bone. This is true for long-bone osteomyelitis in children [24]. After the introduction of antibiotics in clinical medicine, the prevalence of long-bone osteomyelitis further decreased. Between 1980 and 1990, the incidence rate of long-bone osteomyelitis due to *S. aureus* decreased from 5 to 1.8 per 100,000 children below the age of 1 year [4]. The reason for this finding is unclear. It may be due to more rapid diagnosis and better antibiotic treatment of *S. aureus* sepsis in children at a very small age. In contrast, in all other age classes, there was no significant change of incidence over the same time period. Regarding vertebral osteomyelitis, the situation is different. During the same time period, the incidence increased from 0.3 to 1 episode per 100,000 inhabitants above the age of 50, and remained stable in all other age classes [4].

The large French study of Grammatico *et al.* [25] gives the best estimate of the epidemiology of vertebral osteomyelitis. In this study, a nationwide permanent hospital database with information from 2002 and 2003 was used. The authors included all types of vertebral osteomyelitis. Among the 2519 cases, 0.4% were due to *Brucella* spp. and 20.5% due to *M. tuberculosis*. Thus, the data do not exclusively reflect cases with acute osteomyelitis. Overall, 4013 hospitalized patients met their case definition of vertebral osteomyelitis. About 10% of the patients had previous spine surgery, and 90% were cases of probable hematogenous origin. The overall incidence was 2.4 cases per 100,000 inhabitants. Only 3% of all patients were below 20 years of age. The incidence steadily increased up to the age of 80, from 0.3 per 100,000 among persons younger than 20 years of age to 3.5 per 100,000 (50–70 years), to 6.5 per 100,000 among persons older than 70 years of age.

In all studies, a predominance of male patients is reported. In the study of Grammatico *et al.* [25], the male preponderance was observed in all age classes except those below 20 years and above 80 years. Indeed, in 10 studies with patients suffering from different types of vertebral osteomyelitis, between 51 and 70% (mean 58%) of the patients were male [4, 25–33]. In many studies, not only patients with acute osteomyelitis, but also with subacute osteomyelitis, are included, and cannot be analyzed separately. As an exception, Colmenero *et al.* [34] compared the characteristics of patients with acute and subacute osteomyelitis. They found a predominance of 71% males in acute (pyogenic) and 72% in brucellar vertebral osteomyelitis. In contrast, in patients with tuberculosis, the distribution was even (50%). Kim *et al.* [33] observed even a female predominance (64%) in patients with tuberculous vertebral osteomyelitis.

The mean age of patients with vertebral osteomyelitis was 60 years in 12 studies including 972 patients [11, 26, 29–36, 37, 38]. The large majority of patients in these studies had acute osteomyelitis. In the study of Kim *et al.* [33], the mean age of patients was similar in patients with acute as compared to those with tuberculous spondylodiskitis. Similarly, in the study of Colmenero *et al.* [34], patients with acute (pyogenic), tuberculous, and brucellar osteomyelitis had a similar mean age.

Microbiology

In all series reporting microbiology of acute (pyogenic) vertebral osteomyelitis, *S. aureus* is the most frequent single microorganism, accounting for 30–60% of the episodes. Table 15.1 summarizes the data of three large studies with sufficiently precise information on the causing agents and the study population [10–12]. In the study of McHenry *et al.* [10], patients with postoperative vertebral osteomyelitis were included. In contrast, in the studies of Nolla *et al.* [11] and Hadjipavlon *et al.* [12], cases after spine surgery were excluded. In addition, Nolla *et al.* [11] excluded intravenous drug users (IVDU). All 64 episodes were hematogenously acquired. All except one episode of catheter-related infection were community-acquired. In the series of Hadjipavlon *et al.* [12], all 101 patients had primary, that is, hematogenous vertebral osteomyelitis. One-quarter of the patients were IVDU.

In the three studies summarized in Table 15.1, data from 417 episodes with microbiological workup have been reported. Overall, *S. aureus*, *Streptococcus* spp., and *Escherichia coli* were the three most frequent microorganisms accounting for 50.7, 19, and 13.8%, respectively. Thus, more than 80% of all episodes with documented microbiology were caused by these three types of bacteria. Tuberculosis and brucellosis were excluded in all three studies. The fraction of polymicrobial infection varied among the studies between 0% [11], 7.8% [10], and 24.5% [12]. In general, hematogenous infections are monomicrobial with a few exceptions. As an example, in the study of Nolla *et al.* [11], all hematogenous episodes were monomicrobial. In contrast, exogenous infections after spine surgery are frequently polymicrobial. This is especially the case in patients with prolonged secreting wounds. The high fraction of polymicrobial infections in the series of Hadjipavlon *et al.* [12] is unclear. In their population, 25% were IVDU. This could explain the difference to the study of Nolla *et al.* [11], in which IVDU were excluded.

Low-virulence microorganisms such as *Propionibacterium acnes* or coagulase-negative staphylococci almost never cause hematogenous vertebral osteomyelitis in patients without a spinal implant. Indeed, in a series of 29 cases with spondylodiskitis due to

P. acnes, 97% had previous spinal surgery between 0 and 156 months earlier [39]. Similarly, coagulase-negative staphylococci are mainly seen in postoperative vertebral osteomyelitis (confer Chapter 21). However, these microorganisms were also observed in the absence of internal fixation devices in the three series summarized in Table 15.1 [10–12]. In the study of Nolla *et al.* [11], one patient with coagulase-negative staphylococci had an endocarditis. The other patient suffered from liver cirrhosis and had an infected LeVeen shunt. In addition, we observed three cases of spondylodiskitis associated with prolonged bacteremia due to coagulase-negative staphylococci. All three patients had a persisting intravascular device (pacemaker electrode, permanent central venous access device) with a colonized catheter for a prolonged time [40]. The high fraction of coagulase-negative staphylococci in the series of Hadjipavulon *et al.* [12] is not clear, and not analyzed in the publication. The most probable reason is the high fraction of IVDU (25%).

Streptococci are the second most frequent microorganisms causing vertebral osteomyelitis (Table 15.1). It is of clinical importance to know the species. In a recent large series reporting on streptococcal and *S. aureus* spondylodiskitis, concomitant endocarditis was observed in 8% (2/26) of the cases caused by non-viridans, but 50% (16/32) of those caused by viridans streptococci [8].

Risk Factors

Most patients with acute hematogenous vertebral osteomyelitis have underlying conditions that are considered risk factors [5]. Pathogenetically, they can be classified into three different groups. First, underlying diseases impairing host defense increase the risk for bacteremia. Second, local factors favor exogenous inoculation of microorganisms, and adjacent infections enhance the risk for extension to the vertebral column. Third, an infection at a distant site in the body may cause vertebral osteomyelitis by seeding microorganims during bacteremia.

Table 15.2 summarizes the underlying medical conditions. Diabetes mellitus and IV drug use are the two most frequent underlying conditions in most studies reporting risk factors in sufficient detail. Diabetes has been reported as a comorbidity in one quarter of the cases with a range from 14 to 32% in seven different studies with a total of 758 patients [10–12, 34, 37, 38, 41]. Diabetes mellitus is also the predominant risk factor in studies reporting exclusively on *S. aureus* spondylodiskitis [28, 29]. Indeed, diabetes

Table 15.2. Risk factors: underlying medical conditions.

Diabetes mellitus
Intravenous drug use
Immunosuppressive therapy
Malignancy
End-stage renal disease with hemodialysis
Alcohol abuse
Inflammatory rheumatologic conditions
HIV infection
Liver cirrhosis

Table 15.3. Risk factors: local factors favoring exogenous infection or contiguous infection.

Spinal surgery
Penetrating spinal trauma
Diskography
Epidural catheterization or block
Local infiltration
Lumbar puncture
Gun shot or stab wounds
Epidural abscess[a]
Meningitis[a]
Psoas abscess[a]
Adjacent infected aortic graft[a]

[a] Case series reporting these conditions are not sufficiently detailed to differentiate between contiguous primary foci and secondary foci.

mellitus has been shown to increase the risk for bloodstream infection [42]. The relative importance of IV drug use is more difficult to estimate, because this condition was excluded in some studies [11, 37]. Patients with end-stage renal failure have to be considered as immunocompromized [43]. In addition, they are routinely treated with an intravascular access device, which increases the risk for repetitive bacteremia.

Human immunodeficiency virus (HIV) infection is a risk factor for several reasons. First, in this population, IVDU are overrepresented [44]. Second, in untreated HIV infection, a compromised neutrophil function has been observed [45]. Finally, low CD4 counts predispose individuals to infections [46]. Weinstein and Eismont [47] compared the prevalence of spine infection in hospitalized HIV-infected patients with that of HIV-negative patients. The prevalence of acute (pyogenic) vertebral osteomyelitis was 10.9 per 10,000 HIV-positive patients as compared to 4.2 per 10,000 HIV-negative hospitalized patients. As expected, the difference was even more significant when the prevalence of tuberculous vertebral osteomyelitis was assessed (8.2/10,000 HIV-positive versus 0.74/10,000 HIV-negative hospitalized patients). The median CD4 count was 350/μl in HIV patients with pyogenic osteomyelitis and 80/μl in those with tuberculous spondylodiskitis.

Table 15.3 summarizes local factors favoring exogenous vertebral osteomyelitis. Deep wound infection after spinal surgery is the leading risk factor for exogenous infection. This type of infection is presented separately in Chapter 21. In many case series, surgical patients are excluded. Therefore, a meaningful quantitative evaluation is only possible in a minority of the published case series. In the study of McHenry *et al.* [10] where unselected patients were included, 15% of the patients had previous spine surgery or a penetrating spinal trauma. Only 3% of the cases were secondary to an adjacent pyogenic infection. This may occur after epidural infiltration, leading to a local abscess and contiguous osteomyelitis (unpublished observation). However, in most cases, epidural abscesses, meningitis, and psoas abscesses are secondary pyogenic complications of vertebral osteomyelitis.

Table 15.4 shows the most frequent primary sources of hematogenous infection. In about half of the patients with acute vertebral osteomyelitis, a primary distant focus of infection can be determined. In six studies, including a total of 382 patients, primary foci

Table 15.4. Risk factors: distant source of hematogenous infection.

Genitourinary tract infection
Skin and soft tissue infection
Vascular catheter-related infection
Upper respiratory tract infection
Lower respiratory tract infection
Gastrointestinal tract infection
Septic arthritis
Endocarditis
Primary *Staphylococcus aureus* bacteremia

were reported in detail [11, 34, 38, 41, 48, 49]. The three most frequent primary sources were urogenital (15%), skin/soft tissue (13%), and catheter-related (3%) infections. Vertebral osteomyelitis exclusively caused by *S. aureus* shows a similar picture with 46% (62/133) defined primary foci [28]. As suspected, skin infection is by far the most frequent source in this study, accounting for 21% (28/133) of the cases.

Corrah *et al.* [31] looked for the relationship between bacteremia and subsequent development of vertebral osteomyelitis. For this purpose, they analyzed 129 episodes of vertebral osteomyelitis. In 21/74 (28.4%) of the cases with a documented infecting agent, a preceding bacteremia within the previous year was detected. Half of them had recurrent episodes with the same microorganism. All were caused by *S. aureus*. In the same study, the risk for subsequent vertebral osteomyelitis was calculated in 2083 patients with *S. aureus* bacteremia. During a 10-year period, it varied from 0.4 to 3.2% (mean 1.9%). Thus, after an episode of *S. aureus* sepsis, the patients must be carefully followed with regard to possible vertebral osteomyelitis. If recurrent *S. aureus* bacteremia is detected, vertebral osteomyelitis must be actively looked for.

In four studies with a total of 244 patients, 17% suffered from endocarditis [11, 41, 48, 49]. In many cases, it is clinically not clear whether endocarditis preceded or followed vertebral osteomyelitis. However, the following observations indicate that endocarditis is rather the primary focus. Prolonged bacteremia is a risk factor for seeding microorganisms in the vertebral column [40]. In case of subacute endocarditis, patients typically have sustained bacteremia for days to weeks. In contrast, in acute staphylococcal endocarditis, antimicrobial treatment is generally started within hours to days. Indeed, endocarditis was observed in only 5% of the patients with *S. aureus* vertebral osteomyelitis [28]. In contrast, in a recent study, endocarditis was observed in 50% (16/32) of the patients with viridans streptococcal, but only 10% (5/50) with *S. aureus* spondylodiskitis [8]. This clearly favors endocarditis as the primary focus; otherwise, secondary endocarditis would more frequently occur in vertebral osteomyelitis caused by the more virulent *S. aureus*.

Clinical Features

Signs and symptoms are caused either by the primary infection (e.g., urinary tract infection, sepsis) or by the vertebral osteomyelitis and its complications. For obvious reasons, these different clinical features are not separately reported in different publications.

Therefore, pooling of clinical data from different studies is only meaningful for the two leading general symptoms, namely, backache and fever.

Back pain is the most common initial symptom in all case series. In 12 studies including more than 1000 patients, it has been reported in 92% (range: 83–100%) of the cases [10, 11, 26, 28, 29, 32–34, 37, 38, 41, 48]. In contrast, fever is present in only 64% (16–97%) of these episodes. The lowest rate (16%) of fever was reported by Carragee *et al.* [37]. In their study, 29% of the patients were considered as immunosuppressed. The highest rate of fever (97%) was observed by Jensen *et al.* [28]. However, they exclusively reported on *S. aureus* vertebral osteomyelitis.

The location of pain depends on the site of infection. In 14 studies, including greater than 1200 patients in whom spine involvement was reported in sufficient detail, the most common site was the lumbar spine (59%), followed by the thoracic (31%) and the cervical spine (10%) [10–12, 26, 28, 29, 31–34, 37, 41, 48, 50]. In general, two contiguous vertebrae and the interposed disk are involved. Osteomyelitis at multiple levels is observed in less than 5%. Continuous involvement is more frequent than discontinuous involvement at different levels.

Complications of vertebral osteomyelitis may cause the leading signs and symptoms at the time of diagnosis. The main complications are different types of abscesses, causing either persistence of infection or neurologic compromise. According to five studies involving 328 patients, complications occurred in 28% of the patients [11, 34, 41, 48, 50]. This rate is even higher in single studies with detailed information. Turunc *et al.* [32] reported neurologic symptoms (weakness of lower extremities, loss of sensibility, paraparesis, cauda equina syndrome) in 15 out of 30 patients. In the study of McHenry *et al.* [10], detailed information on complications is given. Motor weakness or paralysis is reported in 62/253 (25%) patients. In 14%, the paresis was severe. All neurologic defects were due to abscess formation, either by compression of the spinal cord (13%), cauda equina syndrome (6%), nerve root compression (5%), or compression of the femoral nerve by a psoas abscess (one patient). The motor dysfunction was most severe in patients with cervical, and least severe in those with lumbar osteomyelitis [10].

The most frequent complication is the formation of a spinal epidural abscess. Typical signs and symptoms are back pain, local tenderness, motor weakness, radicular pain, and sensory abnormalities [51]. In case of particularly severe, sharp, or lancinating pain, an epidural abscess should be actively searched with magnetic resonance imaging (MRI). Interestingly, the relative frequency of epidural abscesses is much higher in the cervical than in the thoracic and lumbar regions (29% versus 22% and 12%, respectively) [10]. Nevertheless, lumbar epidural abscesses are more frequent in absolute numbers, since lumbar osteomyelitis is much more frequent than cervical or thoracic involvement. In the same study, paravertebral abscesses were detected in 26% and disk space abscesses in 5% of the cases. Psoas abscesses are more frequent in patients with tuberculous spondylodiskitis [9].

Spinal deformity is rather the exception in patients with acute vertebral osteomyelitis. It mainly occurs in patients with a prolonged course prior to treatment (Figure 15.1). In the comparative study of Colmenero *et al.* [34], only 8% of 72 patients with pyogenic, but 41% of 42 patients with tuberculous spondylodiskitis suffered from spinal deformity. This significant difference is probably due to the more rapid start of treatment in acute osteomyelitis (see Chapter 16).

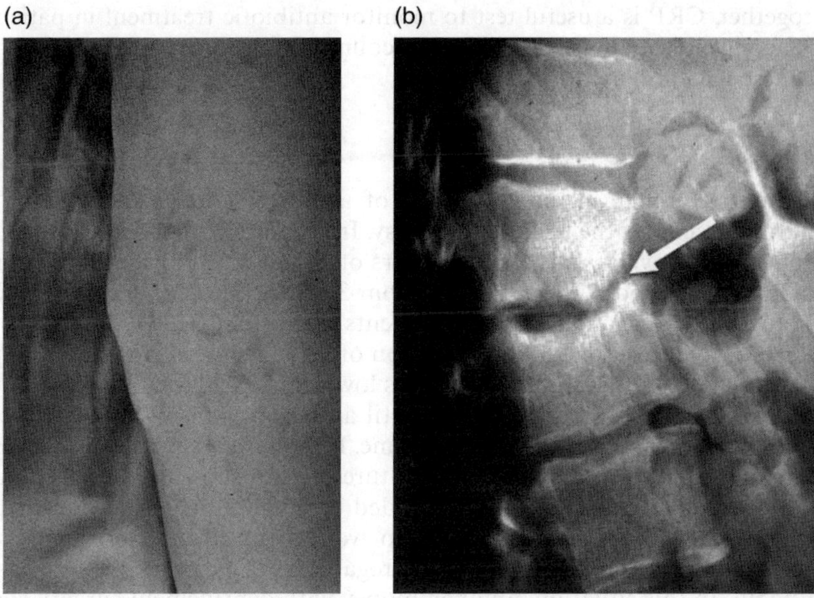

Figure 15.1. A 38-year-old IV drug addict who regularly injected heroin dissolved in lemon juice. He had a history of backache for several weeks when he presented with low-grade fever and immobilizing back pain. Biopsy of vertebra Th10 revealed *Candida albicans* in the culture. (a) gibbus due to severe ventral destruction of vertebra Th10. (b) plain radiograph of thoracic gibbus of the same patient. (*See insert for color representation of the figure.*)

Laboratory Investigation

Hematological and Biochemical Markers

Increased leukocyte or neutrophil counts have a low sensitivity for the diagnosis of osteomyelitis. Both parameters are increased in less than half of the cases [5, 52]. Normochromic anemia has been reported in up to three-quarters of the patients [53]. Erythrocyte sedimentation rate (ESR) is elevated in greater than 90% of the cases [52]. Since different authors use various thresholds, a comparison of different studies is difficult. Hopkinson [53] found an ESR greater than 50mm/h in 91% of their 22 patients with spondylodiskitis. Similarly, Priest and Peacock [29] reported a mean peak ESR of 101 mm/h (range 13 to > 140 mm/h) in 40 patients with staphylococcal hematogenous vertebral osteomyelitis. C-reactive protein (CRP) levels were increased in almost all cases. In the study of Chelsom and Solberg [26], all patients had a value of greater than 10 mg/l. Mete *et al.* [36] compared the characteristics of 100 patients with pyogenic, *Brucella*, or tuberculous spondylodiskitis. They found the highest CRP values in patients with pyogenic infection (mean values: 95 versus 34 versus 67 mg/l, respectively). The CRP level is more closely correlated with the clinical response to therapy than is the ESR, and is therefore the preferred marker to monitor treatment success [54]. While the ESR decreased only by 25% at week 8, CRP values decreased by greater than 60% within 1 week in patients with acute vertebral osteomyelitis [55].

Taken together, CRP is a useful test to monitor antibiotic treatment in patients with vertebral osteomyelitis. In contrast, leukocyte counts and ESR are not useful for the follow-up of these patients.

Blood Cultures

Blood cultures are crucial in the evaluation of acute vertebral osteomyelitis, since a positive culture precludes the need for a biopsy. In patients with a clinical suspicion of acute vertebral osteomyelitis, at least two pairs of blood cultures should be done. The fraction of positive blood cultures ranged from 30 to 72% with a mean of 52% in 11 different studies reporting data from 794 patients [10, 11, 26, 32, 34, 37, 38, 48, 50, 56, 57]. The large differences are due to a selection of patients in several studies reporting on biopsy results (e.g., [32, 57]. In view of this low chance of growth in blood cultures, antibiotic treatment should be restrained until a microorganism has been identified, provided that the patient has no sepsis syndrome. If vertebral osteomyelitis is suspected based on an imaging procedure, but blood cultures do not show the growth of a microorganism, a computed tomography (CT)-guided or open biopsy is needed. If polymicrobial osteomyelitis is suspected (e.g., deep wound infection after wound healing disturbance), a biopsy should be performed regardless of whether the blood cultures are positive [50]. In this situation, an open biopsy with debridement surgery should be performed.

Biopsy

The fraction of positive biopsy cultures ranged from 38 to 100% with a mean of 70% in 17 different studies reporting data from 951 patients [10–12, 26, 32, 34, 37, 38, 48, 50, 55–61]. In three of these studies, the role of empiric antibiotic therapy prior to biopsy was analyzed. In two studies, previous antimicrobial therapy clearly compromised the yield, with a drop from 50 to 25% in one study and from 60 to 23% in the other study [59, 60]. In contrast, in a recent study in which biopsy was performed in 56 patients with and 28 without previous antibiotics, the yield was somewhat higher in pretreated patients (71% versus 54%) [57]. The only rational explanation for this paradox is the possibly higher severity of disease in patients who were empirically treated with antibiotics prior to biopsy. Whether a negative CT-guided biopsy is repeated or followed by an open biopsy depends on the experience in each center. In the study of Marshall *et al.* [57], open biopsy had a much higher sensitivity than CT-guided biopsy (91% versus 53%). Yang *et al.* [62] reported a novel technique for biopsy sampling. They showed that percutaneous endoscopic discectomy had a significantly higher yield of positive cultures than CT-guided biopsy (90% versus 47%, $p < 0.0002$). These results have to be confirmed by other groups before this novel technique can be suggested as a routine procedure.

Bone samples should be cultured for aerobic, anaerobic, and fungal agents and a part of them sent for histopathology. In case of subacute/chronic presentation, suggestive history, or presence of granuloma in the histopathological analysis, mycobacteria and *Brucella* spp. should also be looked for (see Chapter 16). When blood and tissue cultures are negative despite suggestive histopathology, broad-range polymerase chain reaction (PCR) analysis of biopsy specimens or aspirated pus should be considered (see Chapter 2) [63, 64]. With this technique, unusual microorganisms such as *Tropheryma whippelii* can be found [65].

Imaging Procedures

Imaging procedures are the most important tools not only to diagnose osteomyelitis but also to rule out other diseases, such as bone metastases or osteoporotic fractures, which may imitate the clinical picture. They allow to precisely localize the infection and to look for pyogenic complications, such as paravertebral, psoas, epidural, or disk-space abscesses [5]. Plain radiography is a rational first step in patients without neurologic symptoms. It may reveal an alternative diagnosis such as vertebral fracture or bone metastases. However, due to its low sensitivity, it is generally not useful in early diagnosis of acute osteomyelitis.

The gold standard to diagnosing acute osteomyelitis is MRI [66]. It should be the first diagnostic step in patients with neurologic impairment in order to rule out a herniated disk or to rapidly detect pyogenic complications (Figure 15.2). MRI has an accuracy of at least 90% for diagnosing vertebral osteomyelitis [67]. It typically shows high signal intensity within the disk on T2-weighted sequences and loss of the intranuclear cleft. The vertebral end plates are rapidly destroyed, and a high-signal-intensity marrow edema is visible [5, 67]. Typically, the disk space and two adjacent vertebral bodies are involved (see pathogenesis) (Figure 15.3).

CT is less sensitive than MRI. However, it is helpful to guide a percutaneous biopsy.

Three-phase technetium-99m bone scans are typically positive within a few days after the onset of symptoms. However, the findings are nonspecific, and the accuracy only 67% [68].

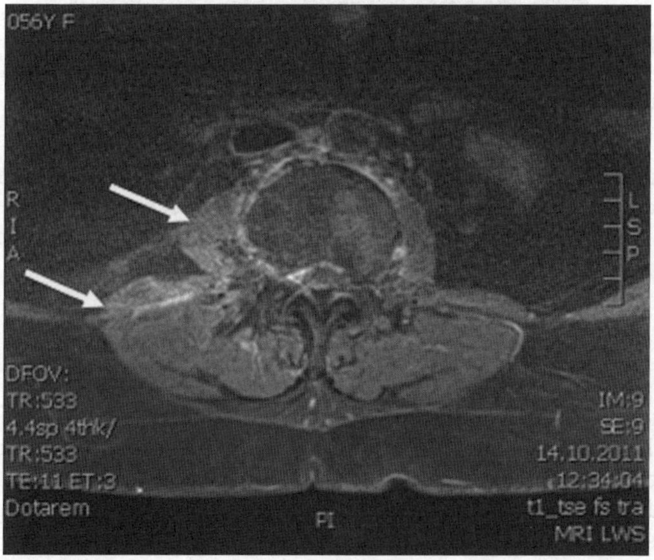

Figure 15.2. This 56-year-old woman was hospitalized with severe lumbar pain and fever. Five days earlier, she suffered from a febrile urinary tract infection treated with nitrofurantoin without consultation of a physician. At presentation, she had 38.2°C fever and severe pain in the L2/L3 region. CRP was at 345 mg/l and blood cultures showed growth of *Escherichia coli*. The MRI (T1 with gadolinium) performed 7 days after the first symptoms revealed a psoas abscess and enhancement of L2/L3 including the intervertebral disk (axial pictures not shown). A diagnosis was made of hematogenous *E. coli* vertebral osteomyelitis with right-side psoas abscess after urosepsis.

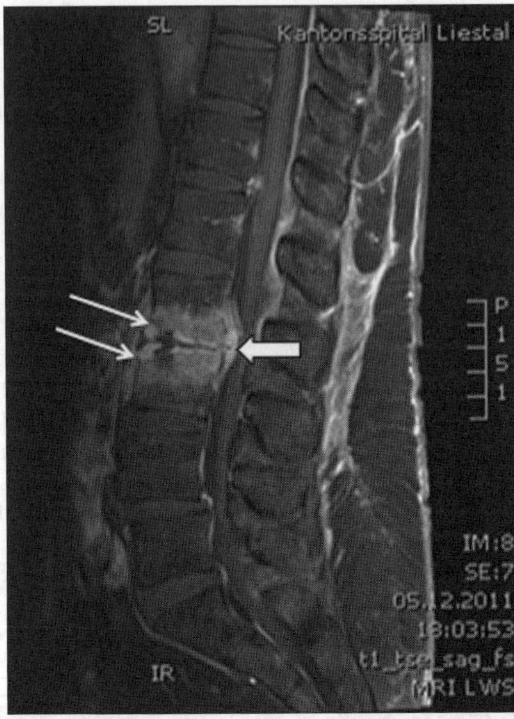

Figure 15.3. MRI (T1 with gadolinium) of the same patient as in Figure 15.2, 7 weeks later. Complete destruction of intervertebral disk, gadolinium enhancement in the bone marrow (thin arrows), and epidural abscess (thick arrow).

In the future, positron emission tomography (PET) scanning with 18F-fluorodeoxyglucose, which has an excellent diagnostic accuracy, may be an alternative imaging procedure in patients with contraindication for MRI. It can especially be considered in patients with implants or with suspicion of several foci [69].

Clinical and Imaging Differential Diagnosis

In a febrile patient with backache, the differential diagnosis is broad. It includes flu-like syndromes, pyelonephritis, pancreatitis, and duodenal ulcer disease, among other causes. In the absence of fever, the differential diagnosis of back pain is even broader, including osteoporotic fracture, spinal stenosis, and disk hernia. There is often a considerable delay between the onset of symptoms and the diagnosis, due to the nonspecific nature of signs and symptoms of vertebral osteomyelitis (see above). The mean interval ranges from 42 to 59 days in four studies including 242 patients [11, 34, 38, 48].

As mentioned earlier, MRI is the diagnostic gold standard for diagnosing acute osteomyelitis. However, even if the MRI pathology is suggestive of vertebral osteomyelitis, alternative diagnoses should be considered, especially if blood cultures are negative, which is the case in half of the patients (see earlier). The most frequent differential diagnosis is erosive osteochondrosis (Figure 15.4) [5]. In addition, gouty spondylodiskitis

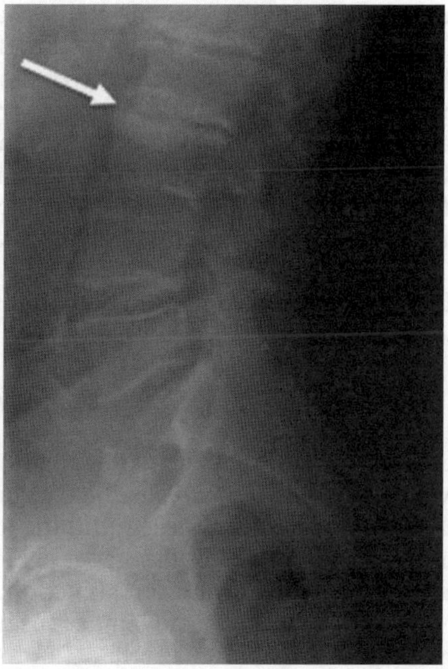

Figure 15.4. This 76-year-old man has been suffering from chronic lumbar backache since several years. He was hospitalized because of exacerbation of the pain and low-grade fever. The plain radiograph suggested vertebral osteomyelitis (arrow). Because the CRP was completely normal and a previous radiograph performed 3 years earlier was almost identical, the most probable diagnosis was erosive osteochondrosis. Therefore, no microbiological workup was performed, and no antibiotics were given. A 3-year follow-up was uneventful, confirming the diagnosis of erosive osteochondrosis.

[70], erosive diskovertebral lesions (Andersson lesions) in ankylosing spondylitis [71], and aseptic bone necrosis [72] may also mimic vertebral osteomyelitis.

Treatment

The aims of therapy are (i) relief of back pain, (ii) elimination of the microorganisms, (iii) protection from further bone loss, (iv) prevention of complications, (v) evacuation of purulent collections, and (vi) stabilization, if needed. In most cases of acute osteomyelitis, surgical treatment is not required.

Conservative Therapy

Nonsurgical management consists of antimicrobial therapy, analgetics, physiotherapy, and initial bed rest [5, 52, 73]. Imperative bed rest is generally not required, except in case of severe positional pain or spinal instability. Complete immobilization with a cast is not used in adults. In case of neurological deficits, intractable back pain, or spinal instability, rapid debridement with internal fixation is indicated [74, 75] (see later). Since directed antimicrobial therapy against a defined microorganism allows a more reliable treatment, microbiological diagnosis should always be performed before starting antibiotics. Empirical antimicrobial therapy, after drawing at least two pairs of blood cultures, should only be considered in hemodynamically unstable patients, or those with neurological symptoms indicating local compression. If blood cultures remain negative, CT-guided biopsy, aspiration of possible pus collection, or open biopsy for microbiology and histopathology should be performed. If empirical therapy is clinically needed, the regimen should cover the most frequent microorganisms, that is, *S. aureus*, streptococci, and gram-negative bacilli.

Data from randomized, controlled trials on antimicrobial therapy of acute osteomyelitis are lacking. In six case series with a total of 581 patients, the survival without relapse was 89% (range: 72–95%) [10, 11, 34, 41, 48, 50]. Only in the large study of McHenry *et al.* [10], the success rate was below 90%. However, in this study, one-third of the patients had nosocomial vertebral osteomyelitis.

Traditionally, bone infections were initially treated by the IV route. However, the need of an IV therapy is not evidence-based. It could well be that oral therapy is equally efficacious, as long as the oral drug and/or the patient fulfill the following requirements: (i) optimal spectrum of the antibiotic (e.g., fluoroquinolones against gram-negative bacilli), (ii) excellent oral bioavailability, (iii) normal intestinal function, and (iv) no vomiting. An initial parenteral therapy of some days may be advantageous against microorganisms at risk for emergence of resistance. This is the case if a fluoroquinolone or a rifampin combination is used.

Table 15.5 provides a summary of suggested antibiotic regimens for the most common microorganisms. Suggestions are based on observational studies, expert opinions, and pharmacokinetic/pharmacodynamic considerations (see Chapter 3) [5].

There are no data from controlled trials on the optimal duration of antimicrobial therapy. The recommendations range from 4 to 6 weeks [76, 77] to 3 months [49, 78]. Most experts suggest a total duration of 6 weeks [5, 66]. However, a longer treatment is recommended in patients with undrained abscesses or with spinal implants (see Chapter 21). Treatment efficacy should be regularly monitored by asking for symptoms (fever, pain), and by checking CRP values. Follow-up MRI controls are only needed in patients with pyogenic complications, since there is a very poor correlation between clinical healing and improvement on MRI [79].

Surgical Treatment

In acute not implant-associated osteomyelitis, surgical intervention is needed only in select cases. First, in patients with negative blood cultures and no growth in the CT-guided biopsy, open biopsy may have a better sensitivity [57]. Second, in patients with large abscesses, which cannot be drained by noninvasive catheter procedures, open debridement allows more rapid control of infection, and shorter antibiotic therapy. Third, in patients with progressive neurological deficits or spinal instability, debridement and internal stabilization may be needed as an urgent intervention [51, 79–84]. Progressive demineralization in the MRI (Figure 15.3) or CT up to about 6 weeks is not an indication of internal stabilization in a patient with vertebral osteomyelitis [66]. However, if bone imaging shows substantial destruction of the bone (Figure 15.1), an orthopedic surgeon should be consulted regarding the use of a fitted back brace or internal fixation. As a rule, surgical debridement is required in patients with spinal implants (cf. Chapter 21) [6].

Table 15.5. Antibiotic therapy of osteomyelitis without implant in adults.

Microorganism	Antimicrobial agent[a]	Dose[b]	Route
***Staphylococcus* spp.**			
Meticillin-susceptible	Nafcillin[c]	2 g every 6 h	IV
	Followed by		
	Rifampin plus	300–450 mg every 12 h	PO
	Levofloxacin	750 mg every 24 h to 500 mg every 12 h	PO
Meticillin-resistant	Vancomycin or	15 mg/kg every 12 h[d]	IV
	daptomycin	>6–8 mg/kg every 24 h	IV
	Followed by		
	Rifampin plus	300–450 mg every 12 h	PO
	Levofloxacin or	750 mg every 24 h to 500 mg every 12 h	PO
	TMP/SMX or	1 forte tablet every 8 h	PO
	Fusidic acid	500 mg every 8 h	PO
Streptococcus spp.	Penicillin G[c] or	5 million U every 6 h	IV
	Ceftriaxone	2 g every 24 h	IV
Enterobacteriaceae (quinolone-susceptible)	Ciprofloxacin	750 mg every 12 h	PO
Enterobacteriaceae (quinonlone-resistant, including ESBL)	Imipenem	500 mg every 6 h	
Pseudomonas aeruginosa	Cefepime or ceftazidime plus aminoglycoside[e] or	2 g every 8 h	IV
	Piperacillin/tazobactam plus		IV
	Aminoglycoside[e]	4.5 g every 8 h	IV
	For 2–4 weeks, followed by		
	Ciprofloxacin[f]	750 mg every 12 h	PO
Anaerobes	Clindamycin	600 mg every 6–8 h	IV
	For 2–4 weeks, followed by		
	Clindamycin[g]	300 mg every 6 h	PO

Reprinted with permission from Ref. 5. © 2010, Massachusetts Medical Society.
PO, orally; IV, intravenously; IM, intramuscularly, forte tablet TMP/SMX: trimethoprim 160 mg plus sulfamethoxazole 800 mg.
[a] The total duration of antimicrobial treatment is generally 6 weeks.
[b] All dosages are for adults assuming normal renal function.
[c] In patients with delayed hypersensitivity, cefuroxime (1.5 g every 6–8 h IV) can be administered. In patients with immediate hypersensitivity, penicillin should be replaced by vancomycin (1 g every 12 h IV).
[d] Vancomycin through concentrations of 15–20 μg/ml.
[e] The need for addition of an aminoglycoside is not proven. However, it may decrease the risk of emergence of resistance to the betalactam.
[f] The rational for starting ciprofloxacin only after pretreatment with a betalactam is the increased risk for emergence of quinolone resistance in the presence of high bacterial load.
[g] Alternatively, penicillin G (5 million U every 6 h IV) or ceftriaxone (2 g every 24 h IV) can be used for gram-positive anaerobes (e.g., *Propionibacterium acnes*), and metronidazole (500 mg every 8 h IV or PO) for gram-negative anaerobes (e.g., *Bacteroides* spp.).

Key Points

- In adults, acute hematogenous osteomyelitis predominantly involves the vertebral column.
- The incidence steadily increases with age, from 0.3 per 100,000 among persons younger than 20 years of age to 6.5 per 100,000 among persons older than 70 years of age.
- *Staphylococcus aureus* causes about half, streptococci one-fifth, and *Escherichia coli* one-seventh of the episodes of acute vertebral osteomyelitis.
- Diabetes mellitus is the most frequent underlying medical condition.
- In about half of the patients, a primary distant focus of infection can be determined, and urogenital and skin/soft tissue infections are the most frequent distant sources of hematogenous vertebral osteomyelitis.
- There is no controlled study on the optimal duration of antimicrobial therapy. Most experts suggest a total duration of 6 weeks.

References

[1] Waldvogel FA, Medoff G, Swartz MN. Osteomyelitis: a review of clinical features, therapeutic considerations and unusual aspects. N Engl J Med 1970;282:198-206.

[2] Widmer A, Zimmerli W. Fatal peripheral catheter phlebitis. Schweiz Med Wochenschr 1988;118(27–28):1053–1055.

[3] Gallie WE. First recurrence of osteomyelitis eighty years after infection. J Bone Joint Surg Br 1951;33-B(1):110–111.

[4] Jensen AG, Espersen F, Skinhoj P, *et al.* Increasing frequency of vertebral osteomyelitis following Staphylococcus aureus bacteraemia in Denmark 1980-1990. J Infect 1997;34(2): 113–118.

[5] Zimmerli W. Clinical practice. Vertebral osteomyelitis. N Engl J Med 2010;362(11): 1022–1029.

[6] Kowalski TJ, Berbari EF, Huddleston PM, *et al.* The management and outcome of spinal implant infections: contemporary retrospective cohort study. Clin Infect Dis 2007;44: 913–920.

[7] Paakkonen M, Peltola H. Bone and joint infections. Pediatr Clin North Am 2013;60(2): 425–436.

[8] Murillo O, Roset A, Sobrino B, *et al.* Streptococcal vertebral osteomyelitis: multiple faces of the same disease. Clin Microbiol Infect 2014;20(1):33–38. DOI:10.1111/1469-0691.12302 [Epub 2013 July 24].

[9] Gehlot PS, Chaturvedi S, Kashyap R, Singh V. Pott's spine: retrospective analysis of MRI scans of 70 cases. J Clin Diagn Res 2012;6(9):1534–1538.

[10] McHenry MC, Easley KA, Locker GA. Vertebral osteomyelitis: long-term outcome for 253 patients from 7 Cleveland-area hospitals. Clin Infect Dis 2002;34(10):1342–1350.

[11] Nolla JM, Ariza J, Gomez-Vaquero C, *et al.* Spontaneous pyogenic vertebral osteomyelitis in nondrug users. Semin Arthritis Rheum 2002;31(4):271–278.

[12] Hadjipavlou A, Mader J, Necessary J, Muffoletto A. Hematogenous pyogenic spinal Infections and their surgical management. Spine 2000;25:1668–1679.

[13] Lio P, Paoletti N, Moni MA, *et al.* Modelling osteomyelitis. BMC Bioinformatics 2012;13(Suppl 14):S12.

[14] Claro T, Widaa A, O'Seaghdha M, *et al.* Staphylococcus aureus protein A binds to osteoblasts and triggers signals that weaken bone in osteomyelitis. PLoS One 2011;6(4):e18748.

[15] Claro T, Widaa A, McDonnell C, *et al.* Staphylococcus aureus protein A binding to osteoblast tumour necrosis factor receptor 1 results in activation of nuclear factor kappa B and release of interleukin-6 in bone infection. Microbiology 2013;159(Pt 1):147–154.

[16] Ahmed S, Meghji S, Williams RJ, *et al.* Staphylococcus aureus fibronectin binding proteins are essential for internalization by osteoblasts but do not account for differences in intracellular levels of bacteria. Infect Immun 2001;69(5):2872–2877.

[17] Nair SP, Williams RJ, Henderson B. Advances in our understanding of the bone and joint pathology caused by Staphylococcus aureus infection. Rheumatology (Oxford) 2000;39(8): 821–834.

[18] Allignet J, Galdbart JO, Morvan A, *et al.* Tracking adhesion factors in Staphylococcus caprae strains responsible for human bone infections following implantation of orthopaedic material. Microbiology 1999;145(Pt 8):2033–2042.

[19] Rasigade JP, Moulay A, Lhoste Y, *et al.* Impact of sub-inhibitory antibiotics on fibronectin-mediated host cell adhesion and invasion by Staphylococcus aureus. BMC Microbiol 2011;11:263.

[20] Sinha B, Francois P, Que YA, *et al.* Heterologously expressed Staphylococcus aureus fibronectin-binding proteins are sufficient for invasion of host cells. Infect Immun 2000;68(12):6871–6878.

[21] Gristina A. Adherent bacterial colonization in the pathogenesis of osteomyelitis. Science 1985;228:990–993.

[22] Spellberg B, Lipsky BA. Systemic antibiotic therapy for chronic osteomyelitis in adults. Clin Infect Dis 2012;54(3):393–407.

[23] Ciampolini J, Harding KG. Pathophysiology of chronic bacterial osteomyelitis. Why do antibiotics fail so often? Postgrad Med J 2000;76(898):479–483.

[24] Gillespie W. The epidemiology of acute haematogenous osteomyelitis of childhood. Int J Epidemiol 1985;14:600–606.

[25] Grammatico L, Baron S, Rusch E, *et al.* Epidemiology of vertebral osteomyelitis (VO) in France: analysis of hospital-discharge data 2002-2003. Epidemiol Infect 2008;136(5): 653–660.

[26] Chelsom J, Solberg CO. Vertebral osteomyelitis at a Norwegian university hospital 1987-97: clinical features, laboratory findings and outcome. Scand J Infect Dis 1998;30(2):147–151.

[27] Bhavan KP, Marschall J, Olsen MA, *et al.* The epidemiology of hematogenous vertebral osteomyelitis: a cohort study in a tertiary care hospital. BMC Infect Dis 2010;10:158.

[28] Jensen AG, Espersen F, Skinhoj P, Frimodt-Moller N. Bacteremic Staphylococcus aureus spondylitis. Arch Intern Med 1998;158(5):509–517.

[29] Priest DH, Peacock JE, Jr. Hematogenous vertebral osteomyelitis due to Staphylococcus aureus in the adult: clinical features and therapeutic outcomes. South Med J 2005;98(9): 854–862.

[30] Sakkas LI, Davas EM, Kapsalaki E, *et al.* Hematogenous spinal infection in central Greece. Spine (Phila Pa 1976) 2009;34(15):E513–E518.

[31] Corrah TW, Enoch DA, Aliyu SH, Lever AM. Bacteraemia and subsequent vertebral osteomyelitis: a retrospective review of 125 patients. QJM 2011;104(3):201–207.

[32] Turunc T, Demiroglu YZ, Uncu H, Colakoglu S, Arslan H. A comparative analysis of tuberculous, brucellar and pyogenic spontaneous spondylodiscitis patients. J Infect 2007;55(2): 158–163.

[33] Kim CJ, Song KH, Jeon JH, *et al.* A comparative study of pyogenic and tuberculous spondylodiscitis. Spine (Phila Pa 1976) 2010;35(21):E1096–E1100.

[34] Colmenero JD, Jimenez-Mejias ME, Sanchez-Lora FJ, *et al.* Pyogenic, tuberculous, and brucellar vertebral osteomyelitis: a descriptive and comparative study of 219 cases. Ann Rheum Dis 1997;56(12):709–715.

[35] Torda AJ, Gottlieb T, Bradbury R. Pyogenic vertebral osteomyelitis: analysis of 20 cases and review. Clin Infect Dis 1995;20(2):320–328.

[36] Mete B, Kurt C, Yilmaz MH, *et al.* Vertebral osteomyelitis: eight years' experience of 100 cases. Rheumatol Int 2012;32(11):3591–3597.

[37] Carragee EJ. Pyogenic vertebral osteomyelitis. J Bone Joint Surg Am 1997;79(6):874–880.

[38] Bateman J, Peuzner M. Spinal osteomyelitis: a review of 10 years' experience. Orthopedics 1995;18:561–565.

[39] Uckay I, Dinh A, Vauthey L, *et al.* Spondylodiscitis due to Propionibacterium acnes: report of twenty-nine cases and a review of the literature. Clin Microbiol Infect 2010;16:353.
[40] Bucher E, Trampuz A, Donati L, Zimmerli W. Spondylodiscitis associated with bacteraemia due to coagulase-negative staphylococci. Eur J Clin Microbiol Infect Dis 2000;19(2): 118–120.
[41] Pigrau C, Almirante B, Flores X, *et al.* Spontaneous pyogenic vertebral osteomyelitis and endocarditis: incidence, risk factors, and outcome. Am J Med 2005;118(11):1287.
[42] Stoeckle M, Kaech C, Trampuz A, Zimmerli W. The role of diabetes mellitus in patients with bloodstream infections. Swiss Med Wkly 2008;138(35–36):512–519.
[43] Knerr K, Füth R, Hemsen P, *et al.* Chronic inflammation and hemodialysis reduce immune competence of peripheral blood leukocytes in end-stage renal failure patients. Cytokine 2005;30:132–138.
[44] Scheidegger C, Zimmerli W. Incidence and spectrum of severe medical complications among hospitalized HIV-seronegative and HIV-seropositive narcotic drug users. AIDS 1996;12: 1407–1414.
[45] Zimmerli W, Borer H, Rüttiman S. Pyomyositis and cryptococcal sepsis in an HIV-infected patient with severe granulocyte defect. AIDS 1992;6:1399–1400.
[46] Cohen MS, Chen YQ, McCauley M, *et al.* Prevention of HIV-1 infection with early antiretroviral therapy. N Engl J Med 2011;365(6):493–505.
[47] Weinstein MA, Eismont FJ. Infections of the spine in patients with human immunodeficiency virus. J Bone Joint Surg Am 2005;87(3):604–609.
[48] Osenbach RK, Hitchon PW, Menezes AH. Diagnosis and management of pyogenic vertebral osteomyelitis in adults. Surg Neurol 1990;33(4):266–275.
[49] Peronne C, Saba J, Behloul ZS-C, *et al.* Pyogenic and tuberculous spondylodiskitis (vertebral osteomyelitis) in 80 adult patients. Clin Infect Dis 1994;19:746–750.
[50] Patzakis MJ, Rao S, Wilkins J, *et al.* Analysis of 61 cases of vertebral osteomyelitis. Clin Orthop Relat Res 1991;264:178–183.
[51] Sendi P, Bregenzer T, Zimmerli W. Spinal epidural abscess in clinical practice. QJM 2008;101(1):1–12.
[52] Gouliouris T, Aliyu SH, Brown NM. Spondylodiscitis: update on diagnosis and management. J Antimicrob Chemother 2010;65(Suppl 3):iii11–iii24.
[53] Hopkinson N, Stevenson J, Benjamin S. A case ascertainment study of septic discitis: clinical, microbiological and radiological features. QJM 2001;94(9):465–470.
[54] Khan MH, Smith PN, Rao N, Donaldson WF. Serum C-reactive protein levels correlate with clinical response in patients treated with antibiotics for wound infections after spinal surgery. Spine J 2006;6(3):311–315.
[55] Yoon SH, Chung SK, Kim KJ, *et al.* Pyogenic vertebral osteomyelitis: identification of microorganism and laboratory markers used to predict clinical outcome. Eur Spine J 2010;19(4):575–582.
[56] Zarrouk V, Feydy A, Sallè F, *et al.* Imaging does not predict the clinical outcome of bacterial vertebral osteomyelitis. Rheumatology (Oxford) 2007;46:292–295.
[57] Marschall J, Bhavan KP, Olsen MA, *et al.* The impact of prebiopsy antibiotics on pathogen recovery in hematogenous vertebral osteomyelitis. Clin Infect Dis 2011;52(7):867–872.
[58] Enoch DA, Cargill JS, Laing R, *et al.* Value of CT-guided biopsy in the diagnosis of septic discitis. J Clin Pathol 2008;61(6):750–753.
[59] de Lucas EM, Gonzalez Mandly A, Gutierrez A, *et al.* CT-guided fine-needle aspiration in vertebral osteomyelitis: true usefulness of a common practice. Clin Rheumatol 2009;28(3): 315–320.
[60] Rankine JJ, Barron DA, Robinson P, *et al.* Therapeutic impact of percutaneous spinal biopsy in spinal infection. Postgrad Med J 2004;80(948):607–609.
[61] Chew FS, Kline MJ. Diagnostic yield of CT-guided percutaneous aspiration procedures in suspected spontaneous infectious diskitis. Radiology 2001;218(1):211–214.

[62] Yang SC, Fu TS, Chen LH, *et al.* Identifying pathogens of spondylodiscitis: percutaneous endoscopy or CT-guided biopsy. Clin Orthop Relat Res 2008;466(12):3086–3092.
[63] Fenollar F, Roux V, Stein A, *et al.* Analysis of 525 samples to determine the usefulness of PCR amplification and sequencing of the 16S rRNA gene for diagnosis of bone and joint infections. J Clin Microbiol 2006;44(3):1018–1028.
[64] Fuursted K, Arpi M, Lindblad BE, Pedersen LN. Broad-range PCR as a supplement to culture for detection of bacterial pathogens in patients with a clinically diagnosed spinal infection. Scand J Infect Dis 2008;40(10):772–777.
[65] Weber U, Morf MH, Gubler JG, *et al.* Spondylodiscitis as the first manifestation of Whipple's disease—a removal worker with chronic low back pain. Clin Rheumatol 2003;22:443–446.
[66] Berbari E, Kanj S, Kowalski TJ, *et al.* IDSA Clinical practice guidelines for the diagnosis and treatment of native vertebral osteomyelitis in adults. Clin Infect Dis 2014 (in press).
[67] Palestro CJ, Love C, Miller TT. Infection and musculoskeletal conditions: Imaging of musculoskeletal infections. Best Pract Res Clin Rheumatol 2006;20(6):1197–1218.
[68] Love C, Patel M, Lonner BS, *et al.* Diagnosing spinal osteomyelitis: a comparison of bone and Ga-67 scintigraphy and magnetic resonance imaging. Clin Nucl Med 2000;25(12): 963–977.
[69] Strobel K, Stumpe K. PET/CT in musculoskeletal infection. Semin Musculoskelet Radiol 2007;11:353–364.
[70] Rufener J, Schulze CC, Tanzler K, *et al.* Sterile spondylodiscitis. Lancet 2012;379(9828):1850.
[71] Park YS, Kim JH, Ryu JA, Kim TH. The Andersson lesion in ankylosing spondylitis: distinguishing between the inflammatory and traumatic subtypes. J Bone Joint Surg Br 2011;93(7): 961–966.
[72] Lin CL, Lin RM, Huang KY, *et al.* MRI fluid sign is reliable in correlation with osteonecrosis after vertebral fractures: a histopathologic study. Eur Spine J 2013;22(7):1617–1623.
[73] Jordan MC, Kirby WM. Pyogenic vertebral osteomyelitis. Treatment with antimicrobial agents and bed rest. Arch Intern Med 1971;128(3):405–410.
[74] Redfern RM, Miles J, Banks AJ, Dervin E. Stabilisation of the infected spine. J Neurol Neurosurg Psychiatry 1988;51(6):803–807.
[75] Kulowski J. The orr treatment of pyogenic osteomyelitis. Annals of Surgery 1936;103: 622–623.
[76] Roblot F, Besnier JM, Juhel L, *et al.* Optimal duration of antibiotic therapy in vertebral osteomyelitis. Semin Arthritis Rheum 2007;36(5):269–277.
[77] Gilbert D, Moellering R, Eliopoulos G, *et al.* The SANFORD Guide to Antimicrobial Therapy. 39th edition. Sperryville: Antimicrobial Therapy; 2009, p. 66.
[78] Livorsi DJ, Daver NG, Atmar RL, *et al.* Outcomes of treatment for hematogenous *Staphylococcus aureus* vertebral osteomyelitis in the MRSA ERA. J Infect 2008;57(2): 128–131.
[79] Kowalski TJ, Berbari EF, Huddleston PM, *et al.* Do follow-up imaging examinations provide useful prognostic information in patients with spine infection? Clin Infect Dis 2006;43(2):172–179.
[80] Dai LY, Chen WH, Jiang LS. Anterior instrumentation for the treatment of pyogenic vertebral osteomyelitis of thoracic and lumbar spine. Eur Spine J 2008;17(8):1027–1034.
[81] Chen WH, Jiang LS, Dai LY. Surgical treatment of pyogenic vertebral osteomyelitis with spinal instrumentation. Eur Spine J 2007;16:1307–1316.
[82] Ito M, Abumi K, Kotani Y, *et al.* Clinical outcome of posterolateral endoscopic surgery for pyogenic spondylodiscitis: results of 15 patients with serious comorbid conditions. Spine (Phila Pa 1976) 2007;32(2):200–206.
[83] Yang SC, Fu TS, Chen LH, *et al.* Percutaneous endoscopic discectomy and drainage for infectious spondylitis. Int Orthop 2007;31(3):367–373.
[84] Robinson Y, Tschoeke SK, Kayser R, *et al.* Reconstruction of large defects in vertebral osteomyelitis with expandable titanium cages. Int Orthop 2009;33(3):745–749.

Chapter 16

Subacute Osteomyelitis: Tuberculous and Brucellar Vertebral Osteomyelitis

Juan D. Colmenero and Pilar Morata

Introduction

Classically, vertebral osteomyelitis has been divided into acute and subacute/chronic presentation depending on the onset of symptoms and their duration before diagnosis. Consequently, and somewhat artificially, the term acute vertebral osteomyelitis is erroneously considered as a synonym of pyogenic vertebral osteomyelitis. This entity is generally caused by *Staphylococcus aureus*, *Streptococcus* spp., and Gram-negative bacilli. On the other hand, the term subacute or chronic vertebral osteomyelitis has been related with specific bacterial etiologies, such as *Mycobacterium tuberculosis*, *Brucella* spp., and different types of fungi.

Current clinical practice, however, shows that this earlier-mentioned classification is not always correct. Vertebral osteomyelitis caused by basically aggressive and pyogenic pathogens can also proceed with a prolonged and indolent course. On the contrary, subacute/chronic osteomyelitis may cause a rapid and even destructive infection in an immunocompromised host. Thus, even knowing the infective agent, the course of disease is not reliably predictable, since it depends not only on the microorganism, but also on the host.

In community-acquired vertebral osteomyelitis, the subacute or chronic course necessitates ruling out specific etiologies, such as infection by *M. tuberculosis* or *Brucella* spp. This chapter deals with these two particular infections, which, given their granulomatous nature and subacute course, are clinically, radiologically, and histologically difficult to differentiate from each other.

Epidemiology

Brucellosis is an emerging and reemerging chronic debilitating infectious disease, highly endemic in many countries in Africa, Central and South America, the Middle East, and the Indian subcontinent [1]. Brucellosis remains the world's most common bacterial

Bone and Joint Infections: From Microbiology to Diagnostics and Treatment, First Edition.
Edited by Werner Zimmerli.

zoonosis, with over half a million new cases annually and prevalence rates in some countries exceeding 10 cases per 100,000 population [2]. There is plenty of evidence to support the conclusion that in countries without strong health systems, official data likely underestimate the true burden [2].

A lower disease incidence is seen in developed countries as compared to low- and middle-income countries. Nevertheless, brucellosis clearly remains a disease of public health importance, even in developed countries that share borders with countries with the highest incidence. As an example, in the United States, the incidence of brucellosis in border states near Mexico is eight times higher than in nonborder states [3]. Finally, multiple data seem to confirm that in the past 15 years the epidemiology of human brucellosis has evolved, reflecting changes resulting from the phenomenon of globalization and international tourism [4].

Tuberculosis is a major global cause of death and disease. The global burden of tuberculosis remains enormous. Recent data from the World Health Organization (WHO) Global Tuberculosis Report 2012 confirm that tuberculosis remains a major infectious killer. In 2011, there were an estimated 8.7 million new cases and 1.4 million people died from tuberculosis [5]. The increased incidence of tuberculosis in some countries in the Western hemisphere at the end of the 1980s and early 1990s was associated with important changes in the epidemiological and clinical features of the disease. This increase was due to socioeconomic factors, the HIV epidemic, and the rising number of patients with severe debilitating diseases or immunosuppression [6].

The incidence of tuberculosis in different European regions varies among and within the countries, from less than 1 tuberculosis case per 100,000 population to over 200 tuberculosis cases per 100,000 in others. The 53 countries in the WHO European region account for around 4.4% of the world's cases, representing an estimated 380,000 individuals with a new episode of tuberculosis or 42 cases per 100,000 population [7]. Both tuberculosis and brucellosis are systemic granulomatous diseases that may affect almost all organs and systems [8, 9].

In the largest series of brucellosis published to date, 20–40% of cases involved at least one focal complication. Among the various focal forms, osteoarticular complications are, without doubt, the most common [10]. In adults, bone and joint infections generally affect the axial skeleton, with vertebral osteomyelitis accounting for 35–50% of all osteoarticular complications [11].

Tuberculosis can affect any organ or system. Approximately one-fifth of all patients with tuberculosis have extrapulmonary involvement [12]. Moreover, in developed countries, the incidence of extrapulmonary tuberculosis has not declined at the same rate as pulmonary tuberculosis [12–14]. This difference cannot be explained solely by the high prevalence of HIV infection; it has also been associated with reactivation of latent tuberculosis infection in immigrants [13, 15]. Bone and joint involvement occurs in 6–12% of all cases of extrapulmonary tuberculosis [12–14, 16], and spinal involvement is certainly the most common form of osteoarticular tuberculosis [13, 17].

Clinical Features

Brucellar vertebral osteomyelitis (BVO) can appear in the context of an acute infection, as an isolated focal complication, or even as the expression of disease relapse [18–20]. Likewise, tuberculous vertebral osteomyelitis (TVO) can accompany pulmonary tuberculosis, with or without miliary dissemination, or be the expression of a late extrapulmonary reactivation [17, 21, 22].

Unlike nonsurgical pyogenic vertebral osteomyelitis, which usually affects patients beyond 60 years of age with comorbidity, immunocompromised persons, patients with focal infection at a distant site or previous bacteremia, BVO and TVO usually affect patients below 60 years of age without relevant medical history. In a recent analysis of our cases, the average age of patients with BVO was 52.3 ± 14.0 years and 48.5 ± 17.8 years for those with TVO [23].

BVO predominates in rural males, while TVO affects both men and women in urban areas. In Western countries, TVO has generally been considered to be a disease mainly affecting persons older than 50 years of age; however, in our experience, the age of patients with TVO has a bimodal distribution, with two peaks, one between 20 and 30 years and another between 60 and 70 years [23].

Backache is the key symptom of vertebral osteomyelitis. It is present in over 85% of cases and almost always has inflammatory characteristics, meaning that it is not relieved by rest. Spinal pain is a very common symptom in clinical practice, and physicians will need to face its initial management in many patients. Although vertebral osteomyelitis should be included in the differential diagnosis of any back pain, it is much less frequent than other spine diseases, which explains why physician do not always initially consider vertebral osteomyelitis in a patient with backache. Unlike pyogenic vertebral osteomyelitis, BVO and TVO have a lower clinical expressivity and a more subacute course. This, coupled with the absence of fever in a high proportion of patients, especially in cases of TVO, leads clinicians to not consider the possibility of infection, thereby resulting in a greater diagnostic delay. In a study by our group, diagnostic delays in BVO and TVO were significantly higher than in hematogenous pyogenic vertebral osteomyelitis, 14.3 and 22.9 weeks versus 7.1 weeks [23]. Table 16.1 shows the most relevant clinical characteristics.

Table 16.1. Main clinical characteristics of patients with tuberculous and *Brucella* vertebral osteomyelitis.

	TVO[a]	BVO[b]
	Number of cases, 110	Number of cases, 97
	Number of patients (%)	
Sex (male/female)	59, (53.6)/51 (46.6)	69, (71.1)/28 (28.9)
Fever	41 (37.3)	89 (91.8)
Chills/rigor	25 (22.7)	75 (77.3)
Constitutional symptoms[c]	52 (47.3)	61 (62.9)
Inflammatory spinal pain[d]	96 (87.3)	92 (94.8)
Increase in local kyphosis	40 (36.4)	6 (6.2)
Paravertebral tenderness	64 (58.2)	75 (77.3)
Neurological deficits	68 (61.9)	30 (30.9)
Motor weakness or paralysis	45 (40.9)	7 (7.9)

[a]TVO, tuberculous vertebral osteomyelitis.
[b]BVO, brucellar vertebral osteomyelitis.
[c]Two or more of anorexia, adynamia, or malaise.
[d]Spontaneous spinal pain unrelieved by rest.

Table 16.2. Frequency of vertebral level involved according to etiological group.

	TVO[a]	BVO[b]
	Number of cases (%)	
Cervical	3 (2.7)	9 (8.2)
Thoracic	**51 (46.4)**	18 (18.6)
Thoraco-lumbar hinge	11 (10)	2 (2.1)
Lumbar	35 (31.8)	**56 (57.7)**
Lumbosacral	5 (4.5)	10 (10.3)
Multiple levels	5 (4.5)	3 (3.1)

[a]TVO, tuberculous vertebral osteomyelitis.
[b]BVO, brucellar vertebral osteomyelitis.

Although any level of the spine can be affected, BVO most often affects the lumbar level and TVO the thoracic level (Table 16.2). The prevalence of neurological deficits is high in both BVO and TVO, ranging from 20 to 30% in BVO and 22 to 76% in TVO.

Contrary to what happens in brucellosis or pyogenic infections, in TVO, bone regeneration is very low, resulting in a large destructive effect that may typically lead to vertebral collapse. This slow consolidation, coupled with the fact that most cases of TVO affect the thoracic level or thoracolumbar hinge, explains the progressive kyphosis that is a common clinical finding in TVO [24]. The increased tendency to produce vertebral collapse, the frequent involvement of the thoracic level, and the high diagnostic delay explain the higher prevalence of severe motor deficits, paraplegia, and paraparesis in TVO than in BVO [17].

Laboratory Investigation

The usual hematological and biochemical parameters are of little value in the diagnosis of vertebral osteomyelitis [23, 25, 26]. This diagnostic difficulty is observed in any case of vertebral osteomyelitis; however, it is even more frequent in patients with TVO and BVO. In our experience, corroborated by other authors [17, 21, 22], the absolute number of leukocytes and differential counts tend to have values within the normal range in patients with TVO. The same is true for patients with BVO. Only 10–15% of them have mild leukocytosis, with most patients having normal or even low leukocyte counts [18–20].

Most patients, in both TVO and BVO, have mild normocytic and normochromic anemia and mild hypoalbuminemia, with the other usual biochemical parameters within the normal range [23].

C-reactive protein is the only biochemical parameter that is somewhat different in the various types of vertebral osteomyelitis. The mean level of C-reactive protein is usually lower in BVO than in pyogenic vertebral osteomyelitis. In patients with TVO the mean level is between the one of pyogenic vertebral osteomyelitis and of BVO. The combined presence of leukocytosis, neutrophilia, and very high values of the erythrocyte sedimentation rate (ESR) and C-reactive protein strongly suggests pyogenic vertebral osteomyelitis [23]. Table 16.3 shows the main hematological and biochemical data present in our patients.

Table 16.3. Summary of hematological and biochemical data.

	TVO[a]	BVO[b]
	Mean ± standard deviation	
Leukocytes (×10³ cells/μl)	8.1 ± 3.6	6.9 ± 2.7
Neutrophils (×10³ cells/μl)	5.6 ± 3.1	4.2 ± 2.1
Hemoglobin (g/dl)	12.9 ± 8.3	12.1 ± 1.9
Hematocrit (%)	35.1 ± 10.9	38.3 ± 5.7
ALT (U/l)	44.0 ± 46.1	40.9 ± 32.0
Alkaline phosphatase (U/l)	199.6 ± 133.9	242.9 ± 129.0
Total protein (g/l)	58.2 ± 26.4	68.9 ± 6.5
Albumin (g/l)	27.2 ± 15.8	36.6 ± 6.0
C-reactive protein (mg/l)	42.7 ± 84.4	63.0 ± 50.7

[a]TVO, tuberculous vertebral osteomyelitis.
[b]BVO, brucellar vertebral osteomyelitis.

Because there are no hematological or biochemical data to help us in the diagnosis of TVO and BVO, diagnosis must be based on microbiological data. Bacteremia is common in patients with brucellosis [8, 27–29]. The diagnostic yield of blood cultures is quite high among patients with BVO, although it is slightly inferior to that obtained among patients with acute brucellosis. In our experience, almost 50% of patients with BVO had positive blood cultures, a rate within the 33–74% range reported by other authors [19, 30, 31]. Therefore, in all patients with suspected BVO, two sets of blood cultures should be performed, even in patients without fever.

Despite the important advances made in the diagnosis of human brucellosis following the general introduction of new semi-automated methods for blood culture processing [32], diagnosis of this disease is still based mostly on the demonstration of specific antibodies by means of different serological techniques. This is mainly because the greatest incidence of brucellosis is found in countries with limited technical resources, as well as the fact that it tends to occur in rural communities.

Several serological tests are available for the diagnosis of human brucellosis, including the Rose Bengal test, standard agglutination, the Coombs anti-Brucella test, immunocapture–agglutination test, and the enzyme-linked immunosorbent assay (ELISA). All of these have good sensitivity. However, they lack the desired specificity in people with previous contact with *Brucella* spp., such as those living in highly endemic areas, those who are occupationally exposed, or those with a recent history of brucellosis. The most widely used strategy for the serological diagnosis of brucellosis consists of the combination of a rapid screening test, such as the Rose Bengal test, and a confirmation test, such as standard agglutination or ELISA. However, the sensitivity is low in patients with focal complications of brucellosis, such as BVO, in which the presence of incomplete antibodies is common. In our experience, standard agglutination can provide titers lower than 1/160 in 30–35% of patients with BVO [18]. For this reason, in patients with suspected BVO, at least two serological tests, such as standard agglutination and Coombs anti-Brucella test or immunocapture agglutination test, should be used to increase sensitivity [33, 34]. In conclusion, patients with suspected BVO should always have blood cultures and two specific serological tests. This approach allows the diagnosis in more than 90% of cases, which makes vertebral biopsy unnecessary.

Unfortunately, there is no noninvasive microbiological test confirming the etiologic diagnosis of TVO. Therefore, vertebral biopsy is generally needed for microbiological confirmation of the diagnosis. The usefulness of the tuberculin skin test (TST) in the diagnosis of tuberculosis has been questioned due to interreader variability, cross-reactivity with nontuberculous mycobacteria, and false-positive results in patients vaccinated with Bacillus Calmette-Guérin (BCG). Furthermore, the TST also has a low sensitivity in immunosuppressed patients. Recently, interferon-gamma release assays (IGRAs) for tuberculosis have overcome most of these limitations. These immunoassays detect in vitro interferon-gamma secreted by peripheral blood mononuclear cells in response to specific antigens of *M. tuberculosis*. Currently, multiple data show that IGRA tests are equally sensitive but more specific than the TST in diagnosing latent tuberculosis infection and active tuberculosis [35, 36]. Unfortunately, in countries with a high prevalence of tuberculosis, the usefulness of TST and IGRA in diagnosis is very limited, as these tests cannot differentiate between latent infection and active disease. However, their negative predictive value is very high, regardless of the prevalence of tuberculosis. For these reasons, it seems advisable to perform an IGRA test in any patient with suspected TVO because, regardless of the prevalence of tuberculosis, a negative IGRA test makes the diagnosis of spinal tuberculosis very unlikely [37]. Unlike BVO, the etiologic diagnosis of TVO requires a vertebral biopsy in a high percentage of cases. In our experience, around 80% of patients with suspected TVO needed a biopsy to achieve the diagnosis [17]. Of these, one-third underwent percutaneous vertebral biopsy and the remaining two-thirds open surgical biopsy. All samples from patients with suspected TVO sent to a microbiology laboratory should be processed for direct microscopical examination and culture.

In contrast to the situation in pulmonary tuberculosis, the bacterial density is much lower in extrapulmonary samples, explaining the low diagnostic yield of microscopy in spinal tuberculosis. In none of the published studies, the positivity of the microscopic examination of the vertebral or paravertebral samples exceeded 36% [21]. Traditionally, solid media such as Lowenstein–Jensen have been used for culturing mycobacteria. This method is very slow, requiring between 3 and 8 weeks for isolation. Currently, liquid media such as Middlebrook can reduce the time required for isolation to 2 or 3 weeks. Several studies have compared the performance of the different culture systems available. The automatic system for mycobacterial culture BACTEC 460 remains the fastest and most sensitive, followed by the mycobacterial detection system BACTEC-MGIT, with the solid media systems being the slowest [37, 38]. Culture remains the gold standard for the diagnosis of tuberculosis and brucellosis. However, as both *Brucella* spp. and *M. tuberculosis* are slow-growing pathogens, cultures are labor-intensive, which may result in unacceptable delay in diagnosis.

Molecular techniques have contributed substantially to the improvement of the diagnosis in the many fields of infectious diseases, especially if fastidious microorganisms are the etiologic agent. Polymerase chain reaction (PCR)–based methods have proven to be more sensitive than conventional methods in both extrapulmonary tuberculosis and focal complications of brucellosis. The use of real-time PCR technology reduces the time to identification of bacterial DNA directly from clinical samples. Additionally, considerable time and effort can be saved by simultaneously amplifying multiple sequences in a single reaction. This strategy, named multiplex PCR, has proven to be very useful in different clinical scenarios. Our group has recently developed a single-tube multiplex real-time

PCR able to amplify *M. tuberculosis* complex and *Brucella* spp. simultaneously [39]. In a recent study that included 15 vertebral samples from patients with TVO or BVO and 9 from pyogenic and nontuberculous mycobacteria, multiplex PCR correctly identified 14 of the 15 samples from patients with TVO and BVO and was negative in all the control samples. Thus, the overall sensitivity and specificity of multiplex PCR were 93.3 and 90%, respectively, with an accuracy of 92%. These results suggest that multiplex real-time PCR is far more sensitive than conventional cultures, and this, together with its speed, makes this technique a very practical approach for the rapid differential diagnosis between TVO and BVO [40].

Imaging Procedures

Two possible clinical scenarios exist in the diagnosis of TVO or BVO. The patient consults within 2 to 4 weeks after the onset of symptoms. Alternatively, due to the oligosymptomatic course of disease, the patient consults only after several weeks, even months. In the former situation, plain radiographs do not reveal any apparent abnormalities. This may mislead the clinician to rule out the diagnosis of vertebral osteomyelitis. Therefore, a high index of suspicion is needed, in order to avoid an inappropriate diagnostic delay. Our experience confirms the limitations of conventional radiological studies in the diagnosis of TVO and BVO. Despite the long duration of symptoms at presentation, plain radiography showed no findings compatible with vertebral osteomyelitis in 10 and 41% of patients with TVO and BVO, respectively. On the other hand, the classic radiological finding of "osteolysis of two contiguous vertebral bodies with the destruction of the intervertebral disk" is easily recognized but nonspecific, as it fails to differentiate between pyogenic osteomyelitis, TVO, BVO, and even noninfectious diseases, such as erosive osteochondrosis (see Chapter 15).

Multiple studies have demonstrated that magnetic resonance imaging (MRI) is more sensitive than plain radiography in the diagnosis of vertebral osteomyelitis of any etiology, resulting in suggestive signal changes just 1 week after onset of symptoms. Therefore, in any patient with suspected TVO or BVO, an MRI should be performed, even if the initial plain radiography has been normal [41–44].

No image, whatever the technique used, is specific for BVO, and depending on the duration of the diagnostic delay, the lesions are so heterogeneous that they range from mild erosion of an end plate (Figure 16.1) to severe destructive lesions with large paravertebral masses.

Although some studies suggest that the presence of a psoas abscess is very uncommon in brucellosis [43, 45], in our experience 10% of patients with BVO in fact had this complication (Table 16.4). Accordingly, the presence of collapse or a large psoas abscess should not exclude a diagnosis of brucellosis.

The granulomatous inflammation caused by *M. tuberculosis* results in a marked osteolytic effect on vertebral tissue, with scarce involvement of the intervertebral disk. In tuberculous osteomyelitis, foci of caseous necrosis tend to coalesce to form abscesses that spread via the subligamentous path. Contrary to what happens in other granulomatous or pyogenic infections, in TVO bone, regeneration is very low, resulting in a large destructive effect that leads to vertebral collapse. This characteristic, coupled with the fact that most cases of TVO affect the thoracic segment or thoracolumbar hinge, explains the progressive kyphosis that is a common clinical finding in TVO [46].

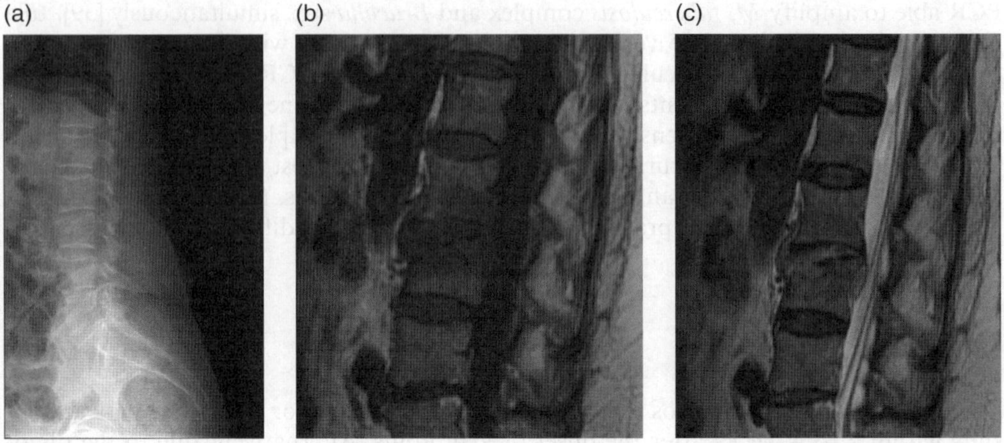

Figure 16.1. L3–L4 vertebral osteomyelitis due to *Bucella melitensis*. (a), Decreased height of the intervertebral disk L3–L4, with L4 anterior superior edge epiphysitis. (b), T1-weighted magnetic resonance (MR) sagittal image showing decreased signal with poor definition of the upper end plate of L4. (c), Sagittal T2-weighted MR image showing increased disk signal with small paravertebral mass and mild epidural abscess.

Table 16.4. Comparative image findings in tuberculous and brucella vertebral osteomyelitis.

	TVO[a]	BVO[b]
Number of affected vertebrae, mean (range)	2.47 (1–11)	2.05 (1–4)
Disk involvement (%)	95.5	81.1
Vertebral body osteolysis (%)		
Anterior	28.2	52.6
Posterior	3.6	12.6
Both anterior and posterior	68.2	34.7
Paravertebral masses (%)	76.4	45.6
Psoas abscesses (%)	28.7	10.0
Ruptured posterior wall (%)	59.1	21.1
Epidural masses (%)	67.9	28.9
Multiple levels involvement (%)	4.5	3.1

[a]TVO, tuberculous vertebral osteomyelitis.
[b]BVO, brucellar vertebral osteomyelitis.

Atypical presentations of TVO have been reported frequently and include involvement of multiple noncontinuous vertebral levels (Figure 16.2), soft tissue mass with calcifications or bony fragments, vertebral collapse with relative preservation of the intervertebral disk, and the exclusive involvement of posterior elements [47–50] (Figure 16.3). It is important to highlight that these findings are highly suggestive though not specific for TVO.

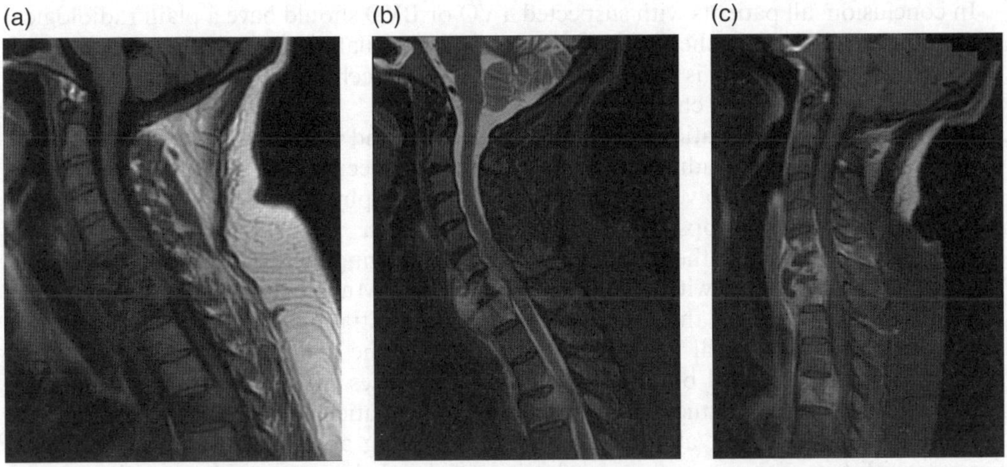

Figure 16.2. Multilevel tuberculous vertebral osteomyelitis. (a), T1-weighted magnetic resonance (MR) sagittal image shows great osteolysis of C6, C7, and T1 bodies, with paravertebral and epidural extension. (b), Increased signal in T2-weighted sequence with simultaneous involvement of T3–T4 and epidural involvement at this level. (c), After gadolinium administration, the paraspinal extension of C6–T1 paravertebral abscess is clearly demonstrated.

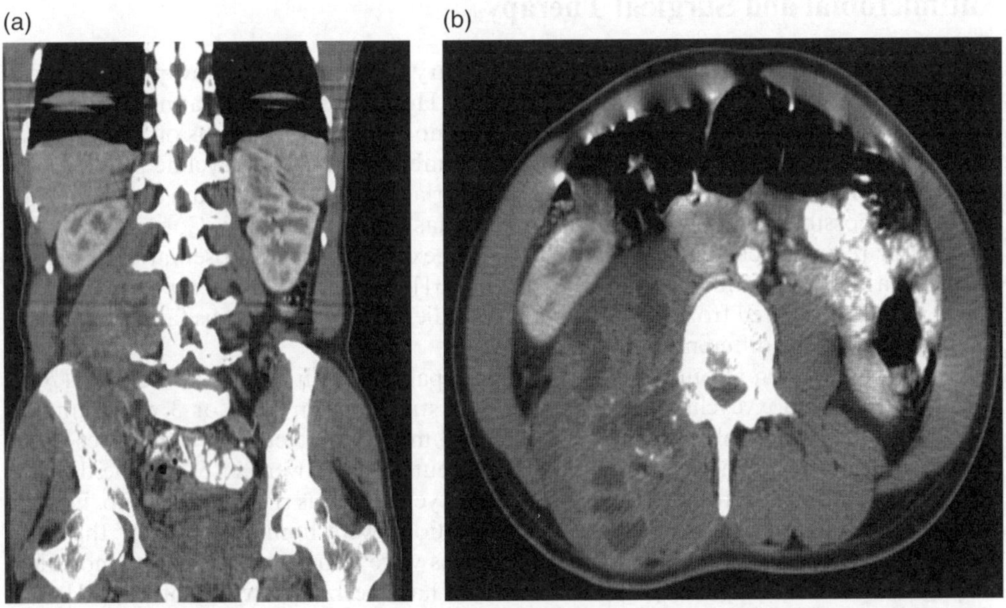

Figure 16.3. Tuberculous vertebral osteomyelitis with involvement of the posterior elements. (a), Coronal computed tomography (CT) image showing heterogeneous mass in the right psoas muscle and total osteolysis of the L4 transverse process. (b), Transverse CT image with a huge right iliopsoas, erector spinae, and quadratus lumborum muscles' abscess containing heterogeneous calcifications.

In conclusion, all patients with suspected TVO or BVO should have a plain radiological study plus an MRI study, the former for its common availability and the latter for its high sensitivity. In addition, it is the most useful imaging technique to evaluate paraspinal tissues and the intraspinal channel.

In order to avoid duplication of imaging techniques and inefficient use of resources, we believe nuclear imaging studies should be reserved for specific cases especially with suspicion of multiple-level involvement. Computed tomography (CT) is indicated for guided percutaneous vertebral biopsy or drainage of an abscess.

There is a tendency for the inappropriate use of imaging methods, especially MRI, in the follow-up of patients with vertebral osteomyelitis. Much evidence demonstrates the large discordance between the clinical evolution after initiation of treatment and the disappearance of radiological abnormalities, especially the changes in T1 and T2 MRI sequences in the vertebral bodies. In fact, in many cases the MRI signal may be even worse despite obvious clinical improvement of the patient. So, we only recommend repeating the MRI study, about 4 weeks after starting appropriate treatment, just in patients with large masses or paravertebral or epidural abscesses or those with unfavorable evolution. In patients with good clinical progress, the return to normal of altered biological parameters, especially C-reactive protein levels, is a marker of good evolution. The persistence of abnormal bone or disk MRI findings alone does not represent therapeutic failure. Similar findings have been reported in patients with acute vertebral osteomyelitis [51, 52]

Antimicrobial and Surgical Therapy

The treatments for acute brucellosis and pulmonary tuberculosis are now perfectly standardized, based on multiple randomized studies. However, the same cannot be said for most of the extrapulmonary sites of tuberculosis nor for the focal forms of brucellosis.

The treatment of brucellosis is based on the combination of doxycycline plus streptomycin or rifampicin. To date, though, no clinical trial designed to evaluate the efficacy of the various possible treatment regimens for BVO has been undertaken. Consequently, the therapeutic regimens currently used are based on experience from observational studies involving a limited number of patients [19, 20, 31]. However, an attempt to unify the experience accumulated from these individual studies in a meta-analysis failed due to the heterogeneity of the different regimens used [53].

A study undertaken at our center involving 96 patients with BVO, of whom 71 (74%) were treated with doxycycline for 3 months plus streptomycin for 2 or 3 weeks and 16 (17%) with doxycycline plus rifampicin, both for 3 months, detected no differences in the efficacy of the two regimens. The rates of therapeutic failure were 15.5% in the doxycycline plus streptomycin group and 18.7% in the doxycycline plus rifampicin group. Despite the different sizes of the two groups, the populations were homogeneous for the main prognostic variables, and no significant differences were found in the rates of mortality, relapse, or functional sequelae [18]. Similar results have been reported by others [31].

In an attempt to reduce this high rate of therapeutic failure, Giannitsioti *et al.* [54] proposed using combined treatment with doxycycline plus rifampicin and a quinolone or sulfamethoxazole/trimethoprim for a period of not less than 6 months. Using this regimen, these authors had no therapeutic failures or relapses, though as the study included only 25 patients, no firm conclusions could be drawn. Thus, until data from larger studies are available, the treatment of choice for BVO is now doxycycline for 3 months plus an aminoglycoside for 2 or 3 weeks, the alternative being doxycycline plus rifampicin for 3 months when aminoglycosides are contraindicated. Although streptomycin has been the most widely used aminoglycoside, gentamicin has proven to be an effective alternative [55].

For TVO, while some authors until recently maintained that caseous necrotic sequestrum may hinder penetration of the drug into bone tissue [56, 57], isoniazid, rifampicin, and pyrazinamide are found in pus and granulation tissues above the minimum inhibitory concentration [58].

The Medical Research Council Working Party on Tuberculosis of the Spine, in their 12th report, confirmed that short treatment regimens of 6–9 months were effective and as successful as an 18-month regimen, even in severe forms of TVO [59]. In a recent study by our group that included 78 TVO patients treated with 9- to 12-month schedules, 5 (6.4%) patients died, but in no case was death related to vertebral tuberculosis; 4 (5.1%) relapsed during the 12 months following conclusion of the therapy; and the remaining 69 evolved favorably. Thus, the cure rate was 94.5% [17]; similar results have been reported by other authors [60].

A special situation exists with TVO due to multidrug-resistant *M. tuberculosis* (i.e., those cases in which simultaneous resistance is demonstrated to both isoniazid and rifampicin). In these cases, the microorganisms are less sensitive and use of second-line drugs implies prolonging the treatment for 18 to 24 months. When these cases of multidrug-resistant TVO are adequately treated, the response seems satisfactory. A recent study by Pawar *et al.* [61] involving 25 cases reports a cure rate of 76%.

Despite the demonstrated efficacy of medical treatment in acute brucellosis and tuberculosis, vertebral osteomyelitis, regardless of its etiology, often requires surgical treatment. There is a general consensus that surgical treatment should be considered in patients with cord or radicular compression, large soft tissue masses, abscesses in which percutaneous drainage is not possible, and highly destructive bone lesions producing spinal instability or in those patients suffering therapeutic failure despite adequate medical treatment.

In a recent review of seven studies including 1008 cases of pyogenic vertebral osteomyelitis, Mylona *et al.* reported that 48% of the patients required some type of surgical therapy [62]. The figures are lower for BVO [53], though they may be even higher in TVO [63].

There is controversy in the literature about the need for surgery in BVO. In our opinion, this could in part be related with the scarcity of data available and the heterogeneity of the centers at which the studies have been carried out [53]. In a study conducted at our center that included 96 patients with BVO, one-third of the patients required surgical treatment [18], a percentage similar to that reported by Lifeso *et al.* [19], although far higher than that reported by others [30, 31]. It is possible that surgery was indicated more frequently in those centers that have access to experienced spinal surgery teams. This is the case at our center, which has a reference unit for spinal orthopedic surgery, and which thus receives a greater number of severe cases compared with other hospitals. Nonetheless, multiple data support the conclusion that in many cases, especially in cases involving a great diagnostic delay, vertebral osteomyelitis is potentially a very serious complication of brucellosis, frequently requiring surgical treatment.

Regarding the need for surgical treatment in TVO, the consensus is far greater than in the case of BVO, especially after the publication of the results of the eighth Report of the Medical Research Council Working Party on Tuberculosis of the Spine, in which the authors stressed that, despite the high efficacy of current medical therapy, a large body of opinion advocates aggressive surgery in TVO [64]. Anterior spinal debridement, decompression, and fusion produce better results in those patients fulfilling the surgical indications proposed therein. Radical debridement enables removal of necrotic tissue and contributes to the early resolution of neurological deficits and the rapid relief of pain. Patients have more rapid abscess resolution, earlier and more frequent bone fusion, and a reduction in the degree of residual kyphosis. Therefore, in our opinion, currently shared by others, ambulatory outpatient chemotherapy for TVO should be reserved for those cases in which surgery is not indicated or operative expertise and facilities are unavailable [65, 66].

Key Points

- In tuberculosis as well as in brucellosis, vertebral osteomyelitis is the most frequent osteoarticular complication. BVO predominates in rural adult males, while TVO affects both men and women in urban areas.
- TVO and BVO often suffer a high diagnosis delay, because of the subacute course and the frequent absence of fever. Therefore, in any patient with suspected TVO or BVO, an MRI should be performed, even if the initial plain radiography has been normal.
- The diagnosis of TVO and BVO requires microbiological confirmation. The combined use of blood cultures and two serological tests permits diagnosis in greater than 90% of cases of BVO. Needle aspiration or vertebral biopsy is needed for culture and histology to diagnose TVO.
- The current medical treatment of BVO with doxycycline plus aminoglycoside or doxycycline plus rifampin for 3 months usually gets good results.
- In TVO, standard antituberculous treatment should be given for 6–9 months.
- Despite good efficacy of medical treatment, surgical treatment should be considered in patients with cord or radicular compression, large soft tissue masses, abscesses in which percutaneous drainage is not possible, and highly destructive bone lesions producing spinal instability or those patients suffering therapeutic failure.

References

[1] Dean AS, Crump L, Greter H, *et al*. Global burden of human brucellosis: a systematic review of disease frequency. PLoS Negl Trop Dis 2012;6(10):e1865.

[2] Franco MP, Mulder M, Gilman RH, Smits HL. Human brucellosis. Lancet Infect Dis 2007;7(12):775–786.

[3] Doyle TJ, Bryan RT. Infectious disease morbidity in the US region bordering Mexico, 1990–1998. J Infect Dis 2000;182(5):1503–1510.

[4] Pappas G, Papadimitriou P, Akritidis N, *et al*. The new global map of human brucellosis. Lancet Infect Dis 2006;6(2):91–99.

[5] Global tuberculosis report 2012. Geneva: World Health Organization. www.who.int/tb/publications/global_report. Accessed on June 9, 2014.

[6] Cegielski JP, Chin DP, Espinal MA *et al*. The global tuberculosis situation: progress and problems in the 20th century, prospects for the 21st century. Infect Dis Clin North Am 2002;16(1):1–58.

[7] Anonymous. Tuberculosis surveillance and monitoring in Europe 2013. Surveillance Report. European Centre for Disease Prevention and Control; 2013, p. 3. http://www.ecdc.europa.eu/en/publications/surveillance_reports/tuberculosis. Accessed on June 1, 2014.

[8] Colmenero JD, Reguera JM, Martos F, *et al.* Complications associated with *Brucella melitensis* infection: a study of 530 cases. Medicine (Baltimore) 1996;75(4):195–211.
[9] Moussa AR, Elhag KM, Khogali M, Marafie AA. The nature of human brucellosis in Kuwait: study of 379 cases. Rev Infect Dis 1988;10(1):211–217.
[10] Colmenero JD, Reguera JM, Fernandez-Nebro A, Cabrera-Franquelo F. Osteoarticular complications of brucellosis. Ann Rheum Dis 1991;50(1):23–26.
[11] González-Gay MA, García-Porrúa C, Ibañez D, García-País MJ. Osteoarticular complications of brucellosis in an Atlantic area of Spain. J Rheumatol 1999;26(1):141–145.
[12] Peto HM, Pratt RH, Harrington TA, *et al.* Epidemiology of extrapulmonary tuberculosis in the United States, 1993–2006. Clin Infect Dis 2009;49(9):1350–1357.
[13] Kruijshaar ME, Abubakar I. Increase in extrapulmonary tuberculosis in England and Wales 1999–2006. Thorax 2009;64(12):1090–1095.
[14] Te Beek LA, van der Werf MJ, Richter C, Borgdorff MW. Extrapulmonary tuberculosis by nationality, The Netherlands, 1993–2001. Emerg Infect Dis 2006;12(9):1375–1382.
[15] Kipp AM, Stout JE, Hamilton CD, Van Rie A. Extrapulmonary tuberculosis, human immunodeficiency virus, and foreign birth in North Carolina, 1993–2006. BMC Public Health 2008;4(8):107.
[16] García-Rodríguez JF, Álvarez-Díaz H, Lorenzo-García MV, *et al.* Extrapulmonary tuberculosis: epidemiology and risk factors. Enferm Infecc Microbiol Clin 2011;29(7):502–509.
[17] Colmenero JD, Jiménez-Mejías ME, Reguera JM,*et al.* Tuberculous vertebral osteomyelitis in the new millennium: still a diagnostic and therapeutic challenge. Eur J Clin Microbiol Infect Dis 2004;23(6):477–483.
[18] Colmenero JD, Ruiz-Mesa JD, Plata A, *et al.* Clinical findings, therapeutic approach, and outcome of brucellar vertebral osteomyelitis. Clin Infect Dis 2008;46(3):426–433.
[19] Lifeso RM, Harder E, McCorkell SJ. Spinal brucellosis. J Bone Joint Surg Br 1985;67(3): 345–351.
[20] Tekkök IH, Berker M, Ózcan OE, *et al.* Brucellosis of the spine. Neurosurgery 1993;33(5): 838–844.
[21] Pertuiset E, Beaudreuil J, Lioté F *et al.* Spinal tuberculosis in adults. A study of 103 cases in a developed country, 1980–1994. Medicine (Baltimore) 1999;78(5):309–320.
[22] Alothman A, Memish ZA, Awada A *et al.* Tuberculous spondylitis: analysis of 69 cases from Saudi Arabia. Spine 2001;26(24):e565–e570.
[23] Colmenero JD, Jiménez-Mejías ME, Sánchez-Lora FJ *et al.* Pyogenic, tuberculous, and brucellar vertebral osteomyelitis: a descriptive and comparative study of 219 cases. Ann Rheum Dis 1997;56(12):709–715.
[24] Rajasekaran S. The problem of deformity in spinal tuberculosis. Clin Orthop Relat Res 2002;398:85–92.
[25] Perronne C, Saba J, Behloul Z, *et al.* Pyogenic and tuberculous spondylodiskitis (vertebral osteomyelitis) in 80 adult patients. Clin Infect Dis 1994;19(4):746–750.
[26] Buchelt M, Lack W, Kutschera HP, *et al.* Comparison of tuberculous and pyogenic spondylitis. An analysis of 122 cases. Clin Orthop 1993;296:192–199.
[27] Corbel MJ. Brucellosis an overview. Emerg Infect Dis 1997;3(2):213–221.
[28] Pappas G, Akritidis N, Bosilkovski M, Tsianos E. Brucellosis. N Engl J Med 2005;352(22): 2325–2336.
[29] Young EJ. An overview of human brucellosis. Clin Infect Dis 1995;21(2):283–289.
[30] Ariza J, Gudiol F, Valverde J, *et al.* Brucellar spondylitis: a detailed analysis based on current findings. Rev Infect Dis 1985;7(5):656–664.
[31] Solera J, Lozano E, Martinez-Alfaro E, *et al.* Brucellar spondylitis: review of 35 cases and literature survey. Clin Infect Dis 1999;29(6):1440–1449.
[32] Yagupsky P. Detection of brucellae in blood cultures. J Clin Microbiol 1999;37(11):3437–3442.
[33] Orduña A, Almaraz A, Prado A, *et al.* Evaluation of an immunocapture agglutination test (brucellacapt) for the serodiagnosis of human brucellosis. J Clin Microbiol 2000;38(11):4000–4005.

[34] Ariza J, Pellicer T, Pallares R, *et al.* Specific antibody profile in human brucellosis. Clin Infect Dis 1992;14(1):131–140.
[35] Rangaka MX, Wilkinson KA, Glynn JR, *et al.* Predictive value of interferon-c release assays for incident active tuberculosis: a systematic review and meta-analysis. Lancet Infect Dis 2012; 12(1):45–55.
[36] Dosanjh DP, Hinks TS, Innes JA *et al.* Improved diagnostic evaluation of suspected tuberculosis. Ann Intern Med 2008;148(5):325–336.
[37] Colmenero JD, Ruiz-Mesa JD, Sanjuan-Jimenez R, *et al.* Establishing the diagnosis of tuberculous vertebral osteomyelitis. Eur Spine J 2013;22(Suppl 4):579–586.
[38] Watterson SA, Drobniewski FA. Modern laboratory diagnosis of mycobacterial infections. J Clin Pathol 2000;53(10):727–732.
[39] Queipo-Ortuño MI, Colmenero JD, Bermudez P, *et al.* Rapid differential diagnosis between extrapulmonary tuberculosis and focal complications of brucellosis using a multiplex real-time PCR assay. PLoS One 2009;4(2):e4526
[40] Colmenero JD, Morata P, Ruiz-Mesa JD, *et al.* Multiplex real-time polymerase chain reaction: a practical approach for rapid diagnosis of tuberculous and brucellar vertebral osteomyelitis. Spine (Phila Pa 1976). 2010;35(24):E1392–E1396.
[41] Modic MT, Feiglin DH, Piraino DW, *et al.* Vertebral osteomyelitis: assessment using MR. Radiology 1985;157(1):157–166.
[42] Stabler A, Reiser MF. Imaging of spinal infection. Radiol Clin North Am 2001;39(1): 115–135.
[43] Smith AS, Weinstein MA, Mizushima A, *et al.* MR imaging characteristics of tuberculous spondylitis vs. vertebral osteomyelitis. Am J Roentgenol 1989;153(2):399–405.
[44] Kim NH, Lee HM, Suh JC. Magnetic resonance imaging for the diagnosis of tuberculous spondylitis. Spine 1994;19(21):2451–2455.
[45] Sharif HS, Aideyan OA, Clark DC, *et al.* Brucellar and tuberculous spondylitis: comparative imaging features. Radiology 1989;171(2):419–425.
[46] Tuli SM. Severe kyphotic deformity in tuberculosis of the spine. Int Orthop. 1995;19(5): 327–331.
[47] Sharif HS, Morgan JL, Al Shahed MS, Al Thagafi MYA. Role of CT and MR imaging in the management of tuberculosis spondylitis. Radiol Clin North Am 1995;33(4):787–804.
[48] Sankaran-Kutty M. Atypical tuberculous spondylitis. Int Orthop 1992;16(1):69–74.
[49] Almadi J, Bajaj A, Destian S, *et al.* Spinal tuberculosis. Atypical observation at MR imaging. Radiology 1993;189(2):489–493.
[50] Arora S, Sabat D, Maini L *et al.* Isolated involvement of the posterior elements in spinal tuberculosis: a review of twenty-four cases. J Bone Joint Surg Am. 2012;94(20):e151.
[51] Kowalski TJ, Berbari EF, Huddleston PM, *et al.* Do follow-up imaging examinations provide useful prognostic information in patients with spine infection? Clin Infect Dis 2006;43(2): 172–179.
[52] Euba G, Narváez JA, Nolla JM, *et al.* Long-term clinical and radiological magnetic resonance imaging outcome of abscess-associated spontaneous pyogenic vertebral osteomyelitis under conservative management 2008;38(1):28–40.
[53] Pappas G, Seitaridis S, Akritidis N, Tsianos E. Treatment of brucella spondylitis: lessons from an impossible meta-analysis and initial report of efficacy of a fluoroquinolone containing regimen. Int J Antimicrob Agents 2004;24(5):502–507.
[54] Giannitsioti E, Papadopoulos A, Nikou P *et al.* Long-term triple-antibiotic treatment against brucellar vertebral osteomyelitis. Int J Antimicrob Agents 2012;40(1):91–93.
[55] Ariza J, Bosilkovski M, Cascio A *et al.* Perspectives for the treatment of brucellosis in the 21st century: the Ioannina recommendations. PLoS Med 2007;4(12):e317.
[56] Dutt AK, Moers D, Stead WW. Short-course chemotherapy for extrapulmonary tuberculosis. Ann Intern Med 1986;104(1):7–12.
[57] Snider DE, Cohn DL, Davidson PT, *et al.* Standard therapy for tuberculosis. Chest 1985;87 (2 Suppl):117S–124S.

[58] Tuli SM, Kumar K, Sen PC. Penetration of antitubercular drugs in clinical osteoarticular tubercular lesions. Acta Orthop Scand 1977;48(4):362–368.
[59] Anonymous. Controlled trial of short-course regimens of chemotherapy in the ambulatory treatment of spinal tuberculosis. Results at three years of a study in Korea. Twelfth report of the Medical Research Council Working Party on Tuberculosis of the Spine. J Bone Joint Surg Br 1993;75(2):240–248.
[60] Ramachandran S, Clifton IJ, Collyns TA, *et al.* The treatment of spinal tuberculosis: a retrospective study. Int J Tuberc Lung Dis 2005;9(5):541–544.
[61] Pawar UM, Kundnani V, Agashe V, *et al.* Multidrug-resistant tuberculosis of the spine–is it the beginning of the end? A study of twenty-five culture proven multidrug-resistant tuberculosis spine patients. Spine (Phila Pa 1976) 2009;34(22):E806.
[62] Mylona E, Samarkos M, Kakalou E, *et al.* Pyogenic vertebral osteomyelitis: a systematic review of clinical characteristics. Semin Arthritis Rheum 2009;39(1):10–17.
[63] Nussbaum ES, Rockswold GL, Bergman TA, *et al.* Spinal tuberculosis: a diagnostic and management challenge. J Neurosurg 1995;83(2):243–247.
[64] Anonymous. A 10-year assessment of a controlled trial comparing debridement and anterior spinal fusion in the management of tuberculosis of the spine in patients on standard chemotherapy in Hong Kong. Eighth Report of the Medical Research Council Working Party on Tuberculosis of the Spine. J Bone Joint Surg Br 1982;64(4):393–398.
[65] Mak KC, Cheung KM. Surgical treatment of acute TB spondylitis: indications and outcomes. Eur Spine J. 2013;22(Suppl 4):603–611.
[66] Jain AK. Tuberculosis of the spine: a fresh look at an old disease. J Bone Joint Surg Br 2010;92(7):905–913.

Chapter 17
Chronic Osteomyelitis in Adults

Silvin Lito, Antoine Lomessy, Pierre Vaudaux, and Ilker Uçkay

Introduction

Chronic osteomyelitis implies an infection of the bone, which may or may not involve the marrow, cortex, periosteum and surrounding tissue, in sequence or at the same time [1]. In clinical practice, the terms osteomyelitis and osteitis are interchangeable. Even though chronic osteomyelitis is generally of infectious origin, rare immunological inflammatory disorders of the bone, such as SAPHO (synovitis, acne, pustulosis, hyperostosis, and osteitis) syndrome or recurrent multifocal osteomyelitis, do exist [2–4]. Osteomyelitis is often associated with "foreign bodies" or "devices," but implant-associated chronic bone infections are presented in another chapter (see Chapter 20). Classification of osteomyelitis according to its duration is widely used. The different forms are labeled as "acute," "subacute," or "chronic," a differentiation that is rarely useful for the clinician [1, 5, 6]. In general, chronic osteomyelitis is arbitrarily defined as bone infection with duration of symptoms for at least 6 weeks to 3 months [1]. A second criterion is the presence of established bone pathology in an imaging procedure. In surgical practice, chronic osteomyelitis implies an infection requiring invasive intervention, characterized by sequestra and deformities. Other classifications based on pathogenetic mechanisms are subdivided into hematogenous as opposed to continuous (*per continuitatem*) osteomyelitis. Vertebral, periarticular, as well as pediatric osteomyelitis are typically hematogenously acquired (see Chapters 5, 6, and 14) [1, 7]. In contrast, long bone, sacral, and foot osteomyelitis often represent infections acquired *per continuitatem* after trauma, surgery [7, 8], or soft tissue ulceration as a result of micro- and macrovascular disease, for example, typically in diabetic [9] patients (Figure 17.1) (see Chapters 18 and 20). Long bone osteomyelitis caused by hematogenous seeding is rare in adults [1, 10].

Several surgical classifications have been proposed. The most frequently used are the Cierny–Mader classification [11] (Table 17.1) for long bone osteomyelitis, and the Perfusion, Extent, Depth, Infection, Sensation (PEDIS) classification for diabetic foot osteomyelitis [12–14] (see Chapter 18). The Cierny–Mader classification takes into

Bone and Joint Infections: From Microbiology to Diagnostics and Treatment, First Edition.
Edited by Werner Zimmerli.

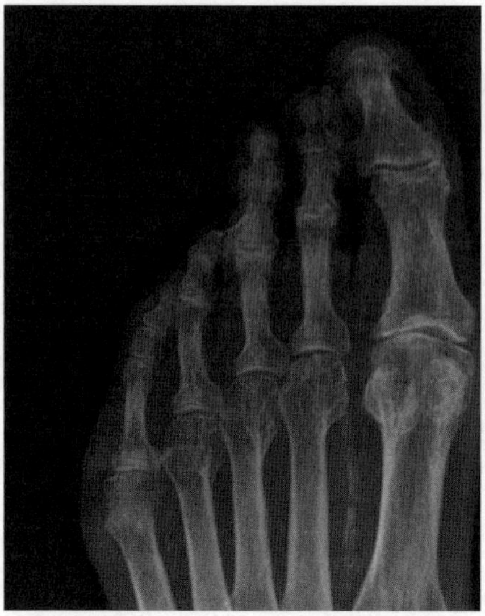

Figure 17.1. Chronic osteomyelitis *per continuitatem* in an 82-year old man with long-standing diabetes mellitus. Osseous lysis on the lateral side of the head of the fifth metatarsal bone. Image is the property of Geneva University Hospitals and is displayed with the permission of the patient.

Table 17.1. The Cierny–Mader staging system for long bone osteomyelitis.

Anatomical type		Physiological class	
Type I	Medullary osteomyelitis	A host	Normal immunity and conserved tissue absorption
Type II	Nonmedullary osteomyelitis	B host	Immunocompromised locally (i.e., overlaying sore or skin breach) [B^L] or systemically [B^S]—AIDS, or any inborn or acquired, humoral, or cellular functional deficiency
Type III	Confined focus osteomyelitis	C host	Not classifiable—invasive treatment discouraged; consider suppressive or no treatment at all. Risks by far exceed possible benefits
Type IV	Widespread (circumferential) osteomyelitis		

Adapted from Ref. 11.

account the quality of the tissues, anatomy, and prognosis. It is especially useful for planning surgical management (see Chapter 13).

Regarding microbiology, the vast majority of chronic osteomyelitis is caused by bacterial infections. Of lesser importance are fungi, which are mainly observed in intravenous drug users (IVDU) [15], in skull-base osteomyelitis [16], and in the immunocompromised host. Parasites (i.e., echinococcosis) rarely cause osteomyelitis, and no case of viral bone infection has been described until now. A majority of cases are due to *Staphylococcus aureus* [1, 5, 7, 17–20], followed by *Streptococcus* spp. [18, 19] and Gram-negative pathogens including non-fermenting pathogens, such as *Pseudomonas aeruginosa* [19]. A polymicrobial etiology is frequent after trauma and in case of long-lasting ulcerations [20]. Anaerobes remain rare [18, 21], and coagulase-negative staphylococci are detected almost exclusively in device-associated infections [22] (see Chapter 20). *Kingella kingae* is mainly seen in children under the age of 4 years in poor countries [23]. Worldwide, but excessively in poor countries, there is endemicity of subacute and chronic osteomyelitis due to tuberculosis [24] and brucellosis (see Chapter 16).

Epidemiology and Pathogenesis

Both prevalence and incidence of chronic osteomyelitis are likely influenced by age, geographical location, socioeconomic status, and comorbidities. Concrete data are either not available or only explored in particular settings. Of all patients suffering from diabetic foot ulcers, 15–20% eventually develop osteomyelitis [1, 20] (see Chapter 18). Osteomyelitis resulting from surgical site infections is an infrequent complication occurring in 1–4% of all interventions [25].

Some bacterial species, in particular *S. aureus*, have the ability to bind to bone matrix through specific fibronectin receptors and other surface matrix-binding proteins [17, 22]. Host defenses can be overcome by either intracellular invasion [22] or biofilm production [11, 22, 26]. Vascular tunnels are obliterated by severe inflammation, and ischemic bone necrosis follows. Bone segments devoid of blood supply become separated (sequestra) and provide an excellent substratum for bacterial growth [1]. Abscess formation may occur as early as 48 h after infection [17]. At the infarction edges, reactive hyperemia causes excess osteoclastic activation, resulting in osteopenia [1]. In the meantime, vigorous osteoblastic activity generates zones of periosteal apposition and new bone called involucrum (Figure 17.2). Ample evidence suggests the contribution of growth factors, drugs, cytokines, and hormones in regulating osteoblastic and osteoclastic activity [1, 17].

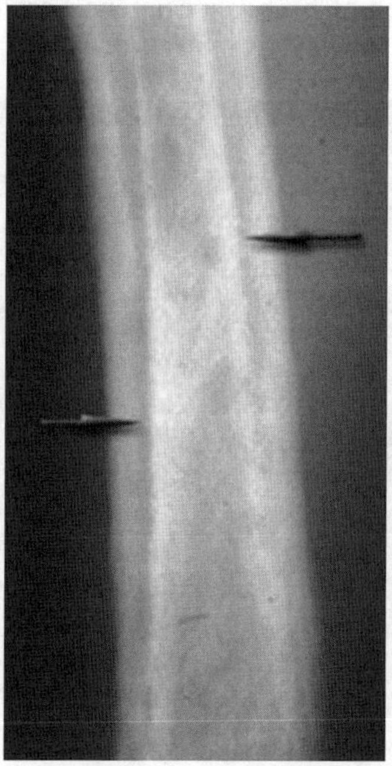

Figure 17.2. Involucrum with "new" bone formation inside an infected tibia. Image is the property of Geneva University Hospitals and is displayed with the permission of the patient.

The passage of pus through cloacae, which are holes in the involucrum, may spread the exudates into the surrounding soft tissue. Next, sequestra and/or involucra may become fibrotic, resulting in sclerosis, indicating a persistence of infection of generally at least 1 month [8]. Biofilms play a crucial role not only in foreign body infections but also in implant-free osteomyelitis [9, 22] (for implant-related biofilms, see Chapter 20).

Diagnosis

Clinical and Laboratory Diagnostics

The typical clinical presentation of chronic osteomyelitis includes pain in the affected region, frequently accompanied by local inflammatory signs such as swelling, warmth, localized fluctuation, impaired function, or a sinus tract (Figure 17.3) with purulent discharge. The latter is almost pathognomonic for infectious chronic osteomyelitis. However, the ultimate diagnostic proof is the association of a compatible clinical presentation with two separate positive microbiological deep bone biopsy samples [26]. Eubacterial polymerase chain reaction (PCR) has a high specificity, but generally a low sensitivity. In addition, it provides only very limited antibiotic susceptibility data (e.g., methicillin or rifampin resistance). Nevertheless, specific PCR are useful for slow-growing germs that are falsely interpreted as culture-negative in most microbiological media. Among these microorganisms are *Bartonella* spp., *Brucella* spp., *Coxiella* spp., *K. kingae*, *Mycobacterium tuberculosis* [27], or *Mycobacterium ulcerans* (Buruli disease) [28]. In culture- and PCR-negative osteomyelitis, the diagnosis of infection is established by the combination of clinical, laboratory (C-reactive protein), and radiological signs, as well as by the histological examination of biopsy specimens. Serology is very rarely helpful [29].

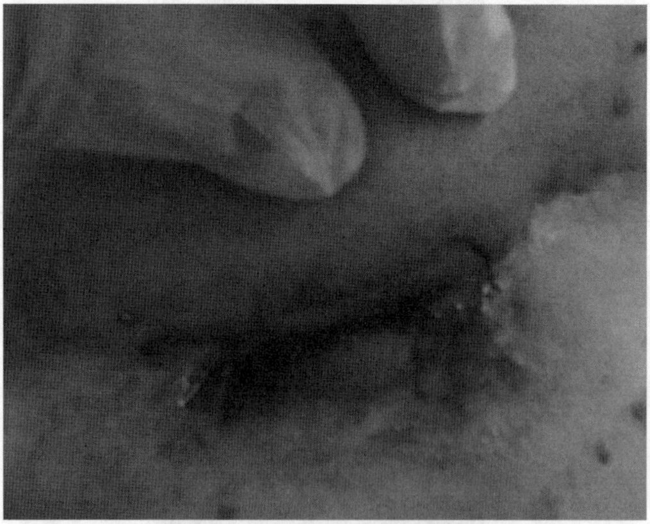

Figure 17.3. External orifice of a sinus tract on the calve/femur with maceration of the skin. Image is the property of Geneva University Hospitals and is displayed with the permission of the patient. (*See insert for color representation of the figure.*)

Probe-to-bone test does not prove diagnosis of underlying chronic osteomyelitis [20], because it may indicate only superficial bone contamination. Many trials have addressed this issue. In one well-designed, representative study, bone probing for osteomyelitis had a sensitivity of 66%, a specificity of 85%, a positive predicting value of 89%, and a negative predicting value of 56% [30]. The diagnostic value of pathogen identification on surface bone swabs is unclear. The same problem applies to specimens retrieved from any communicating aperture between the bone and the outer environment, as sinus tract cultures for instance. Mackowiak *et al.* [31] compared sinus tract samples with intraoperative bone biopsy and found concordance in only 44% of cases. We cannot emphasize enough the value of several bone biopsy cultures. A microbiological diagnosis of underlying bone infection by two identical consecutive samples of the sinus tract is an approach performed especially in France [32, 33], but needs confirmation in larger studies.

Radiological Diagnosis

Soft tissue swelling, periosteal thickening/elevation, and focal osteopenia are the first changes that can be seen on plain radiographs. Unfortunately, these changes only appear when the majority (> 50–75%) of the bone matrix is lost, which typically takes at least 2 weeks [20]. Classical pathognomonic radiological signs in plain radiographs are the presence of involucra and/or sequestra, which are, however, only seen in a minority of patients. More useful tools for early diagnosis are computed tomography (CT) scans (Figure 17.4) or magnetic resonance imaging (MRI). The latter, in particular, has shown an excellent sensitivity and a very high specificity before surgical treatment. Nevertheless, in those cases where there is concomitant "overuse syndromes," such as gout, neuropathic osteoarthropathy, fibrosis, and bone remodeling after previous surgery, the specificity of

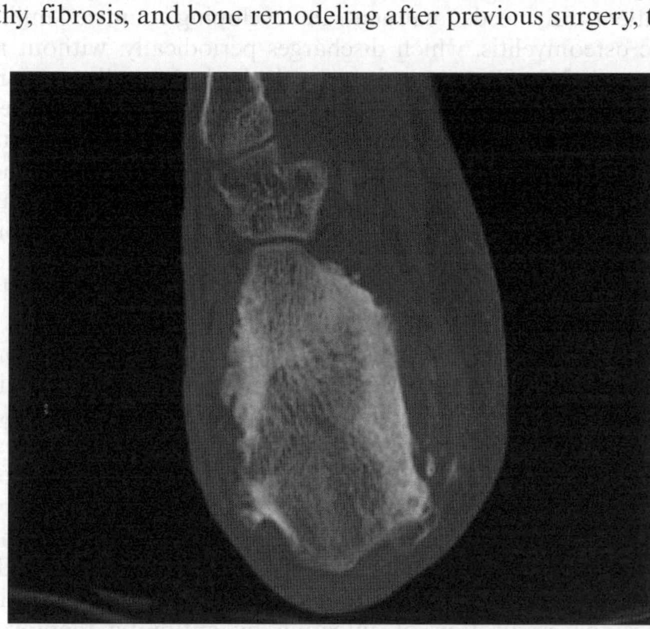

Figure 17.4. Bone lesions and sclerosis in a calcaneum of a 29-year-old man (posttraumatic chronic osteomyelitis). Image is the property of Geneva University Hospitals and is displayed with the permission of the patient.

MRI is diminished [34]. According to Kaim *et al.* [35], in cases of chronic posttraumatic osteomyelitis, the reported sensitivity, specificity, and accuracy of MRI are 100, 69, and 78%, respectively. Nowadays, advanced CT scans combine good sensitivity and an overall more acceptable price–quality ratio than MRI. Furthermore, they offer additional comfort, providing a faster diagnostic procedure as well as a faster and easier setting, in particular for claustrophobic patients. An even more accurate diagnostic imaging technique might be positron emission tomography (PET). Nanni *et al.* [36] reported a sensitivity of 100%, a specificity of 76%, and an overall accuracy of 90% for the ^{68}Ga–citrate PET–CT. However, differentiating between postsurgery tissue remodeling and persistent infection at a cutoff of 6 weeks may still prove to be a problem, as the only studies available on this topic have been performed in rabbits [37]. The technical difficulty of setting up a PET–CT scanner and the high operative costs still confine the routine use of this technology to selected tertiary healthcare centers worldwide. As far as scintigraphy is concerned, its role is diminishing due to the prevalence of more accurate methods [38, 39]. Despite the fact that imaging procedures are very precious diagnostic tools in the hands of an experienced clinician, radiological modalities remain of limited usefulness for directing treatment, as they cannot provide any clue regarding the causal pathogen.

Treatment

Treatment of chronic osteomyelitis is a real challenge, and requires a multidisciplinary team including orthopedic surgeons, infectious disease specialists, nurses, and physiotherapists, in order to achieve the best possible outcome. The first question to be answered is whether the patient truly needs treatment. In developing countries, millions of patients live with chronic osteomyelitis, which discharges periodically, without impairing their everyday activities [7, 26]. The patient's general physical condition, comorbidity, age, eventual allergies, and possible drug interactions as well as overall life expectancy must be taken into account before beginning any treatment. We strongly discourage beginning any long-term antibiotic treatment in patients who are stable, nonfebrile, and asymptomatic or, for any reason, would not be able to complete a full course of therapy. For these patients, we suggest an etiology-driven (if applicable) suppressive therapy of intermittent duration in case of symptomatic flares and exacerbations of chronic osteomyelitis (e.g., purulent discharge, pain), with the purpose of diminishing the bacterial loading temporarily.

When deciding on a curative approach, two major rules should be considered. First, chronic osteomyelitis cannot be cured, if the underlying reason for the infection is not identified and reversed. In these situations, infection is only an epiphenomenon, the tip of the iceberg, of a much more serious problem. Good examples of this situation are sacral osteomyelitis in tetraplegic patients or diabetic foot osteomyelitis in patients with Charcot deformation. If the underlying problem cannot be solved, every attempt for sustained cure is futile. Even if there is cure of the current episode, recurrence of chronic osteomyelitis will occur, because the pathogenetic reasons are still present. Second, every chronic osteomyelitis in adult patients is a surgical disease with only few exceptions [1, 10, 11]. There is evidence for eradication of infection by antibiotic therapy alone only in hematogenous childhood osteomyelitis, skull and maxillar osteomyelitis, spondylodiscitis, and diabetic toe osteomyelitis in selected patients. The presence of sequestra, biofilm, and tissue necrosis compromises the effectiveness of antibiotic therapy. Thus, debridement and drainage are compulsory to achieve a satisfactory therapeutic result. The patient must accept at least one if not several surgical interventions and a corresponding hospital stay of one (amputation) or several weeks (cure and reconstruction).

Surgical Treatment

In this chapter, we will not go into the detailed review of surgical techniques. Various techniques have been described [10, 26, 40]. State-of-the-art surgical management includes sequestrectomy, resection of scarred and fibrous tissue [1, 11], restoration of effective blood supply, adequate soft tissue coverage, and dead space obliteration as well as bone mechanical stability [5, 11, 40]. For better visualization, sinus tracts can be injected with methylene blue, which facilitates excision. The intramedullary canal should be reopened through reaming from the site of infection toward both metaphyses to restore vascular flow. Not sampling biopsies for microbiological culture is an error, since antimicrobial therapy should be based on a defined microorganism with a known susceptibility pattern. Ideally, several bone biopsies, not superficial bone swabs or liquid samples, should be dispatched to microbiology and pathology. Aggressive debridement pays off, as shown by a prospective study of surgical resections in the setting of chronic osteomyelitis. For example, Simpson *et al.* [41] showed that resection margins wider than 5 mm had an overall better outcome. If blood supply is insufficient, blood flow must be restored proximally to the affected region through vascular bypass and/or intravascular stenting. If such a procedure is not feasible, amputation must be considered. Only appropriate tissue coverage allows cure of osteomyelitis and prevents recurrence. Large dead spaces are filled with surrounding tissue to discourage infection and favor stability. In cases where surrounding tissue is insufficient, the cavity can be filled using a local muscle flap [10] or free tissue transfer [42]. The choice of soft tissue coverage by autologous graft is left to the plastic reconstructive surgeon [40]. Some surgeons use antibiotic-loaded beads for both local anti-infective therapy and space filling, although neither indication is evidence-based in terms of cure. In cases of bone instability, a two-step procedure may be advised [43], consisting of preliminary debridement followed by external fixation and dressing. After a few weeks of systemic antibiotic treatment, a second intervention involving a new debridement is performed; the eventually filling material is removed and replaced by bone grafts and bone stabilized by internal fixation. The use of vacuum-assisted closure (VAC) is controversial [44]. While some surgeons renounce it, Tan *et al.* [45] showed that a series of patients treated by VAC had a recurrence rate that was 18% lower than without. In addition, 30% fewer patients needed flap surgery. Additional benefits were increased bacterial clearance and substantial lower costs compared to control. The pro and contra points of view regarding VAC therapy are the subject of ongoing research.

Antibiotic Treatment

Pharmacokinetic and Pharmacodynamic Considerations

Traditionally, the spectrum of activity and the in vitro susceptibility to antimicrobials have been the cornerstones in the choice of antimicrobial regimens. However, there is little doubt that the pharmacokinetic (PK) characteristics of a drug and the optimal pharmacodynamic (PD) exposure, not only in plasma but also at the infection site, are essential parameters to consider for optimal antimicrobial usage [46]. Another problem is the penetration of antibiotics in different compartments, and the potential influence of underlying disease on antimicrobial penetration into bone (see Chapter 3). Patients with chronic osteomyelitis and disorders of peripheral vessels may have impaired blood flow circulation at the site of infection. Bone is less vascularized than other tissues and is functionally composed of two distinct parts, namely, cortical bone and cancellous bone. Despite gradually improved knowledge on pharmacological characteristics of antimicrobials, current clinical pharmacological data on the treatment of chronic osteomyelitis are still inadequate for determining the best agent, route, or duration of antibiotic therapy [47]. Taken together, PK studies of bone penetration can provide important information, but they cannot replace large-scale clinical effectiveness trials [46].

Duration of Antibiotic Treatment

As far as duration of therapy after surgery is concerned, no evidence-based data are available to support a particular duration. At the present time, there are no clues that this situation will change in the near future. Too many known and unknown factors can influence the "fine-tuning" of pharmacological therapy. Therefore, guidelines covering a wide range of scenarios of chronic osteomyelitis seem unrealistic. For a vast majority of pathogens, the traditional approach for treatment duration is to start with several weeks [1, 26] of IV therapy, followed by oral pharmacotherapy for some weeks or months [8, 26]. In contrast, recent data from retrospective studies suggest the noninferiority of regimens including early compared to late oral switch involving prolonged IV administration [5, 26, 48]. While bioavailability is set by default at 100% in IV administration, long-term IV therapy may lead to a rate of complication of 15%. Recent data suggest that 2 weeks of initial IV regimen, followed by oral medication for a total of roughly 6 weeks, may be nearly as effective but safer [26, 49]. Thus, in most situations, it provides a clinically convenient and cost-effective trade-off [50]. Eyichukwu and Anyaehie [51] achieved arrest of chronic osteomyelitis by using surgery and short-term (2–3 days) IV therapy based on microbial sensitivity data, followed by oral administration. Ross and Cole [52] reported remarkable results for 91% of their pediatric population treated by 2 days of IV therapy, followed by 6 weeks of oral antibiotics. In a similar way, the total duration of anti-infective therapy that was proposed in the traditional approach of chronic osteomyelitis therapy has also been recently challenged. Some recent data suggest that antibiotic therapy could be limited to a total of 6 weeks, regardless of the type of osteomyelitis [26, 50], which is in line with recent research developments in the field of osteoarticular infections. Treatment durations longer than 6 weeks were not associated with any significant benefit. The significance of this latter study for current clinical practice should be reconsidered, in view of the recent improvements in surgical methods, diagnostic tools, and new antibiotics developed in the meantime. Antibiotics with high oral bioavailability have demonstrated their efficacy in clinical practice [26, 50]. Typical examples are fluoroquinolones, linezolid, clindamycin, and fusidic acid in combination with rifampin. Several trials have shown excellent results with early switch to oral medication. In a recent review, Haidar *et al.* [48] listed a number of small-site studies performed in animal models and clinical trials, which achieved remission of osteomyelitis with treatment lengths of 2–4 weeks. A recent retrospective study performed at our institution showed by a multivariate analysis that 1 week of IV therapy had the same remission rate as 2–3 weeks (0.2, 0.1–1.9) or ≥3 weeks of IV therapy (0.3, 0.1–2.4). In addition, more than 6 weeks were not significantly different from less than 6 weeks of total antibiotic treatment (0.8, 0.1–5.2) in terms of recurrence [49]. Current literature data and expert opinion indicate that the total duration of antibiotic administration is not influenced by the nature of the pathogen, by its antibiotic resistance profile, or by the evolution of serum inflammatory markers. The few exceptions indicating a longer treatment by default are chronic osteomyelitis due to tubercular and nontubercular mycobacteria (*M. ulcerans, Mycobacterium leprae*), fungi, *Coxiella burneti*, *Nocardia* spp., or *Brucella* spp. [26].

Choice of Appropriate Antibiotic (see also Chapter 6)

Factors that need to be taken into account before starting any antimicrobial treatment are susceptibility of the isolated pathogen, bone penetration, and oral bioavailability (Table 17.2). Detailed information on the different antimicrobial agents is given in Chapter 6. The indications for the different agents are not different in chronic osteomyelitis.

Local Antibiotic-Releasing Delivery Systems

Antibiotic-containing bone cement is utilized for the treatment and prevention of prosthetic-related infections [53–55], but could also be used for implant-free chronic osteomyelitis. However, the most frequent local vehicles in patients with chronic osteomyelitis are antibiotic-containing beads. These systems can achieve local drug concentrations up to 1000 times in excess of the minimum inhibitory concentrations (MICs)

Table 17.2. Bone penetration of selected antibiotics in the literature.

Antibiotic agent	Range of mean bone to serum ratio
Gram-positive pathogens	
Flucloxacillin	0.16–1.2
Cefazolin	0.179
Vancomycin	0.26
Rifampicin	0.57
Fusidic acid	0.46–0.93
Clindamycin	0.21–0.45
Linezolid	0.4–0.51
Gram-negative pathogens	
Ceftriaxone	0.07–0.17
Ceftazidime	0.04–0.06
Ciprofloxacin	0.27–1.1
Mixed Gram-positive and Gram-negative pathogens	
Cefuroxime	0.04–0.06
Amoxicillin/clavulanic acid	0.22/0.11–0.14
Cefepime	0.46–0.75
Levofloxacin	0.36–1.0
Piperacillin/tazobactam	0.2–0.3/0.3
Ertapenem	0.13–0.19

Adapted from Ref. 46.

for the most frequently encountered pathogens. This provides the potential benefits of local delivery, namely, high local activity without adverse systemic effects [53]. Limited data suggests a remission rate of 78% in chronic osteomyelitis in patients treated by antibiotics beads alone [55], but, as far as we know, there is no hard data suggesting that local plus systemic antibiotic prescription would be superior to systemic administration alone. Unfortunately, time-dependent drug release is unpredictable and difficult to ascertain in vivo, limiting the reliability and effectiveness of the local antibiotic therapy. Currently, the most frequent antibiotic-laden cements contain tobramycin, gentamicin, vancomycin, fluoroquinolones, cephalosporins, amphotericin B, or fluconazole. Biodegradable materials are theoretically preferred, as they do not require surgical removal. However, in clinical practice surgeons combine the removal of undegradable beads for a second look about 10–20 days after insertion.

Hyperbaric Oxygen Therapy

Hyperbaric oxygen therapy is relatively expensive and should be considered as the last armentarium against chronic osteomyelitis, if ever [56]. Hyberbaric oxygen may contribute to faster wound healing and better tissue vascularization by promoting angiogenesis, osteogenesis, and collagen production. However, so far no convincing data have shown its utility in reducing infection recurrence per se [57].

Particular Types of Chronic Osteomyelitis

Diabetic Foot Osteomyelitis (see Chapter 18)

Foot infections are one of the most common complications in diabetes [20]. Chronic osteomyelitis has to be suspected in patients with nonhealing ulcers despite state-of-the-art local management, including "off-loading" [58]. In addition, a positive probe-to-bone test, and cortical lysis in X-rays may indicate, but not entirely prove [30], diabetic foot osteomyelitis. Serial foot radiographs provide a good and inexpensive diagnostic tool. Clinically, there may be local signs like foul-smelling, purulent discharge, swelling, and dry/wet necrosis. Systemic complication such as fever or sepsis is uncommon but can be dramatic. In these cases, necrotizing fasciitis and other complications needing emergency surgery have to be considered. Unless there is amputation, antibiotic treatment is compulsory. International expert agreement is empirical treatment for Gram-positive organisms (mainly staphylococci/streptococci) and additional coverage of Gram-negative *Enterobacteriaceae* in severe or pretreated cases [20, 58]. It is important to stress that methicillin-resistant *S. aureus* (MRSA) skin colonization is not an indication of treatment by default [59]. While treatment duration of 6 weeks is recommended for acute osteomyelitis, some prefer duration of 12 weeks for chronic diabetes foot osteomyelitis without surgical debridement or amputation [20, 58, 60]. In selected cases, remission of diabetic chronic osteomyelitis by antibiotics, after local debridement and off-loading, has been achieved in up to 70% of case series [60] (for antibiotic choice see Table 17.3).

Brucella Osteomyelitis (see Chapters 7 and 16)

Osteoarticular infection is a common presentation of brucellosis. The most typical locations are the sacroiliac joints and the large joints of the lower extremities [61]. Sacroiliitis is a common form of disease encountered in around 25% of cases. Common

Table 17.3. Usual targeted antibiotic therapy for chronic osteomyelitis after surgical debridement (according to Geneva University Hospitals).

Pathogen(s)	Initial therapy	Alternative therapy	Oral therapy
S. aureus (MSSA)	Cefazolin 2 g (IV) tid 0–2 weeks	Fluoclaxillin 2 g (IV) qd 0–2 weeks	Rifampin 600 mg qd + ciprofloxacin 750 mg bid 4–6 weeks
			Clindamycin 600 mg qd 4–6 weeks
S. aureus (MRSA)	Vancomycin 1 g (IV) bid 0–2 weeks		Rifampin 600 mg qd 4–6 weeks + fusidic acid 500 mg tid 4–6 weeks
	Daptomycin 6–8 mg/kg (IV) qd 0–2 weeks		Clindamycin 600 mg qd 4–6 weeks (is susceptible)
			Linezolid 600 mg (PO) bid 4–6 weeks
B hemolytic streptococci	Penicillin G (IV) 6× 4 million Units	Ceftriaxone 2 (IV) qd	Clindamycin 600 mg qd 4–6 weeks
Enterobacteriaceae	Ceftriaxone 2 g (IV) qd, 0–2 weeks	Ceftazidime 2 g (IV) tid	Ciprofloxacin 750 mg qd for 4–6 weeks, eventually trimethoprim/ sulfamethoxazole
	Ertapenem 1 g (IV) qd, 0–2 weeks	Cefepime 2 g (IV) tid	
Brucella spp.	Streptomycin 1 g (IM), 21 days		Doxycycline 100 mg bid + Rifampin 600 qd 6–12 weeks

treatment involves antibiotic bitherapy for an overall duration of 12 weeks [62, 63]. Due to the usual lack of sequestra, a surgical approach is less frequent than in staphylococcal chronic osteomyelitis. Surgical treatment may be advised in case of spinal cord compression, paravertebral and epidural abscess, and, finally, structural instability. The infection responds well to oral doxycycline 100 mg bid coupled to streptomycin 1 g intramuscularly for the first 14–21 days, before switching to oral doxycycline combined with rifampin (600–900 mg qd) [63]. An alternative combination would be ciprofloxacin (750 mg bid) plus rifampin [64].

Osteomyelitis Associated to Sickle Cell Disease

Sickle cell disease is an autosomal recessive disorder and provokes anemia starting in childhood [65]. The disease also leads to bone necrosis caused by microvascular occlusion [66]. Its infectious hallmark is Gram-negative pathogens [65], for example, *Salmonella* spp., although the association with *Salmonella* spp. is poorly defined. Lifetime incidence of osteomyelitis in severe homozygote disease is estimated at 3% [65]. Treatment is that of classical osteomyelitis.

Sacral Osteomyelitis

This disease is chronic and related to decubitus in patients with multiple comorbidities and/ or neurological disorders. It is particularly difficult to treat, since there is no remission, if the reason for chronic osteomyelitis cannot be reversed. In these chronic decubitus patients, the infected sacral bone often cannot be excised and the patient cannot be improved neurologically. Prevention is of utmost importance. A thorough daily nursing care and debridement is the key to success. In improved cases, plastic surgeons may graft on the naked bone. The ideal duration of antibiotic administration is unknown. The aim is often not to eradicate osteomyelitis, but to control it. More data are needed in this field of osteomyelitis.

Jaw Osteomyelitis (see Chapter 19)

Chronic mandibular osteomyelitis occurs after dental procedures, trauma, or in very poor settings of noma disease. There has not been much research in the past. The causative pathogens are often polymicrobial and stem from the oral flora. *Actinomyces* spp. are particular pathogens that may mimic neoplasm. Treatment consists of maxilla–facial surgery, often repeated, and of long-lasting oral antibiotic therapy for which the choice of amoxicillin/clavulanate covers the majority of the oral flora.

Clinical Follow-Up during Treatment

Osteomyelitis patients should be regularly followed up throughout the treatment, for early detection of complications, adverse events, and control of wounds. However, it is an unresolved issue, whether regular follow-up really enhances remission rates and shortens treatment. Outpatient visits upon request of the patient are an alternative to scheduled regular follow-up controls. Again, there are no data comparing both possibilities of follow-up. A substantial improvement would be the use of diagnostic imaging to judge how long therapy remains necessary. Indeed, the duration of antibiotic administration for chronic osteomyelitis is usually decided from the start and maintained thereafter; independent of the individual case and markers. C-reactive protein is widely used in the follow-up of patients with localized bone and implant infections, but trauma or surgery may also result in its transient elevation. An additional problem is the fact that many cases with chronic osteomyelitis have a normal C-reactive protein even before treatment. Ideally, the question of serological markers to monitor treatment duration needs a prospective, randomized trial.

Future Perspectives

New antibiotic drugs and new approaches such as phage therapy are clearly needed and existing ones should be studied in-depth. Promising medications are dalbavancin, telavancin, oritavancin, and others [22]. Yet, we do not expect any revolution in the next few years, but rather remission rates in vivo comparable with actual available drugs. Studying specific bacterial genes coding virulence factors promoting invasiveness might bring to light new possible targets for future therapy. Finally, great effort must be invested in multicenter, blinded, large, randomized trials for the study of many unanswered questions in the domain of chronic osteomyelitis.

Key Points

- Chronic osteomyelitis in adults is a frequent disease requiring surgical debridement for cure (very few exceptions).
- The recommended duration of concomitant antibiotic therapy is scheduled for roughly 6 weeks.

- There is no difference between an oral and an IV antimicrobial treatment, as long as the blood levels of the drugs are identical.
- Few antibiotic agents reveal sufficient bone penetration and oral bioavailability for the treatment of chronic osteomyelitis, for example, clindamycin, linezolid, fluoroquinolones, rifampin.

Acknowledgments

We would like to thank the Orthopedic Surgery and the Infectious Disease Division of the Geneva University Hospitals for their help.

References

[1] Lew DP, Waldvogel FA. Osteomyelitis. Lancet 2004;364(9431):369–379.
[2] Jansson A, Renner ED, Ramser J, *et al.* Classification of non-bacterial osteitis: retrospective study of clinical, immunological and genetic aspects in 89 patients. Rheumatology (Oxford, England) 2007;46(1):154–160.
[3] Karadağ-Saygı E, Gündüz OH, Gümrükcü G, Akyüz G. SAPHO syndrome: misdiagnosed and operated. Acta Reumatol Port 2008;33(4):460–463.
[4] Shimizu M, Tone Y, Toga A, *et al.* Colchicine-responsive chronic recurrent multifocal osteomyelitis with MEFV mutations: a variant of familial Mediterranean fever? Rheumatology (Oxford, England) 2010;49(11):2221–2223.
[5] Lazzarini L, Lipsky BA, Mader JT. Antibiotic treatment of osteomyelitis: what have we learned from 30 years of clinical trials? Int J Infect Dis 2005;9(3):127–138.
[6] Waldvogel FA, Medoff G, Swartz MN. Treatment of osteomyelitis. N Engl J Med 1970;283(15):822–823.
[7] Beckles VL, Jones HW, Harrison WJ. Chronic haematogenous osteomyelitis in children: a retrospective review of 167 patients in Malawi. J Bone Joint Surg Br 2010;92(8):1138–1143.
[8] Auh JS, Binns HJ, Katz BZ. Retrospective assessment of subacute or chronic osteomyelitis in children and young adults. Clin Pediatr (Phila) 2004;43(6):549–555.
[9] Hannan CM, Attinger CE. Special considerations in the management of osteomyelitis defects (diabetes, the ischemic or dysvascular bed, and irradiation). Semin Plast Surg 2009;23(2):132–140.
[10] Parsons B, Strauss E. Surgical management of chronic osteomyelitis. Am J Surg 2004;188(1):57–66.
[11] Cierny G, 3rd, Mader JT, Penninck JJ. A clinical staging system for adult osteomyelitis. Clin Orthop Relat Res 2003;414:7–24.
[12] Byren I, Peters EJ, Hoey C, *et al.* Pharmacotherapy of diabetic foot osteomyelitis. Expert Opin Pharmacother 2009;10(18):3033–3047.
[13] Schaper NC. Diabetic foot ulcer classification system for research purposes: a progress report on criteria for including patients in research studies. Diabetes Metab Res Rev 2004;20(1):90–95.
[14] Calhoun JH, Manring MM. Adult osteomyelitis. Infect Dis Clin North Am 2005;19(4): 765–786.
[15] Legout L, Assal M, Rohner P, *et al.* Successful treatment of *Candida parapsilosis* (fluconazole-resistant) osteomyelitis with caspofungin in a HIV patient. Scand J Infect Dis 2006;38(8):728–730.
[16] Blyth CC, Gomes L, Sorrell TC, *et al.* Skull-base osteomyelitis: fungal vs. bacterial infection. Clin Microbiol Infect 2011;17(2):306–311.

[17] Ciampolini J, Harding KG. Pathophysiology of chronic bacterial osteomyelitis. Why do antibiotics fail so often? Postgrad Med J 2000;76(898):479–483.

[18] Beals RK, Bryant RE. The treatment of chronic open osteomyelitis of the tibia in adults. Clin Orthop Relat Res 2005;433:212–217.

[19] Karamanis EM, Matthaiou DK, Moraitis LI, Falagas ME. Fluoroquinolones versus beta-lactam based regimens for the treatment of osteomyelitis: a meta-analysis of randomized controlled trials. Spine 2008;33(10):297–304.

[20] Lipsky BA, Berendt AR, Deery HG, *et al.* Diagnosis and treatment of diabetic foot infections. Plast Reconstr Surg 2006;117(7):212–238.

[21] Uçkay I, Dinh A, Vauthey L, *et al.* Spondylodiscitis due to *Propionibacterium acnes*: report of twenty-nine cases and a review of the literature. Clin Microbiol Infect 2010;16(4):353–358.

[22] Uçkay I, Pittet D, Vaudaux P, *et al.* Foreign body infections due to *Staphylococcus epidermidis*. Ann Med 2009;41(2):109–119.

[23] Ceroni D, Cherkaoui A, Ferey S, *et al. Kingella kingae* osteoarticular infections in young children: clinical features and contribution of a new specific real-time PCR assay to the diagnosis. J Pediatr Orthop 2010 30(3):301–304.

[24] Peto HM, Pratt RH, Harrington TA, *et al.* Epidemiology of extrapulmonary tuberculosis in the United States, 1993-2006. Clin Infect Dis 2009;49(9):1350–1357.

[25] Uçkay I, Hoffmeyer P, Lew D, Pittet D. Prevention of surgical site infections in orthopaedic surgery and bone trauma: state-of-the-art update. J Hosp Infect 2013;84(1):5–12.

[26] Uçkay I, Jugun K, Gamulin A, *et al.* Chronic osteomyelitis. Curr Infect Dis Rep 2012;14(5):566–575.

[27] Colmenero JD, Morata P, Ruiz-Mesa JD, *et al.* Multiplex real-time polymerase chain reaction: a practical approach for rapid diagnosis of tuberculous and brucellar vertebral osteomyelitis. Spine (Phila Pa 1976) 2010;35(24):1392–1396.

[28] Abgueguen P, Pichard E, Aubry J. Buruli ulcer or *Mycobacterium ulcerans* infection. Med Mal Infect 2010;40(2):60–69.

[29] Uçkay I, Ferry T, Stern R, *et al.* Use of serum antistreptolysin O titers in the microbial diagnosis of orthopedic infections. Int J Infect Dis 2009;13(4):421–424.

[30] Grayson ML, Gibbons GW, Balogh K, *et al.* Probing to bone in infected pedal ulcers. A clinical sign of underlying osteomyelitis in diabetic patients. JAMA 1995;273(9):721–723.

[31] Mackowiak PA, Jones SR, Smith JW. Diagnostic value of sinus-tract cultures in chronic osteomyelitis. JAMA 1978;239(26):2772–2775.

[32] Bernard L, Uçkay I, Vuagnat A, *et al.* Two consecutive deep sinus tract cultures predict the pathogen of osteomyelitis. Int J Infect Dis 2010;14(5):390–393.

[33] Bernard L, Assal M, Garzoni C, Uçkay I. Predicting the pathogen of diabetic toe osteomyelitis by two consecutive ulcer cultures with bone contact. Eur J Clin Microbiol Infect Dis 2011; 30(2):279–281.

[34] Vartanians VM, Karchmer AW, Giurini JM, Rosenthal DI. Is there a role for imaging in the management of patients with diabetic foot? Skeletal Radiol 2009;38(7):633–636.

[35] Kaim A, Ledermann HP, Bongartz G, *et al.* Chronic post-traumatic osteomyelitis of the lower extremity: comparison of magnetic resonance imaging and combined bone scintigraphy/immunoscintigraphy with radiolabelled monoclonal antigranulocyte antibodies. Skeletal Radiol 2000;29(7):378–386.

[36] Nanni C, Errani C, Boriani L, *et al.* 68Ga-citrate PET/CT for evaluating patients with infections of the bone: preliminary results. J Nucl Med 2010;51(12):1932–1936.

[37] Koort JK, Makinen TJ, Knuuti J, *et al.* Comparative 18F-FDG PET of experimental *Staphylococcus aureus* osteomyelitis and normal bone healing. J Nucl Med 2004;45(8): 1406–1411.

[38] Sönmezoğlu K, Sönmezoğlu M, Halaç M, *et al.* Usefulness of 99mTc-ciprofloxacin (infecton) scan in diagnosis of chronic orthopedic infections: comparative study with 99mTc-HMPAO leukocyte scintigraphy. J Nucl Med 2001;42(4):567–574.

[39] Schauwecker DS. The scintigraphic diagnosis of osteomyelitis. AJR Am J Roentgenol 1992;158(1):9–18.

[40] Papineau LJ, Alfageme A, Dalcourt JP, Pilon L. Chronic osteomyelitis: open excision and grafting after saucerization. Int Orthop 1979;3(3):165–176.

[41] Simpson AH, Deakin M, Latham JM. Chronic osteomyelitis. The effect of the extent of surgical resection on infection-free survival. J Bone Joint Surg Br 2001;83(3):403–407.

[42] Fitzgerald RH Jr, Ruttle PE, Arnold PG, *et al.* Local muscle flaps in the treatment of chronic osteomyelitis. J Bone Joint Surg Am 1985;67(2):175–185.

[43] Catagni MA, Ottaviani G, Camagni M. Treatment of massive tibial bone loss due to chronic draining osteomyelitis: fibula transport using the Ilizarov frame. Orthopedics 2007;30(8):608–611.

[44] Kumar S, O'Donnell ME, Khan K, *et al.* Successful treatment of perineal necrotising fasciitis and associated pubic bone osteomyelitis with the vacuum-assisted closure system. World J Surg Oncol 2008;6(67):1–4.

[45] Tan Y, Wang X, Li H, *et al.* The clinical efficacy of the vacuum-assisted closure therapy in the management of adult osteomyelitis. Arch Orthop Trauma Surg 2011;131(2):255–259.

[46] Landersdorfer CB, Bulitta JB, Kinzig M, *et al.* Penetration of antibacterials into bone. Pharmacokinetic, pharmacodynamic and bioanalytical considerations. Clin Pharmacokinet 2009;48(2):89–124.

[47] Joukhadar C, Muller M. Microdialysis: current applications in clinical pharmacokinetic studies and its potential role in the future. Clin Pharmacokinet 2005;44(9):895–913.

[48] Haidar R, Der Boghossian A, Atiyeh B. Duration of post-surgical antibiotics in chronic osteomyelitis: empiric or evidence-based? Int J Infect Dis 2010;14(9):752–758.

[49] Rod-Fleury T, Dunkel N, Assal M, *et al.* Duration of post-surgical antibiotic therapy for adult chronic osteomyelitis: a single-centre experience. Int Orthop 2011;35(11):1725–1731.

[50] Daver NG, Shelburne SA, Atmar RL, *et al.* Oral step-down therapy is comparable to intravenous therapy *for Staphylococcus aureus* osteomyelitis. J Infect 2007;54(6):539–544.

[51] Eyichukwu GO, Anyaehie UE. Outcome of management of chronic osteomyelitis at National Orthopaedic Hospital, Enugu. Niger J Med 2009;18(2):194–198.

[52] Ross ER, Cole WG. Treatment of subacute osteomyelitis in childhood. J Bone Joint Surg Br 1985;67(3):443–448.

[53] Kanellakopoulou K, Giamarellos-Bourboulis EJ. Carrier systems for the local delivery of antibiotics in bone infections. Drugs 2000;59(6):1223–1232.

[54] Cui Q, Mihalko WM, Shields JS, *et al.* Antibiotic-impregnated cement spacers for the treatment of infection associated with total hip or knee arthroplasty. J Bone Joint Surg Am 2007;89(4):871–882.

[55] Walenkamp GH, Kleijn LL, de Leeuw M. Osteomyelitis treated with gentamicin-PMMA beads: 100 patients followed for 1-12 years. Acta Orthop Scand 1998;69(5):518–522.

[56] Berendt AR. Counterpoint: hyperbaric oxygen for diabetic foot wounds is not effective. Clin Infect Dis 2006;43(2):193–198.

[57] Andel H, Felfernig M, Andel D, *et al.* Hyperbaric oxygen therapy in osteomyelitis. Anaesthesia 1998;53(2):68–69.

[58] Uçkay I, Gariani K, Pataky Z, Lipsky BA. Diabetic foot infections: state-of-the-art. Diabetes Obes Metab 2013;16(4):305–316. DOI:10.1111/dom.12190.

[59] Reber A, Moldovan A, Dunkel N, *et al.* Should the methicillin-resistant *Staphylococcus aureus* carriage status be used as a guide to treatment for skin and soft tissue infections? J Infect 2012;64(5):513–519.

[60] Valabhji J, Oliver N, Samarasinghe D, *et al.* Conservative management of diabetic forefoot ulceration complicated by underlying osteomyelitis: the benefits of magnetic resonance imaging. Diabet Med 2009;26(11):1127–1134.

[61] Geyik MF, Gür A, Nas K, *et al.* Musculoskeletal involvement of brucellosis in different age groups: a study of 195 cases. Swiss Med Wkly 2002;132(7):98–106.

[62] Colmenero JD, Ruiz-Mesa JD, Plata A, *et al.* Clinical findings, therapeutic approach, and outcome of brucellar vertebral osteomyelitis. Clin Infect Dis 2008;46(3):426–433.
[63] Pappas G, Akritidis N, Bosilkovski M, Tsianos E. Brucellosis. N Engl J Med 2005;352(22):2325–2336.
[64] Alp E, Koç RK, Durak AC, *et al.* Doxycycline plus streptomycin versus ciprofloxacin plus rifampicin in spinal brucellosis. BMC Infect Dis 2006;6:72–82.
[65] Al-Tawfiq JA. *Bacteroides fragilis* bacteremia associated with vertebral osteomyelitis in a sickle cell patient. Int Med (Tokyo, Japan) 2008;47(24):2183–2185.
[66] Hernigou P, Daltro G, Flouzat-Lachaniette CH, *et al.* Septic arthritis in adults with sickle cell disease often is associated with osteomyelitis or osteonecrosis. Clin Orthop Relat Res 2010;468(6):1676–1681.

Chapter 18
Diabetic Foot Osteomyelitis

Eric Senneville and Sophie Nguyen

Introduction

Over half of chronic diabetic foot wounds are complicated by soft tissue infection that may involve underlying osteoarticular structures in 20–60% of the cases according to the infection severity [1–4]. These osteoarticular infections, designated by the term diabetic foot osteomyelitis (DFO), are difficult to treat and are associated with increased risk of relapsing infection, hospitalization episodes, and foot amputations [5]. Suspicion of DFO is based on clinical and radiological elements, while definite diagnosis requires microbiological and histological criteria [6]. In addition to the classical limitations of the treatment of chronic osteomyelitis (i.e., bone necrosis and difficulties in achieving high local concentrations of anti-infective agents), peripheral arterial disease and impaired leukocyte function associated with diabetes increase the risk of poor outcome especially if they are simultaneously present in a patient.

Many of the limb losses related to diabetes are due to severe soft tissue infection and osteomyelitis, and much has still to be clarified regarding the diagnosis and management of DFO, despite significant advances in this field during the last few decades.

The aim of this chapter is to review the currents advances in the knowledge of DFO regarding diagnosis and treatment.

Classification

Osteomyelitis is defined as an inflammatory process in bone that is caused by infection, which is responsible for osseous fragmentation and overall destruction of the skeletal architecture [7]. DFO is an osteomyelitis that affects, by definition, any part of the foot located under the malleoli in a patient with diabetes. DFO is mostly the consequence of a soft tissue infection that spreads to the underlying osteoarticular structures.

Bone and Joint Infections: From Microbiology to Diagnostics and Treatment, First Edition.
Edited by Werner Zimmerli.

Hematogenous origin of DFO leading to a primary infection of the bone marrow is almost never encountered, but DFO can result from either penetrating injury or ischemic soft tissue injury [8]. Puncture of wound may result in direct inoculation of medullar bone or joints. However, the first osseous structure generally involved during DFO is the cortical bone and periosteum infection leading to subsequent affection of the bone marrow. According to Hofmann *et al.* [9], this kind of "centripetal" infection is defined as an *osteitis* rather than an *osteomyelitis*.

Only a few authors have studied the histological findings of DFO. Chantelau *et al.* [10] reported the results of routine histopathology assessment of bone specimens taken in 43 patients, including 30 with peripheral ischemic vascular disease (PIVD) who had been amputated. Among 22 clinical cases of DFO, 20 cases of gangrene, and 3 cases of pressure ulcer, histomorphological examination showed osteomyelitis ($n = 29$), bone necrosis ($n = 1$), myelofibrosis ($n = 8$), and normal bone ($n = 7$) at the affected sites, as compared to normal bone ($n = 26$), myelofibrosis ($n = 12$), and osteoporosis ($n = 7$) at the osteotomy sites. In cases of clinical gangrene, bones were also affected by osteomyelitis, but less so than in cases of clinical osteomyelitis (8/18 versus 22/22; $P < 0.001$). Interestingly, bone tissue at the osteotomy sites was normal, with some myelofibrosis in both conditions. The authors concluded that histomorphology of unaffected foot bone appears mostly normal in diabetic patients with neuropathy and PIVD.

The Cierny–Mader classification is based on both the anatomy of bone infection and the physiology of the host [11, 12]. However, it has been suggested that this classification would be less useful for classifying osteomyelitis in diabetic patients, especially in case of involvement of the small bones in the foot than in patients with posttraumatic osteomyelitis [7]. According to a recent study reported by Aragon-Sanchez *et al.* [13], three types of osteomyelitis, namely, acute, chronic, and acute exacerbation of chronic osteomyelitis, can be found in the diabetic foot. Acute osteomyelitis is defined by necrosis, destroyed bone, and infiltrations of polymorphonuclear granulocytes at cortical sites and inside the bone marrow usually associated with congestion or thrombosis of medullary or periosteal small vessels. Chronic osteomyelitis is characterized by destroyed bone and infiltrations of lymphocytes, histiocytes, and/or plasmatic cells at cortical sites and inside the bone marrow. Acute exacerbation of chronic osteomyelitis corresponds to a background of chronic osteomyelitis, upon which infiltration of polymorphonuclear granulocytes is present. In all cases of osteomyelitis, areas of fibrosis in variable forms as well as medullar edema were described [13].

More recently, the same team proposed to separate DFO into four classes. Class I corresponds to DFO without ischemia and without soft tissue involvement, class II to DFO with ischemia without soft tissue involvement, class III to DFO with soft tissue involvement, and class IV to DFO with ischemia and soft tissue involvement [14]. This study showed a statistically significant trend toward increased severity (i.e., from class I to class IV) and increased amputation and mortality rates. However, as stated by the authors, the diagnosis of deep tissue infection associated with DFO may be difficult to achieve before surgery, which makes this classification more useful for conducting clinical research studies, than to help the physician choose the most appropriate treatment strategy for a given patient.

The International Working Group on the Diabetic Foot (IWGDF) has recently proposed a classification of diabetic foot wounds that uses the acronym "PEDIS" (i.e., perfusion of the foot, extent and depth of the foot wound, infection, and sensation [neuropathy]) [15]. The severity of each item is graded semiquantitatively. An infection involving the osteoarticular structures underlying the foot ulcer is graded 3 and in the presence of signs defining the systemic inflammatory response syndrome (SIRS) is graded 4. The infection part of the classification differs only in small details from the classification proposed by the Infectious Diseases Society of America (IDSA) (i.e., grades 1, 2, 3, and 4 of the IWGDF classification correspond to absence, mild, moderate, and severe diabetic foot infection (DFI), respectively, in the IDSA classification) [16].

Diagnostic criteria for DFO have been proposed by the IWGDF in 2008 based on the same model as in the Duke University classification of infectious endocarditis. The expert panel of this group stratified DFO into four categories based on the results of clinical, imaging, and bone examination (ranging from unlikely [<10% post-test probability], through possible [10–50%], probable [51–90%], and certain [>90%]) [17]. According to IWGDF, DFO is certain in case of positive culture of a bone sample (taken with proper precautions in order to minimize contamination by bacteria belonging to the colonizing flora) associated with histopathological findings consistent with the diagnosis of osteomyelitis (i.e., acute or chronic inflammation and necrosis); all categories reported in this classification are detailed in Table 18.1.

It appears, however, that these definition criteria may be more useful for clinical research rather than for the physician in the setting of routine practice. Overall, there is currently no classification of DFO that may help select the most appropriate therapeutic approach of the patients and predict the outcome regarding the risk of amputation of the limb or recurrence of DFO.

Table 18.1. Diagnostic criteria of diabetic foot osteomyelitis (IWGDF).

Diagnosis	%	Criteria
Certain	>90	Positive bone culture + positive histopathology or
		Intra-osseous pus seen intraoperatively or
		Sequestra through the ulcer or
		Intra-osseous abscess seen on MRI
Probable	51–90	Medullar bone exposed or
		Signs consistent with osteomyelitis on MRI or
		Positive bone culture and unavailable or negative histopathology
Possible	10–50	Cortical osteolysis on plain X-ray or
		Osseous edema on MRI or
		Cortical bone exposed or
		ESR >70 mm/h with no other identified origin or
		Delayed ulcer healing on a well-perfused foot and adequate off-loading of the wound >6 weeks or infected foot ulcer of duration >2 weeks
Excluded	<10	Absence of signs of clinical and radiological signs of infection and time <2 weeks of infection and superficial foot wound and normal MRI or bone scan normal

Modified from Ref. 17.

Microbiology

Spectrum of Microorganisms in DFO

The microbiology of DFO is usually polymicrobial [18], and in almost all the reported series, *Staphylococcus aureus* is the most common pathogen cultured from bone samples [19, 20]. Other Gram-positive cocci frequently isolated from bone samples are *Staphylococcus epidermidis*, beta-hemolytic streptococci, and diphteroids. Among the Enterobacteriaceae, *Escherichia coli*, *Klebsiella pneumoniae*, and *Proteus* spp. are the most common pathogens followed by *Pseudomonas aeruginosa* (Table 18.2). The number of obligate anaerobes (mostly *Finegoldia magna* and other anaerobic peptococci) depends on the method of sampling and transportation of the bone fragments.

The increased prevalence of multiresistant bacteria (especially methicillin-resistant *S. aureus* [MRSA] and extended spectrum beta-lactamase [ESBL]–producing Gram-negative bacilli) has been reported in patients with DFO [21]. In a recent retrospective study of 130 patients with DFO, the presence of MRSA was associated with adverse outcome (53.3% versus 21.1%, $P=0.04$), which was defined as death, amputation, and failure to heal [22].

Culture and Histopathology of Bone Biopsy

The concordance between cultures from soft tissue swab cultures and bone is generally low, well below 50%. In numerous studies, it has been shown that cultures of bone specimens provide more accurate microbiological data than those of soft tissue specimens for patients with DFO [18, 19, 23]. Needle aspiration has recently been found more accurate than superficial samples (swabs) for microbiological diagnosis of DFO [24]. However, the overall concordance between microbiological results from needle aspiration and bone biopsy has been assessed in a recent retrospective study of diabetic patients with low-grade infection of the foot and suspected osteomyelitis, and it was only 23.9%, with *S. aureus* showing the strongest correlation (46.7%), as already established by Mackowiack *et al.* [25] in another setting [26].

Table 18.2. Microbiology of diabetic foot osteomyelitis.

	Swab samples	Bone biopsy samples
No. of isolates	109	125
No. (%) of isolates, by pathogen		
Gram-positive cocci	70 (64.2)	93 (74.4)
Staphylococcus aureus	36 (33.0)	33 (26.4)
Coagulase-negative staphylococci	5 (4.6)	32 (25.6)
Streptococcus spp.	22 (20.2)	15 (12.0)
Enterococcus spp.	5 (4.6)	10 (8.0)
Other	2 (1.8)	3 (2.4)
Corynebacterium spp.	8 (7.3)	3 (2.4)
Gram-negative bacilli	28 (25.7)	23 (18.4)
Obligate anaerobes	3 (2.8)	6 (4.8)

Modified from Ref. 19.

The main advantage of bone biopsy is that it provides reliable data on the organisms responsible for the infection and determines their profile of susceptibility to antimicrobial agents [27]. In only one retrospective multicenter study, bone culture–guided antibiotic treatment was associated with a significantly better clinical outcome than treatment guided by soft tissue culture results [28].

On the basis of current medical literature, bone biopsy performed with an appropriate procedure appears to be the most appropriate method for definite diagnosis and microbiological documentation of DFO [29]. In order to reduce the risk of false-negative culture results, it is preferable to perform bone biopsy after an antibiotic-free period of at least 2 weeks given the prolonged release of some antibiotics from bones [16, 30], provided the infection is not severe and the patient needs no urgent antibiotic therapy.

Percutaneous bone biopsy performed through a noninfected skin area in order to reduce the contamination by bacteria not involved in the bone infection appears to be the best technique to obtain a reliable microbiological documentation of DFO (Figure 18.1) [5]. Using fluoroscopic or computed tomography (CT) guidance may help to reduce the risk of error mapping and false-negative results. However, the realization of transcutaneous bone biopsies is a little complex, as it requires a multidisciplinary team with orthopedic surgeons interested in diabetic foot problems. In addition, for a minority of patients without severe loss of protective sensory, an anesthetist is required. By using a 11-gauge needle (bone-cutting needles, such as Jamshidi [Perfectum Corporation; distributed by Propper and Sons] or Ostycut Bard Products [distributed by Angiomed]), it is possible to obtain a large amount of bone that can be cut into one part for microbiological culture and another for histopathological examination [16]. Ideally, it seems best to obtain at

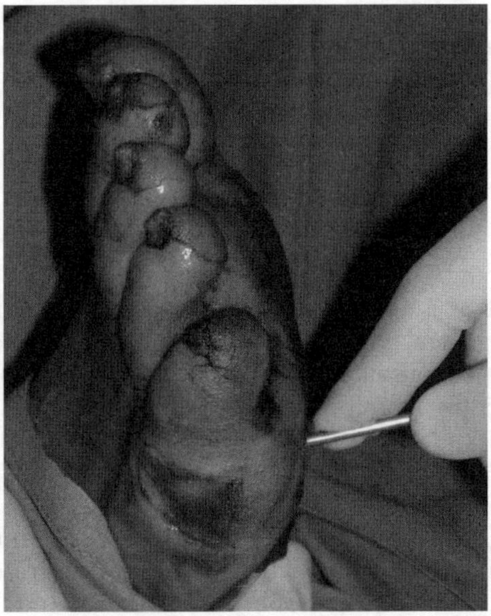

Figure 18.1. Transcutaneous bone biopsy of the metatarso-phalangeal joint of the fifth ray on the right foot in a diabetic foot patient; the trocard is introduced through a noninfected skin area opposite to the ulcer. Courtesy of Dr. Eric Beltrand, Department of Orthopedic Surgery Dron Hospital Tourcoing F-59200 France. (*See insert for color representation of the figure.*)

least two bone samples for culture. Histological examination may be helpful in interpreting the results of bone culture especially in case of negative culture or identification of bacteria belonging to commensal skin flora (especially *S. epidermidis*, *Dermabacter hominis*, *Propionibacterium acnes*, *Corynebacteria*, etc.).

In some cases, it may be possible to aspirate only a few bony spicules. This is especially the case if the infected bone is liquefied or from small toe bones (especially the distal phalanx). Of note, it has recently been shown that bone culture and histopathological examination have the same value for diagnosing DFO. However, bone culture has the obvious advantage of identification of the pathogen(s) and its/their susceptibility profile to antibiotics [31].

Skin and Soft Tissue Cultures

Skin cultures should not be performed to identify bone pathogens. However, when bone biopsy is not available, they may provide some information on the bacterial flora potentially involved in the underlying bone infection especially in case of acute infection where *S. aureus* is likely to be involved. In these situations, a tissue specimen should be obtained for culture by scraping with a sterile scalpel or curettage/biopsy from the base of a foot ulcer that has been debrided [16]. Deep tissue cultures may be of interest when they are taken preoperatively in those patients with severe DFI associated with DFO that require surgical debridement. Multiple samples may increase the chance of identifying the pathogens.

Blood Cultures

In case of severe infection of the foot and/or signs of sepsis or other signs associated with sepsis (hypotension, confusion, vomiting, or evidence of metabolic disturbances, such as acidosis, severe hyperglycemia, and new-onset azotemia), blood cultures should be performed ideally before any systemic antibiotic for culture.

Risk Factors

Risk factors for developing DFO have been assessed by Lavery *et al.* [32] who prospectively followed a cohort of 1666 persons with diabetes presenting to a Managed Care Diabetes Disease Management Program during a mean period of 27.2 months. Among 151 patients who developed foot infections, 30 (19.9%) had DFO confirmed by a positive bone culture. Independent risk factors for having foot osteomyelitis were (i) wounds that extended to bone or joint (relative risk [RR] = 23.1), (ii) previous history of a wound prior to enrollment (RR = 2.2), and (iii) recurrent or multiple wounds during the study period (RR = 1.9) [32].

Aragón-Sánchez *et al.* [33] found in a retrospective study that independent risk factors for developing a DFO due to Gram-negative bacteria (alone or combined with a Gram-positive isolate) were glycated hemoglobin less than 7% (odds ratio [OR] = 2.0; 95% confidence interval [CI] = 1.1–3.5) and a wound caused by traumatic injury (OR = 2.0; 95% CI = 1.0–3.9). All factors that contribute to the development of a foot ulcer in a diabetic patient, including deformity secondary to neuropathy or trauma, inappropriate footwear, and PIVD, are indirect risk factors for DFO. The patient's general conditions (social, economic, etc.) are other parameters that are likely to influence the patient's adherence to measures taken to shorten the healing duration of a foot ulcer and as a result the risk of DFO.

Clinical Features

Underlying osteomyelitis should be suspected when an ulcer does not heal after at least 6 weeks of appropriate wound care and off-loading on a foot with good blood supply. Both bone exposure and ulcer area larger than $2\,cm^2$ make osteomyelitis more likely [34]. DFO typically affects bones underlying sites where ulcers usually develop (i.e., toes, metatarsal heads, and calcaneum). Midfoot bones are less commonly involved, except when deformities have created hyperpressure ulcerations as can be seen in patients with advanced forms of neuropathic osteoarthropathy (Charcot foot). The ulcer depth is often not clinically apparent, and, as a consequence, any foot wound should be carefully explored at each consultation with a sterile blunt metal probe (the probe-to-bone [PTB] test) [16]. Exposed bone through the ulcer and/or a positive PTB test (i.e., showing hard and gritty contact) enhance the likelihood of osteomyelitis underlying the ulcer [35, 36]. The ability of the PTB test to predict or to exclude osteomyelitis is, however, largely influenced by the pretest prevalence of DFO in the studied population [3]. Overall, a positive PTB test is strongly associated with DFO when the wound is infected, and negative PTB is associated with a low probability of DFO in a noninfected ulcer [37]. However, positive and negative PTB tests do not constitute definite criteria for confirming/eliminating the existence of an underlying DFO.

It has been suggested that a combination of clinical and laboratory findings (ulcer depth >3 mm or C-reactive protein >32 mg/l, ulcer depth >3 mm or erythrocyte sedimentation rate [ESR] >60 mm/h) may help to differentiate osteomyelitis from soft tissue infection [38]. Exposed bone and toe involvement, particularly when it is erythematous and indurated (the so-called sausage toe), increase the likelihood of DFO [39]. Neither the presence of signs of infection of the wound nor an elevated white blood cell count seems to influence the likelihood of osteomyelitis [16]. Interestingly, the likelihood ratio (LR) for the clinical suspicion of DFO is high (positive = 5.5, negative = 0.54) [40, 41].

Given its potential severity, it is important to keep in mind that DFO is not a rare event in patients with diabetic foot wounds and that although it is not an emergency sensu stricto, it should be diagnosed and treated without delay.

Inflammatory Parameters

ESR has been identified as a biological marker associated with the presence of an osteomyelitis underlying a diabetic foot wound when it is elevated more than 70 mm/h (positive and negative LR values at, respectively, 11 and 0.34) and no other origin has been identified [34, 40, 42]. Some authors have also suggested that elevated values of C-reactive protein and procalcitonin may also be predictive of the presence of osteomyelitis [43]. Elevated white blood cell count does not influence the likelihood of osteomyelitis [44].

Except ESR, there are currently no other biological markers that may help physicians identify patients with DFO.

Imaging Procedures

Plain radiographs of the foot are first-line diagnostic tools when DFO is suspected, as they are widely available, repeatable, and relatively inexpensive. Bone abnormalities, especially cortical erosion, periosteal reaction, and sometimes fractures and even bone fragmentation and complete disappearance of osteoarticular structures underlying a soft tissue ulcer, should lead physicians to consider the diagnosis of DFO (Figure 18.2a and b). However, a recent meta-analysis reported a pooled sensitivity of only 0.54, and pooled specificity of only 0.68 of plain radiographs in patients with DFO [41]. The low negative predictive value of plain radiographs is because it may take 2–4 weeks after the onset of osteomyelitis to become visible [35, 45, 46]. Serial plain radiographs repeated after 2–4 weeks may have greater sensitivity and specificity to reliably identify bone changes at the site where the foot ulcer is present, but this has never been studied in this setting [37]. The continued absence of any bony abnormalities on repeated radiographs performed in an interval of 1 month or more probably excludes osteomyelitis. Another limitation of plain radiographs, which is not specific to this technique, is the difficulty to distinguish neuro-osteoarthropathy from osteomyelitis origin of bone abnormalities.

Magnetic resonance imaging (MRI) is currently considered the most accurate imaging technique for the diagnosis of bone infection [47–49]. When compared to other imaging techniques, MRI provides the best sensitivity and specificity for the detection of DFO with values as high as 90% and 79–82.5%, respectively, established in two recent meta-analyses [41, 47]. Findings consistent with DFO are similar to those observed in other types of osteomyelitis (e.g., decreased intensity of osteomyelitic areas on T1-weighted

(a) (b)

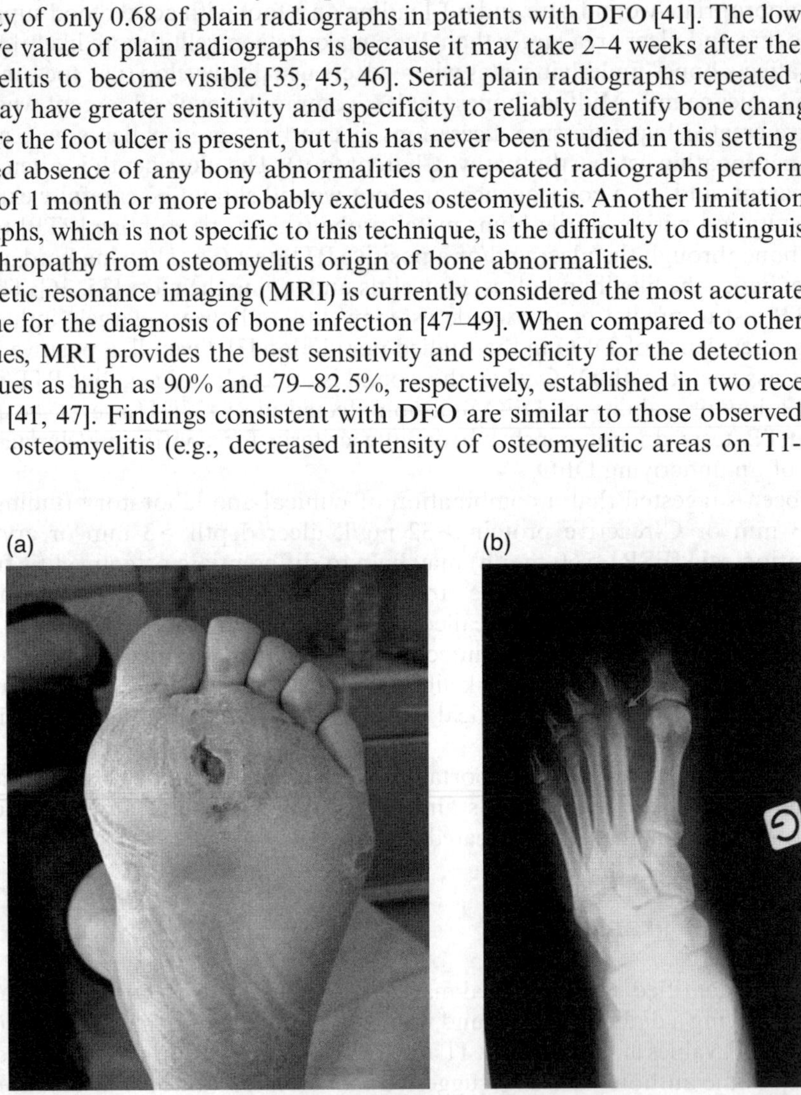

Figure 18.2. (a) Clinical picture of a diabetic patient with a chronic ulcer of the plantar surface with regard to the second metatarsal head. (b) Radiological signs of osteomyelitis on plain X-rays of the left foot in the same patient. The red arrow indicates joint destruction. (*See insert for color representation of the figure.*)

sequence and enhanced intensity on T2-weighted sequence and on postcontrast images). MRI may not be able to distinguish DFO from neuro-osteoarthropathy; clinical clues supporting neuro-osteoarthropathy in this context include midfoot location and absence of a soft tissue wound, while those favoring osteomyelitis include the presence of an overlying ulcer and the presence of a fistula communicating from the ulcer to the affected bone [50, 51]. However, MRI use may be limited by its low availability in some centers and difficulties in interpreting the findings [29].

Radioisotope scans (leukocyte or antigranulocyte scan) are more sensitive than radiographs for detecting early osteomyelitis, but they are rather nonspecific [45]. Their specificity is clearly below that of MRI [46]. The best techniques are either labeled leukocyte (with ^{99m}Tc or ^{111}In) or antigranulocyte Fab fragments (e.g., sulesomab). However, these techniques are limited by their need for blood manipulation (labeled leukocytes), and their availability, complexity (it may take >2 days to be completed), and cost [48].

In preliminary studies, fluorodeoxyglucose positron emission tomography (^{18}F-FDG–PET) has better accuracy for confirming or excluding the diagnosis of chronic osteomyelitis than MRI, but its role in diabetic patients is not yet established [52]. In a recent comprehensive literature search of studies on ^{18}F-FDG–PET and PET/CT that comprised 299 patients with DFO, the meta-analysis of four selected studies provided a sensitivity of 74% (95% CI: 60–85%), a specificity of 91% (95% CI: 85–96%), a positive LR of 5.56 (95% CI: 2.02–15.27), a negative LR of 0.37 (95% CI: 0.10–1.35), and a diagnostic OR of 16.96 (95% CI: 2.06–139.66) [53].

MRI is currently the most accurate imaging technique for the diagnosis of DFO. However, despite its low sensitivity, plain radiographs remain important tools in routine practice to identify patients with DFO. The UK National Institute for Health and Care Excellence (NICE) guidelines recommend the use of MRI or white blood cell scintigraphy when DFO is suspected but not confirmed by initial plain radiographs [54]. MRI has the advantage over other imaging techniques in providing data on soft tissue infection, including sinus tracts and tissue necrosis, that may help guide the surgical intervention [51].

Management

General Aspects

The management of DFO is one of the most conflicting aspects of DFI [55–59]. As for other types of osteomyelitis, the main limitation of the management of such infections is the very large diversity of the situations encountered by physicians. As a result, the management must be tailored to the characteristics of each patient and type of DFI. Despite this, recommendations for the management of DF including osteomyelitis have been proposed by international working groups [16, 29]. These recommendations are mostly based on expert opinions as the few available published clinical studies on the management of DFO do not provide definitive helpful conclusions. Indeed, published studies are heterogeneous, especially as regards the definition of the criteria of bone infection and cure and the existence of posttreatment follow-up. None of these studies were comparative, so that no conclusions can be drawn about the most effective antimicrobial regimens.

IDSA guidelines recommend obtaining a plain radiograph of the foot when DFO is suspected in a patient with DFI. If the radiographs do not sustain the suspicion of associated DFO, the soft tissue infection should be treated with antibiotics for 2–3 weeks, associated with surgical debridement in some severe cases [16]. Radiographs are repeated 2–4 weeks after the initial ones. If they remain normal, the likelihood of DFO is low. Such patients should be followed for 3 months, in order to assess the outcome. Nonhealing of the ulcer despite appropriate off-loading and wound care is an argument for the existence of an underlying osteomyelitis [37]. If bone changes appear on the new radiograph, antibiotic treatment guided by bone biopsy or deep tissue culture results if bone biopsy is unavailable should be considered.

The question of whether DFO can be treated without removing the infected bone is still being debated. It has long been thought that removal of the infected bone is necessary for arresting chronic bone infection, because the success rates reported in older reports using antimicrobial therapy alone were disappointing [55]. Ha Van *et al.* [60] showed that limited resection of the infected phalanx or of the metatarsal bone under the wound, together with removal of the ulcer site, was more effective than antibiotic therapy alone for shortening the duration of wound healing. Tan *et al.* [61] reported a lower rate of above-ankle amputation and a shorter hospital stay in patients who underwent debridement or local limited amputation than in patients treated with antibiotic therapy alone. However, these studies did not comprise prolonged clinical and radiological follow-up designed to evaluate the rate of relapse of osteomyelitis, as usually recommended. In addition, the choice of the antibiotic was not optimal for chronic bone infection. Other authors have suggested that prompt toe amputation is a cost-effective solution [62], but this approach has not been unambiguously evaluated. Nevertheless, the main objective of the management of such patients is to minimize the number of amputations, even minor ones, which may be responsible for recurrent DFO in relation with subsequent biomechanical abnormalities [63]. In addition, the dividing line between infected and noninfected bone is difficult to evaluate, even by sophisticated imaging assessment including MRI.

The main goal of surgical intervention in DFI is to control deep infection by draining pus and removing all necrotic tissue, in order to reduce the bacterial inoculum, which is essential for the antibiotic agents to be active. Because of the unique nature of these infections, which simultaneously require control of the vascular status of the limb, determination of the best level of amputation, and the possibility of covering the dehiscence due to resection, an interested and experienced team of surgeons (orthopedic, vascular, and plastic reconstructive surgeons) is needed.

The persistence of infected bone that has not been treated with suitable antibiotics is likely to be responsible for recurrent infections at the same site or contiguously. Interestingly, some recent clinical studies have shown that DFO can be arrested by using highly bioavailable oral agents especially fluoroquinolones, clindamycin, and rifampicin without bone resection [64–68]. In another retrospective study of 147 patients with DFO, Game and Jeffcoate [69] treated 113 patients with antibiotics alone. Ninety-three (82.3%) of these 113 patients achieved remission without surgery. The remaining 34 patients underwent minor amputation (28 patients, including 22 with remission) or major amputation (6 patients). However, the apparent satisfactory results established by these studies are limited by their retrospective design and the doubt that persists regarding the mode of selection of patients who were included. It must be taken into account that prolonged antibiotics are associated with a significant risk of drug-related side effects, including *Clostridium difficile* diarrhea, and the selection of antibiotic-resistant organisms, in particular MRSA and resistant Gram-negative bacilli.

In extreme situations (modest changes of bone on imaging assessment or, at the opposite end, major bone destruction), trained clinicians can easily decide whether or not surgical debridement of the infected bone or amputation is needed. However, there is currently no valid method to determine whether medical or surgical strategy should be used in a given patient with moderate bone destruction on imaging studies. The 2012 IDSA guidelines [16] propose that nonsurgical management of DFO can be considered in the following situations: (i) if surgery results in unacceptable loss of tissue and/or function, (ii) in patients with unreconstructable vascular disease who refuse amputation, (iii) in case of forefoot infection with minimal soft tissue loss, and (iv) in patients for whom surgery is not a reasonable option. In nonsurgical patients, the question arises as to whether bone culture is necessary. Transcutaneous bone biopsy is not implemented in routine practice for the management of DFO in most diabetic foot clinics. However, it is generally considered when the diagnosis of osteomyelitis remains uncertain despite clinical and imaging evaluations. In addition, it may be useful in case of ambiguous data from soft tissue cultures, when the infection has failed to respond to initial empirical antibiotic therapy or when considering an antibiotic regimen with a higher potential for emergence of resistant organisms (e.g., rifampin, fluoroquinolones).

Antibiotic Regimens

The IWGDF has reported the results of a systematic review showing that no data support the superiority of any antibiotic agent or treatment strategy for DFO. This review also highlighted the paucity of evidence-based data that may help physicians rationalize their prescriptions of antibiotics in patients with DFO [29].

Antibiotic therapy, chosen on the basis of nonbone tissues (swabs or wound biopsies) and prescribed for longer durations than for soft tissue infection, has been classically used by physicians involved in the management of such patients with apparently good results [67, 68]. However, these studies were retrospective and did not include at least 1 year posttreatment follow-up, which would allow detecting late relapsing episodes of DFO.

Another approach is to consider that DFO should be treated like any other chronic osteomyelitis. It is widely accepted that reliable bacterial documentation should be obtained in patients with chronic osteomyelitis (other than DFO) in order to prescribe proper conditions for antibiotic regimens comprising agents with high bone/biofilm penetration and sustained activity against bacteria in the stationary-growth phase as seen in these chronic infections. Antibiotics exhibiting such properties are limited to rifampicin against staphylococci, fluoroquinolones against Gram-negative bacilli, and clindamycin against anaerobes. The propensity of these antibiotics to select the resistant mutants present naturally in the offending bacteria population requires a prescription of these agents in combination to prevent the emergence of these mutants. This represents a strong argument for obtaining a reliable identification of the offending pathogens and their susceptibility profile to antibiotics when the prescription of such antibiotics is considered. Nevertheless, despite attempts to provide such patients with the most adapted antibiotic treatment, it has not been clearly demonstrated that this approach improves the outcome of patients with DFO.

The most appropriate duration of therapy for any type of DFI depends on the presence of any residual dead or infected bone and the state of the soft tissues. If all infected bone tissues have been removed, the duration must not exceed 1–2 weeks, as the surgical debridement must have notably reduced the bacterial inoculum inside the infected soft tissues. If infected bone or soft tissues remain despite surgery, continued treatment is needed for 6–12 weeks. In patients for whom the resection of infected and necrotic bone could not be done for any reasons, the treatment might be prolonged for 12 weeks or more [16]. Of note, most recommendations on the antibiotic treatment duration come from expert opinion, since no study has assessed this question so far. Parenteral therapy may be delivered in the outpatient setting, where available [70].

Each case requires an individualized approach because of the numerous variables that may impact the final decision, which should be best taken in multidisciplinary teams including senior physicians with skilled experience in this field.

There is no evidence that the use of adjunctive therapies like hyperbaric oxygen therapy, growth factors, including granulocyte-stimulating factor, maggots (larvae), or topical negative pressure therapy (e.g., vacuum-assisted closure) is beneficial in the management of DFO [29, 71].

Prevention

The best way to prevent DFO is to reduce the occurrence of foot ulcers in persons with diabetes. This implies prevention of the two major diabetes-related complications, that is, peripheral neuropathy and vascular disease (PVD), that contribute to the formation of foot ulcers. Glycemic control is the corner stone for long-term prevention of these complications. For patients with installed neuropathy and/or PVD, education about feet care and regular surveillance of the feet by podiatrists are essential for the prevention of DFO. In addition, prompt evaluation and treatment of any suspicion of DFO may help reduce the risk of relapsing infections and the need for amputation.

References

[1] Lavery LA, Armstrong DG, Wunderlich RP, Mohler MJ, Wendel CS, Lipsky BA. Risk factors for foot infections in individuals with diabetes. Diab Care 2006;29:1288–1293.

[2] Prompers L, Huijberts M, Apelqvist J, *et al.* High prevalence of ischaemia, infection and serious comorbidity in patients with diabetic foot disease in Europe. Baseline results from the Eurodiale study. Diabetologia 2007;50:18–25.

[3] Shone A, Burnside J, Chipchase S, Game F, Jeffcoate W. Probing the validity of the probe-to-bone test in the diagnosis of osteomyelitis of the foot in diabetes. Diab Care 2006;29:945.

[4] Lipsky BA. A report from the international consensus on diagnosing and treating the infected diabetic foot. Diabetes Metab Res Rev 2004;20Suppl 1:S68–S77.

[5] Mutluoglu M, Sivrioglu AK, Eroglu M, Uzun G, Turhan V, Ay H, Lipsky BA. The implications of the presence of osteomyelitis on outcomes of infected diabetic foot wounds. Scand J Infect Dis 2013;45(7):497–503.

[6] Lipsky BA, Peters EJ, Berendt AR, Senneville E, Bakker K, Embil JM, Lavery LA, Urbančič-Rovan V, Jeffcoate WJ; International Working Group on Diabetic Foot. Specific guidelines for the treatment of diabetic foot infections 2011. Diabetes Metab Res Rev 2012;28(Suppl 1): 234–235.

[7] Lew DP, Waldvogel FA. Osteomyelitis. Lancet 2004;364:369–379.

[8] Lavery LA, Harkless LB, Ashry HR, Felder-Johnson K. Infected puncture wounds in adults with diabetes: risk factors for osteomyelitis. J Foot Ankle Surg 1994;33(6):561–566.

[9] Hofmann G, Gonschorek O, Hofmann GO, Bühren V. Stabilisierungsverfahren bei Osteomyelitis. Osteosyn Intern 1997;5:226–231.

[10] Chantelau EA, Wolf A, Ozdemir S, Hachmöller A, Ramp U. Bone histomorphology may be unremarkable in diabetes mellitus. Med Klin (Munich) 2007;102(6):429–433.
[11] Cierny G 3rd, Mader JT, Penninck JJ. A clinical staging system for adult osteomyelitis. Clin Orthop Relat Res 2003;414:7–24.
[12] Cierny G 3rd, Mader JT. Approach to adult osteomyelitis. Orthop Rev 1987;16:259–270.
[13] Aragón-Sánchez FJ, Cabrera-Galván JJ, Quintana-Marrero Y, *et al.* Outcomes of surgical treatment of diabetic foot osteomyelitis: a series of 185 patients with histopathological confirmation of bone involvement. Diabetologia 2008;51:1962–1970.
[14] Aragón-Sánchez J. Clinical-pathological characterization of diabetic foot infections: Grading the severity of osteomyelitis. Int J Low Extrem Wounds 2012;11:107–112.
[15] Schaper NC. Diabetic foot ulcer classification system for research purposes: a progress report on criteria for including patients in research studies. Diabetes Metab Res Rev 2004;20(Suppl 1): S90–S95.
[16] Lipsky BA, Berendt AR, Cornia PB, *et al.* Infectious Diseases Society of America Clinical Practice guideline for the diagnosis and treatment of diabetic foot infections. Clin Infect Dis 2012;54(12):132–173.
[17] Peters EJ, Lipsky BA, Berendt AR, *et al.* A systematic review of the effectiveness of interventions in the management of infection in the diabetic foot. Diabetes Metab Res Rev 2012;28(Suppl 1):142–162.
[18] Wheat J. Diagnostic strategies in osteomyelitis. Am J Med 1985;78:218–224.
[19] Senneville E, Melliez H, Beltrand E, *et al.* Culture of percutaneous bone biopsy specimens for diagnosis of diabetic foot osteomyelitis: concordance with ulcer swab cultures. Clin Infect Dis 2006;42:57–62.
[20] Lesens O, Desbiez F, Vidal M, *et al.* Culture of per-wound bone specimens: a simplified approach for the medical management of diabetic foot osteomyelitis. Clin Microbiol Infect 2011;17:285–291.
[21] Game F, Jeffcoate W. MRSA and osteomyelitis of the foot in diabetes. Diabet Med 2004; 21(Suppl 4):16–19.
[22] Acharya S, Soliman M, Egun A, Rajbhandari SM. Conservative management of diabetic foot osteomyelitis. Diabetes Res Clin Pract 2013;101:18–20.
[23] Elamurugan TP, Jagdish S, Kate V, Chandra Parija S. Role of bone biopsy specimen culture in the management of diabetic foot osteomyelitis. Int J Surg 2011;9:214–216.
[24] Kessler L, Piemont Y, Ortega F, *et al.* Comparison of microbiological results of needle puncture vs. superficial swab in infected diabetic foot ulcer with osteomyelitis. Diabet Med 2006;23:99–102.
[25] Mackowiak PA, Jones SR, Smith JW. Diagnostic value of sinus-tract cultures in chronic osteomyelitis. JAMA 1978;239:2772–2775.
[26] Senneville E, Morant H, Descamps D, *et al.* Needle puncture and transcutaneous bone biopsy cultures are inconsistent in patients with diabetes and suspected osteomyelitis of the foot. Clin Infect Dis 2009;48:888–893.
[27] Ertugrul MB, Baktiroglu S, Salman S, *et al.* The diagnosis of osteomyelitis of the foot in diabetes: microbiological examination vs. magnetic resonance imaging and labelled leucocyte scanning. Diabet Med 2006;23:649–653.
[28] Senneville E, Lombart A, Beltrand E *et al.* Outcome of diabetic foot osteomyelitis treated non-surgically: a retrospective cohort study. Diab Care 2008;31:637–642.
[29] Berendt AR, Peters EJ, Bakker K, *et al.* Diabetic foot osteomyelitis: a progress report on diagnosis and a systematic review of treatment. Diabetes Metab Res Rev 2008;24(Suppl 1): S145–S161.
[30] Witso E, Persen L, Loseth K, Bergh K. Adsorption and release of antibiotics from morselized cancellous bone: in vitro studies of 8 antibiotics. Acta Orthop Scand 1999;70(3):298–304.
[31] Weiner RD, Viselli SJ, Fulkert KA, Accetta P. Histology versus microbiology for accuracy in identification of osteomyelitis in the diabetic foot. J Foot Ankle Surg 2011;50:197–200.

[32] Lavery LA, Peters EJ, Armstrong DG, Wendel CS, Murdoch DP, Lipsky BA. Risk factors for developing osteomyelitis in patients with diabetic foot wounds. Diabetes Res Clin Pract 2009;83:347–352.
[33] Aragón-Sánchez J, Lipsky BA, Lázaro-Martínez JL. Gram-negative diabetic foot osteomyelitis: risk factors and clinical presentation. Int J Low Extrem Wounds 2013;12:63–68.
[34] Newman LG, Waller J, Palestro CJ, *et al.* Unsuspected osteomyelitis in diabetic foot ulcers. Diagnosis and monitoring by leukocyte scanning with indium in 111 oxyquinoline. JAMA 1991;266:1246–1251.
[35] Grayson ML, Gibbons GW, Balogh K, Levin E, Karchmer AW. Probing to bone in infected pedal ulcers. A clinical sign of underlying osteomyelitis in diabetic patients. JAMA 1995; 273:721–723.
[36] Lavery LA, Armstrong DG, Peters EJ, Lipsky BA. Probe-to-bone test for diagnosing diabetic foot osteomyelitis: reliable or relic? Diabetes Care 2007;30:270–274.
[37] Hartemann-Heurtier A, Senneville E. Diabetic foot osteomyelitis. Diabetes Metab 2008;34:87–95.
[38] Fleischer AE, Didyk AA, Woods JB, Burns SE, Wrobel JS, Armstrong DG. Combined clinical and laboratory testing improves diagnostic accuracy for osteomyelitis in the diabetic foot. J Foot Ankle Surg 2009;48(1):39–46.
[39] Rajbhandari SM, Sutton M, Davies C, Tesfaye S, Ward JD. 'Sausage toe': a reliable sign of underlying osteomyelitis. Diabet Med 2000;17(1):74–77.
[40] Butalia S, Palda VA, Sargeant RJ, Detsky AS, Mourad O. Does this patient with diabetes have osteomyelitis of the lower extremity? JAMA 2008;299(7):806–813.
[41] Dinh MT, Abad CL, Safdar N. Diagnostic accuracy of the physical examination and imaging tests for osteomyelitis underlying diabetic foot ulcers: metaanalysis. Clin Infect Dis 2008; 47(4):519–527.
[42] Kaleta JL, Fleischli JW, Reilly CH. The diagnosis of osteomyelitis in diabetes using erythrocyte sedimentation rate: a pilot study. J Am Podiatr Med Assoc 2001;91(9):445–450.
[43] Dinh T, Snyder G, Veves A. Current techniques to detect foot infection in the diabetic patient Int J Low Extrem Wounds 2010;9(1):24–30.
[44] Armstrong DG, Lavery LA, Sariaya M, Ashry H. Leukocytosis is a poor indicator of acute osteomyelitis of the foot in diabetes mellitus. J Foot Ankle Surg 1996;35(4):280–283.
[45] Sella, E.J. and D.M. Grosser. Imaging modalities of the diabetic foot. Clin Podiatr Med Surg, 2003;20(4):729–740.
[46] Shults, DW, Hunter GC, McIntyre KE, Parent FN, Piotrowski JJ, Bernhard VM. Value of radiographs and bone scans in determining the need for therapy in diabetic patients with foot ulcers. Am J Surg 1989;158(6):525–529.
[47] Kapoor A, Page S, Lavalley M, Gale DR, Felson DT. Magnetic resonance imaging for diagnosing foot osteomyelitis: a meta-analysis. Arch Intern Med 2007;167(2):125–132.
[48] Schweitzer ME, Daffner RH, Weissman BN, *et al.* ACR appropriateness criteria on suspected osteomyelitis in patients with diabetes mellitus. J Am Coll Radiol 2008;5(8):881–886.
[49] Chatha DS, Cunningham PM, Schweitzer ME. MR imaging of the diabetic foot: diagnostic challenges. Radiol Clin North Am 2005;43(4):747–759.
[50] Berendt AR,Lipsky BA. Is this bone infected or not? Differentiating neuro-osteoarthropathy from osteomyelitis in the diabetic foot. Curr Diab Rep 2004;4(6):424–429.
[51] Tan PL, The J. MRI of the diabetic foot: differentiation of infection from neuropathic change. Br J Radiol 2007;80(959):939–948.
[52] Termaat MF, Raijmakers PG, Scholten HJ, Bakker FC, Patka P, Haarman HJ. The accuracy of diagnostic imaging for the assessment of chronic osteomyelitis: a systematic review and meta-analysis. J Bone Joint Surg Am 2005;87(11):2464–2471.
[53] Treglia G, Sadeghi R, Annunziata S, *et al.* Diagnostic performance of Fluorine-18-Fluorodeoxyglucose positron emission tomography for the diagnosis of osteomyelitis related to diabetic foot: A systematic review and a meta-analysis. Foot (Edinb) 2013;23:140–148.

[54] Tan T, Shaw EJ, Siddiqui F, Kandaswamy P, Barry PW, Baker M. Inpatient management of diabetic foot problems: summary of NICE guidance; Guideline Development Group. BMJ 2011;342:d1280. doi:10.1136/bmj.d1280.
[55] Jeffcoate, WJ and BA Lipsky. Controversies in diagnosing and managing osteomyelitis of the foot in diabetes. Clin Infect Dis 2004;39(Suppl 2):S115–S122.
[56] Lipsky BA. Osteomyelitis of the foot in diabetic patients. Clin Infect Dis 1997;25(6):1318–1326.
[57] Snyder RJ, Cohen MM, Sun C, Livingston J. Osteomyelitis in the diabetic patient: diagnosis and treatment. Part 2: medical, surgical, and alternative treatments. Ostomy Wound Manage 2001;47(3):24–30.
[58] Lipsky BA, Peters EJ, Senneville E, *et al.* Expert opinion on the management of infections in the diabetic foot. Diabetes Metab Res Rev 2012;28(Suppl 1):163–178.
[59] Embil JM. The management of diabetic foot osteomyelitis. The Diabetic Foot 2000;3:76–84.
[60] Ha Van G, Siney H, Danan JP, Sachon C, Grimaldi A. Treatment of osteomyelitis in the diabetic foot. Contribution of conservative surgery. Diabetes Care 1996;19:1257–1260.
[61] Tan JS, Friedman NM, Hazelton-Miller C, Flanagan JP, File TM Jr. Can aggressive treatment of diabetic foot infections reduce the need for above-ankle amputation? Clin Infect Dis 1996;23(2):286–289.
[62] Kerstein MD, Welter V, Gahtan V, Roberts AB. Toe amputation in the diabetic patient. Surgery 1997;122(3):546–547.
[63] Molines-Barroso RJ, Lázaro-Martínez JL, Aragón-Sánchez J, García-Morales E, Beneit-Montesinos JV, Alvaro-Afonso FJ. Analysis of transfer lesions in patients who underwent surgery for diabetic foot ulcers located on the plantar aspect of the metatarsal heads. Diabet Med 2013;30(8):973–976.
[64] Venkatesan P, Lawn S, Macfarlane RM, Fletcher EM, Finch RG, Jeffcoate WJ. Conservative management of osteomyelitis in the feet of diabetic patients. Diabet Med 1997;14:487–490.
[65] Senneville E, Yazdanpanah Y, Cazaubiel M, *et al.* Rifampicin-ofloxacin oral regimen for the treatment of mild to moderate diabetic foot osteomyelitis. J Antimicrob Chemother 2001;48:927–930.
[66] Pittet D, Wyssa B, Herter-Clavel C, Kursteiner K, Vaucher J, Lew PD. Outcome of diabetic foot infections treated conservatively: a retrospective cohort study with long-term follow-up. Arch Intern Med 1999;159:851–856.
[67] Yadlapalli N, Vaishnar A, Sheehan P. Conservative management of diabetic foot ulcers complicated by osteomyelitis. Wounds 2002;14:31–35.
[68] Embil JM, Rose G, Trepman E, *et al.* Oral antimicrobial therapy for diabetic foot osteomyelitis. Foot Ankle Int 2006;27:71–79.
[69] Game FL, Jeffcoate WJ. Primarily non-surgical management of osteomyelitis of the foot in diabetes. Diabetologia 2008;51(6):962–967.
[70] Tice AD, Hoaglund PA, Shoultz DA. Outcomes of osteomyelitis among patients treated with outpatient parenteral antimicrobial therapy. Am J Med 2003;114:723–728.
[71] Ubbink DT, Westerbos SJ, Evans D, Land L, Vermeulen H. Topical negative pressure for treating chronic wounds. Cochrane Database Syst Rev 2008; 3:CD001898.

Chapter 19
Osteomyelitis of the Jaws

Werner Zimmerli

Introduction

Osteomyelitis of the jaws has become a rare disease after the introduction of antibiotics in clinical medicine [1]. Before antibiotics were available, odontogenic (dental infection–associated) osteomyelitis was quite common. Nowadays, early surgical and antibiotic treatment of dental and periodontal infection dramatically reduces the incidence of bone infection of the jaws. In addition, hematogenous osteomyelitis in neonates and young infants has almost completely disappeared. In contrast, during the past 10 years, bisphosphonate-associated osteonecrosis of the jaws appeared as a new entity [2, 3].

In general, patients with cranio-maxillofacial infections do not consult internists or infectious disease specialists, but either the dentist or the oral and maxillofacial surgeon. Therefore, the clinical experience of most infectious disease consultants in this field is quite limited. The aim of this chapter is to highlight the topic from the perspective of the infectious disease specialist, the microbiologist, and the maxillofacial surgeon. The different types of osteomyelitis of the jaws are not uniformly classified in the literature. In this chapter, we consequently use "the Zurich classification system," which has been proposed by Baltensperger and Eyrich [4].

Classification

Acute and secondary chronic osteomyelitis are caused by many different pathogenetic mechanisms, as shown in Table 19.1 [5–12]. The term "secondary chronic" is synonymous with "suppurative" and the term "primary chronic" with "nonsuppurative" osteomyelitis. It describes the clinical presentation with or without pus. The relative frequency of these various etiologies is very different according to the epidemiology in a defined region. In industrial regions such as Europe and the United States, peri-implantitis became the

Bone and Joint Infections: From Microbiology to Diagnostics and Treatment, First Edition.
Edited by Werner Zimmerli.

Table 19.1. The Zurich classification of osteomyelitis of the jaws [4–8].

Acute osteomyelitis (< 1 month duration) → Secondary chronic osteomyelitis
Neonatal, tooth germ–associated
Trauma-/fracture-related (e.g., open fracture)
Odontogenic
Periodontal
Pulpal
Paranasal sinus–related
Extraction wound–related
Chronic neck infection (actinomycosis, tuberculosis) (contiguous)
Implant-associated (e.g., dental implant or internal fixation device)
Graft-associated (e.g., bone grafts for reconstruction)
Associated with bone pathology and/or systemic disease (e.g., Albers–Schönberg disease) [9]
Not clearly classifiable
Primary chronic osteomyelitis
Early onset (juvenile chronic osteomyelitis) [10]
Adult onset
Radiation-induced avascular bone necrosis
Osteochemonecrosis (corticosteroids, antineoplastic drugs, bisphosphonate therapy)
Syndrome-associated (e.g., chronic recurrent multifocal osteomyelitis [11], Garré's sclerosing osteomyelitis [12], SAPHO = synovitis, acne, pustulosis, hyperostosis, and osteitis) [7]

leading etiology during the last decade [13–15]. In a series of 84 cases from India, radionecrosis-associated osteomyelitis was the most common pathogenesis [16]. Finally, in Nigeria, among 141 cases, 50 were of odontogenic origin, and 18 suffered from necrotizing stomatitis (cancrum oris), leading to osteomyelitis [17].

Bone infection with sequesters or sinus tracts should be classified as secondary chronic osteomyelitis, even if no microorganism can be detected. This group of osteomyelitis includes chronic neck infections, such as actinomycosis or cervical lymph node tuberculosis. Both may cause secondary chronic osteomyelitis of the jaws by contiguous infection [18, 19]. These microorganisms must be actively looked for, since special culture conditions are needed for detection.

The term nonsuppurative osteomyelitis is not uniformly used in the literature. In the Zurich classification, it labels a heterogenous group of chronic inflammatory bone diseases, mostly of unknown etiology. In this type of osteomyelitis, there is no pus formation. In addition, sinus tracts or sequestration are also missing [4]. Primary chronic osteomyelitis typically has an insidious course without a clear-cut acute phase. In the literature, a multitude of different entities are classified (Table 19.1). According to the series of Baltensperger *et al.* [4, 20], primary chronic osteomyelitis has two incidence peaks, one between 11 and 20 years and a second one after the age of 50 years. There is no formal proof for an infectious etiology of these entities. However, some authors postulate low-virulence microorganisms as a trigger [6, 21, 22]. Radiation-induced chronic osteomyelitis is initially not a bone infection, but an avascular necrosis of the bone [23]. Due to this necrosis, and to chronic nonhealing wounds, secondary infection of the bone is frequent [8, 16]. If superinfection occurs, this entity has to be considered as acute or secondary chronic osteomyelitis. Nowadays, bisphosphonate-induced osteochemonecrosis of the jaws has become a frequent form of primary chronic osteomyelitis, especially when used by the intravenous (IV) route [24]. The prevalence is estimated at 0.7–18.6% after IV administration, and 0.01–0.04% after oral treatment [24–26]. Like radiation-induced osteomyelitis, chemonecrosis is also a risk factor for secondary pyogenic osteomyelitis. In this chapter, only acute and secondary chronic osteomyelitis will be covered.

Microbiology

The spectrum of microorganisms causing osteomyelitis of the jaws is much broader than the one in osteomyelitis at other localizations. Another difference is the high frequency of polymicrobial infections [27, 28]. In the series of Calhoun *et al.* [28], 93% of the episodes were polymicrobial with an average of 3.9 microorganisms per patient.

The infecting agents differ among the different types of osteomyelitis, because of the various pathogenesis. In contrast to other types of osteomyelitis, *Staphylococcus aureus* is not the most important pathogen in bone infection of the jaws, except in neonatal osteomyelitis of the maxilla, which is not only tooth germ–associated, but alternatively caused by the hematogenous route [29]. Most other types of osteomyelitis of the jaws are contiguous infections (Table 19.1). As can be suspected, in contiguous infection, microorganisms, which normally colonize the oral cavity, are the potential pathogens. In addition, microorganisms of the transient mouth flora may also play a role. Since *S. aureus* may colonize the paranasal sinuses, it is a typical microorganism of rhinosinusitis-associated osteomyelitis [30].

In chronic neck infection, pathogens such as *Actinomyces* spp. and *Mycobacterium tuberculosis* can be found, provided that appropriate culture techniques are applied [18, 19, 31, 32]. In endemic regions, *Coccidioides immitis*, which causes pulmonary diseases by inhalation, can seed to multiple areas of the bone, including the mandibule [33]. Other fungi, such as *Aspergillus* spp. or *Mucor* spp. preferentially, but not exclusively, cause osteomyelitis of the jaws in immunocompromised patients [34–37].

In patients with internal fixation devices, *S. aureus* and coagulase-negative staphylococci are by far the most frequent pathogens at all locations except the jaws. In the latter type of osteomyelitis, staphylococci play a role only if the fracture communicates with extraoral soft tissue and skin, which is mainly the case after severe facial trauma.

In patients with dental implants, the pathogens can only be identified indirectly, since removed implants are heavily contaminated with the flora of the oral cavity. In a study of 17 patients with 98 dental implants, Hultin *et al.* [38] analyzed which microorganisms in subgingival samples were predictive for peri-implantitis. Paradontal pathogens such as *Actinobacillus actinomycetemcomitans*, *Porphyromonas gingivalis*, *Prevotella intermedia*, *Bacteroides forsythus*, and *Treponema denticola* were cultivated in high concentrations in patients with peri-implantitis, but not in controls without implant pathology, suggesting that these microorganisms may be crucial for the infectious loosening of dental implants.

In the retrospective case series of Baltensperger [39], 48 patients with acute osteomyelitis have been observed. Five bone biopsies and 29 soft tissue/pus specimens have been analyzed. Viridans streptococci were by far the most frequent pathogens, with half of them from the *Streptococcus milleri* group. Interestingly, in 7/34 specimens, Gram-negative bacilli (*Escherichia coli*, *Enterobacter* spp., *Klebsiella* spp., *Serratia marcescens*, and *Proteus* species) have been isolated [40]. These microorganisms only transiently colonize the oral cavity, mainly during antimicrobial therapy, in critically ill patients or in patients of old age [41, 42]. In a series reported by Calhoun *et al.* [28], anaerobic bacteria were cultivated in half of the patients (30/60).

In 203 patients with secondary chronic osteomyelitis, 46 bone biopsy specimens and 96 specimens from adjacent soft tissue or pus have been sampled from 203 patients [39]. The most frequent pathogens were again viridans streptococci in bone biopsies (21/46 = 46%), as well as in soft tissue/pus samples (54/96 = 56%). Half of the viridans streptococci were identified as of the *S. milleri* group. Mixed infections, mainly with anaerobic microorganisms of the local flora, were frequent in oral abscesses. *Enterobacteriaceae*, mainly *E. coli*, were isolated from 17/142 (12%) specimens, despite the fact that these bacteria only transiently colonize the oral cavity [41, 42]. *Haemophilus* spp. and *Actinomyces* spp. were both isolated in 15 cases. Interestingly, *S. aureus* was isolated from only 2/46 (4%) bone samples [40].

Risk Factors

There are several underlying conditions that can be considered as risk factors for the occurrence of osteomyelitis of the jaws. The most important conditions are dental and periodontal infection [43], chronic sinusitis [16, 30], dental implants [13–15], fracture with or without internal fixation [44, 45], and facial infection [46, 47]. In addition, other underlying conditions, such as previous radiation [23, 48], bisphosphonate therapy [2, 3, 24], alcohol and/or tobacco use [49], malnutrition [17], sickle cell disease [50], osteopetrosis [9], diabetes [16, 37], and acute leukemia [51] predispose to osteomyelitis of the jaws. Interestingly, 74–96% of the episodes are located in the mandible, with only a small minority in the maxilla [17, 49, 52]. This is presumably due to the more limited blood supply of the mandible, as compared to the maxilla. A small minority of cases is caused by extension from septic arthritis of the temporomandibular joint [53].

Clinical Features

The clinical presentation depends on the type of osteomyelitis (Table 19.1). Both types of *neonatal osteomyelitis* occur within the first few weeks after birth (1–15 weeks) [29]. Hematogenous as well as tooth germ–associated osteomyelitis of the jaws affect the maxillar bone in most instances [4]. Initially, in hematogenous osteomyelitis, systemic signs of sepsis or of the primary focus may predominate. Clinical signs include swelling around the eye followed by proptosis, chemosis, and ophthalmoplegia. In addition, unilateral nasal discharge and swelling in the oral cavity at the affected side may be observed [29]. In case of delayed treatment, the microorganisms may spread to the dural sinus veins, causing infection in the central nervous system. In addition, facial bone deformities and tooth loss can be observed.

In the different types of acute osteomyelitis of the jaws, the patient suffers from fever, malaise, red and swollen facial skin, trismus, and regional lymphadenopathy [1]. However, in the series of Baltensperger [39], only 6/43 (14%) patients had fever ≥38°C, and 47% had a normal body temperature ≤37°C. A special sign that points toward osteomyelitis of the mandible is hypoesthesia of the lower lip, indicating the involvement of the inferior alveolar nerve, observed in half of the patients. In case of odontogenic, periodontal, pulpal, or extraction wound–related origin, local signs of infection can easily be detected. Typical signs are pus in the gingival sulcus, and sinus tracts ending either in the oral mucosa or less frequently in the facial skin. A fetid foetor, which is caused by anaerobic bacteria, is often present.

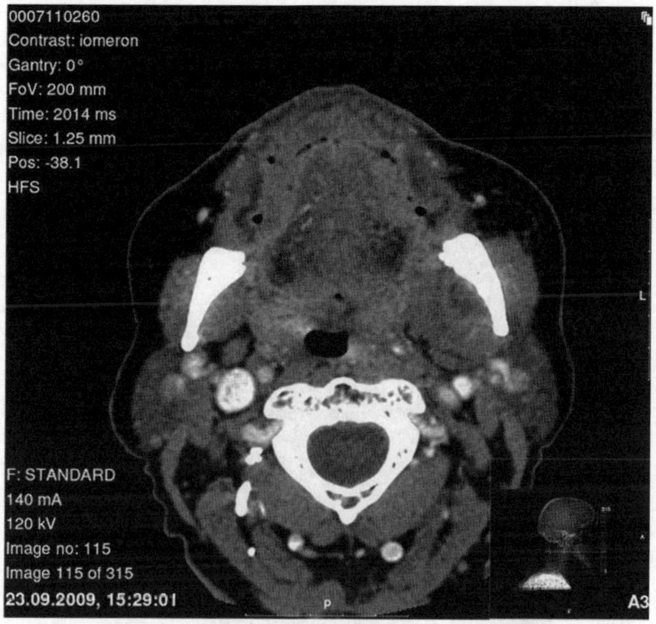

Figure 19.1. A 66-year-old man with pain in the right mandible since 5 months. Transitory improvement after a 3-week course of clindamycin. Second treatment course after recurrence. Computed tomography (CT) scan revealed chronic osteomyelitis of the left lower jaw and an abscess of the *Musculus pterygoideus medialis*. Treatment with surgical debridement of the bone, lavage of the abscess, and antibiotics. Courtesy of Parham Sendi, University Hospital Bern/Switzerland.

Maxillar osteomyelitis was described in 3/13 patients with acute rhinosinusitis in a series from India [30]. However, it is an extremely rare complication in countries with a good healthcare system. Patients with secondary chronic osteomyelitis have less extensive signs and symptoms than those with acute osteomyelitis. In the series of Baltensperger [39], only 4/193 (2%) had fever ≥38°C, and 69% had a normal temperature ≤37°C. The local edema is replaced by a palpable tenderness of high consistency caused by tissue infiltration and periostitis. Sequestrum formation, tooth loss, and pathological fracture of the mandible are reported [17]. Sinus tracts, draining either in the oral cavity or through the skin, are frequent in patients with chronic osteomyelitis, which is not rapidly and adequately treated [8, 54]. In addition, in patients with long-lasting osteomyelitis of the mandible, large abscesses in the muscles of jaw and neck can be detected (Figure 19.1).

Actinomycosis and tuberculosis rarely cause osteomyelitis of the jaws. Both are chronic neck infections, with different presentations. Cervicofacial actinomycosis is generally acquired after disruption of the mucosal barrier integrity, for example, after tooth extraction. It causes chronic inflammation, which is characterized by a tender or a painless indurated mass of woody consistency in the face or neck, typically followed by one or several sinus tracts (Figure 19.2). Cervicofacial actinomycosis may extend to the underlying mandible or facial bone, leading to osteomyelitis [55–57]. In general, actinomycosis is diagnosed and treated before causing osteomyelitis. Untreated cervical lymph node tuberculosis can also extend to the bone. However, this is a very rare event, which has not been observed in 203 patients with chronic osteomyelitis reported by Baltensperger [39].

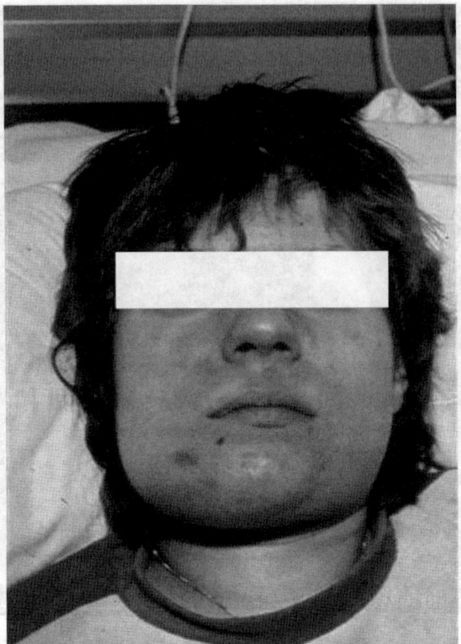

Figure 19.2. A 28-year-old man with a 6-month history of painful swelling of the right lower jaw after tooth extraction. Anaerobic cultures of tissue biopsies revealed *Actinomyces israelii* and *Fusobacterium* sp. (*See insert for color representation of the figure.*)

Dental implant–associated osteomyelitis is not described as an entity in the literature. Clinically, peri-implantitis has to be delineated from peri-implant mucositis. Peri-implant mucositis resembles gingivitis, which reflects a host response to gingival microorganisms. In contrast, peri-implantitis also affects the supporting bone [58]. It is defined as an inflammatory condition characterized by loss of supporting bone in the tissues surrounding the implant [13, 59]. According to Berglundh *et al.* [60], the following features are signs of peri-implantitis: probing depth of >6 mm in combination with bleeding on probing or suppuration, and attachment loss/bone loss of 2.5 mm. In their systematic review of 51 studies, the rate of peri-implantitis was 6.5%.

The different etiologies of primary chronic osteomyelitis are at least initially not due to infecting agents. Therefore, for their clinical characteristics, the pertinent literature cited in Table 19.1 should be consulted.

Laboratory Investigation

As in other types of osteomyelitis, there are no specific laboratory signs proving bone infection. To the best of our knowledge, the diagnostic yield of different inflammatory parameters has not been systematically studied. Data from retrospective observational studies is the best available.

Inflammatory Parameters

Erythrocyte sedimentation rate and C-reactive protein are nonspecific. Therefore, they only allow estimating the degree of inflammation, but not the presence or absence of infection. In the large series of Baltensperger [39], 23/38 (61%) of the patients with acute and 135/183 (74%) of those with secondary chronic osteomyelitis had completely normal leukocyte counts. In the same series, 4/23 (17%) of the patients with acute and 54/124 (44%) with secondary chronic osteomyelitis had a completely normal sedimentation rate. The sensitivity of C-reactive protein was even worse with 33 and 50% normal values (< 5 mg/l) in patients with acute and chronic osteomyelitis, respectively. Thus, these parameters are by far not sensitive enough to exclude osteomyelitis of the jaws.

Microbiological Workup

Osteomyelitis must be treated for several weeks. Therefore, the choice of the antimicrobial agents should be based on relevant culture results. There are several factors that impede the interpretation of microbiological results from the workup of osteomyelitis of the jaws. First, in contrast to other types of osteomyelitis, the risk for specimen contamination is quite high, since, in most cases, during sampling the colonized surface of the oral cavity has to be passed. Second, most infections are polymicrobial. Thus, there is a considerable risk for missing part of the causing pathogens. This is especially the case in microorganisms requiring special growth conditions and needing a long incubation period such as *Actinomyces* spp., *Nocardia* spp., mycobacteria. Finally, the risk for superinfection after starting directed therapy is not negligible, as long as the patient has an intraoral wound. Therefore, it could be argued that an empirical treatment course would be as good as pathogen-directed therapy. To the best of our knowledge, this question has never been addressed. However, there is no regimen active against all pathogens. Treatment based on pathogen identification and susceptibility testing is especially important in *Enterobacteriaceae*, *Pseudomonas aeruginosa*, methicillin-resistant *S. aureus*, fungal agents, *Actinomyces* spp., and *M. tuberculosis.*

In case of abscesses, aspiration through the intact skin surface is the best technique for relevant specimens. If biopsies or pus collections must be sampled through the oral mucosa, the site should be isolated with cotton rolls, dried, and swabbed vigorously with povidone–iodine, which is allowed to remain on the site for at least 1 min before insertion of the needle [40]. Swabs from sinus tracts may be misleading, because of irrelevant contaminating microorganisms. Biopsy material is better than swabs, because the number of microorganisms is much higher and the risk for drying is lower. It should be considered that a delay of as little as 15 min decreases the yield of anaerobic bacteria [40]. Therefore, transport in an anaerobic transport agar tube is advisable. If specimens are rapidly processed in the lab, biopsy samples can also be transported in a sterile tube after adding sterile physiological saline.

In case of chronic osteomyelitis, several biopsies should be sampled. From each biopsy, one part should be processed for microbiological workup, the other one for histopathology. This workup has several advantages. First, the comparison of the results of these pairs better allows detecting irrelevant specimens with contamination. Second, histopathology is important for the diagnosis of actinomycosis, tuberculosis, or fungal infection. Third, primary chronic osteomyelitis (presumably sterile inflammation) can be differentiated from secondary chronic osteomyelitis by the type of histology (lack of neutrophils, predominant sclerosis, and periosteal new bone formation) [6].

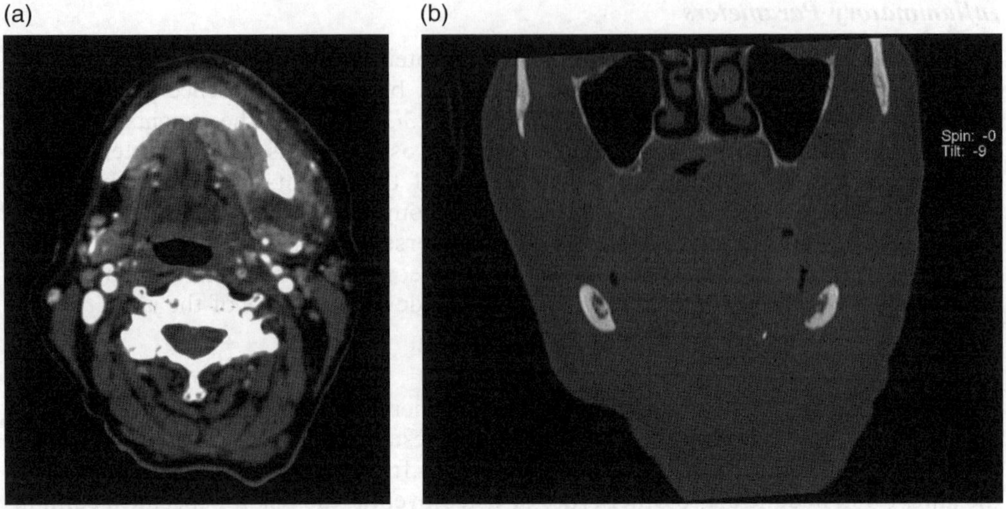

Figure 19.3. A 79-year-old edentulous woman suffering from a pressure sore by her dental prosthesis on the left lower jaw since several weeks. Increasing pain during several weeks and inability to eat since about 1 week. A spontaneous perforation was observed in the region of the former tooth 34. C-reactive protein at 108 mg/l, leukocyte counts at 12.5 giga/l. Abscess drainage showed growth of *S. milleri* group. No clinical signs of sepsis. (a) Computed tomography (CT) scan transversal section: Perimandibular abscess and osteomyelitis of the left jaw. (b) Coronary section: bone erosion. Courtesy of Peter Graber, Basel University Medical Clinic, Liestal/Switzerland.

Imaging Procedures

Conventional orthopantomography is still the first-line imaging procedure to assess not only the dental status, but also the bony structure [61]. However, this technique is limited by superposition and artifacts. Computed tomography (CT) allows better assessment of bone morphology, including signs of osteomyelitis such as periosteal reaction, osteolysis (Figure 19.3a and b), sclerosis, and sequesters. In a recent study by Bolouri *et al.* [62], the performance of different imaging procedures was analyzed in a population of 42 patients with suspected or exacerbated previously known osteomyelitis of the jaws. The observers were blinded except for the suspicion of osteomyelitis. In 30 patients, the gold standard for the diagnosis was histopathology from a biopsy; in 12 patients without biopsy, a follow-up of at least 6 months was available. In patients with an uneventful follow-up without specific treatment, the diagnosis of osteomyelitis was excluded. According to the defined standards, osteomyelitis was diagnosed in 35 of 42 patients, that is, the prevalence in the study population was 83%. Single-photon emission computed tomography plus conventional CT (SPECT/CT) had the best performance with a sensitivity of 100%, a specificity of 86%, and an accuracy of 98%, followed by planar bone scintigraphy with 100% sensitivity, 71% specificity, and 95% accuracy. CT and conventional orthopantomography performed significantly poorer with sensitivities of 77 and 59%, specificities of 86 and 100%, and accuracies of 79 and 66%, respectively. Thus, in this study, SPECT/CT was the best modality for the diagnosis of osteomyelitis of the jaws. However, the accuracy of SPECT/CT was not significantly better than the one of planar bone scintigraphy. Therefore, the use of the less costly bone scintigraphy as first-line procedure seems reasonable.

Management

General Aspects

The goals of osteomyelitis therapy of the jaws are focus eradication by meticulous debridement including removal of dead bone and unstable internal fixation devices, combined with correct empirical and pathogen-directed antimicrobial therapy [63]. If an empirical therapy is started after sampling for microbiology, the spectrum should include the microorganisms of the normal oral flora, as mentioned earlier. Amoxicillin/clavulanic acid as well as clindamycin fulfill this requirement. Antimicrobial therapy differs between the different types of osteomyelitis of the jaws. The most important entities are presented in detail.

Neonatal Osteomyelitis

In case of early diagnosis, in most cases, no surgical treatment is required. If teeth are involved, which is by definition the case in tooth germ–associated osteomyelitis, extraction of the dental focus is advisable. If adequate treatment is delayed, additional surgical drainage of abscesses or removal of sequesters may be needed. Before starting antibiotics, at least two pairs of blood cultures should be drawn. Starting with amoxicillin/clavulanic acid (30–50 mg/kg i.v. every 12 h) or clindamycin (5 mg/kg i.v. every 12 h) covers the spectrum of the most frequent microorganisms (see earlier). If hematogenous osteomyelitis is suspected, the empirical choice should be directed against *S. aureus*. In regions with a high prevalence of methicillin-resistant *S. aureus*, vancomycin (12.5 mg/kg i.v. every 12 h) should be preferred. If the primary focus is meningitis, or if septic thrombosis of a sinus vein is suspected, cefotaxime (50 mg/kg every 12 h) is suggested. As soon as the microorganisms are known, treatment should be optimized according to the susceptibility of the isolates. Neonatal osteomyelitis has to be treated by the IV route for 3–6 weeks, since during the first couple of weeks after birth, oral therapy would not be reliable.

Acute Trauma-/Fracture-Related Osteomyelitis

In most cases, this infection is polymicrobial, including microorganisms of the oral cavity (see earlier). If osteomyelitis occurs after internal fixation of the bone, the spectrum of microorganisms is not different, except in case of communication of the fracture with extraoral soft tissue and skin. In such cases, *S. aureus* and coagulase-negative staphylococci play a major role. The treatment is identical to the one of bone and joint infections at other locations (see tables in Chapters 9 and 15). The treatment duration is 6 weeks in the absence and 3 months in the presence of an internal fixation device.

Trauma-/Fracture-Related Secondary Chronic Osteomyelitis

In general, secondary chronic osteomyelitis cannot be treated with antibiotics alone. Before antibiotic treatment, meticulous debridement surgery, usually decortication of the bone, including the excision of sequesters and lavage/drainage of pus, is required accompanied by adequate (re)fixation of the fracture. After sampling of several biopsies for histology and culture, treatment should be started with an antibiotic acting on microorganisms of the oral cavity. Amoxicillin/clavulanic acid (2.2 g i.v. every 8 h) or, in case of delayed-type penicillin allergy, ceftriaxone (2 g i.v. once daily) cover the appropriate spectrum. If a patient has a history of an immediate-type penicillin allergy, clindamycin (600 mg i.v. every 8 h) is a rational alternative. This treatment should be optimized as soon as the culture and susceptibility results are available. The duration of treatment is not standardized. Experts suggest 6 weeks to 3 months, depending on the clinical and laboratory response to treatment and the presence of a fixation device.

Acute Odontogenic Osteomyelitis

The management does not differ from the one of posttraumatic osteomyelitis. When soft tissue necrosis or abscesses have not yet developed, tooth extraction combined with antimicrobial therapy (see earlier) is suggested. The duration of treatment is 4–6 weeks.

Odontogenic Secondary Chronic Osteomyelitis

After a duration of 1 month, osteolysis, bone sequesters, and periosteal can be documented with a CT scan or MR imaging. Meticulous surgical debridement including dental extraction combined with antimicrobial therapy (see earlier) is needed for curative treatment. The duration of antibiotic treatment is suggested between 6 weeks and 6 months. In the large series of Kim and Jang [64], the outcome of 49 patients with secondary chronic osteomyelitis of the jaws was analyzed. The success rate in 39 patients with surgery and 8 weeks of antibiotics was 94.9%, as compared to 60% in 10 patients with surgery alone.

Chronic Contiguous Neck Infection

The two main entities are cervicofacial actinomycosis and cervical lymph node tuberculosis with contiguous bone infection. Both entities must be actively searched, since specific culture conditions are required. Surgical treatment includes sinus tract excision, removal of necrotic bone and soft tissue, and sampling of at least three biopsies before starting specific antimicrobial therapy. In case of actinomycosis, penicillin G (20 million units/day i.v. in four doses) is the treatment of choice. In case of penicillin allergy, clindamycin or an IV cephalosporin (e.g., ceftriaxone 2 g/day i.v.) are alternatives. After a duration of 2 weeks, it can be switched to an oral regimen, which should be continued for at least 6 months [55].

Bone tuberculosis is treated with a standard quadruple regimen during the first 2 months (pyrazinamide 15–30 mg/kg/day plus isoniazid 5 mg/kg/day plus rifampin 10 mg/kg/day plus ethambutol 15–25 mg/kg/day) [65]. After this initial phase, therapy is continued with isoniazid plus rifampin, provided that the microorganisms are susceptible to these two standard drugs. In case of resistance, an individual regimen with alternative drugs has to be chosen [66]. Generally, tuberculous osteomyelitis is treated for 9 months [67] (see Chapter 16).

Peri-implantitis

Dental implants are exposed to and colonized by microorganisms from the normal mouth flora and by putative periodontal pathogens. The biofilm on the surface of dental implants cannot be eliminated by antimicrobial therapy. Therefore, the aim of antimicrobial therapy is the decrease of the local inflammation and the infection around the device. Unfortunately, there are no controlled clinical trials regarding antibiotic therapy of peri-implantitis [59]. In most cases, antimicrobial therapy has to be combined with surgical debridement [13]. Surgical management includes removal of granulation tissue, surface cleaning of the tooth, and bone resection if needed [15, 68]. In two comprehensive reviews, various aspects of the treatment of peri-implant infections are summarized [69, 70]. The antimicrobial therapies in the different studies were so different that no standardized regimen can be deduced from these systematic reviews. In 9 out of 10 studies analyzed by Javed *et al.* [59], local or systemic antibiotic treatment resulted in reduced gingival bleeding, suppuration, and peri-implant pocket. However, none of these studies was a randomized placebo-controlled trial. Thus, the role of antimicrobial therapy remains unclear. If systemic antimicrobial therapy is given, the choice of the drug should consider the microbial spectrum in severe peri-implantitis. Either amoxicillin/clavulanic acid (625 mg p.os every 8 h) or clindamycin (300 mg p.os every 6 h) for 7–10 days are rational choices [63].

Key Points

- Osteomyelitis of the jaws can be classified as acute (<1 month duration), secondary chronic, and primary chronic osteomyelitis. Acute and secondary chronic osteomyelitis are caused by microorganisms, whereas the different types of primary chronic osteomyelitis do not have a proven infectious etiology.
- Osteomyelitis of the jaws is generally polymicrobial and typically caused by microorganisms from the mouth flora.
- Osteomyelitis of the jaws is most often caused by dental or periodontal infection. Peri-implantitis is of increasing importance. IV bisphosphonate therapy may cause osteonecrosis, which can be followed by secondary infection.
- SPECT/CT and planar bone scintigraphy are the most accurate imaging procedures.
- The treatment concept is surgical focus eradication combined with antimicrobial therapy.

References

[1] Hudson JW. Osteomyelitis of the jaws: a 50-year perspective. J Oral Maxillofac Surg 1993; 51(12):1294–1301.

[2] Beninati F, Pruneti R, Ficarra G. Bisphosphonate-related osteonecrosis of the jaws (Bronj). Med Oral Patol Oral Cir Bucal 2013;18(5):e752–e758.

[3] Ruggiero SL, Mehrotra B, Rosenberg TJ, Engroff SL. Osteonecrosis of the jaws associated with the use of bisphosphonates: a review of 63 cases. J Oral Maxillofac Surg 2004;62(5): 527–534.

[4] Baltensperger M, Eyrich G. Osteomyelitis of the jaws: definition and classification. In: Baltensperger MM, Eyrich GK, editors. *Osteomyelitis of the Jaws*, 1st edition. Berlin: Springer-Verlag; 2009. pp. 5–56.

[5] Bernier S, Clermont S, Maranda G, Turcotte JY. Osteomyelitis of the jaws. J Can Dent Assoc 1995;61(5):441–442, 445–448.

[6] Eyrich GK, Baltensperger MM, Bruder E, Graetz KW. Primary chronic osteomyelitis in childhood and adolescence: a retrospective analysis of 11 cases and review of the literature. J Oral Maxillofac Surg 2003;61(5):561–573.

[7] Eyrich GK, Harder C, Sailer HF, *et al.* Primary chronic osteomyelitis associated with synovitis, acne, pustulosis, hyperostosis and osteitis (SAPHO syndrome). J Oral Pathol Med 1999;28(10):456–464.
[8] Chowdri NA, Sheikh S, Gagloo MA, *et al.* Clinicopathological profile and surgical results of nonhealing sinuses and fistulous tracts of the head and neck region. J Oral Maxillofac Surg 2009;67(11):2332–2336.
[9] Trivellato AE, Ribeiro MC, Sverzut CE, *et al.* Osteopetrosis complicated by osteomyelitis of the maxilla and mandible: light and electron microscopic findings. Head Neck Pathol 2009;3(4):320–326.
[10] Theologie-Lygidakis N, Schoinohoriti O, Iatrou I. Surgical management of primary chronic osteomyelitis of the jaws in children: a prospective analysis of five cases and review of the literature. Oral Maxillofac Surg 2011;15(1):41–50.
[11] Probst FP. Chronic recurrent multifocal osteomyelitis. Case report and overview as a contribution to the knowledge about the disease. Radiologe 1984;24(1):24–30.
[12] Batcheldor GD Jr., Giansanti JS, Hibbard ED, Waldron CA. Garre's osteomyelitis of the jaws: a review and report of two cases. J Am Dent Assoc 1973;87(4):892–897.
[13] Klinge B, Meyle J. Peri-implant tissue destruction. The Third EAO Consensus Conference 2012. Clin Oral Implants Res 2012;23(Suppl 6):108–110.
[14] Mombelli A, Muller N, Cionca N. The epidemiology of peri-implantitis. Clin Oral Implants Res 2012;23(Suppl 6):67–76.
[15] Muthukuru M, Zainvi A, Esplugues EO, Flemmig TF. Non-surgical therapy for the management of peri-implantitis: a systematic review. Clin Oral Implants Res 2012;23(Suppl 6):77–83.
[16] Prasad KC, Prasad SC, Mouli N, Agarwal S. Osteomyelitis in the head and neck. Acta Otolaryngol 2007;127(2):194–205.
[17] Adekeye EO, Cornah J. Osteomyelitis of the jaws: a review of 141 cases. Br J Oral Maxillofac Surg 1985;23(1):24–35.
[18] Finley AM, Beeson MS. Actinomycosis osteomyelitis of the mandible. Am J Emerg Med 2010;28:118.e1–118.e4.
[19] Sheikh S, Pallagatti S, Gupta D, Mittal A. Tuberculous osteomyelitis of mandibular condyle: a diagnostic dilemma. Dentomaxillofac Radiol 2012;41(2):169–174.
[20] Baltensperger M, Gratz K, Bruder E, *et al.* Is primary chronic osteomyelitis a uniform disease? Proposal of a classification based on a retrospective analysis of patients treated in the past 30 years. J Craniomaxillofac Surg 2004;32(1):43–50.
[21] Marx RE. Chronic osteomyelitis of the jaws. Int J Oral Maxillofac Surg 1991;3:367–381.
[22] Eyrich GK, Langenegger T, Bruder E, *et al.* Diffuse chronic sclerosing osteomyelitis and the synovitis, acne, pustolosis, hyperostosis, osteitis (SAPHO) syndrome in two sisters. Int J Oral Maxillofac Surg 2000;29(1):49–53.
[23] Marx RE, Tursun R. Suppurative osteomyelitis, bisphosphonate induced osteonecrosis, osteoradionecrosis: a blinded histopathologic comparison and its implications for the mechanism of each disease. Int J Oral Maxillofac Surg 2012;41(3):283–289.
[24] Yamazaki T, Yamori M, Yamamoto K, *et al.* Risk of osteomyelitis of the jaw induced by oral bisphosphonates in patients taking medications for osteoporosis: a hospital-based cohort study in Japan. Bone 2012;51(5):882–887.
[25] Walter C, Al-Nawas B, Grotz KA, *et al.* Prevalence and risk factors of bisphosphonate-associated osteonecrosis of the jaw in prostate cancer patients with advanced disease treated with zoledronate. Eur Urol 2008;54(5):1066–1072.
[26] Coleman R, Woodward E, Brown J, *et al.* Safety of zoledronic acid and incidence of osteonecrosis of the jaw (ONJ) during adjuvant therapy in a randomised phase III trial (AZURE: BIG 01-04) for women with stage II/III breast cancer. Breast Cancer Res Treat 2012;127(2): 429–438.
[27] Arunkumar JS, Naik AS, Prasad KC, Santhosh SG. Role of nasal endoscopy in chronic osteomyelitis of maxilla and zygoma: a case report. Case Rep Med 2011;2011:802964.

[28] Calhoun KH, Shapiro RD, Stiernberg CM, *et al.* Osteomyelitis of the mandible. Arch Otolaryngol Head Neck Surg 1988;114(10):1157–1162.

[29] Loh FC, Ling SY. Acute osteomyelitis of the maxilla in the newborn. J Laryngol Otol 1993;107(7):627–628.

[30] Ali A, Kurien M, Mathews SS, Mathew J. Complications of acute infective rhinosinusitis: experience from a developing country. Singapore Med J 2005;46(10):540–544.

[31] Chaudhary S, Kalra N, Gomber S. Tuberculous osteomyelitis of the mandible: a case report in a 4-year-old child. Oral Surg Oral Med Oral Pathol Oral Radiol Endod 2004;97(5): 603–606.

[32] Sharkawy AA. Cervicofacial actinomycosis and mandibular osteomyelitis. Infect Dis Clin North Am 2007;21(2):543–556, viii.

[33] Sieber OF Jr., Larter W, Smith PJ, *et al. Coccidioides immitis* osteomyelitis of the mandible in an infant. J Oral Surg 1977;35(9):721–725.

[34] Bathoorn E, Escobar Salazar N, Sepehrkhouy S, *et al.* Involvement of the opportunistic pathogen *Aspergillus* tubingensis in osteomyelitis of the maxillary bone: a case report. BMC Infect Dis 2013;13:59.

[35] Hovi L, Saarinen UM, Donner U, Lindqvist C. Opportunistic osteomyelitis in the jaws of children on immunosuppressive chemotherapy. J Pediatr Hematol Oncol 1996;18(1):90–94.

[36] Dimitrakopoulos I, Lazaridis N, Asimaki A. Craniofacial invasive aspergillosis in an immunocompetent patient: a case report. J Oral Maxillofac Surg 2005;63(6):845–848.

[37] Oswal NP, Gadre PK, Sathe P, Gadre KS. Mucormycosis of mandible with unfavorable outcome. Case Rep Dent 2012;2012:257940.

[38] Hultin M, Gustafsson A, Hallstrom H, *et al.* Microbiological findings and host response in patients with peri-implantitis. Clin Oral Implants Res 2002;13(4):349–358.

[39] Baltensperger M. A retrospective analysis of 290 osteomyelitis cases treated in the past 30 years at the Department of cranio-Maxillofacial Surgery Zürich with special recognition of the classification Medical thesis, Zürich, 2003.

[40] Zbinden R. Microbiology. In: Baltensperger MM, Eyrich GK, editors. *Osteomyelitis of the Jaws*, 1st edition. Berlin: Springer-Verlag; 2009. pp. 135–143.

[41] Frandah W, Colmer-Hamood J, Mojazi Amiri H, *et al.* Oropharyngeal flora in patients admitted to the medical intensive care unit: clinical factors and acid suppressive therapy. J Med Microbiol 2013;62(Pt 5):778–784.

[42] Hull MW, Chow AW. Indigenous microflora and innate immunity of the head and neck. Infect Dis Clin North Am 2007;21(2):265–282, v.

[43] Tomaselli DL Jr., Feldman RS, Krochtengel AL, Fernandez P. Osteomyelitis associated with chronic periodontitis in a patient with end-stage renal disease: a case report. Periodontal Clin Investig 1993;15(2):8–12.

[44] Lukosiunas A, Kubilius R, Sabalys G, *et al.* An analysis of etiological factors for traumatic mandibular osteomyelitis. Medicina (Kaunas) 2011;47(7):380–385.

[45] Bormann KH, Wild S, Gellrich NC, *et al.* Five-year retrospective study of mandibular fractures in Freiburg, Germany: incidence, etiology, treatment, and complications. J Oral Maxillofac Surg 2009;67(6):1251–1255.

[46] Bertocchi M, Hamel-Teillac D, Emond S, *et al.* Facial cellulite associated with mandibular osteomyelitis in an infant. Ann Dermatol Venereol 2002;129(4 Pt 1):405–407.

[47] Koorbusch GF, Deatherage JR. Osteomyelitis of the jaw. How can we diagnose and treat osteomyeltisi of the jaws as early as possible. Oral Maxillofacial Surg Clin N Am 2011;23: 557–567.

[48] Chrcanovic BR, Reher P, Sousa AA, Harris M. Osteoradionecrosis of the jaws—a current overview—part 1: physiopathology and risk and predisposing factors. Oral Maxillofac Surg 2010;14(1):3–16.

[49] Koorbusch GF, Fotos P, Goll KT. Retrospective assessment of osteomyelitis. Etiology, demographics, risk factors, and management in 35 cases. Oral Surg Oral Med Oral Pathol 1992; 74(2):149–154.

[50] Khullar S, Khullar SM, Tvedt D, *et al.* Sixty cases of extreme osteonecrosis and osteomyelitis of the mandible and maxilla in a West African population. Int J Oral Maxillofac Surg 2012;41:978–985.
[51] Santos FA, Pochapski MT, Pilatti GL, *et al.* Severe necrotizing stomatitis and osteomyelitis after chemotherapy for acute leukaemia. Aust Dent J 2009;54(3):262–265.
[52] Chen L, Li T, Jing W, *et al.* Risk factors of recurrence and life-threatening complications for patients hospitalized with chronic suppurative osteomyelitis of the jaw. BMC Infect Dis 2013;13:313.
[53] Kito S, Hirashima S, Yoshioka I, *et al.* A case of chronic infectious arthritis of the temporomandibular joint associated with osteomyelitis without malocclusion. Open Dent J 2010; 4:29–32.
[54] Pigrau C, Almirante B, Rodriguez D, *et al.* Osteomyelitis of the jaw: resistance to clindamycin in patients with prior antibiotics exposure. Eur J Clin Microbiol Infect Dis 2009; 28:317–323.
[55] Smego RA Jr., Foglia G. Actinomycosis. Clin Infect Dis 1998;26(6):1255–1261; quiz 62–63.
[56] Bartkowski SB, Zapala J, Heczko P, Szuta M. Actinomycotic osteomyelitis of the mandible: review of 15 cases. J Craniomaxillofac Surg 1998;26(1):63–67.
[57] Robinson JL, Vaudry WL, Dobrovolsky W. Actinomycosis presenting as osteomyelitis in the pediatric population. Pediatr Infect Dis J 2005;24(4):365–369.
[58] Lang NP, Berglundh T. Periimplant diseases: where are we now?—Consensus of the Seventh European Workshop on Periodontology. J Clin Periodontol 2011;38(Suppl 11):178–181.
[59] Javed F, Alghamdi AS, Ahmed A, *et al.* Clinical efficacy of antibiotics in the treatment of peri-implantitis. Int Dent J 2013;63(4):169–176.
[60] Berglundh T, Persson L, Klinge B. A systematic review of the incidence of biological and technical complications in implant dentistry reported in prospective longitudinal studies of at least 5 years. J Clin Periodontol 2002;29(Suppl 3):197–212; discussion 32–33.
[61] Schuknecht BF, Carls FR, Valavanis A, Sailer HF. Mandibular osteomyelitis: evaluation and staging in 18 patients, using magnetic resonance imaging, computed tomography and conventional radiographs. J Craniomaxillofac Surg 1997;25(1):24–33.
[62] Bolouri C, Merwald M, Huellner MW, *et al.* Performance of orthopantomography, planar scintigraphy, CT alone and SPECT/CT in patients with suspected osteomyelitis of the jaw. Eur J Nucl Med Mol Imaging 2013;40(3):411–417.
[63] Zimmerli W. Osteomyelitis antibiotic therapy. In: Baltensperger MM, Eyrich GK, editors. *Osteomyelitis of the Jaws*, 1st edition. Berlin: Springer-Verlag; 2009. pp. 179–190.
[64] Kim SG, Jang HS. Treatment of chronic osteomyelitis in Korea. Oral Surg Oral Med Oral Pathol Oral Radiol Endod 2001;92(4):394–398.
[65] Neff M. ATS, CDC, and IDSA update recommendations on the treatment of tuberculosis. Am Fam Physician 2003;68(9):1854, 1857–1858, 1861–1862.
[66] Pontali E, Matteelli A, Migliori GB. Drug-resistant tuberculosis. Curr Opin Pulm Med 2013;19(3):266–272.
[67] Editorial. Controlled trial of short-course regimens of chemotherapy in the ambulatory treatment of spinal tuberculosis. Results at three years of a study in Korea. Twelfth report of the Medical Research Council Working Party on Tuberculosis of the Spine. J Bone Joint Surg Br 1993;75:240–248.
[68] Schou S, Berglundh T, Lang NP. Surgical treatment of peri-implantitis. Int J Oral Maxillofac Implants 2004;19(Suppl):140–149.
[69] Klinge B, Gustafsson A, Berglundh T. A systematic review of the effect of anti-infective therapy in the treatment of peri-implantitis. J Clin Periodontol 2002;29(Suppl 3):213–225; discussion 32–33.
[70] Roos-Jansaker AM, Renvert S, Egelberg J. Treatment of peri-implant infections: a literature review. J Clin Periodontol 2003;30(6):467–485.

Chapter 20

Implant-Associated Osteomyelitis of Long Bones

Martin McNally and Parham Sendi

Introduction

In 2008, Donaldson *et al.* [1] calculated a fracture incidence of 3.6 fractures per 100 people per year in England. The lifetime fracture prevalence exceeded 50% in middle-aged men and 40% in women over the age of 75 years. Fractures most commonly involved feet and hands, followed by long bones (i.e., humerus, radius/ulna, femur, tibia/fibula). Implant-associated osteomyelitis (IAOM) of long bones is a feared complication after bone fixation.

Darouiche [2] estimated in 2004 that two million fracture fixation devices were implanted per year in the United States, and the projected number resulted in 100,000 (5%) infections. The extent of trauma influences the incidence. Open trauma implies a communication between fracture and wound and therefore a contamination of the bone. The more extensive the open trauma is, the higher the incidence rate of infection. Moreover, in fixation of open fractures, it may not be possible to close the wound directly. This results in a delay in covering the implant, which may increase the chance of early colonization of fracture implants and IAOM.

The high susceptibility of fixation devices to bacterial and fungal infection does not differ from that of periprosthetic joint infection (PJI). The presence of foreign body material leads to a locally impaired host defense (see Chapter 8) [3]. Thus, during and after bone fixation, the surgical wound containing an implant is highly prone to infection. Foreign devices are rapidly coated by host proteins after implantation. Several host proteins, such as fibronectin and laminin, favor the adherence of staphylococci to foreign surfaces [4]. After bacterial adherence, microorganisms accumulate, produce exopolysaccharides, and form complex communities resembling multicellular organisms [5–7]. This so-called biofilm resists not only natural host defenses but also antibiotics [3, 8]. Instability, infection, and disturbed endosteal blood supply can lead to impaired bone consolidation, resulting in nonunion [3, 9, 10].

Bone and Joint Infections: From Microbiology to Diagnostics and Treatment, First Edition.
Edited by Werner Zimmerli.

Treatment concepts of IAOM and PJI are often considered to be similar. This is only partially true, because there are significant differences between the management of PJI and IAOM. First, in PJI, the implant has to be maintained for appropriate limb function. In IAOM, it may be possible to remove the implant material after fracture healing. Second, exogenous infection following surgical treatment for open fractures is more frequent than in PJI. Third, a large variety of hardware can be implanted. Fracture fixation devices include plates, screws, intramedullary nails, rods, and pins from external fixation devices. Thus, there is often a heterogeneous population when studies on IAOM are analyzed.

The definition of IAOM requires at least one of the following criteria:

- Pus surrounding the implant
- Sinus tract communicating with the implant
- Evidence of an identical microorganism in at least two samples (biopsies, exudate, sonication fluid)
- Presence of ≥5 granulocytes per 10 high-power fields in a biopsy

The histopathological criterion is typically used in chronic infections. Additional signs of infection are bone loss and fluid accumulation around the implant, which is visible in any imaging procedure.

Classification and Risk Factors

Internal fixation of a closed fracture has no higher risk of infection as compared to joint replacement. The risk of IAOM depends on the type and site of injury, particularly the soft tissue damage. It is especially high in patients with open fractures of the tibia, multiple injuries, high-energy trauma, vascular injury, and late admission to a trauma treatment center [11–15]. Open fractures are commonly classified according to Gustilo–Anderson (Table 20.1) [16], with risk of infection increasing according to the degree of tissue injury. In a systematic review comprising 32 articles reporting on 3060 open tibia fractures, the deep infection rate for grade I fracture was 1.8% (range 0–3.6%), grade II 3.3% (range 0–11%), grade IIIA 5% (range 0–28.6%), grade IIIB 12.3% (9.4–15.1%), and grade IIIC 16.1% (10.2–22%) [11].

There are also patient-related risk factors. For instance, smoking increases the risk of infection in open tibia fractures. In one study, Castillo *et al.* [17] separated patients with

Table 20.1. Open fracture classification according to Gustilo *et al.* [16].

Fracture type	Description
Type I	The wound is <1 cm long with a clean piercing puncture
Type II	The laceration is >1 cm long with no extensive soft tissue damage, and the puncture wound is moderately contaminated
Type IIIA	Extensive laceration with adequate coverage of the fractured bone
Type IIIB	Extensive loss of soft tissue with periosteal stripping and exposure of bone resulting in massive contamination
Type IIIC	Open fracture with an arterial injury regardless of the degree of soft tissue injury

unilateral open tibia fractures into three groups (never smoked [$n=81$], previous smoker [$n=82$], and current smoker [$n=105$]). Current smokers were more than twice as likely to develop an infection ($P=0.05$) and 3.7 times as likely to develop osteomyelitis ($P=0.01$). Patients with multiple comorbidities are also at high risk for infection after open fracture. McPherson *et al.* [18] developed a patient score based on the number of comorbid conditions. The "host" score has three classes: A = no comorbid condition, B = one or two compromising factors, and C = more than three comorbid conditions or one of the following conditions: absolute blood neutrophil counts less than 1000 cells/μl, CD4 cells less than 100 cells/μl, intravenous drug users (IVDU), chronic active infection of another site, dysplasia, or neoplasia. In 146 patients with 162 infections, the incidence of infection in patients with host grade A was 4%, in grade B 15%, and in grade C 31% [14].

There are three routes of infection: (1) exogenous, (2) hematogenous, and (3) contiguous. Exogenous infection mainly occurs during the perioperative period. Hematogenous infection results from seeding during bacteremia [19, 20]. Contiguous infection progresses from septic arthritis or soft tissue infection to the neighboring bone fixation device. In contrast to PJI, the exogenous route is much more frequent in bone fixation-associated infection, since the surgical field after open trauma is not sterile at the time of intervention.

A classification according to the time interval from index surgery to infection manifestation is helpful for clinical practice, because it considers three features of pathogenesis: typical infection route, typical microorganisms, and findings on clinical examination (see Section "Clinical Features"). It includes "early infection," which is diagnosed within 2 weeks after implantation; "delayed" infection between the third and the 10th week; and "late infection" more than 10 weeks after implantation (Table 20.2) [21, 22]. Until the 1970s, chronic late infections were a frequent manifestation of osteomyelitis acquired during the preantibiotic era or due to insufficient initial surgical management. Nowadays, these infections are mainly observed in patients with a postoperative wound healing disturbance that has not been adequately treated.

Table 20.2. Open fracture classification according to onset of symptoms after implantation [21, 22].

Time of onset of symptoms	Characteristics
Early infection (within <2 weeks after implantation)	*Clinical picture:* signs of wound infection such as persistent fever, pain, erythema, swelling, wound healing disturbances
	Typical microorganisms: S. aureus, group A streptococci, Gram-negative bacilli
Delayed infection (3–10 weeks after implantation)	*Clinical picture*: persistent pain, low-grade fever, mechanical instability, sinus tract
	Typical microorganisms: low-virulence microorganisms such as coagulase-negative staphylococci or mixed skin flora in the case of a sinus tract
Late infection (>10 weeks after implantation)	*Clinical picture:* (a) **acute** hematogenous infection, sepsis syndrome, local pain, and signs of inflammation; (b) **chronic** (delayed infection or recurrence of incorrectly treated early infection), signs of infection after interval with bridging symptoms such as pain, wound healing disturbances, nonunion
	Typical microorganisms: (a) *S. aureus* and *E. coli*, (b) any microorganism, including polymicrobial infection

Microbiology

Virtually any bacterial or fungal agent can cause IAOM, including mycobacteria and fungi [23–27]. Even microorganisms of low virulence, which are generally considered as commensals (e.g., *Corynebacterium* species), have been described in patients with internal fixation devices [27]. Table 20.3 summarizes the most frequent isolates in patients with IAOM [23, 24, 27]. Up to three-quarters of the cases are caused by staphylococci, with *Staphylococcus aureus* being most frequent. *Propionibacterium acnes* is not only a common pathogen in periprosthetic shoulder infection (see Chapter 10) but is also observed after internal fixation of proximal humeral fractures [28]. Multidrug-resistant Gram-negative bacilli, fungal agents, and polymicrobial infections are mainly observed in patients with internal fixation of grade III open fractures or in those with prolonged wound healing disturbance and continuous empiric antibiotic therapy [29, 30].

Importantly, epidemiological data about MRSA and multidrug-resistant Gram-negative bacilli vary considerably between different regions. Therefore, each institution should constantly evaluate the distribution of the isolated pathogens and their antimicrobial susceptibility patterns. These data together with clinical features are helpful for an empirical antimicrobial treatment.

Clinical Features

The clinical presentation depends on (i) the duration of infection, (ii) the type of microorganism, (iii) the anatomic localization, (iv) the mechanism of the trauma leading to

Table 20.3. Frequency of microbiological isolates in patients with bone infection and metalwork *in situ* at sampling.

Study/number of patients	[24] % (*n* = 58)	[23] % (*n* = 777)	[27] % (*n* = 247)
Monomicrobial culture	57	[a]	[b]
Polymicrobial culture	22	[a]	[b]
No growth	12[c]		
Staphylococcus aureus	32	44	43
Coagulase-negative staphylococci	28	33	30
Enteric Gram-negative bacilli	22	5	7
Streptococcus species	9	9	8
Pseudomonas species	7	2	10
Enterococcus species	5	4	6
Anaerobes	5	0.1	
Diphtheroids	2		0 (1)
Fungi	2	1	
Other	2		1

Column adds to more than 100% because isolates within polymicrobial infections are also presented. Numbers are rounded.
[a]The study included 777 monomicrobial and 125 polymicrobial infections. Only pathogens of monomicrobial infections are listed.
[b]Only results from monoculture are presented (infections with internal fixation devices).
[c]In addition to 12% no growth, no sample for culture in 9%.

fracture, and (v) the type of surgery that has been performed. Regarding therapeutic management, the duration of infection is the most relevant parameter to be considered.

Early Infection

Early infection is often, but not always, acute and manifests itself within 2 weeks after bone fixation. These infections are typically caused by virulent pathogens (e.g., *S. aureus*, group A *Streptococcus*, Gram-negative bacilli). Patients often report pain after initial bone fixation. In addition, fever (or at least elevated body temperature) is common. The wound is characterized by erythema, hyperthermia, swelling, or wound discharge of turbid fluid or even pus. Prolonged secretion, wound edge necrosis, and hematoma are not only risk factors for IAOM but may also be signs of already established infection. Therefore, these clinical signs should lead to rapid surgical review for diagnostic and therapeutic purposes, which increases the chance for implant retention until fracture consolidation. In the early phase after fracture fixation, the diagnosis "superficial wound infection" should be used with caution, as it may delay investigation and treatment of a deep implant infection.

Delayed Infection

This type of infection is generally acquired during the perioperative period but is diagnosed with a delay of more than two and up to 10 weeks. The diagnostic delay is due to (i) low virulence of the infecting agent, (ii) previous empiric antimicrobial therapy without diagnostic work-up, or (iii) misinterpretation of nonspecific symptoms. Coagulase-negative staphylococci are typical microorganisms involved. If there is a sinus tract, polymicrobial flora is often found in microbiological samples. The key symptom is pain, but occasionally patients report low-grade fever.

Late Infection

There are two types of late infections. Most late infections are also acquired during the perioperative period. These infections are characterized either by continuous symptoms or by recurrent exacerbations in cases of insufficient prior treatment. Typical clinical features of these chronic infections are pain and wound healing disturbances. Alternatively, late infections can be acquired via hematogenous seeding [19, 20]. Typical clinical features are an uneventful postoperative course and a sudden onset of symptoms, mostly pain and swelling. Infection occurs via silent bacteremia, or occasionally, an antecedent focus can be found (e.g., skin and soft tissue infection).

Laboratory Investigation

Blood Tests

Early infections (<2 weeks) are mainly diagnosed clinically. Blood tests such as C-reactive protein (CRP), erythrocyte sedimentation rate (ESR), and white blood cells (WBC) are not helpful, but can point toward the severity of the systemic inflammatory response. In delayed and late infections, the diagnosis is mainly clinical because the key symptom is pain in the area of the implant, although in fractures with delayed union or aseptic nonunion, it can be difficult to rule out late infection. These patients present with pain, swelling, and tenderness. Laboratory tests may be helpful, as aseptic causes of pain will not usually elevate the CRP or WBC.

In patients with acute symptoms such as high fever and chills, blood cultures must be obtained because they may identify the causative organism.

Microbiology and Histopathology

Surgical exploration and microbiological sampling are the cornerstones of diagnosis and, more important, decisive for antimicrobial treatment. Whenever possible, samples should be taken before empirical antibiotics are started. Only in unstable patients with sepsis syndrome, this may not be possible. Superficial swabs and surface biopsies from open wounds must be avoided, because there is a high chance of culturing bacteria belonging to skin flora colonizing the wound. If this is the case, the differentiation between contaminants and the IAOM-causative pathogen is difficult. Diagnostic material should be obtained from deep tissues adjacent to the implant. These can be taken during surgery or under image-guided percutaneous biopsy, though we recommend the former method. At least three samples each should be taken for microbiological culture and histopathological examination, although we recommend obtaining 5 or 6 samples for each investigation. Suitable sample material includes free pus, membranes around plates, necrotic tissue, and dead bone fragments. The samples should be numbered and labeled with the anatomic localization during harvesting.

As mentioned earlier, an optimal sampling technique facilitates the differentiation between colonizing wound flora and IAOM-causative organisms. The "no skin touch" technique of sampling is applied successfully in Oxford, United Kingdom. Each sample is harvested with separate instruments (forceps, knife, curette, or bone nibbler) to avoid contamination across the samples (Fig. 20.1). It is important that the instruments do not touch the patient's skin. Another sampling method that may aid differentiation between contaminants and IAOM-causative pathogens is parallel sampling for microbiological culture and histopathological examination. This method is applied in Liestal, Switzerland, particularly in patients with chronic infections. Each single sample is paralleled by a second sample from the identical anatomical localization during harvesting. The samples are designated with numbers and letters (e.g., 1A and 1B, 2A and 2B, and so on; As are sent to the microbiology laboratory and Bs to the institute for histopathology). With this method, each microbiological result can be correlated with the histopathological finding.

Irrespective of the method of sample harvesting, it is important to think of potentially involved species prior to the intervention and to have good collaboration with the bacteriology laboratory. Anaerobic bacteria, for example, should be sought in cases with dead spaces, extensive tissue necrosis, and tissue with poor vascularization. These bacteria may require special media and rapid transport from the operating theater to the microbiology laboratory [31] (see Chapter 2).

Removed hardware can be cultured in enrichment broth. This technique is prone to contamination. Sonication of removed hip and knee arthroplasties has shown good sensitivity and specificity [32]. However, the method has shown variable results of success. In early infections, the diagnostic accuracy of sonication seems not to be increased when compared to sample culture results [33]. In addition, data specifically on internal fixation devices are scarce [34]. Finally, in patients with open wounds, multiple organisms may adhere to the foreign material. This can lead to an overestimate of involved pathogens for the antimicrobial

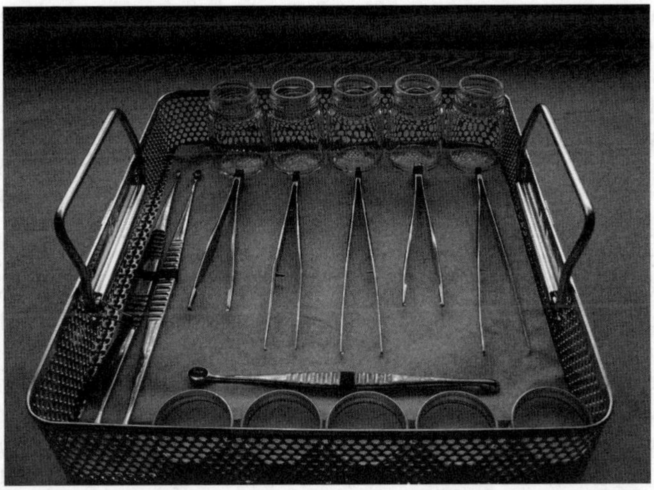

Figure 20.1. Prepared instrument set for biopsy sampling. Separate instruments (forceps, knife, curette or bone nibbler) and transport tubes are used for each biopsy sample. (*See insert for color representation of the figure.*)

treatment of IAOM. Thus, in patients with IAOM after having had open fractures and severely damaged soft tissue, the diagnostic value of sonication results is limited.

Imaging Procedures

Plain radiographs in serial follow-ups are required to monitor bone healing, formation of a nonunion, implant loosening, bone loss around screws, and migration of devices. Changes seen on plain X-rays are not specific for infection. Implant loosening can be due to mechanical fixation failure, but loosening early after implantation is suspicious for infection.

In our experience, ultrasound plays only a minor role in the assessment of IAOM. It may show fluid collection in the proximity of a device, and if so, it helps guiding puncture for microbiological sampling. In our institutions, computed tomography (CT) is frequently used to estimate the extent of inflamed tissue (e.g., up to intramedullary), allows detection of bone necrosis, and demonstrates infected fracture fragments (i.e., sequestra). Isotope bone scanning has little place in the diagnosis of IAOM. Magnetic resonance imaging (MRI) may be valuable in late infections, but often images are degraded by artifacts due to the implant. Moreover, it tends to overestimate the extent of osteomyelitis.

Positron emission tomography (PET) and PET-CT are used only for selected cases in which the infection suspicion persists and can neither be excluded nor diagnosed by other means.

Management

General Aspects

The goals of the therapeutic management of IAOM are consolidation of bone and cure of osteomyelitis. These goals must be fulfilled in a way that allows rehabilitation of the limb, including weight-bearing and joint movement.

It is essential to establish the diagnosis of IAOM early. If infection is suspected, anti-inflammatory drugs should be avoided, because they can reduce symptoms and delay diagnosis. Empirical antibiotic therapy without diagnostic work-up should never be considered. In acute infections with sepsis syndrome, blood cultures must be obtained before starting rational empiric antimicrobial treatment. In the absence of severe systemic inflammatory response syndrome, biopsy samples must be obtained (percutaneous or open surgical). All results from diagnostic investigations must be evaluated carefully prior to surgical intervention. This is necessary to develop a management plan and a final treatment strategy. The evaluation includes the patient's comorbidities, the neurovascular function of the extremity, the duration of disease before diagnosis was established, imaging results, the condition of the surrounding soft tissue, and—if available—the causative pathogen and its susceptibility patterns.

Comorbidities, such as diabetes, smoking, and poor nutrition must be actively addressed to improve the likelihood of adequate wound healing. Injuries of the neurovascular system affect the functional outcome. The longer the disease duration, the more established the biofilm formation becomes and the less likely the implant can be retained. Imaging studies should be reviewed to assess loosening of the device, the extent of infection, and the integrity of the bone (e.g., nonunion). This assessment helps in decision making for the surgical intervention (e.g., debridement or resection of the bone, removing or retaining the implant) and the bone stabilization method. The quality and integrity of the soft tissue must be evaluated for postoperative wound closure and the necessity of flap coverage. Some pathogens are extremely difficult to eradicate (e.g., small-colony variants of bacteria, fungi), even with the correct antimicrobial treatment. They have a strong adherence to the foreign material, making implant retention difficult.

In principle, stable fractures are less susceptible to infection than unstable bone components [10, 35]. Thus, the fracture must be stable during the treatment for infection until consolidation of the bone is achieved. The decision to retain or exchange the fixation device may be difficult. It must be based on a careful assessment of fracture stability. Another factor to consider is the likelihood of the fracture achieving union within the time that the implant can be expected to survive without fatigue failure. Infected diaphyseal fractures may be very slow to heal, placing prolonged stresses on the fixation device. If an infected fracture implant is retained, it must be assumed that there is biofilm present. As a consequence, antimicrobial suppression with activity against biofilm-forming bacteria is required until bone union (e.g., ≥3 months). After union, the decision to retain or remove the foreign material must be considered. In cases of early infection or acute hematogenous infection, the implant may be retained. It is a precondition that rapid and adequate surgery be performed. In a study by Zimmerli *et al.* [36] that included 10 patients with staphylococcal IAOM, the cure rate with debridement and a 3-month course of antimicrobials (fluoroquinolone/rifampin combination) was high. Hence, in these cases, the implant must not necessarily be removed after consolidation. In contrast, in patients with delayed infection, the biofilm may persist on the implant. Thus, in these cases, all foreign material should be removed after bone union. Thereafter, antimicrobial therapy can be stopped in a short while (i.e., within 1 week).

Surgical Interventions

Surgery will be required for almost all cases of IAOM of the long bones. A suppressive regimen of antibiotics alone is only considered in selected patients with severe comorbidity, who are not fit for surgery or who have a short life expectancy. Surgical management of an IAOM is complex and often requires close collaboration between orthopedic and plastic surgeons, infectious disease physicians, and microbiologists. If expertise and experience in treating infected fractures is not available, the patient should be transferred to a team that regularly manages such cases. The involvement of a dedicated multidisciplinary team has been shown to improve outcomes in posttraumatic IAOM [37].

Timing of Treatment

The urgency of treatment is variable. In early infection, it is important to operate rapidly to try to preserve the stabilization device before infection causes loosening, bone loss, and instability. In this situation, urgent drainage of abscesses and excision of infected tissue can be limb saving. However, in late infections, particularly if the fracture has developed an established infected nonunion, it is better to fully evaluate the patient, complete investigations, and optimize general health before beginning a major reconstructive surgical intervention.

Influence of IAOM Classification on Surgical Intervention

The classification of IAOM is helpful for considering the typical infection route and the typical microorganisms in association with findings on clinical examination. For the decision on surgical management, other criteria are more important (e.g., fracture stability, implant loosening, bone healing). Only in early infections can the time of infection (i.e., within 2 weeks after fracture fixation) be reliably estimated. The majority of early infections, but not all, can be treated with debridement and implant retention. This is also true for patients with acute symptoms after an uneventful postoperative course. For example, sudden systemic reaction of a silently growing bone abscess requires a different surgical approach than typically used for acute infections. Similarly, not all late infections must have major bone resections and reconstruction. Therefore, the distinction between early and late infections is not decisive for the nature of the surgery required. As mentioned earlier, the classification may be helpful in the decision to remove the implant after bone healing or to retain it after completing the antimicrobial treatment.

Soft Tissue Management

In IAOM, the infection will only be cured and the fracture united if the soft tissue envelope is well vascularized and intact. The provision of good soft tissues increases the blood flow to the area, delivering systemic antimicrobial agents, immune cells, and antibodies.

In the thigh and in the upper limb, it is often possible to close the skin directly. Femur, humerus, and forearm bones are all surrounded by muscles providing a healthy cover. In contrast, in the tibia, high-energy fractures will disrupt the limited soft tissues, producing major compromise. Even if the skin is closed, the underlying vascularity will be poor. The tibia will be surrounded by scar tissue, which can allow bacterial adherence and biofilm formation. This tissue may prevent adequate perfusion with antimicrobial agents. The concept of *the reconstructive ladder* defines a progression from simple wound closure through split skin grafting to local rotation of skin and muscle flaps up to free microvascular tissue transfer [38]. In the tibia, it is often necessary to transfer muscle flaps. The upper third of the tibia can be covered by a local musculus gastrocnemius flap. More distal infections require a free flap with a gracilis or latissimus dorsi muscle [39]. There are many types of soft tissue reconstruction available, but muscle flaps have the advantage of being highly vascular and resistant to colonization by microorganisms. They have been shown to deliver higher oxygen concentrations around the bone and thus aid elimination of bacteria when compared with random-pattern skin flaps [40]. Microvascular tissue transfer can be combined with complex bone reconstructions by using the Ilizarov method, with excellent results [41]. Rarely, in severe infections with a systemically unwell patient, it may be necessary to excise and drain infected tissues with extensive areas of inflammation around the surgical wound. In such cases, it is appropriate to leave the wound open and to reinspect the tissues after 48 hours. This open wound can be managed with occlusive dressings or a vacuum-assisted dressing [42]. However, the wound should be closed as soon as it is safe to do so to prevent secondary infection, poor mobilization, and fracture nonunion. Prolonged vacuum-assisted wound therapy has been shown to allow persistence of bacteria in the wound in patients with IAOM [43].

Consolidated Fracture with Surrounding Infection

In this situation, surgery is required to obtain microbiological diagnosis, with sampling as described in Section "Microbiology and Histopathology," and to remove the infected material. The implant can be safely removed. It is also essential to inspect the surgical field and remove any dead bone or compromised soft tissue. If a plate and screws have been removed for infection, there is usually a layer of dead cortical bone under the plate that can be shaved off with an osteotome. Screw holes should be overdrilled. Removal of an infected intramedullary nail should always be followed by reaming of the diaphysis and overdrilling of the locking screw holes. If there is a cavity in the metaphysis (proximal or distal), the bone should be opened through a small window and the lining of the cavity removed with curettes. After revision, the bone is washed to reduce contamination. We advocate 0.05% aqueous chlorhexidine. Normal saline is used, if a joint surface has been exposed. In deep cavities, suction catheters can be used to wash from deep to superficial, preventing accumulation of material in the medullary canal. After washing, parenteral antibiotics are given depending on the local antibiotic policy.

If the resection of dead bone has been significant, there may be a bone defect that will act as a dead space. Not covering this space adequately will result in fluid collection within the bone and an increased risk of infection recurrence. Good soft tissue cover will address superficial defects in cortical bone, but medullary dead space requires filling with a material that will resist microbial adherence. The subject of antibiotic-loaded carriers is discussed in the section "Antimicrobial Treatment."

Unhealed Fracture with a Stable Implant

In many early infections and in some late cases, the infection can present with an unhealed bone, but there is no major bone lysis or implant loosening. Because maintenance of stability is critical for fracture healing and to control infection, it is appropriate to retain a stable implant to achieve these goals. It should be remembered that fractures will heal even in the presence of suppressed infection [10, 35]. However, the principles of treatment must not be compromised. Adequate sampling and excision of dead tissue should be performed. The decision to retain or remove the implant is not only based on bone stability but also on the soft tissue conditions and risk factors for delayed wound healing.

The difficulty of achieving stability without the implant is also to be considered. In complex intra-articular fractures, it may be impossible to stabilize the bone without retaining the internal fixation. Conversely, a transverse tibial fracture can be easily stabilized by a simple monolateral external fixator.

If the implant is retained, careful clinical follow-up with regular radiology is mandatory. Signs of implant loosening, loss of fracture alignment, or new discharge of purulent fluid or wound breakdown are indicators of treatment failure. Early reintervention to remove the failing fixation before major bone loss occurs may allow limb salvage.

Unhealed Fracture with Unstable Fixation

There is little merit in keeping unstable implants in IAOM. Surgery must be focused on removing all foreign material together with the dead and compromised tissue in order to produce a healthy zone for bone healing. Once the implant is removed, the bone can be fully inspected and the need for radical debridement assessed. In early infections, there may be extensive inflammation but little dead bone. Fracture edges may need to be trimmed back, and loose fragments should be removed unless they have good soft tissue attachment. New bone formation around the fracture should be preserved whenever possible. Vital bone has a good blood supply and may aid early bone bridging of the fracture.

In open fractures, there may be extensive periosteal stripping with large areas of dead bone. These should be removed, as they will be colonized with bacteria and have established biofilm. Such excision will render the fracture grossly unstable and produce a major bone defect. After debridement and sampling, the bone is washed and IV antibiotics are given. The fracture must be stabilized. It is not acceptable to splint the limb in a cast or external support when there is established infection and a mobile fracture. The standard method of stabilization involves application of an external fixator with pins fixed to the bone above and below the fracture, outside the zone of infection. This is a safe technique that can allow rehabilitation and soft tissue healing. External fixators are versatile devices that can be used in many situations. However, they are cumbersome for the patient and have the risks of pin-tract infections with loss of fracture stability. Nevertheless, they remain the method of choice for most unstable infected fractures.

In recent years, there has been interest in using internal fixation after removal of infected implants. This can be done immediately or after an interval with an external fixator. In 1986, Klemm [44] showed that infected femoral and tibial fractures could be stabilized with an intramedullary nail, but union was achieved in only 89% of femurs and 62% of tibias. Nailing after infected plates were removed from tibia had a very high failure rate. More recently, Prasarn *et al.* [45] managed femoral infected nonunions with a single-stage protocol, exchanging internal fixation in one operation. Ten of the 11 patients healed their fractures, but 5 required revision surgery. The authors stressed the need for radical debridement of the infected fracture.

The use of narrow unreamed nails has been a recent trend for intramedullary nailing of fractures. However, when infection occurs, these will not provide sufficient stability. We advocate nail removal, excision of the medullary canal with the Reamer–Irrigator–Aspirator [46], and renailing with a larger reamed nail. There may be an advantage in inserting a new nail that has been coated with antibiotic-loaded polymethyl methacrylate (PMMA) cement [47].

In all cases of unstable, unhealed fractures, the soft tissues must be fully restored. These fractures will often be slow to heal. They require good blood supply to deliver antibiotics and stem cells to effect bone healing.

Infected Nonunion and Bone Defects

Nonunion is said to have occurred when the fracture will not progress to union without intervention. It is most often diagnosed many weeks or months after fracture, but the condition may be present from an early episode of infection causing bone death and failure of healing. Nonunions can be classified as viable (all bone being alive) or nonviable (containing a portion of dead bone). In all infected nonunions, there is an area of dead bone preventing fracture bridging and allowing persistence of infection. This may be surrounded by new bone (callous and involucrum), producing the typical "elephant's foot" appearance. Such cases provide a major reconstructive challenge.

Infected nonunions require excision of all dead bone, stabilization, soft tissue cover, and management of dead space around the bone defect.

The Ilizarov Method

Stabilization can be achieved by using a circular external fixator (Ilizarov frame), which facilitates correction of deformity and compression or distraction of the bone and allows full weight-bearing. However, this method is not used in every center. Segmental excision of an infected nonunion will produce a bone defect that can be managed in a single procedure or in stages. If the defect is small (< 2 cm), the limb can be shortened to give good bone contact and allow skin closure. If the defect is up to 4 cm in the tibia and 6 cm in the femur, the limb can be acutely shortened to give bone contact, but this will produce a significant limb length discrepancy that must be addressed. Within the Ilizarov fixator, it is possible to relengthen the bone through a separate corticotomy, away from the site of infection [48]. In severe purulent infection, the corticotomy can be delayed until 4 weeks after segmental resection to reduce the risk of infection in the lengthening regenerate.

In larger defects, gradual defect filling by Ilizarov bone transport has been shown to be a safe and effective method of securing union and eradicating infection [49, 50]. However, this is a very time-consuming technique with long fixator times (often over 1 year), with various problems and complications. It requires careful surgical technique and meticulous aftercare to ensure good results.

Ilizarov bone transport can be combined with staged internal fixation to reduce the fixator time [51].

Transfer of Vascularized Bone Segment

An alternative single-stage reconstruction involves transfer of a vascularized segment of bone (fibula or pelvic ilium) to bridge the defect. This can be stabilized by an external fixator or internal fixation [52, 53]. The experience of free fibular grafts is good in the upper limb with well-consolidated grafts and good function. The fibula is almost the same size as the humerus and forearm bones, and so little remodeling is needed. In the lower limb, the clinical experience with fibular grafts was less promising during past years. Certain centers have reported a significant risk of developing nonunions and recurrent infections. Biomechanically, the weight-bearing can lead to stress fractures or malunion in the remodeling period. Thus, patients must protect their graft for many months. In addition, because of the functional ankle instability, there is a risk of secondary ankle osteoarthritis in the donor leg [54].

Staged Reconstruction

This procedure is now more popular in large bone defects. The initial surgery is aimed at eradication of infection with the production of a bone defect free of microorganisms. This can then be filled with delayed bone grafting once the soft tissues are well healed. The dead space left after resection can be managed with an absorbable antibiotic carrier or with a temporary PMMA cement block. In Masquelet's technique, the cement block is said to produce an inductive membrane around it. After some weeks, the cement can be removed and bone graft packed into the defect inside the carefully preserved membrane. If the bone remains stable, this graft may consolidate and remodel over many months [55].

Antibiotic Cement-Coated Nails

The use of this method with or without staged bone grafting has offered a new approach to segmental bone defect management. This method potentially allows early weight-bearing and avoidance of external fixation [47]. As with all methods involving internal fixation after infection, it is dependent on adequate bone excision, careful sampling, and appropriate antimicrobial therapy.

IAOM with External Fixation Pin Sites

Effective use of an external fixator relies on stable bone–pin interfaces. Acute pin-site infection in the skin around the fixator pin is common, but extension to the bone is less frequent. If deep bone infection occurs, the fixation pin will loosen and the fracture stability will be lost. This is more common in external fixation for infected fractures than for other conditions. Pin infection should be rapidly treated with local cleaning and oral antibiotics. A drug with good antistaphylococcal activity and high bone bioavailability should be chosen (see Chapter 3, Table 9.2). The pin sites should be reviewed and regular radiographs taken to identify pin loosening early. Loose pins may need to be removed if the infection does not settle within 1 week of antibiotic therapy. In the first few weeks after fracture stabilization, loose pins will need to be replaced to restore adequate fixation. Toward the end of bone healing, loss of one or two pins may not be critical, as there will be some stability within the healing fracture. With established pin-site bone infection, the surrounding bone will die, forming a ring sequestrum. Simply removing the infected pin will not resolve this infection. A ring sequestrum must be excised.

Amputation

Most published series of IAOM treatment report a number of amputations, either as the initial treatment for severe infection or as salvage after failed surgery. Amputation for upper limb IAOM should be avoided at all costs. In the last 13 years in Oxford, United Kingdom, below-knee or above-knee amputation was performed in 0.77% (7 of 914 patients) with IAOM of the long bones. In three cases, the infection was treatable with surgery and antibiotics, as described earlier, but the patients did not wish to have a prolonged treatment program with further time off work. In one case, we decided that the infection was not treatable without amputation. In the other three cases, we failed to eradicate infection and secure union (two cases), or the patient had intractable pain (one case) [56]. It is essential to discuss the place of amputation at the outset of treatment. Patients should be aware that it offers a good chance of infection cure, but it may have severe quality-of-life disadvantages, particularly with above-knee amputation. Elderly patients will rarely manage heavy above-knee prosthesis and will spend much of their time in a wheelchair. Among the worst-case scenarios, below-knee is better tolerated than above-knee amputation.

Summary of Surgical Interventions

In conclusion, IAOM of the long bones can present throughout the period of fracture healing. It requires a multidisciplinary approach from clinicians dedicated to the management of bone and joint infection. The principles of general health optimization for the patient, early microbiological diagnosis from tissue sampling, appropriate antimicrobial therapy, adequate surgical debridement, fracture stability, dead-space management, and bone defect reconstruction are key steps in successful treatment. The methods available to deliver each element will vary and evolve, but the principles remain.

Antimicrobial Treatment

Concepts of IV and oral antimicrobial treatment are discussed elsewhere (Chapters 8–10 and 21). In general, most concepts derive from treatment concepts for PJI. Nevertheless, there are no recommendations nor comparative studies defining the length of treatment specifically for IAOM. We administer long-term antibiotic treatment, if hardware is maintained. In early infection or acute hematogenous infection that occurred (a) after an uneventful postoperative period and (b) was treated without delay, we propose 3 months of antimicrobial therapy [36, 57]. As in all types of biofilm infections, the cure rate with implant retention depends on the duration of infection [36, 58]. This is at least partially due to the maturation of the biofilm, which makes it more resistant to antimicrobial agents [59]. Therefore, in cases of delayed or late infection with chronic symptoms, treatment duration is to be prolonged as long as the device is retained. Bone consolidation must be controlled at regulatory intervals (see Case 2 in the section "Instructive Cases"). After bone union, the foreign material should be removed and antimicrobial therapy stopped within 1 week.

Local Antimicrobial Therapy

In IAOM, the management of a dead space, in particular when a high bacterial load is present, includes the filling of antibiotic-loaded beads or implantation of spacers. With this method, poorly vascularized bone segments are treated with antimicrobial agents. The drug release has its peak within two to three days and rapidly decreases thereafter [60]. An aminoglycoside (e.g., gentamicin, tobramycin)—either alone or in combination with vancomycin—is the standard antimicrobial agent in spacers and beads, but virtually any compound can be added. PMMA cement is the standard material on which antibiotics are impregnated [61]. The beads or spacers have to be removed as early as possible in order to enable subsequent bone grafting. This usually requires an additional surgical intervention several weeks after insertion. In contrast, a bioabsorbable material as a vehicle for local antibiotic delivery does not require another surgical stage and may allow bone regrowth. Studies on bioabsorbable bone substitutes (BBS) impregnated with antibiotics for the management of IAOM have been recently reported [62–64]. Alpha-hemihydrate calcium sulfate is used as a carrier for aminoglycosides (e.g., tobramycin and gentamicin) [65]. The carrier material has been proposed to enhance bone growth [66]. In a prospective clinical trial, McKee *et al.* [63] used PMMA cement beads in 15 patients and BBS pellets in another 15 patients for the treatment of chronic osteomyelitis and infected nonunion. Both materials were impregnated with tobramycin. The infection eradication rate at a mean follow-up of 38 months was 86% in both groups. However, conditioned by the nature of the carrier material, there were significantly more reoperations in the PMMA group than in the BBS group (15 vs. 7, $p = 0.04$). The practice to fill dead space within long bones with BBS tobramycin pellets was then continued in a series of 198 patients in Oxford, United Kingdom [64]. The results were encouraging with a success rate of over 90% having resolved the infection. If these results are confirmed in multiple centers, including a similar large number of patients with IAOM, BBS-impregnated antibiotic pellets may eventually replace PMMA cement beads as the standard carrier.

Instructive Cases

Case 1: Chronic Implant-Associated Osteomyelitis after a Gustilo–Anderson Grade IIIB Open Tibia Fracture

A 22-year-old man suffered a Gustilo–Anderson grade IIIB open tibia fracture in a motorcycle road traffic accident. He underwent immediate wound excision, stabilization with an intramedullary nail (Fig. 20.2a), and delayed soft tissue cover with a free microvascular latissimus dorsi muscle flap and split skin graft. The wound did not heal, and he developed persistent discharge. The patient remained systemically well with normal WBC and CRP. A surface swab that was taken from his wound 3 weeks after injury grew fully sensitive *S. aureus*, and he was given oral flucloxacillin for 2 weeks.

On referral 7 months later, he had obvious wound discharge with compromised soft tissues (Fig. 20.2b) and a painful unstable fracture. He was a smoker, but had no other comorbidities. At surgery, the nail was loose with free pus in the medullary canal. Deep samples were taken. The nail was removed and the medullary canal reamed. The medial cortex of the upper tibia was exposed and a dead fracture fragment excised. The dead space was filled with BBS calcium sulfate pellets with gentamicin and the skin closed. IV vancomycin and meropenem were given. The fracture was stabilized with an Ilizarov external fixator (Fig. 20.2c). This was used to compress the fracture for 3 weeks followed by gradual distraction of 1 mm per day for 10 days to stimulate bone union. Full weight-bearing was encouraged. Microbiological culture revealed a sensitive *S. aureus* in 4 of 6 specimens. He was treated with oral ciprofloxacin and rifampin until union.

The fixator was removed at 21 weeks with a well-healed tibia and no recurrence of infection. At follow-up 34 months after surgery, he remains infection free with excellent limb function (Fig. 20.2d).

Learning Points

- This patient presented with clinical signs (poor wound healing despite muscle flap reconstruction) of an early IAOM in the first few weeks after injury. He did not have appropriate initial investigation. Therefore, definitive treatment was delayed for many months.
- The delay in diagnosis with short-course oral antibiotics allowed progression of the infection, causing implant loosening over 7 months. This removed the possibility of

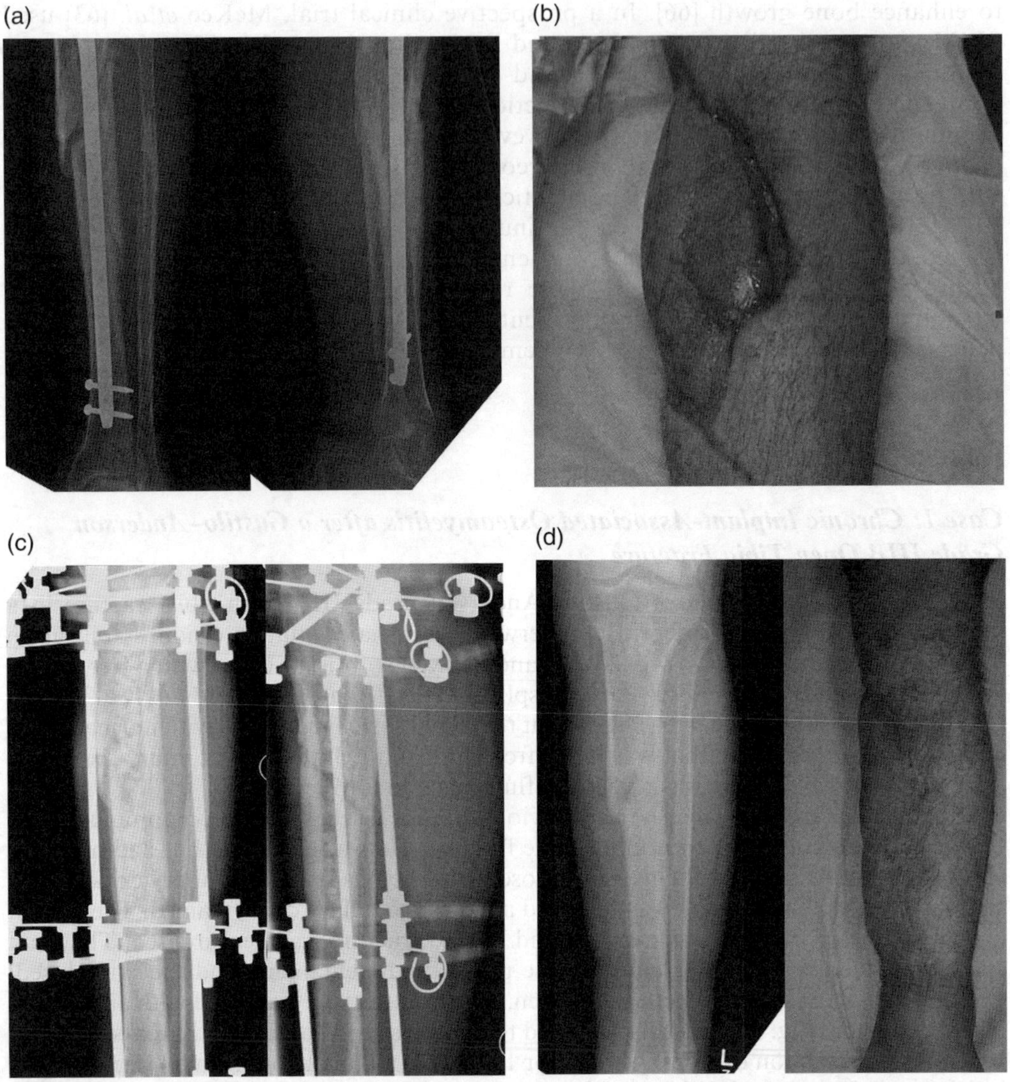

Figure 20.2. A 22-year-old man with chronic implant-associated osteomyelitis after a Gustilo–Anderson grade IIIB open tibia fracture. (a) Anteroposterior and lateral radiographs of a fragmented tibial fracture with an intramedullary nail *in situ*, taken 14 weeks after injury. (b) Wound discharge, indicating chronic infection, 7 months after primary intervention. (c) Stabilization of the fracture with an Ilizarov external fixator. (d) Final X-ray and clinical assessment at 34-month follow-up. (*See insert for color representation of the figure.*)

early debridement and implant retention. The patient then required a much more invasive surgical strategy with prolonged Ilizarov external fixation.

- Fortunately, his early use of antibiotics without adequate surgical treatment did not produce antibiotic resistance, but this is a high risk in this type of mismanagement.

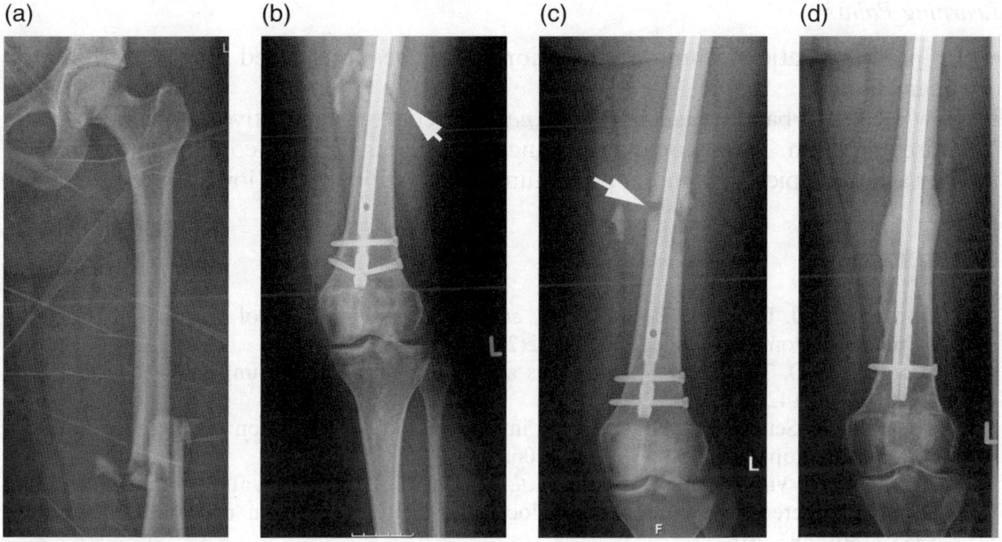

Figure 20.3. A 46-year-old woman with a septic nonunion after a grade II open femur fracture. (a–d) Serial anteroposterior radiographs of the left femur showing no bone healing until a second intramedullary nail exchange was performed and correct antimicrobial treatment administered.

Case 2: Septic Nonunion after a Grade II Open Femur Fracture

A 46-year-old woman suffered a grade II open left femur fracture in a motorcycle road traffic accident (Fig. 20.3a). She had no comorbidities. The fracture was immediately stabilized with an intramedullary nail. The postoperative course was uneventful, and no weight-bearing was allowed for 6 weeks. In the following months, she reported persistent pain. In follow-up controls, blood tests (CRP, WBC) revealed normal values, but no sign of bone union was observed in serial radiographs (Fig. 20.3b). Fifteen months later, aseptic nonunion of the femoral shaft fracture was postulated and exchange nailing performed. Three biopsy samples were obtained, and the intramedullary nail was sent for sonication and microbiological culture of the sonicated fluid. The postoperative course was uneventful and the patient discharged. One specimen grew *Propionibacterium* spp., while microbiological culture of sonicated fluid revealed coagulase-negative staphylococci with a low number of colony-forming units (100 cfu/ml). The results were interpreted as contamination. However, 6 months later, still no sign of bone formation was observed (Fig. 20.3c). At this time, the previous microbiological results were reconsidered, and chronic low-grade infection diagnosed. The nonunion was decorticated, the intramedullary nail exchanged, and the bone grafted with pelvic autograft. Ten samples were obtained, five each for microbiological and histopathological investigation. The removed nail was sent for microbiological culture of the sonicated fluid. IV treatment with vancomycin was administered. Results from histopathology were consistent with chronic infection. *P. acnes* grew in three samples and coagulase-negative staphylococci in two samples. The latter also grew in sonicated fluid culture. IV antibiotics were switched to doxycycline plus rifampin for a 6-month course, and the patient was discharged. At follow-up, 9 months after surgery, successful bone union was documented (Fig. 20.3d), and full weight-bearing was possible.

Learning Points

- In posttraumatic nonunion, infection must be considered in the differential diagnosis.
- Low-virulence bacteria, such as *P. acnes* and coagulase-negative staphylococci, can cause nonunion. They should not be uncritically considered as contamination, even if they are not typically found in posttraumatic infections of the lower extremities.

References

[1] Donaldson LJ, Reckless IP, Scholes S, *et al.* The epidemiology of fractures in England. J Epidemiol Community Health 2008;62(2):174–180.
[2] Darouiche RO. Treatment of infections associated with surgical implants. N Engl J Med 2004;350(14):1422–1429.
[3] Zimmerli W, Sendi P. Pathogenesis of implant-associated infection: the role of the host. Semin Immunopathol 2011;33(3):295–306.
[4] Herrmann M, Vaudaux PE, Pittet D, *et al.* Fibronectin, fibrinogen, and laminin act as mediators of adherence of clinical staphylococcal isolates to foreign material. J Infect Dis 1988;158(4):693–701.
[5] Zimmerli W, Moser C. Pathogenesis and treatment concepts of orthopaedic biofilm infections. FEMS Immunol Med Microbiol 2012;65(2):158–168.
[6] Moons P, Michiels CW, Aertsen A. Bacterial interactions in biofilms. Crit Rev Microbiol 2009;35(3):157–168.
[7] Harris LG, Richards RG. Staphylococci and implant surfaces: a review. Injury 2006;37 Suppl 2: S3-S14.
[8] Anderl JN, Zahller J, Roe F, Stewart PS. Role of nutrient limitation and stationary-phase existence in Klebsiella pneumoniae biofilm resistance to ampicillin and ciprofloxacin. Antimicrob Agents Chemother 2003;47(4):1251–1256.
[9] Velard F, Laurent-Maquin D, Guillaume C, *et al.* Polymorphonuclear neutrophil response to hydroxyapatite particles, implication in acute inflammatory reaction. Acta Biomater 2009;5(5):1708–1715.
[10] Schmidt AH, Swiontkowski MF. Pathophysiology of infections after internal fixation of fractures. J Am Acad Orthop Surg 2000;8(5):285–291.
[11] Papakostidis C, Kanakaris NK, Pretel J, *et al.* Prevalence of complications of open tibial shaft fractures stratified as per the Gustilo-Anderson classification. Injury 2011;42(12): 1408–1415.
[12] Dellinger EP, Miller SD, Wertz MJ, *et al.* Risk of infection after open fracture of the arm or leg. Arch Surg 1988;123(11):1320–1327.
[13] Pollak AN, Jones AL, Castillo RC, *et al.* The relationship between time to surgical debridement and incidence of infection after open high-energy lower extremity trauma. J Bone Joint Surg Am 2010;92(1):7–15.
[14] Bowen TR, Widmaier JC. Host classification predicts infection after open fracture. Clin Orthop Relat Res 2005;(433):205–211.
[15] Dunkel N, Pittet D, Tovmirzaeva L, *et al.* Short duration of antibiotic prophylaxis in open fractures does not enhance risk of subsequent infection. Bone Joint J 2013;95-B(6): 831–837.
[16] Gustilo RB, Merkow RL, Templeman D. The management of open fractures. J Bone Joint Surg Am 1990;72(2):299–304.
[17] Castillo RC, Bosse MJ, MacKenzie EJ, Patterson BM. Impact of smoking on fracture healing and risk of complications in limb-threatening open tibia fractures. J Orthop Trauma 2005; 19(3):151–157.

[18] McPherson EJ, Woodson C, Holtom P, *et al.* Periprosthetic total hip infection: outcomes using a staging system. Clin Orthop Relat Res 2002;403:8–15.

[19] Lalani T, Chu VH, Grussemeyer CA, *et al.* Clinical outcomes and costs among patients with Staphylococcus aureus bacteremia and orthopedic device infections. Scand J Infect Dis 2008; 40(11–12):973–977.

[20] Murdoch DR, Roberts SA, Fowler Jr VG, *et al.* Infection of orthopedic prostheses after Staphylococcus aureus bacteremia. Clin Infect Dis 2001;32(4):647–649.

[21] Willenegger H, Roth B. Treatment tactics and late results in early infection following osteosynthesis. Unfallchirurgie 1986;12(5):241–246.

[22] Trampuz A, Zimmerli W. Diagnosis and treatment of infections associated with fracture-fixation devices. Injury 2006;37(Suppl 2):S59–S66.

[23] Lipsky BA, Weigelt JA, Gupta V, *et al.* Skin, soft tissue, bone, and joint infections in hospitalized patients: epidemiology and microbiological, clinical, and economic outcomes. Infect Control Hosp Epidemiol 2007;28(11):1290–1298.

[24] Sheehy SH, Atkins BA, Bejon P, *et al.* The microbiology of chronic osteomyelitis: prevalence of resistance to common empirical anti-microbial regimens. J Infect 2010;60(5): 338–343.

[25] Suy F, Carricajo A, Grattard F, *et al.* Infection due to Mycobacterium thermoresistibile: a case associated with an orthopedic device. J Clin Microbiol 2013;51(9):3154–3156.

[26] Spinner RJ, Sexton DJ, Goldner RD, Levin LS. Periprosthetic infections due to Mycobacterium tuberculosis in patients with no prior history of tuberculosis. J Arthroplasty 1996;11(2):217–222.

[27] Arciola CR, An YH, Campoccia D, *et al.* Etiology of implant orthopedic infections: a survey on 1027 clinical isolates. Int J Artif Organs 2005;28(11):1091–1100.

[28] Athwal GS, Sperling JW, Rispoli DM, Cofield RH. Acute deep infection after surgical fixation of proximal humeral fractures. J Shoulder Elbow Surg 2007;16(4):408–412.

[29] Carsenti-Etesse H, Doyon F, Desplaces N, *et al.* Epidemiology of bacterial infection during management of open leg fractures. Eur J Clin Microbiol Infect Dis 1999;18(5):315–323.

[30] Gomez-Garces JL, Saez D, Almagro M, *et al.* Osteomyelitis associated to CTX-M-15-producing Aeromonas hydrophila: first description in the literature. Diagn Microbiol Infect Dis 2011;70(3):420–422.

[31] Larsen LH, Lange J, Xu Y, Schonheyder HC. Optimizing culture methods for diagnosis of prosthetic joint infections: a summary of modifications and improvements reported since 1995. J Med Microbiol 2012;61(Pt 3):309–316.

[32] Trampuz A, Piper KE, Jacobson MJ, *et al.* Sonication of removed hip and knee prostheses for diagnosis of infection. N Engl J Med 2007;357(7):654–663.

[33] Puig-Verdie L, Alentorn-Geli E, Gonzalez-Cuevas A, *et al.* Implant sonication increases the diagnostic accuracy of infection in patients with delayed, but not early, orthopaedic implant failure. Bone Joint J 2013;95-B(2):244–249.

[34] Esteban J, Sandoval E, Cordero-Ampuero J, *et al.* Sonication of intramedullary nails: clinically-related infection and contamination. Open Orthop J 2012;6:255–260.

[35] Worlock P, Slack R, Harvey L, Mawhinney R. The prevention of infection in open fractures: an experimental study of the effect of fracture stability. Injury 1994;25(1):31–38.

[36] Zimmerli W, Widmer AF, Blatter M, *et al.* Role of rifampin for treatment of orthopedic implant-related staphylococcal infections: a randomized controlled trial. Foreign-Body Infection (FBI) Study Group. JAMA 1998;279(19):1537–1541.

[37] Ziran BH, Rao N, Hall RA. A dedicated team approach enhances outcomes of osteomyelitis treatment. Clin Orthop Relat Res 2003;(414):31–36.

[38] Levin LS. The reconstructive ladder. An orthoplastic approach. Orthop Clin North Am 1993;24(3):393–409.

[39] Small JO, Mollan RA. Management of the soft tissues in open tibial fractures. Br J Plast Surg 1992;45(8):571–577.

[40] Mathes SJ, Alpert BS, Chang N. Use of the muscle flap in chronic osteomyelitis: experimental and clinical correlation. Plast Reconstr Surg 1982;69(5):815–829.
[41] Hollenbeck ST, Woo S, Ong S, *et al.* The combined use of the Ilizarov method and microsurgical techniques for limb salvage. Ann Plast Surg 2009;62(5):486–491.
[42] Hou Z, Irgit K, Strohecker KA, *et al.* Delayed flap reconstruction with vacuum-assisted closure management of the open IIIB tibial fracture. J Trauma 2011;71(6):1705–1708.
[43] Diefenbeck M, Mennenga U, Guckel P, *et al.* Vacuum-assisted closure therapy for the treatment of acute postoperative osteomyelitis. Z Orthop Unfall 2011;149(3):336–341.
[44] Klemm KW. Treatment of infected pseudarthrosis of the femur and tibia with an interlocking nail. Clin Orthop Relat Res 1986;(212):174–181.
[45] Prasarn ML, Ahn J, Achor T, *et al.* Management of infected femoral nonunions with a single-staged protocol utilizing internal fixation. Injury 2009;40(11):1220–1225.
[46] Zalavras CG, Sirkin M. Treatment of long bone intramedullary infection using the RIA for removal of infected tissue: indications, method and clinical results. Injury 2010;41(Suppl 2):S43–S47.
[47] Thonse R, Conway JD. Antibiotic cement-coated nails for the treatment of infected nonunions and segmental bone defects. J Bone Joint Surg Am 2008;90(Suppl 4):163–174.
[48] Rozbruch RS, Weitzman AM, Watson TJ, *et al.* Simultaneous treatment of tibial bone and soft-tissue defects with the Ilizarov method. J Orthop Trauma 2006;20(3):197–205.
[49] Cierny G, 3rd, DiPasquale D. Treatment of chronic infection. J Am Acad Orthop Surg 2006;14(10 Spec No.):S105–S110.
[50] Papakostidis C, Bhandari M, Giannoudis PV. Distraction osteogenesis in the treatment of long bone defects of the lower limbs: effectiveness, complications and clinical results; a systematic review and meta-analysis. Bone Joint J 2013;95-B(12):1673–1680.
[51] Oh CW, Apivatthakakul T, Oh JK, *et al.* Bone transport with an external fixator and a locking plate for segmental tibial defects. Bone Joint J 2013;95-B(12):1667–1672.
[52] Yajima H, Tamai S, Mizumoto S, Inada Y. Vascularized fibular grafts in the treatment of osteomyelitis and infected nonunion. Clin Orthop Relat Res 1993;293:256–264.
[53] Levin LS. Vascularized fibula graft for the traumatically induced long-bone defect. J Am Acad Orthop Surg 2006;14(10 Spec No.):S175–S176.
[54] Farhadi J, Valderrabano V, Kunz C, *et al.* Free fibula donor-site morbidity: clinical and biomechanical analysis. Ann Plast Surg 2007;58(4):405–410.
[55] Masquelet AC, Begue T. The concept of induced membrane for reconstruction of long bone defects. Orthop Clin North Am 2010;41(1):27–37.
[56] Kugan R, Aslam N, Bose D, McNally MA. Outcome of arthrodesis of the hindfoot as a salvage procedure for complex ankle pathology using the Ilizarov technique. Bone Joint J 2013;95-B(3):371–377.
[57] Sendi P, Zimmerli W. Antimicrobial treatment concepts for orthopaedic device-related infection. Clin Microbiol Infect 2012;18(12):1176–1184.
[58] Sendi P, Graber P, Zimmerli W. Risk factors associated with acute hip prosthetic joint infections and outcome of treatment with a rifampin-based regimen. Acta Orthop 2008;79(3):454.
[59] Stewart PS, Costerton JW. Antibiotic resistance of bacteria in biofilms. Lancet 2001; 358(9276):135–138.
[60] Sendi P, Zimmerli W. Challenges in periprosthetic knee-joint infection. Int J Artif Organs 2011;34(9):947–956.
[61] Kanellakopoulou K, Giamarellos-Bourboulis EJ. Carrier systems for the local delivery of antibiotics in bone infections. Drugs 2000;59(6):1223–1232.
[62] McKee MD, Wild LM, Schemitsch EH, Waddell JP. The use of an antibiotic-impregnated, osteoconductive, bioabsorbable bone substitute in the treatment of infected long bone defects: early results of a prospective trial. J Orthop Trauma 2002;16(9):622–627.

[63] McKee MD, Li-Bland EA, Wild LM, Schemitsch EH. A prospective, randomized clinical trial comparing an antibiotic-impregnated bioabsorbable bone substitute with standard antibiotic-impregnated cement beads in the treatment of chronic osteomyelitis and infected nonunion. J Orthop Trauma 2010;24(8):483–490.

[64] Ferguson JY, Riley N, Dudareva M, *et al.* The use of a biodegradeable antibiotic-loaded calcium sulphate carrier containing tobramycin for the treatment of osteomyelitis: a series of 198 cases. Bone Joint J 2014; 96B:829–836.

[65] Thomas DB, Brooks DE, Bice TG, *et al.* Tobramycin-impregnated calcium sulfate prevents infection in contaminated wounds. Clin Orthop Relat Res 2005;(441):366–371.

[66] Urban RM, Turner TM, Hall DJ, *et al.* Increased bone formation using calcium sulfate-calcium phosphate composite graft. Clin Orthop Relat Res 2007;(459):110–117.

Chapter 21
Implant-Associated Vertebral Osteomyelitis

Todd J. Kowalski

Introduction

Rates of spinal fusion surgery with instrumentation are rapidly increasing. Indications for spinal fusion surgery include scoliosis and fracture, but the majority of surgeries are performed for expanding indications such as spinal stenosis, spinal degeneration, and disk disorders [1]. In the United States, there were 492,000 hospital stays for spinal fusion surgery in 2010, a 115% increase from 1997 [2]. From 1997 to 2009, the average annual cost of spinal fusion procedures increased more than that of any other procedure in the United States [3].

Deep surgical site infection (SSI) after spinal fusion with implants, hereafter referred to as implant-associated vertebral osteomyelitis (IAVO), is one of the most common and serious complications of spinal fusion surgery. In addition to the substantial morbidity associated with IAVO, this complication substantially increases the cost of care [4]. IAVO is a particularly challenging bone and joint infection to manage because the need to maintain spinal stability must be balanced with the need to treat the infection effectively. In the early postoperative period, implant removal can lead to spinal instability, placing the patient at risk of neurologic injury.

As with other implant-associated infections, the key virulence feature in IAVO is the presence of a biofilm. Biofilms are microbial cells in a polysaccharide matrix that, in the case of IAVO, adhere to the spinal implants. Within this milieu of extracellular polymeric substances, bacteria may enter a stationary phase, rendering them resistant to antibiotic therapy [5]. Microorganisms commonly associated with medically relevant biofilms include *Staphylococcus aureus*, coagulase-negative staphylococci, *Enterococcus* spp., *Pseudomonas aeruginosa*, and *Propionibacterium acnes*, all of which are common causes of IAVO [5, 6].

Diagnosis and treatment strategies for IAVO are guided by the time from instrumentation to the onset of infection, categorized as early, delayed, or late onset. Most published

Bone and Joint Infections: From Microbiology to Diagnostics and Treatment, First Edition.
Edited by Werner Zimmerli.

data and clinical experience are with early-onset infection, albeit variably defined in the literature as ≤30 days, <3 months, <6 months, or <12 months from instrumentation. The incidence and recognition of late-onset infection, however, is increasing, perhaps due to increased prevalence of *in situ* implants and enhanced awareness and culturing methods. Throughout this chapter, IAVO onset from instrumentation is defined as early (≤30 days), delayed (1–12 months), and late (>12 months).

Epidemiology, Risk Factors, and Preventive Measures

IAVO occurs in 0.22–20% of patients who undergo spinal fusion surgery, with higher infection rates in patients with neuromuscular scoliosis or more extensive surgical procedures [7, 8]. Two-thirds of infections are identified within the first 30 days [9]. Most early-onset IAVO are acquired intraoperatively, although a minority of infections may be acquired at any time from a persistently draining or dehisced postoperative wound. Relatively virulent pathogens, such as *S. aureus*, beta-hemolytic streptococci, or Gram-negative rods, cause local or constitutional symptoms that lead to the diagnosis.

Late-onset IAVO is estimated to occur in 0.2–8.3% of patients who undergo spinal fusion surgery [10–12]. The true prevalence of late-onset IAVO is difficult to estimate because it depends upon how it is defined and the intensity with which a microbiologic diagnosis is sought in patients with late complications such as pseudoarthrosis or pain. As in early-onset IAVO, most cases of late-onset IAVO are acquired intraoperatively. Occasionally, presentation of virulent pathogens, such as *S. aureus*, is delayed, but the microorganisms that cause late-onset IAVO tend to be less virulent, such as coagulase-negative staphylococci, *Corynebacterium* spp., and *P. acnes*. Although common in prosthetic joint infections, in our experience, hematogenous seeding is exceedingly rare in spinal implants.

Data are inconsistent regarding risk factors for SSI following spine surgery [13]. Diabetes mellitus and obesity, typically defined as a body mass index (BMI, calculated as weight in kilograms divided by height in meters squared) ≥30, are the most consistently sited risk factors for spine SSI [13]. The risk of SSI, however, may be more closely associated with the distribution of adipose tissue than with BMI [14]. Many other factors have inconsistently been reported to increase the risk of SSI after spine surgery, including a surgical indication of trauma, older age, malnutrition, American Society of Anesthesiologists (ASA) score of 2 or 3, smoking, alcohol abuse, posterior surgical approach, blood transfusion, perioperative hyperglycemia, and duration of surgery [15–18]. In one large study malignancy, weight loss, anemia, and smoking were reported to increase the risk of infection following spinal fusion [19]. In another report, BMI greater than 35, hypertension, and renal disease were three host factors independently associated with infection [20]. Patients with neuromuscular scoliosis who have either cerebral palsy or myelomeningocele may have a particularly high rate of infection [21, 22]. A spine surgical invasiveness index correlates with infection risk [20]. Other procedure-related factors associated with infection include multiple-level surgery and staged procedures, such as anterior/posterior procedures on separate days [23]. Notably, the use of allograft has not been consistently shown to be a risk factor for spinal implant infection [15]. In a rabbit model, steel implants have been shown to have higher infection rates than titanium implants [24]. Increasingly novel spinal implants are used in fusion procedures utilizing minimally invasive spine surgery (MISS) techniques, that is, doing spinal procedures through a tubular retractor-type system. Infection rates are reported to be substantially lower in MISS than in open procedures [25]. More rigorous and standardized trials are needed to more completely and accurately determine modifiable risk factors for IAVO.

A number of interventions have been used to reduce the rate of SSI after spine surgery. Appropriately timed administration of preoperative antibiotics and intraoperative redosing of antibiotics are recommended to reduce SSI rates in spine surgery [7, 26]. A first-generation cephalosporin such as cefazolin is suitable to use for most patients. In patients with a beta-lactam allergy, vancomycin and clindamycin are alternatives. Some advocate for enhanced coverage of Gram-negative rods in fusion with instrumentation, which is advisable in pediatric neuromuscular scoliosis patients, especially if the fusion extends into the lumbosacral spine. Antibiotic duration should be less than 24 h, and there is no benefit to extending duration based upon the presence of a drain [7, 27, 28]. Intraoperative wound irrigation with 0.35% povidone–iodine solution has been shown to reduce surgical infection rates in two randomized trials [29, 30] and has also been found beneficial in joint arthroplasty SSI prevention [31]. One retrospective study demonstrated that using greater than 2 l/h of irrigation on average during spine surgery was associated with lower rates of SSI. Based upon this finding, the investigators recommend 500 ml of irrigation every 15 min during extensive spine surgery [32]. Retrospective reports have suggested that topical application of vancomycin powder intraoperatively prior to closure of the incision reduces SSI after spine surgery, though this result should be confirmed in well-designed trials prior to widespread adoption [8, 33]. Placement of gentamicin-impregnated poly(lactic-co-glycolic acid) microspheres in the surgical bed reduced IAVO in a rabbit model [34].

Closed-suction drains are frequently used after spinal fusion to reduce the incidence of hematoma and compression of the cauda equina. Drains are considered to be a potentially helpful intervention to prevent infection, but in randomized trials, closed-suction drains have not been shown to reduce SSI [35, 36]. In fact, prolonged use of closed-suction drains postoperatively was associated with infection in one study [37], and removal of drains as soon as possible has been recommended [38].

Risk factors for late-onset IAVO are not well characterized. Ho *et al.* [39] examined risk factors for delayed-onset infection, which included patients with infections greater than 6 months from the primary surgery. Not using a drain at the time of surgery was associated with a higher infection rate, and when drains were used, increased drainage volume was a risk factor for late infection. Having the most distal fusion level in the lumbar spine rather than in the thoracic spine was associated with a greater risk of infection [39]. Hematogenous seeding of implants has been implicated as the etiology of infection [12, 40]; however, the microbiology and clinical circumstances described in most reports purporting hematogenous seeding are instead far more suggestive of intraoperative seeding.

Microbiology

The microbiology of IAVO varies depending on the patient population, time from implantation to onset of infection, and level of fusion. Tables 21.1 and 21.2 show the microbiology of early-onset infection from selected series in which all infections were identified within 30 days of implantation. In six of the nine series [23, 41–45], *S. aureus* was the most common isolate (Table 21.1). In the three studies [21, 46, 47] in which *S. aureus* was not the most common isolate (Table 21.2), the study populations

Table 21.1. Microorganisms isolated from early-onset (<30 days postimplantation) implant-associated vertebral osteomyelitis (IAVO) patients from cohorts composed primarily of adult patients without neuromuscular scoliosis.

	Published series							
Microorganism	Sierra-Hoffman *et al.* [41]	Mirovsky *et al.* [42]	Kowalski *et al.* [43]	Glassman *et al.* [23]	Dubee *et al.* [44]	Stambough *et al.* [45]	Subtotal *n* (%)	Early-onset summary (Tables 21.1 and 21.2)[a]
S. aureus	17	4	13	9	27	9	79 (41)	87 (26)
Coagulase-negative staphylococcus	0	0	6	3	6	5	20 (10)	26 (14)
Streptococcus spp.	0	1	5	0	1	0	7 (4)	11 (3)
Enterococcus spp.	1	1	0	3	5	0	10 (5)	24 (7)
Escherichia coli	1	0	0	3	7	0	11 (6)	28 (8)
Proteus spp.	0	0	0	1	7	0	8 (4)	23 (7)
Klebsiella spp.	1	0	0	1	1	0	3 (2)	9 (3)
Serratia spp.	0	0	0	0	0	0	0 (0)	4 (1)
Acinetobacter spp.	0	1	0	0	0	0	1 (0)	5 (2)
Enterobacter spp.	0	0	0	3	8	2	13 (7)	18 (5)
Pseudomonas spp.	1	2	0	1	5	2	11 (6)	18 (5)
Gram-negative rods not otherwise specified	0	0	10	0	0	4	14 (7)	16 (5)
Propionibacterium spp.	1	1	2	0	2	0	6 (3)	11 (3)
Peptostreptococcus spp.	0	0	2	3	0	0	5 (3)	16 (5)
Bacteroides spp.	1	0	0	1	1	0	3 (1)	21 (6)
Others[b]	0	0	1	0	2	0	3 (1)	14 (4)

IAVO, implant-associated vertebral osteomyelitis.

[a]Percentages will not total 100% due to rounding.

[b]Other microorganisms isolated in the early-onset adult non-neuroscoliosis cohorts were *Corynebacterium* spp. (2) and *Morganella morganii* (1).

Table 21.2. Microorganisms isolated from early-onset (<30 days postimplantation) implant-associated vertebral osteomyelitis (IAVO) patients from cohorts composed primarily of pediatric or neuromuscular scoliosis patients.

	Published series				
Microorganism	Brook *et al.* [46]	Sponseller *et al.* [21]	Brook *et al.* [47]	Subtotal *n* (%)[a]	Early-onset summary (Tables 21.1 and 21.2)[a]
S. aureus	2	3	3	8 (6)	87 (26)
Coagulase-negative staphylococcus	0	6	0	6 (4)	26 (14)
Streptococcus spp.	1	1	2	4 (3)	11 (3)
Enterococcus spp.	3	7	4	14 (10)	24 (7)
Escherichia coli	6	3	8	17 (12)	28 (8)
Proteus spp.	5	3	7	15 (11)	23 (7)
Klebsiella spp.	3	0	3	6 (4)	9 (3)
Serratia spp.	1	1	2	4 (3)	4 (1)
Acinetobacter spp.	0	2	2	4 (3)	5 (2)
Enterobacter spp.	1	3	1	5 (4)	18 (5)
Pseudomonas spp.	2	0	5	7 (5)	18 (5)
Gram-negative rods not otherwise specified	0	2	0	2 (1)	16 (5)
Propionibacterium spp.	2	0	3	5 (4)	11 (3)
Peptostreptococcus spp.	5	0	6	11 (8)	16 (5)
Bacteroides spp.	9	0	9	18 (13)	21 (6)
Others[b]	3	1	7	11 (8)	14 (4)

IAVO, implant-associated vertebral osteomyelitis.
[a]Percentages will not total 100% due to rounding.
[b]Other microorganisms isolated in the early-onset neuroscoliosis cohorts were *Clostridium* spp. (3), *Veillonella* spp. (3), *Corynebacterium* spp. (3), *Providencia* spp. (1), and *Micrococcus* spp. (1).

consisted predominantly of pediatric patients with neuromuscular disorders such as cerebral palsy, muscular dystrophy, myelomeningocele, or spinal cord injury from trauma. The microbiology in these three series was very different from that of the other six, with a small number of infections due to *S. aureus*. Enteric Gram-negative rods such as *Escherichia coli*, *Proteus* spp., and *Klebsiella* spp., as well as the Gram-positive *Enterococcus* spp., were commonly isolated, as were anaerobes such as *Bacteroides* species and *Peptostreptococcus*. These findings may reflect higher rates of enteric skin colonization and wound contamination in patients with fecal and urinary incontinence, limited soft tissue availability for wound coverage, or lack of sensation that may predispose to peri-incisional pressure ulceration of skin [21, 22]. Gram-negative rods and anaerobic bacteria are more commonly isolated in patients who undergo lumbar or lumbosacral procedures than in those who have procedures at a higher vertebral level [48].

The microbiology found in series of late-onset implant-associated infections [10, 11, 40, 49–52], defined as symptoms and diagnosis onset greater than 12 months from

Table 21.3. Microorganisms isolated from patients with late-onset (> 1 year postimplantation) implant-associated vertebral osteomyelitis (IAVO).

Microorganism	Published series							Total *n* (%)[a]
	Viola *et al.* [10]	Heggenes *et al.* [40]	Richard *et al.* [49]	Clark *et al.* [52]	Richards *et al.* [11]	Hahn *et al.* [50]	DiSilvestre *et al.* [51]	
S. aureus	0	3	1	1	0	0	4	9 (12)
Coagulase-negative staphylococcus	6	1	4	6	3	0	8	28 (38)
Streptococcus spp.	0	1	0	0	0	0	0	1 (1)
Enterococcus spp.	0	0	0	2	0	0	0	2 (3)
Serratia spp.	0	0	0	0	0	0	1	1 (1)
Propionibacterium spp.	1	1	12	3	5	6	2	30 (41)
Others[b]	0	0	1	0	1	0	0	2 (3)

IAVO, implant-associated vertebral osteomyelitis.

[a] Percentages will not total 100% due to rounding.

[b] The other microorganism isolated from the late-onset IAVO cohorts was *Micrococcus* spp. (2).

the index procedure, is markedly different from that of early-onset infections (Table 21.3). *P. acnes* and coagulase-negative staphylococci are the most commonly isolated pathogens in late-onset infections. Both are biofilm-producing organisms [5, 6]. *P. acnes* demonstrate fastidious growth, and prolonged incubation times enhance their culture yield. When standardized tissue sampling and optimized culture techniques were used, *P. acnes* was the most common etiology of late infections and also was isolated in 50% of early-onset infections [53]. Therefore, *P. acnes* may be a more frequent etiologic agent of IAVO than is recognized. Gram-negative bacteria are rare causes of late-onset infection.

The microbiology of delayed-onset infection is insufficiently described in the literature but shares features of that of both early- and late-onset infections. The microbiology of infections with onset in the first 6 months closely mirrors that of early-onset infections, while onset after 6 months more closely resembles that of late-onset infections.

Clinical Features

Patients with early-onset IAVO most commonly seek care between 1 and 3 weeks postoperatively [21, 41, 43, 44]. Neck, back, or hip pain is the most common symptom and is present in approximately 80% of patients [41, 54]. Wound drainage, which can be serous, cloudy, or purulent, is present in 73–100% of patients [41, 45, 54]. On examination, fever is present in most patients, although a substantial minority may not have fever. The surgical wound often demonstrates erythema, induration, dehiscence, or fluctuation. It can be difficult to determine the depth of infection in patients with early-onset SSI after spinal fusion [54]. Surgical exploration with sampling biopsies for culture and histology is the only reliable method of determining depth of infection and should be promptly performed when infection is suspected. Mistaking infection that involves implants for a superficial infection inevitably leads to delayed infectious complications.

Diagnosis of early-onset IAVO in the absence of either constitutional symptoms or wound breakdown is challenging and requires a high index of suspicion. There may be deep infection below the fascia, and progressive, often severe, pain may be the patient's only symptom. Patients may report a transient period of improvement in the pain or radicular symptoms that prompted the spinal fusion procedure, followed by recurrent and increasing pain one to 4 weeks postoperatively.

The clinical presentation of late-onset IAVO is often subtle and differs substantially from that of early-onset IAVO. Most patients have mild, poorly localized pain [11, 50]. Fever is usually absent [11, 50, 55]. The most common findings on examination are localized swelling over the spine or the development of a sinus tract, signs that may be present in up to 70% of patients [11, 50, 55].

The clinical features of delayed-onset IAVO are less well characterized in the literature. Clinical presentation may be variable, depending on the host, the etiologic pathogen, and the time since surgery. Patients may have progressive back pain and either wound drainage or a sinus tract, evidence of a deep fluid collection on imaging, or fever. A high index of suspicion for infection should be maintained in any patient with spinal implants and persistent or progressive neck or back pain.

Diagnostic Procedures

Early-Onset IAVO

The most important diagnostic procedure, when fever, wound drainage, and pain suggest early-onset IAVO, is surgical exploration to determine the depth and extent of involvement. Appropriate specimens may be acquired and sent for culture at this time, as well. Without operative exploration, the depth of infection is often difficult to determine [54]. In cases of periprosthetic joint infection, it is recommended that at least three and optimally 5 or 6 tissue samples be acquired to maximize the chance of obtaining a microbiologic diagnosis [56]; a similar approach is warranted in IAVO because there are often multiple levels and multiple implants that may be involved. Tissue samples are strongly preferred to wound swabs because their microbiologic yield is higher.

Tissue samples should be cultured aerobically and anaerobically. Prolonged culture duration up to 14 days may enhance yield, particularly for *Propionibacterium* spp. [57, 58]. Sonication of orthopedic implants has been shown to enhance the ability to make a microbiologic diagnosis in patients who have received antibiotics preoperatively [59]. A protocol that employed vortexing and sonication of spinal implants followed by culture of the sonicate fluid was more sensitive than peri-implant tissue cultures for the microbiologic diagnosis of IAVO [60]. Polymerase chain reaction (PCR)-based molecular techniques continue to be developed but are not yet widely available. 16S rRNA PCR has been shown to have better sensitivity than traditional culture-based methods in patients undergoing spinal biopsy for culture but at this time is still considered adjunctive [61].

Laboratory tests are adjunctive in establishing the diagnosis of early-onset IAVO. Leukocyte counts, erythrocyte sedimentation rates (ESR), and C-reactive protein (CRP) concentrations are usually elevated in early-onset IAVO, but these tests are nonspecific [39, 43, 62]. In patients who have undergone spinal fusion, postoperative CRP concentrations peak approximately 2–3 days after surgery and typically peak higher as complexity of the surgery increases. CRP values slowly return to normal over 14–21 days in patients without infection [63]. A rising CRP beyond postoperative day 8 is associated with infection [63]. A second rise in the CRP, or failure to decrease as expected, was found to be 82% sensitive but only 48% specific for infectious complications [64]. ESR results after spine surgery do not accurately discriminate between benign postoperative elevation and infectious complications [63, 64]. Blood cultures should be obtained prior to antibiotic administration and have been shown to be positive for growth 26–43% of the time [43, 44].

Imaging is of limited value in early-onset IAVO, because the diagnosis of infection can often be confirmed clinically. Plain radiographs establish the appropriate position of the implants but may take up to 8 weeks to detect any evidence of deep infection [65]. Typically no changes suggestive of osteomyelitis are noted in early-onset IAVO. Ultrasonography may be used to assess for fluid collections that may be amenable to aspiration in patients with swelling. If signs or symptoms that suggest deep infection exist but clear indications for surgical exploration are absent, magnetic resonance imaging (MRI) is preferred over other imaging modalities, owing to its superior soft tissue resolution [66]. MRI can reveal fluid collections with enhancement suggestive of abscess, evidence of collapsed disk space, endplate destruction osteomyelitis, or a peri-implant soft tissue mass [67, 68]; however, it can sometimes be difficult to distinguish inflammatory changes caused by infection from postoperative inflammation [69]. Computed tomography (CT) scans may be limited by beam artifact, which causes difficulty in interpreting soft tissue findings [66]. ^{18}F-fluoro-D-deoxyglucose positron emission tomography (PET)/CT scans show great promise as a highly sensitive diagnostic test for spine infection although specificity is reduced when spinal implants are present [70].

Delayed- and Late-Onset IAVO

Delayed- and late-onset IAVO can be challenging to diagnose, but laboratory tests and imaging may be helpful. Leukocyte counts are normal in most patients [10, 51]. Infection, primarily with *P. acnes*, was unexpectedly found at the time of delayed spinal implant removal in 25% of patients in a series of 74 patients [53]. In this series, 17% of CRP results, 45% of ESR results, and 97% of white blood cell (WBC) counts were normal despite the microbiologically and histopathologically confirmed diagnosis of infection. However, in another study of 54 patients undergoing spinal implant removal, preoperative CRP greater than 46 mg/l had a sensitivity of 79% and specificity of 68% to detect IAVO [71]. Blood cultures are not often obtained in patients with late infection given the lack of fever and constitutional symptoms.

Plain radiographs may demonstrate pseudoarthrosis or nonunion, implant loosening, or implant failure, all of which may be associated with late-onset IAVO [68]. These findings suggest infection, but are not specific enough to establish the diagnosis. Currently, MRI provides the best assessment for the presence of infection, but interference from implants may limit resolution [68]. CT is more sensitive than plain radiography for bone abnormalities and assessment of arthrodesis. CT-guided aspiration of spine tissue suspicious for infection can establish a microbiologic diagnosis. This procedure is most commonly employed in patients with native spine pyogenic vertebral osteomyelitis, but in IAVO patients without a sinus tract, it may occasionally be helpful in establishing a diagnosis and treatment plan.

Although WBC-labeled imaging is useful in diagnosis of other orthopedic implant-associated infections, it is a poor test in the spine. Bone scintigraphy with gallium imaging is both sensitive and specific for the diagnosis of pyogenic vertebral osteomyelitis in the absence of recent surgery. However, data on its accuracy in the postoperative setting are limited, and the testing protocol is cumbersome and takes days to complete [65]. Gallium has been reported to accumulate in normally healing wound beds months after surgery. PET has shown promise as a highly sensitive modality for the diagnosis of IAVO, although false-positive results can occur with fluorodeoxyglucose (FDG) uptake in patients with implants [69]. Specificity was only 65% among patients with spinal implants. The high sensitivity, lack of interference from the implants, and rapid results are positive features of PET, but additional clinical validation of its accuracy is needed prior to widespread utilization in diagnosing IAVO.

Management

Management of IAVO is dependent upon host factors, the onset of infection, the microbiologic etiology, the fixation of the spinal implants, and the stability of the spine at the time of diagnosis. Figure 21.1 depicts an overview of the management of IAVO.

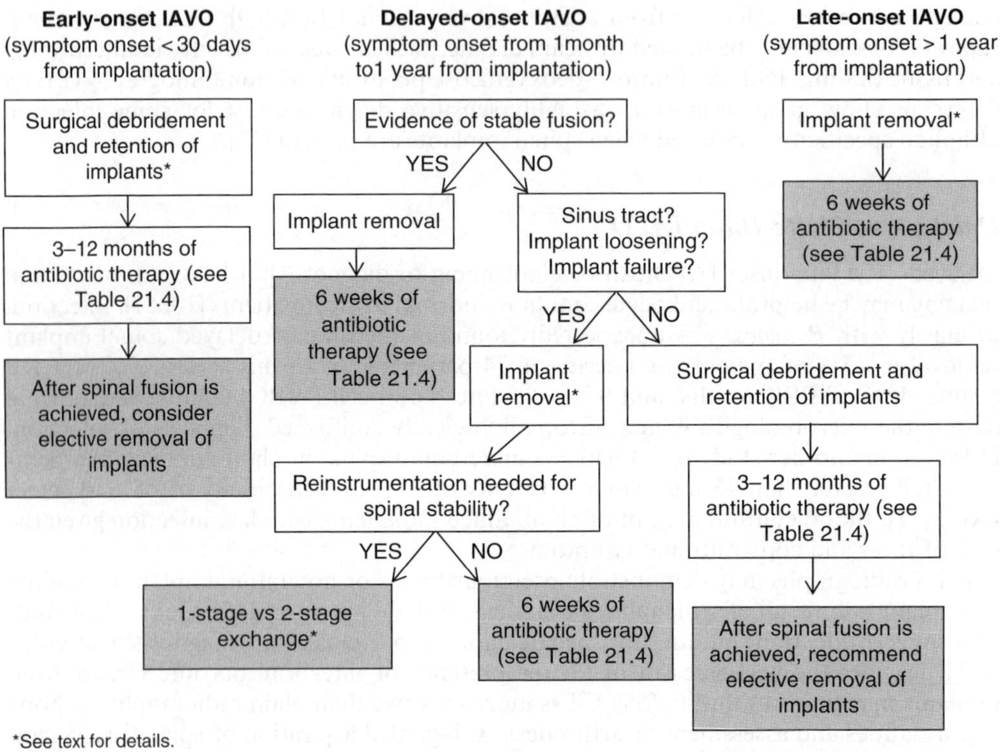

Figure 21.1. Algorithm for the management of implant-associated vertebral osteomyelitis (IAVO).

Early-Onset IAVO

Early-onset IAVO occurs within 30 days of surgery, and symptoms of infection are typically present for only a brief time prior to diagnosis [43, 44]. When early-onset IAVO is suspected, an aggressive treatment approach is necessary, the foundation of which is early and thorough surgical debridement of all infected, purulent, and nonviable necrotic materials. Except in the rare circumstance, in which only a superficial cellulitis is present, wounds should be explored both superficial and deep to the fascia, and multiple samples of tissue suspicious for infection should be obtained for culture. Well-fixed and functional spinal implants should be retained in patients with early-onset IAVO diagnosed within 1 month of implant placement [23, 43–45, 64, 72, 73]. Some advocate removal of bone graft; others leave well-fixed bone graft *in situ* and debride any loose particles [23, 44]. In most instances, the wound may be packed open or a negative pressure wound dressing applied, unless complete control of the infection with healthy tissues is achieved during the first operation [73, 74]. A number of authors have advocated for either closed-suction irrigation systems or the placement of antibiotic-impregnated beads into the wound [74, 75]. Repeat incision and drainage approximately every 48–72 h is warranted until the wound appears healthy and ready for delayed primary closure.

Although rare, removal of spinal implants is sometimes necessary for early-onset IAVO. Indications for removal are inability to control sepsis with debridement and antibiotic therapy and loose and poorly fixed implants. In such cases, implants may need to be replaced in order to maintain correction and spinal integrity, either in a single-stage exchange or after a short period of treatment with intravenous (IV) antibiotics. Collapse of disk space, loss of lordosis, and spine nonfusion occurred in all 14 patients who had implants removed for early-onset IAVO, highlighting the long-term sequelae that can occur with explantation in early-onset infection [76]. Implant removal for infection within the first year of placement was a risk factor for spinal deformity progression [22]. Occasionally, patients, particularly pediatric patients with neuromuscular conditions, will require musculocutaneous flap placement to facilitate implant coverage and wound closure.

Empiric IV antibiotics should be administered once intraoperative culture specimens have been obtained, or sooner if the patient has sepsis or there is concern for neurologic deficits related to an abscess. While awaiting culture results, empiric antibiotics should be directed at Gram-negative rods and *Staphylococcus* species, including methicillin-resistant *S. aureus* (MRSA) if the local prevalence of MRSA is high. Reasonable empiric regimens include vancomycin plus either ceftriaxone, ciprofloxacin, or a carbapenem, depending upon local antibiotic resistance patterns.

There are two primary approaches to antibiotic therapy duration in patients with early-onset IAVO: (1) suppressive antibiotic therapy until the spine is fused, typically 9–12 months postimplantation, and (2) prolonged defined duration treatment, typically for 3–4 months. In most reported series, pathogen-directed IV or highly bioavailable oral antibiotics are used for approximately 6 weeks after surgical debridement [13, 23, 41–43, 54, 64, 77], though shorter durations of IV antibiotics have also been employed with apparent success [45, 78]. Antibiotic recommendations by pathogen for initial therapy and for subsequent therapies, either prolonged defined duration or suppressive, are listed in Table 21.4. Antibiotic suppression is suggested for the treatment of early-onset IAVO [22, 43, 54, 64]. The goal of therapy is a stable, fused, pain-free spine and the prevention of chronic osteomyelitis. Antibiotic suppression of infection to allow time for spinal fusion to occur may help accomplish these goals. In one retrospective study, the use of oral suppressive antibiotics was associated with lower rates of treatment failure in early-onset IAVO [43]. Functional and clinical outcomes in infected patients treated with this approach were similar to those of patients who did not have infection [43, 54, 64].

Prolonged defined treatment duration has been proven effective in other orthopedic device infections [79–81]. For this approach, it is essential that the diagnosis of infection be made early and that rifampin-based dual antibiotic regimens be used [82] against staphylococci. Published data on the treatment of early-onset IAVO do not include cohorts or treatment protocols that routinely employed rifampin in the antibiotic program of staphylococcal infections, with the following notable exception. A cohort of 50 patients with early-onset IAVO in whom 33 patients had staphylococcal infection was reported in 2012 [44]. Patients were treated with IV antibiotics for 2 weeks and then, whenever possible, with oral antibiotics to complete 3 months of therapy total. Of the 33 patients with staphylococcal infection, 22 received rifampin as part of their antibiotic program. Outcomes were excellent, with only 6% of patients having relapse or reinfection after 2 years of follow-up.

Fusion of the spine typically occurs 6–9 months after surgery in patients without infection complications. Spinal fusion may be delayed in patients with early-onset IAVO [83]. Once spinal fusion is achieved, elective removal of the implants can be considered [53]. Collins *et al.* [53] reported that 9 of 15 (40%) patients with retained implants who were treated with debridement followed by suppressive antibiotics had evidence of infection

Table 21.4. Antibiotic treatment of common organisms causing implant-associated vertebral osteomyelitis (IAVO).

Microorganism	Initial parenteral/highly bioavailable antimicrobial therapy when implants retained or the definitive treatment course when implants are removed. Typical duration is 6 weeks[a,b]	Subsequent defined duration (3 mo) or suppressive (up to 1 y) oral antimicrobial therapy in early- or delayed-onset IAVO with retained implants[a,b]	Comments
MSSA	Nafcillin[c] 2 g IV q 4–6 h or Flucloxacillin[c] 2 g IV q 6 h or Cefazolin[c] 2 g IV q 8 h *plus* Rifampin 450 mg PO q 12 h if implants retained	Levofloxacin 750 mg PO q 24 h plus rifampin 450 mg PO q 12 h or Cefadroxil 1 g PO q 12 ± rifampin 450 mg PO q 12 h or Doxycycline/minocycline 100 mg PO q 12 h ± rifampin 450 mg PO q 12 h	If implants are retained and defined duration of 3 months is planned, rifampin combination should be used if isolate is sensitive
MRSA	Vancomycin 15 mg/kg IV q 12 h[d] or Daptomycin 6 mg/kg IV q 24 h PLUS Rifampin 450 mg PO q 12 h if implants retained	Levofloxacin 750 mg PO q 24 h plus rifampin 450 mg PO q 12 h or Doxycycline/minocycline 100 mg PO q 12 h ± rifampin 450 mg PO q 12 h or TMP/SX 160/800 mg PO q 12 h ± rifampin 450 mg PO q 12 h or Clindamycin 300 mg PO q 6 h ± rifampin 450 mg PO q 12 h	If implants are retained and defined duration of 3 months is planned, rifampin combination should be used if isolate is sensitive
Enterococcus spp., penicillin susceptible	Ampicillin sodium 2 g IV q 4 h or Penicillin G 20–24 million units IV over 24 h	Amoxicillin 1 g PO q 8 h	Aminoglycoside optional to use during initial 2–4 weeks of treatment
Enterococcus spp., penicillin resistant	Vancomycin 15 mg/kg IV q 12 h[d]	n/a	Aminoglycoside optional to use during initial 2–4 weeks of treatment

Streptococcus spp.	Penicillin G 20–24 million units IV over 24 h or Ceftriaxone 2 g IV q 24 h	Amoxicillin 1 g PO q 8 h	
Enterobacteriaceae	Ceftriaxone 2 g IV q 24 h or Ciprofloxacin 750 mg PO q 12 h	Ciprofloxacin 750 mg PO q 12 h	
P. aeruginosa	Cefepime 2 g IV q 12 h or Meropenem 1 g IV q 8 h or Ciprofloxacin 750 mg PO q 12 h	Ciprofloxacin 750 mg PO q 12 h	Consider double coverage with either ciprofloxacin or an aminoglycoside during initial 2–4 weeks of treatment
Enterobacter spp.	Cefepime 2 g IV q 12 h or Ertapenem 1 g IV q 24 h or Ciprofloxacin 750 mg PO q 12 h	Ciprofloxacin 750 mg PO q 12 h	
P. acnes	Penicillin G 20–24 million units IV over 24 h or Ceftriaxone 2 g IV q 24 h	Penicillin V 1000 mg PO q 8 h	
Gram-positive anaerobes	Penicillin G 20–24 million units IV over 24 h or Ceftriaxone 2 g IV q 24 h	Penicillin V 1000 mg PO q 8 h or Clindamycin 300 mg PO q 6 h	
Gram-negative anaerobes	Metronidazole 500 mg IV/PO q 8 h	Amoxicillin/clavulanate 875 mg PO q 12 h	Neuropathy may occur with prolonged metronidazole use

Adapted from Refs. [82] and [56].

[a] IAVO, implant-associated vertebral osteomyelitis; MSSA, methicillin-sensitive *S. aureus*; MRSA, methicillin-resistant *S. aureus*; mo, months; y, year; h, hours.

[b] For details of antibiotic treatment strategies and duration, see text.

[c] Doses for patients without normal renal and hepatic function may need to be adjusted. Antibiotic susceptibility to recommended agents should be confirmed.

[d] For patients with severe beta-lactam allergy, vancomycin 15 mg/kg IV q 12 h (goal trough levels of 15–20) may be substituted. Goal vancomycin steady-state trough levels are 15–20 μg/ml.

at subsequent implant removal. Antibiotics should be stopped at least 2 weeks prior to elective removal of implants to avoid compromising intraoperative tissue culture results. If cultures are positive for growth at the time of implant removal, antibiotics should be administered for 6 weeks to prevent chronic osteomyelitis [84]. In patients treated with antibiotic suppression and for whom elective removal of implants is not anticipated, antibiotics can be stopped after successful fusion. A high index of suspicion for subtle symptoms that may be attributable to persistent infection should be maintained for years. If subsequent evidence of infection develops, the implants should be removed. For most patients, removal does not result in loss of correction or spinal instability [53], although this may occur more frequently in pediatric scoliosis patients and may occasionally require reinstrumentation [39, 55].

Late-Onset IAVO

The primary surgical treatment of late-onset IAVO is complete removal of the spinal implants. Treatment failure occurs at a high rate if implants are not removed, regardless of antibiotic therapy [11, 12, 39, 43, 51, 53]. Surgical debridement should include removal of all implants and bone graft, as well as excision of sinus tracts and any inflamed, necrotic, or infected material. Multiple intraoperative tissue specimens should be obtained, and cultures should be optimized to yield *P. acnes*. Careful inspection of the fusion mass should be performed to assess for pseudoarthrosis, which may not have been apparent on preoperative radiologic images [21]. In pediatric patients with neuromuscular disorders or scoliosis, some authors advise that single-stage replacement of implants be considered to minimize loss of correction [55]; however, it is not clear which patients will develop loss of correction after implant removal, and data on the long-term results of single-stage exchange procedures in IAVO are limited. Loss of correction after implant removal does not necessarily correlate with poor patient outcomes [85].

Patients with late-onset IAVO whose implants are removed should receive 6 weeks of pathogen-directed antibiotics as shown in Table 21.4. Laboratory monitoring of patients on IV antibiotics should be performed in accordance with published guidelines [86].

Delayed-Onset IAVO

The surgical and medical management of delayed-onset IAVO is highly variable and depends on the onset and severity of infection, integrity of the implants, and fusion status of the spine. Published data to inform management decisions in this population are limited. Figure 21.1 outlines factors that influence surgical decisions when managing delayed-onset IAVO. Cases of delayed-onset IAVO are generally stratified based on clinical findings as being more suggestive of either an early-onset infection, in which implants are retained, or of a late-onset infection, in which implants are removed. In most cases, an assessment of the stability of the spine and adequacy of fusion at the time of diagnosis guides the decision for implant removal or retention. If implants are retained in delayed-onset IAVO, implant removal should be strongly considered after spinal stability and fusion are achieved because eradication of subacute or chronic implant infection is unlikely. In select circumstances, instrument exchange may be desired to eradicate infection but also maintain spinal stability. The principles that guide surgical treatment of prosthetic joint infection may be helpful in guiding the decision for a single- versus a two-stage exchange in a patient in whom reinstrumentation is necessary for spinal stability or to prevent loss of correction [56]. Activity limitation, spinal bracing, and delayed recovery time for patients with spinal instability undergoing two-stage exchange procedures are detriments to this approach. Furthermore, optimal timing of reimplantation has not been established.

Medical therapy of delayed-onset IAVO is dependent upon whether implants were retained. In cases where implants are retained to facilitate spinal fusion, antibiotics should be given as for early-onset IAVO. However, because of the higher rate of expected treatment failure with implant retention in delayed-onset IAVO, if the microbiology is amenable, oral suppressive antibiotic therapy should be used until spinal fusion is obtained. Likewise, if implants are exchanged in a single stage, consideration should be given to extending antibiotic therapy until fusion has occurred. If implants are removed, 6 weeks of pathogen-directed antibiotics is sufficient.

Because late treatment failure of IAVO can occur many years after the inciting infection, patients should be followed periodically after they have completed treatment for infection. Pain may be the only symptom in patients with residual infection [53]. Patients should be counseled to watch for the symptoms and signs of late-onset IAVO, including progressive pain, localized swelling, or the development of a draining sinus tract. Patients with late spinal pain for whom implant removal is planned should have intraoperative samples sent for culture to assess for occult infection [53]. Pseudoarthrosis has been reported as an outcome of IAVO in up to 25% of patients [22].

Instructive Cases

Case 1: Early-onset IAVO mistaken as superficial surgical site infection

An 8-year-old boy with congenital L2 hemivertebra abnormality underwent L1–L3 fusion with instrumentation and posterolateral allograft bone grafting for the indication of spinal instability with bladder incontinence. Twelve days after surgery, he had a fever, cloudy wound drainage, and peri-incisional pain. Surgical incision and drainage was performed. Purulence was noted superficial to the fascia, but the fascia was deemed healthy and intact. Therefore, exploration and debridement to the level of the implants was not performed. Intraoperative cultures grew methicillin-sensitive *S. aureus*. He was treated with IV cefazolin for 2 weeks, followed by 4 weeks of oral cefadroxil.

Postoperatively, he complained of vague mild bilateral flank discomfort over many months and was not gaining weight. He had no fever and developed no localized swelling or sinus tracts over his incision. Plain radiographs and a CT scan 9 months after implant placement demonstrated incomplete fusion and radiolucency around the pedicle screws (Figure 21.2). MRI demonstrated increased signal on T2 images and contrast enhancement in the posterior paraspinal musculature suggesting inflammation. CRP concentration was normal at 4 mg/l, and ESR was mildly elevated at 30 mm/h (reference range 0–24 mm/h).

For the indication of nonunion after spinal fusion with associated pain, the patient underwent a spinal implant revision procedure. Intraoperative findings included peri-implant purulence, gross loosening of implants, and a mildly unstable spine at L1–L2. All implants were removed. Intraoperative cultures were positive for *S. aureus* with the identical susceptibility pattern as the initial strain. The patient received 6 weeks of IV cefazolin therapy, during which time he was maintained in a brace. Approximately 11 months

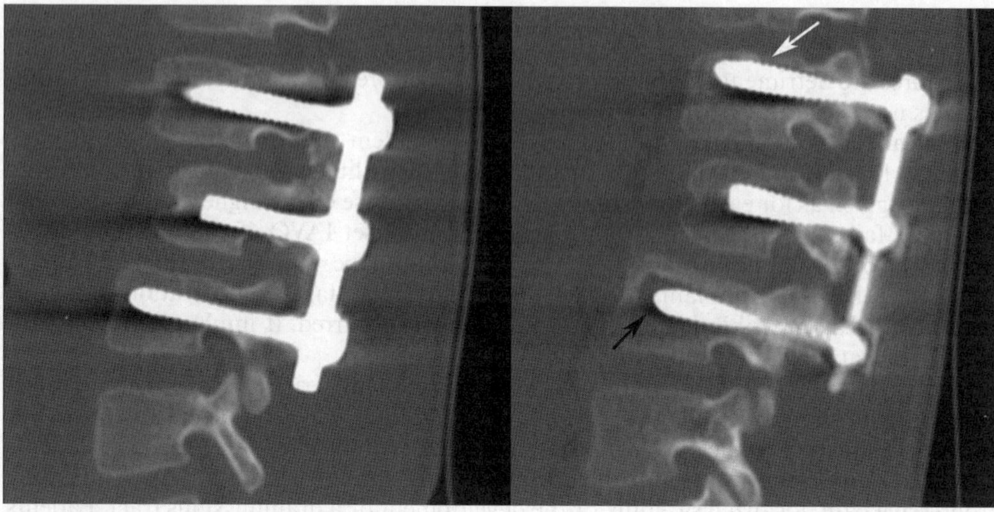

Figure 21.2. Computed tomography findings on the right demonstrate radiolucency around a pedicle screw (black arrow) and evidence of superior migration of a pedicle screw (white arrow) 9 months after the fusion procedure compared with immediate postinstrumentation images (shown on the left). Purulence about the implants was noted upon subsequent operative exploration.

after his index procedure, and 6 weeks after implant removal, he underwent anterior spinal fusion with interbody cages and anterior instrumentation from L1 to L3 using iliac crest autograft. Subsequently, the patient has done well.

Learning Points

This case demonstrates the importance of thorough operative exploration and debridement in early-onset IAVO. The infection was wrongly presumed to be superficial, and surgical and medical therapy was accordingly less aggressive. It is possible that early aggressive debridement to the level of the implants followed by either a rifampin-based prolonged defined duration antibiotic approach or suppressive antibiotic therapy may have controlled the infection enough to facilitate a stable fusion, thereby obviating the need for additional instrumentation. Additionally, it highlights the need to have a very low index of suspicion for persistent infection in patients who have vague complaints following early-onset IAVO. Finally, it highlights the insensitivity of inflammatory markers for the detection of infection outside of early-onset infections.

Case 2: Polymicrobial early-onset IAVO in a patient with neuromuscular scoliosis

A 34-year-old woman with myelomeningocele and neuromuscular scoliosis underwent a thoracolumbosacral posterior fusion with Harrington rods many years previously. The patient was incontinent of stool but had an ileal conduit minimizing urinary contamination. She underwent revision spinal fusion from T2 to sacrum with instrumentation and allograft placement for the indication of implant failure with prominent migration and pseudoarthrosis. Two weeks after surgery, she returned with fever, turbid wound drainage, and wound dehiscence. She underwent incision and drainage with nine liters of irrigation via pulse lavage with removal of necrotic tissue and all nonadhered bone chips.

Intraoperative cultures grew multiple microorganisms, including *S. aureus*, *Enterococcus faecalis*, coagulase-negative staphylococci, *Acinetobacter lwoffii*, and *Bacteroides fragilis*. She underwent three additional incision and drainage procedures every 2 days, until the wound was clean with healthy granulation tissue. Twenty-four days following her revision procedure, she underwent closure of the wound with bilateral trapezius, latissimus dorsi, and gluteus maximus advancement flaps. She was treated with ampicillin/sulbactam 3 G intravenously every 6 h and levofloxacin 500 mg orally daily for 6 weeks, followed by 6 months of oral clindamycin 300 mg three times daily plus levofloxacin 500 mg daily. She demonstrated a radiographically stable fusion 9 months after revision surgery. Eight years later, swelling and a sinus tract with purulent drainage appeared over her lumbosacral spine. The wound probed to the spinal implant on examination. The patient underwent surgical exploration and removal of the implants. Operative cultures grew *S. aureus*, *E. faecalis*, and coagulase-negative staphylococci. She was treated with 4 weeks of IV vancomycin followed by 6 weeks of oral amoxicillin/clavulanate. She has done well, without clinical evidence of infection recurrence for 3 years.

Learning Points

This case demonstrates many salient features of IAVO. First, it highlights the propensity for polymicrobial infection with Gram-negative rods in early-onset IAVO, particularly in patients with neuromuscular scoliosis. Second, it demonstrates that repeat incision and drainage in early-onset IAVO is often required but that implants can usually be retained even in the setting of severe localized infection. Third, with prolonged pathogen-directed antibiotics, most patients will go on to successful fusion. Fourth, it highlights the possibility of very late infection relapse—in this case 8 years after apparently successful treatment. Additionally, it raises the question of whether an initial rifampin-based prolonged defined duration treatment regimen during the initial infection would have been more effective in preventing very late treatment failure. Finally, though the patient technically went on to have late microbiologic treatment failure, the management of the initial infection should be considered a success overall, since it allowed control of sepsis and stabilization of the spine, from which implants could later be removed without progression of deformity or other sequelae.

References

[1] Deyo RA, Nachemson A, Mirza SK. Spinal-fusion surgery—the case for restraint. N Engl J Med 2004;350(7):722–726.

[2] Pfunter A, Wier LM, Stocks C. Most frequent procedures performed in U.S. hospitals, 2010. HCUP statistical brief #149. Available at: http://www.hcup-us.ahrq.gov/reports/statbriefs/sb149.pdf. Accessed on May 29, 2014.

[3] Pfunter A, Levit K, Elixhauser A. Components of cost increases for inpatient hospital procedures, 1997–2009. HCUP statistical brief #133. Available at: http://www.hcup-us.ahrq.gov/reports/statbriefs/sb133.pdf. Accessed on May 29, 2014.

[4] Calderone RR, Garland DE, Capen DA, Oster H. Cost of medical care for postoperative spinal infections. Orthop Clin North Am 1996;27(1):171–182.

[5] Donlan RM. Biofilms: microbial life on surfaces. Emerg Infect Dis 2002;8(9):881–890.

[6] Ramage G, Tunney MM, Patrick S, Gorman SP, Nixon JR. Formation of *Propionibacterium acnes* biofilms on orthopaedic biomaterials and their susceptibility to antimicrobials. Biomaterials 2003;24(19):3221–3227.

[7] Bratzler DW, Dellinger EP, Olsen KM, *et al.* Clinical practice guidelines for antimicrobial prophylaxis in surgery. Am J Health Syst Pharm 2013;70(3):195–283.

[8] Molinari RW, Khera OA, Molinari WJ, 3rd. Prophylactic intraoperative powdered vancomycin and postoperative deep spinal wound infection: 1,512 consecutive surgical cases over a 6-year period. Eur Spine J 2012;21(Suppl 4):S476-S482.
[9] Mackenzie WG, Matsumoto H, Williams BA, *et al.* Surgical site infection following spinal instrumentation for scoliosis: a multicenter analysis of rates, risk factors, and pathogens. J Bone Joint Surg Am 2013;95(9):800,6, S1-2.
[10] Viola RW, King HA, Adler SM, Wilson CB. Delayed infection after elective spinal instrumentation and fusion. A retrospective analysis of eight cases. Spine (Phila Pa 1976) 1997;22(20):2444,50; discussion 2450–1.
[11] Richards BS. Delayed infections following posterior spinal instrumentation for the treatment of idiopathic scoliosis. J Bone Joint Surg Am 1995;77(4):524–529.
[12] Bose B. Delayed infection after instrumented spine surgery: case reports and review of the literature. Spine J 2003;3(5):394–399.
[13] Pull ter Gunne AF, Cohen DB. Incidence, prevalence, and analysis of risk factors for surgical site infection following adult spinal surgery. Spine (Phila Pa 1976) 2009;34(13):1422–8.
[14] Mehta AI, Babu R, Karikari IO, *et al.* 2012 Young Investigator Award winner: the distribution of body mass as a significant risk factor for lumbar spinal fusion postoperative infections. Spine (Phila Pa 1976) 2012;37(19):1652–1656.
[15] Schuster JM, Rechtine G, Norvell DC, Dettori JR. The influence of perioperative risk factors and therapeutic interventions on infection rates after spine surgery: a systematic review. Spine (Phila Pa 1976) 2010;35(9 Suppl):S125-37.
[16] Fang A, Hu SS, Endres N, Bradford DS. Risk factors for infection after spinal surgery. Spine (Phila Pa 1976) 2005;30(12):1460–5.
[17] Wimmer C, Gluch H, Franzreb M, Ogon M. Predisposing factors for infection in spine surgery: a survey of 850 spinal procedures. J Spinal Disord 1998;11(2):124–128.
[18] Klein JD, Hey LA, Yu CS, *et al.* Perioperative nutrition and postoperative complications in patients undergoing spinal surgery. Spine (Phila Pa 1976) 1996;21(22):2676–82.
[19] Veeravagu A, Patil CG, Lad SP, Boakye M. Risk factors for postoperative spinal wound infections after spinal decompression and fusion surgeries. Spine (Phila Pa 1976) 2009; 34(17):1869–72.
[20] Cizik AM, Lee MJ, Martin BI, *et al.* Using the spine surgical invasiveness index to identify risk of surgical site infection: a multivariate analysis. J Bone Joint Surg Am 2012;94(4): 335–342.
[21] Sponseller PD, LaPorte DM, Hungerford MW, *et al.* Deep wound infections after neuromuscular scoliosis surgery: a multicenter study of risk factors and treatment outcomes. Spine (Phila Pa 1976) 2000;25(19):2461–6.
[22] Cahill PJ, Warnick DE, Lee MJ, *et al.* Infection after spinal fusion for pediatric spinal deformity: thirty years of experience at a single institution. Spine (Phila Pa 1976) 2010;35(12): 1211–7.
[23] Glassman SD, Dimar JR, Puno RM, Johnson JR. Salvage of instrumental lumbar fusions complicated by surgical wound infection. Spine (Phila Pa 1976) 1996;21(18):2163–9.
[24] Arens S, Schlegel U, Printzen G, *et al.* Influence of materials for fixation implants on local infection. an experimental study of steel versus titanium DCP in rabbits. J Bone Joint Surg Br 1996;78(4):647–651.
[25] O'Toole JE, Eichholz KM, Fessler RG. Surgical site infection rates after minimally invasive spinal surgery. J Neurosurg Spine 2009;11(4):471–476.
[26] Watters WC, 3rd, Baisden J, Bono CM, *et al.* Antibiotic prophylaxis in spine surgery: an evidence-based clinical guideline for the use of prophylactic antibiotics in spine surgery. Spine J 2009;9(2):142–146.
[27] Shaffer WO, Baisden J, Fernand R, Matz P. Antibiotic Prophylaxis in Spine Surgery/North American Spine Society (NASS) Evidence-Based Clinical Guideline. Rev. ed. Burr Ridge: North American Spine Society; 2013.

[28] Takemoto RC, Park J, Ricart-Hoffiz PA, *et al.* Prospective, randomized study of surgical site infections with the use of perioperative antibiotics for 24 hours vs. the duration of a drain after spinal surgery [abstract 31]. Presented at the Scoliosis Research Society 46th Annual Meeting and Course, Louisville, Kentucky, September 14–17, 2011.

[29] Chang FY, Chang MC, Wang ST, *et al.* Can povidone-iodine solution be used safely in a spinal surgery? Eur Spine J 2006;15(6):1005–1014.

[30] Cheng MT, Chang MC, Wang ST, *et al.* Efficacy of dilute betadine solution irrigation in the prevention of postoperative infection of spinal surgery. Spine (Phila Pa 1976) 2005;30(15): 1689–93.

[31] Brown NM, Cipriano CA, Moric M, *et al.* Dilute betadine lavage before closure for the prevention of acute postoperative deep periprosthetic joint infection. J Arthroplasty 2012;27(1): 27–30.

[32] Watanabe M, Sakai D, Matsuyama D, *et al.* Risk factors for surgical site infection following spine surgery: efficacy of intraoperative saline irrigation. J Neurosurg Spine 2010;12(5): 540–546.

[33] Strom RG, Pacione D, Kalhorn SP, Frempong-Boadu AK. Lumbar laminectomy and fusion with routine local application of vancomycin powder: decreased infection rate in instrumented and non-instrumented cases [published online ahead of print April 23, 2013]. Clin Neurol Neurosurg; 115(9):1766–1769. DOI:S0303-8467(13)00129-7.

[34] Stall AC, Becker E, Ludwig SC, *et al.* Reduction of postoperative spinal implant infection using gentamicin microspheres. Spine (Phila Pa 1976) 2009;34(5):479–83.

[35] Brown MD, Brookfield KF. A randomized study of closed wound suction drainage for extensive lumbar spine surgery. Spine (Phila Pa 1976) 2004;29(10):1066–8.

[36] Payne DH, Fischgrund JS, Herkowitz HN, Gelb D, Poelstra KA. Efficacy of closed wound suction drainage after single-level lumbar laminectomy. J Spinal Disord 1996;9(5):401–403.

[37] Rao SB, Vasquez G, Harrop J, *et al.* Risk factors for surgical site infections following spinal fusion procedures: a case-control study. Clin Infect Dis 2011;53(7):686–692.

[38] Reiffel AJ, Barie PS, Spector JA. A multi-disciplinary review of the potential association between closed-suction drains and surgical site infection. Surg Infect (Larchmt) 2013;14(3): 244–269.

[39] Ho C, Sucato DJ, Richards BS. Risk factors for the development of delayed infections following posterior spinal fusion and instrumentation in adolescent idiopathic scoliosis patients. Spine (Phila Pa 1976) 2007;32(20):2272–7.

[40] Heggeness MH, Esses SI, Errico T, Yuan HA. Late infection of spinal instrumentation by hematogenous seeding. Spine (Phila Pa 1976) 1993;18(4):492–496.

[41] Sierra-Hoffman M, Jinadatha C, Carpenter JL, Rahm M. Postoperative instrumented spine infections: a retrospective review. South Med J 2010;103(1):25–30.

[42] Mirovsky Y, Floman Y, Smorgick Y, *et al.* Management of deep wound infection after posterior lumbar interbody fusion with cages. J Spinal Disord Tech 2007;20(2):127–131.

[43] Kowalski TJ, Berbari EF, Huddleston PM, *et al.* The management and outcome of spinal implant infections: contemporary retrospective cohort study. Clin Infect Dis 2007;44(7): 913–920.

[44] Dubee V, Lenoir T, Leflon-Guibout V, *et al.* Three-month antibiotic therapy for early-onset postoperative spinal implant infections. Clin Infect Dis 2012;55(11):1481–1487.

[45] Stambough JL, Beringer D. Postoperative wound infections complicating adult spine surgery. J Spinal Disord 1992;5(3):277–285.

[46] Brook I, Frazier EH. Aerobic and anaerobic microbiology of wound infection following spinal fusion in children. Pediatr Neurosurg 2000;32(1):20–23.

[47] Brook I, Frazier EH. Aerobic and anaerobic microbiology of surgical-site infection following spinal fusion. J Clin Microbiol 1999;37(3):841–843.

[48] Olsen MA, Nepple JJ, Riew KD, *et al.* Risk factors for surgical site infection following orthopaedic spinal operations. J Bone Joint Surg Am 2008;90(1):62–69.

[49] Richard JC, Vilain R. Acute septic arthritis of the fingers. A clinical study of 87 cases. Annales de chirurgie de la main: organe officiel des sociétés de chirurgie de la main 1982; 1(3):214–20.
[50] Hahn F, Zbinden R, Min K. Late implant infections caused by Propionibacterium acnes in scoliosis surgery. Eur Spine J 2005;14(8):783–788.
[51] Di Silvestre M, Bakaloudis G, Lolli F, Giacomini S. Late-developing infection following posterior fusion for adolescent idiopathic scoliosis. Eur Spine J 2011;20(Suppl 1):S121–S127.
[52] Clark CE, Shufflebarger HL. Late-developing infection in instrumented idiopathic scoliosis. Spine (Phila Pa 1976) 1999;24(18):1909–12.
[53] Collins I, Wilson-MacDonald J, Chami G, *et al.* The diagnosis and management of infection following instrumented spinal fusion. Eur Spine J 2008;17(3):445–450.
[54] Abbey DM, Turner DM, Warson JS, et al. Treatment of postoperative wound infections following spinal fusion with instrumentation. J Spinal Disord 1995;8(4):278–283.
[55] Muschik M, Luck W, Schlenzka D. Implant removal for late-developing infection after instrumented posterior spinal fusion for scoliosis: reinstrumentation reduces loss of correction. A retrospective analysis of 45 cases. Eur Spine J 2004;13(7):645–651.
[56] Osmon DR, Berbari EF, Berendt AR, *et al.* Diagnosis and management of prosthetic joint infection: Clinical Practice Guidelines by the Infectious Diseases Society of America. Clin Infect Dis 2013;56(1):e1–e25.
[57] Butler-Wu SM, Burns EM, Pottinger PS, *et al.* Optimization of periprosthetic culture for diagnosis of *Propionibacterium acnes* prosthetic joint infection. J Clin Microbiol 2011;49(7): 2490–2495.
[58] Schafer P, Fink B, Sandow D, *et al.* Prolonged bacterial culture to identify late periprosthetic joint infection: a promising strategy. Clin Infect Dis 2008;47(11):1403–1409.
[59] Trampuz A, Piper KE, Jacobson MJ, *et al.* Sonication of removed hip and knee prostheses for diagnosis of infection. N Engl J Med 2007;357(7):654–663.
[60] Sampedro MF, Huddleston PM, Piper KE, *et al.* A biofilm approach to detect bacteria on removed spinal implants. Spine (Phila Pa 1976) 2010;35(12):1218–24.
[61] Gouliouris T, Aliyu SH, Brown NM. Spondylodiscitis: update on diagnosis and management. J Antimicrob Chemother 2010;65(Suppl 3):iii11–24.
[62] Levi AD, Dickman CA, Sonntag VK. Management of postoperative infections after spinal instrumentation. J Neurosurg 1997;86(6):975–980.
[63] Lee JH, Lee JH, Kim JB, *et al.* Normal range of the inflammation related laboratory findings and predictors of the postoperative infection in spinal posterior fusion surgery. Clin Orthop Surg 2012;4(4):269–277.
[64] Mok JM, Pekmezci M, Piper SL, *et al.* Use of C-reactive protein after spinal surgery: comparison with erythrocyte sedimentation rate as predictor of early postoperative infectious complications. Spine (Phila Pa 1976) 2008;33(4):415–21.
[65] Gemmel F, Dumarey N, Palestro CJ. Radionuclide imaging of spinal infections. Eur J Nucl Med Mol Imaging 2006;33(10):1226–1237.
[66] Thakkar RS, Malloy JP, 4th, Thakkar SC, et al. Imaging the postoperative spine. Radiol Clin North Am 2012;50(4):731–747.
[67] Hayashi D, Roemer FW, Mian A, *et al.* Imaging features of postoperative complications after spinal surgery and instrumentation. AJR Am J Roentgenol 2012;199(1):W123–W129.
[68] Allouni AK, Davis W, Mankad K, *et al.* Modern spinal instrumentation. Part 2: multimodality imaging approach for assessment of complications. Clin Radiol 2013;68(1):75–81.
[69] De Winter F, Gemmel F, Van De Wiele C, *et al.* 18-fluorine fluorodeoxyglucose positron emission tomography for the diagnosis of infection in the postoperative spine. Spine (Phila Pa 1976) 2003;28(12):1314–9.
[70] Gemmel F, Rijk PC, Collins JM, *et al.* Expanding role of 18F-fluoro-D-deoxyglucose PET and PET/CT in spinal infections. Eur Spine J 2010;19(4):540–551.

[71] Piper KE, Fernandez-Sampedro M, Steckelberg KE, *et al.* C-reactive protein, erythrocyte sedimentation rate and orthopedic implant infection. PLoS One 2010;5(2):e9358.
[72] Picada R, Winter RB, Lonstein JE, *et al.* Postoperative deep wound infection in adults after posterior lumbosacral spine fusion with instrumentation: incidence and management. J Spinal Disord 2000;13(1):42–45.
[73] Weinstein MA, McCabe JP, Cammisa FP, Jr. Postoperative spinal wound infection: a review of 2,391 consecutive index procedures. J Spinal Disord 2000;13(5):422–426.
[74] Mehbod AA, Ogilvie JW, Pinto MR, *et al.* Postoperative deep wound infections in adults after spinal fusion: management with vacuum-assisted wound closure. J Spinal Disord Tech 2005;18(1):14–17.
[75] Rohmiller MT, Akbarnia BA, Raiszadeh K, *et al.* Closed suction irrigation for the treatment of postoperative wound infections following posterior spinal fusion and instrumentation. Spine (Phila Pa 1976) 2010; 35(6): 642–646.
[76] Kim JI, Suh KT, Kim SJ, Lee JS. Implant removal for the management of infection after instrumented spinal fusion. J Spinal Disord Tech 2010;23(4):258–265.
[77] Ahmed R, Greenlee JD, Traynelis VC. Preservation of spinal instrumentation after development of postoperative bacterial infections in patients undergoing spinal arthrodesis. J Spinal Disord Tech 2012;25(6):299–302.
[78] Perry JW, Montgomerie JZ, Swank S, *et al.* Wound infections following spinal fusion with posterior segmental spinal instrumentation. Clin Infect Dis 1997;24(4):558–561.
[79] Zimmerli W, Widmer AF, Blatter M, *et al.* Role of rifampin for treatment of orthopedic implant-related staphylococcal infections: a randomized controlled trial. Foreign-Body Infection (FBI) Study Group. JAMA 1998;279(19):1537–1541.
[80] Giulieri SG, Graber P, Ochsner PE, Zimmerli W. Management of infection associated with total hip arthroplasty according to a treatment algorithm. Infection 2004;32(4):222–228.
[81] Laffer RR, Graber P, Ochsner PE, Zimmerli W. Outcome of prosthetic knee-associated infection: evaluation of 40 consecutive episodes at a single centre. Clin Microbiol Infect 2006;12(5):433–439.
[82] Trampuz A, Zimmerli W. Diagnosis and treatment of implant-associated septic arthritis and osteomyelitis. Curr Infect Dis Rep 2008;10(5):394–403.
[83] Ali MH, Kowalski T, Berbari E, *et al.* Spinal fusion rate in patients with early-onset spine implant associated infections [abstract]. Presented at the Musculoskeletal Infection Society 2007 Annual Meeting, San Dieg, August 10–11, 2007.
[84] Trampuz A, Zimmerli W. Diagnosis and treatment of infections associated with fracture-fixation devices. Injury 2006;37(Suppl 2):S59–S66.
[85] Rihn JA, Lee JY, Ward WT. Infection after the surgical treatment of adolescent idiopathic scoliosis: evaluation of the diagnosis, treatment, and impact on clinical outcomes. Spine (Phila Pa 1976) 2008;33(3):289–94.
[86] Tice AD, Rehm SJ, Dalovisio JR, *et al.* Practice guidelines for outpatient parenteral antimicrobial therapy. IDSA guidelines. Clin Infect Dis 2004;38(12):1651–1672.

Chapter 22
Postoperative Sternal Osteomyelitis

Parham Sendi, Örjan Friberg, Mihai A. Constantinescu,
and Lars Englberger

Introduction

Sternal osteomyelitis is classified as either primary or secondary. Secondary sternal osteomyelitis follows exogenous inoculation and is usually a complication of median sternotomy. In contrast, the pathogenesis of primary sternal osteomyelitis is based on hematogenous or lymphogenous seeding. It is associated with an antecedent infection (e.g., *Staphylococcus aureus* bacteremia [1]) or lung tuberculosis [2]. Primary sternal osteomyelitis is a rare entity. Secondary infection after cardiac surgery accounts for the vast majority of sternal osteomyelitis. The common cardiac interventions include coronary artery bypass surgery, valve repair or replacement, surgery at the aorta, and heart transplantation.

In postoperative sternal osteomyelitis (PSTOM)—apart from chronic infections—the delineation of the bone to the infected adjacent tissue is difficult. Owing to the pre- and retrosternal anatomical structures, deep sternal wound infections (SWI) or mediastinitis commonly involves the bone. Therefore, the terminology used for PSTOM overlaps with those for the infected adjacent tissue. Several factors suggest that a deep surgical site infection of a sternal wound is equivalent to bacterial inoculation of the bone. The pectoralis muscle tendons are in close proximity to the sternal bone (i.e., insertion sites to the manubrium and corpus sterni). After sternotomy, these anatomical barriers have been abrogated, which facilitates the migration of bacteria to deeper structures. In addition, the fascia is close to the cerclage wires. A low bacterial inoculum is sufficient to cause a biofilm on these foreign body materials [3], contributing to persistent infection. Finally, in the early postoperative period, the pathway of the exogenous infection cannot be estimated. The sternum may have been contaminated during surgery, or bacteria belonging to the skin flora may have migrated through the surgical wound. Therefore, it is reasonable to consider deep SWI as an infection involving the bone. Also, in mediastinitis, the involvement of the sternal bone is anatomically often obvious. As a consequence

Bone and Joint Infections: From Microbiology to Diagnostics and Treatment, First Edition.
Edited by Werner Zimmerli.

of this overlapping terminology (i.e., PSTOM, deep SWI, postoperative mediastinitis), there is little or no differentiation in the evaluation of epidemiology, risk factors, microbiology, clinical features, diagnostics, or treatment concepts for these conditions.

Most epidemiological data on the infection incidence have to be reviewed with caution because they summarize all poststernotomy infections as SWI. Nowadays, patients are discharged early after cardiac surgery (e.g., within 1 week). If 30- and 90-day follow-up examinations are not performed, the infection incidence rate is clearly underreported [4]. The incidence rates of SWI are approximately 5% (3.6–7.6%) 30 days [4–6] and 9% 90 days after surgery [4, 6]. The incidence rates of *deep* SWI from different centers range between 0.8 and 3.2% [7–11]. Little is known about the incidence of chronic sternal infection. A Scandinavian study that included 12,297 patients who underwent sternotomy found a cumulative incidence of sternocutaneous fistulas of 0.23% ($n = 32$) [12]. Several indicators suggest that the incidence rates of deep SWI are increasing. One potential explanation for this observation is that the absolute number of cardiac interventions is overall increasing, while at the same time surveillance systems for complications have become mandatory in many institutions. This may correct previously underreported incidence rates. Moreover, the population of patients with multiple comorbidities who are over 65 years is increasing in developed countries. Consequently, cardiac surgery is increasingly performed in patients with several risk factors for infection.

In the past few decades, several expert groups have proposed various classifications for sternal infections, mainly based on the time of infection presentation after index surgery (Table 22.1) [13–15]. These classifications are associated with typical clinical presentations and have, in essence, two purposes. First, they allow cases to have the same definitions, in order to be comparable intra- and interinstitutionally. Second, they help in the decision making for appropriate therapeutic management. In our experience, it is difficult differentiating various forms of infection within the first 4 weeks after sternotomy. Virulent pathogens (e.g., *S. aureus*, Gram-negative bacteria) can cause purulent deep infections at any time during the postoperative course. On the other hand, depending on the bacterial load inoculated and the host's immune status, PSTOM with low-virulence bacteria (e.g., coagulase-negative staphylococci [CNS]) can present early after surgery. We classify PSTOM in acute (≤4 weeks after surgery) or chronic infection (>4 weeks after surgery). Whether the cutoff time for chronic osteomyelitis is set at 4 or 6 weeks after index surgery is arbitrary for a single patient, because infection presentation varies depending on multiple factors. We prefer 4 weeks to anticipate early detection and rapid treatment.

Risk Factors

Several studies have investigated the variables associated with increased risk for deep SWI or mediastinitis after sternotomy. The risk factors can be categorized as patient related or surgery related. The surgery-related category can be further subdivided into preoperative-, operative-, and postoperative-related risk factors.

Patient-Related Risk Factors

These factors include diabetes mellitus, obesity (≥30 kg/m^2), chronic obstructive pulmonary disease, renal failure, peripheral vascular disease, and smoking [8, 12, 16–18]. Diabetes

Table 22.1. Classifications of postoperative sternal osteomyelitis (PSTOM).

	Infected median sternotomy wounds		Postoperative mediastinitis		Postoperative sternomediastinitis	
			Wound infection associated with sternal osteomyelitis with or without infected retrosternal space			
	Pairolero and Arnold [13]		El Oakley and Wright [14]		Robicsek [15]	
Type	Time interval after sternotomy	Clinical presentation	Time interval after sternotomy	Description	Time interval after sternotomy	Description
I	Within a few days	Wound separations with or without sternal instability Often no mediastinal suppuration Skin pathogens involved	Within 2 weeks	No risk factors[a]	<1 week	Nonpurulent sternomediastinitis and no soft tissue or bone necrosis
II	Within the first few weeks	Fulminant mediastinitis Cellulitis, purulent wound drainage Obvious communication with the sternum and the mediastinum	2–6 weeks	No risk factors[a]	1–3 weeks	Virulent infections with tissue necrosis
III	Several weeks or several months	Chronic, draining sinus tract into the sternum or costochondral arches	IIIA: <2 weeks	One or morerisk factors[a]	One month up to 1 year	Chronic, smoldering infections
III			IIIB: 2–6 weeks	One or more risk factors[a]		
IV			IVA: <2 weeks	Type I, II, IIIA, or IIIB after one failed therapeutic trial		
IV			IVB: 2–6 weeks	Type I, II, IIIA, or IIIB after more than one failed therapeutic trial		
V			>6 weeks	First time presentation		

[a]Risk factors in this classification included diabetes, obesity, and the requirement of immunosuppressive agents. We classify PSTOM in acute (≤4 weeks after surgery) or chronic infection (>4 weeks after surgery).

mellitus and obesity have consistently been reported as risk factors, with odds ratios (ORs) ranging from 2.6 to 5.8 and from 1.2 to 6.5, respectively (reviewed in [19, 20]).

Surgery-Related Risk Factors (Preoperative)

Nasal S. aureus Carriage

Kluytmans *et al.* [21] have shown that preoperative nasal carriage of *S. aureus* is a risk factor for SWI. Later, the same group demonstrated that preoperative elimination of nasal *S. aureus* carriage using mupirocin nasal ointment and chlorhexidine soap reduced the risk of SWI [22]. However, the influence of this preventive measure on PSTOM (i.e., number needed to treat) is not yet firmly quantified.

Timing and Duration of Systemic Antibiotic Prophylaxis

The current data for effectiveness in preventing surgical site infections suggest 30 min (for beta-lactams) or 60 min (for glycopeptides) prior to incision [23]. The prevalence of infection was increased by 1% if beta-lactams were given 60 min instead of 30 min and by 1.4% if glycopeptides were given 75 min instead of 60 min before surgery [24].

An ongoing debate is the optimal duration of antibiotic prophylaxis (i.e., <24 h vs. ≥24 h) in cardiac surgery. A meta-analysis reviewed 12 studies involving 7893 patients [25]. Compared with administering prophylaxis for <24 h, prolonged antibiotic prophylaxis reduced the risk of deep SWI (risk ratio 1.68, 95% CI 1.12–2.53). The authors pointed out that the findings were limited by the heterogeneity of antibiotic regimens used and the risk of bias in the published studies. It is important to underline that adaption of antimicrobial prophylaxis must be according to institutional policies.

Surgery-Related Risk Factors (Intraoperative)

Emergency Surgery and Blood Transfusion

Emergency intervention has been associated with increased risk for deep SWI [26]. These types of interventions may be associated with an increased need for blood products, which itself can be a risk factor [8].

Mobilization of Internal Mammary Artery

The use of single and bilateral internal mammary artery (IMA) has been associated with deep SWI, in particular in patients with diabetes mellitus [16, 27]. The method of IMA harvesting has, therefore, gained importance. Comparisons between pedicled (the artery plus its accompanying vein, fascia, and surrounding tissue) and skeletonized (only the artery) harvest techniques have been made. Pedicled grafts have been associated with reduced sternal blood flow [28] and, as a consequence, delayed wound healing and increased risk for infection [29].

Operation and Bypass Time

Several studies have reported that prolonged surgical as well as longer time on cardiopulmonary bypass (CPB) is associated with an elevated risk for deep SWI [8, 16, 30, 31]. It is conceivable that there is a correlation between the duration of surgery and the risk of infection [32]. Since patients undergoing sternotomy present regularly with a variety of risk factors, determining a precise cutoff time for an elevated risk for infection is difficult and clinically not meaningful. Several studies consider the threshold for an elevated risk of infection as between 180 and 240 min for the entire operation time and as between 100 and 125 min for CPB time [8, 16, 30, 31].

Sternal Bone Stability

A stable sternum has a lower risk for infection [33]. Friberg *et al.* [7] analyzed the incidence of SWI in patients having six or fewer fixation wires (ST group) versus those having seven or more cerclages (XW group). The incidence of deep SWI was 0.4% (1 of 264) in the ST group versus 1.7% (13 of 741) in the XW group. However, it should be taken into consideration when stabilizing the sternum that a large quantity of metal ware requires a very low inoculum of bacteria to adhere and cause infection [3].

Within the past few years, many topical agents have entered the market. Their use is timed at sternal closure with the purpose of enhancing sternal healing, early bone stability and reducing infection rates. These include bone wax, water-soluble polymer wax, Kryptonite (a biocompatible bone adhesive), and thermoreactive nitillium clips for sternal closure, among others [34–36]. As with every new method and compound, relatively small single-center studies report promising results on efficacy and safety, but large, multicenter trials are warranted to convince an unbiased audience.

Local Antibiotic Prophylaxis

At sternal closure, local collagen–gentamicin between the sternal halves can be added [37]. This method has been successfully applied for the reduction of wound complications and superficial SWI in Swedish and German cardiac surgery units [38, 39]. In other institutions, a reduction in SWI could not be confirmed [40]. The quantitative benefit of local collagen–gentamicin for the prevention of PSTOM is not fully determined.

Surgery-Related Risk Factors (Postoperative)

Reexploration is the most consistently reported risk factor for deep SWI [8, 11, 16, 30]. Other risk factors in the postoperative period include a prolonged intensive care unit stay, prolonged ventilator time (> 72 h), the use of cardiac assist devices (including intra-aortic balloon pump), and renal replacement therapy [16, 18, 26].

Microbiology

The microbiological findings of three studies consisting of 321 patients with deep SWI are presented in Table 22.2 [9, 20, 41]. The majority of patients have a monomicrobial infection. This might be different in chronic PSTOM. Tocco *et al.* [42] reported polymicrobial infection in 42.8% of 70 patients with chronic infections. CNS are the causative pathogens in approximately 40–60% of the cases. *S. aureus* is the second most frequent pathogen (10–20%), followed by Gram-negative bacteria (5–10%) and *Propionibacterium acnes* (0–10%). This distribution may vary between institutions and departments, as well as between subsets of analyzed patients. In a French study that included 316 patients in an intensive care unit, *S. aureus* (including methicillin-resistant *S. aureus* (MRSA)) was responsible for 60% of the cases with acute mediastinitis. Gram-negative bacteria caused 16.5% and CNS 13% of the infections [43].

Table 22.2. Microbiological findings in patients with deep sternal wound infections.

Number of patients/study reference	129 [41]	91 [9]	101 [20]
Microorganism	*n* (%)	*n* (%)	*n* (%)
Single microorganism	95 (74)	80 (88)	86 (85)
Multiple microorganisms (mostly involving coagulase-negative staphylococci or *S. aureus*)	NA	11 (12)	15 (15)
Coagulase-negative staphylococci	49 (38)	56 (62)	61 (60)
Staphylococcus aureus	9 (7)	17 (19)	10 (10)
Propionibacterium acnes	13 (10)	–	2 (2)
Gram-negative bacteria	5 (4)	7 (8)	12 (12)[1]
Others	8 (6)[2]	–	1 (1)
Negative or missing	26 (20)	4 (4)	–

Percentages are rounded.
[1] Gram-negative bacteria included Enterobacter spp., Klebiella spp. and Escherichia coli.
[2] Number/proportion includes others or multiple species.

It is possible that certain pathogens are associated with a typical constellation and/or a particular comorbidity. In a study investigating this hypothesis, *S. aureus* was more frequent in patients with a stable sternum than in patients with sternal dehiscence (18 of 38, 47%, vs. 13 of 80, 16%). In patients with reoperations, Gram-negative bacteria were more frequently found when these patients later developed deep SWI [44].

Each center should constantly evaluate their microbiological patterns of PSTOM. These data, together with the risk profile of the patient, may help in the decision regarding both antibiotic prophylaxis and empiric antimicrobial therapy.

Clinical Features

PSTOM ≤4 Weeks after Sternotomy

The diagnosis of early PSTOM is often obvious, because patients present within a few weeks after surgery. Wound dehiscence is almost always present, with either serosanguineous or purulent drainage [41]. The former mode of discharge is often seen within the first 2 weeks, the latter between the second and fourth week after sternotomy. There is, however, no rule for these time estimates. The skin is erythematous and edematous (Figure 22.1), and clinical findings can be consistent with cellulitis. Occasionally, a crepitus can be detected. Chest pain is frequently reported, though the significance of this symptom is difficult to assess after recent sternotomy. Sternal instability together with purulent discharge is a PSTOM until proven otherwise.

The clinical presentation can be acute or subacute. In the case of a prolonged incubation period enabling a significant bacterial load and/or involvement of a virulent pathogen, signs of systemic inflammation are prominent (e.g., fever, tachycardia). Subsequently, bacteremia is either present or will occur within a short time.

PSTOM >4 Weeks after Sternotomy

The range of possible clinical presentations is broad. The surgical wound can be unremarkable or only showing slight redness and discreet edema on the scar, while osteolysis

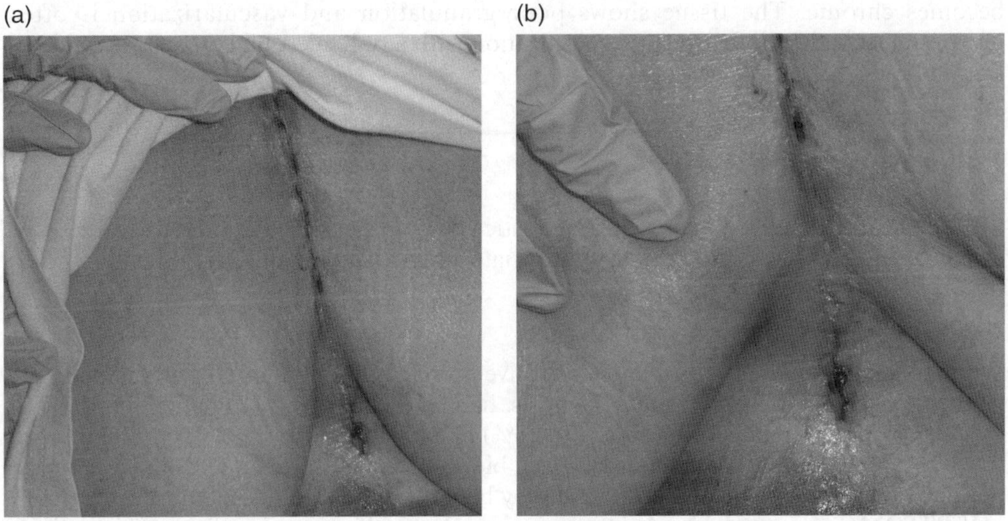

Figure 22.1. (a) Sternal wound 10 days after median sternotomy in a patient who developed fever. (b) After the breasts were moved aside, purulent discharge was visible. (*See insert for color representation of the figure.*)

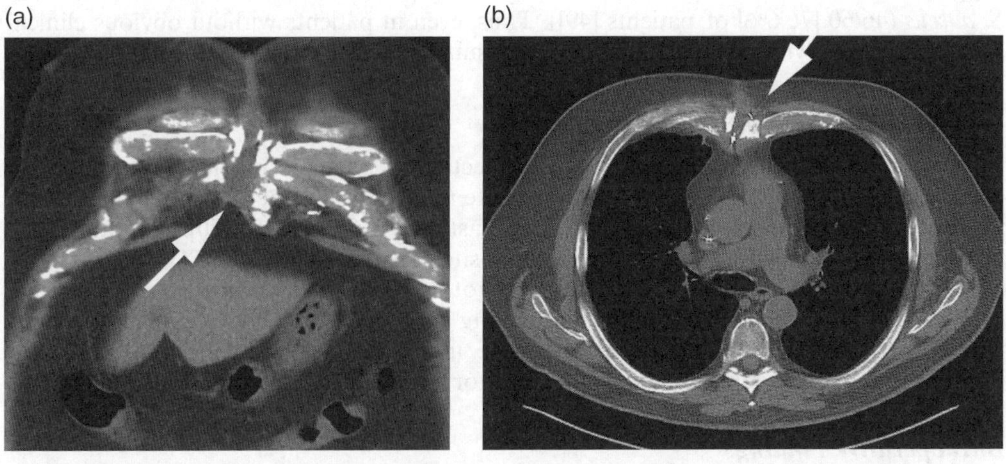

Figure 22.2. CT scan of patient with chronic sternal osteomyelitis. Despite wound closure 14 months previously, he reported sternal pain. (a) Coronal view showing an abscess within the sternum. (b) Axial view showing a sternum without signs of healing and presternal soft tissue inflammation.

of the sternum is progressing (Figure 22.2). Another possible clinical feature is the presence of a sternocutaneous fistula. Patients frequently report having had a previous SWI, either superficial or deep [12].

Among patients with chronic PSTOM, a large proportion show delayed or no wound healing. The diagnosis of deep SWI is made in the postoperative phase, but the disease becomes chronic. The tissue shows poor granulation and vascularization is often impaired. Secondary infection and colonization with a polymicrobial (hospital-acquired) flora follows.

Laboratory Investigation

SWI is mainly clinically diagnosed. Diagnostic work-up steps are required to evaluate the extent of the disease and to initiate appropriate surgical and medical treatment.

Blood Tests

White blood cell (WBC) counts and C-reactive protein (CRP) are usually elevated but are neither specific nor helpful for the diagnosis. In a study including 129 patients with SWI, 55 (42.6%) had deep infections and 21 (16.3%) had a temperature >38.5°C; CRP ranged from 7 to 576 (median 100) mg/l [41]. Thus, blood test results are an add-on piece of the puzzle in the overall clinical picture that may help in estimating the extent of infection.

If PSTOM is suspected, blood cultures are mandatory. Bacteremia is frequent in patients with acute postoperative mediastinitis. Most published studies on the frequency of bacteremia are not recent. The results of these studies, in which patient numbers vary from 21 to 107, noted bacteremia in 27% (12/107) to 57% (12/21) of the patients [45–47]. Positive blood culture results can help in identifying acute PSTOM as the focus of sepsis in a significant number of patients (61/186 [32.8%] to 16/27 [59.3%]) [48, 49]). This is in particular true for *S. aureus* (46/60 [76.6%] of patients [49]). Thus, even in patients without obvious clinical signs of SWI but with positive blood culture results, mediastinitis must be sought.

Preoperative Sampling

This method can be useful if a large fluid collection or purulent mediastinitis is seen on computed tomography (CT). A French institute reported the benefit of sternal puncture for the early diagnosis of poststernotomy mediastinitis in 23 patients [50]. Aspiration was performed with a needle inserted through the site of sternotomy. In PSTOM, this diagnostic method is an alternative if surgery cannot be performed within a reasonable time or if the risk of morbidity and mortality during surgery outweighs the potential benefit with conservative treatment. In our experience, in most cases, rapid surgery is necessary in order to judge the depth of the wound and for adequate microbiological sampling.

Intraoperative Findings

The intraoperative aspect of the wound is crucial to assess the extent of infection. If the fascia is compromised or shows significant inflammation or if cerclages are visible, criteria for deep SWI are fulfilled. If there is macroscopic doubt about the differentiation between "superficial" and "deep" infection, we propose considering it as "deep" SWI.

Microbiological Sampling

If clinically possible, samples must be obtained prior to administering antibiotics. We discourage the use of swabs because of their lower sensitivity in culture results. Moreover, with most available swabs in routine microbiology, it is not possible to perform polymerase chain reaction (PCR) for technical reasons. Therefore, tissue and bone biopsies should be obtained. Bone biopsies are needed for confirming the diagnosis of PSTOM. We recommend at least three but preferably six samples. Biopsy samples should be sent to a microbiological laboratory within a reasonable time (preferably less than 1 h) because some bacteria require both special media and optimal conditions to grow. An extended incubation time (up to 10–14 days) should be requested, if *Propionibacterium* spp. are potentially involved [51]. This might be the case if sternal dehiscence occurs early, the infection is not purulent, and the CRP level is only slightly elevated [41]. The sampling during the first revision surgery is the most important, in particular if the wound is not primarily closed.

In our experience, the number of colonizing bacteria increases with (i) the number of revisions (i.e., debridement and/or vacuum-assisted closure (VAC) exchange) and (ii) the time interval prior to definite secondary wound closure. The pathogenic role of these bacteria is uncertain (i.e., contaminants vs. secondary infection) and causes challenges for antimicrobial treatment. In our institution, microbiological sampling is performed only in the first revision surgery and then again prior to wound closure, unless there are systemic or local signs of secondary infection. This procedure occurs irrespective of the wound closure method (e.g., muscle flap, secondary wound closure). The relevance of microbiological results obtained from the final intervention is assessed on the basis of the organism characteristics and in accordance to the clinical course (i.e., wound healing).

Imaging Procedures

Because the diagnosis of PSTOM is primarily clinical, each imaging procedure must have a specific diagnostic goal. As a principle, in case of emergency (e.g., septic patients with purulent mediastinitis), diagnostic imaging must not delay the surgical intervention.

In mediastinitis, chest radiographs can show mediastinal widening in the anterior view and retrosternal air in the lateral view. In chronic PSTOM, dislocated wires and, rarely, osteolysis of the bone can be detected. To estimate the extent of infection, which has an influence on therapeutic management, a CT is the method of choice. For the diagnosis of mediastinitis, CT has excellent sensitivity but insufficient specificity in the early postoperative period [52]. Of note, with the rapidly developing technique and resolution possibilities of CT, the differentiation of fluid, soft tissue, and bone consistency continues to improve. The sensitivity for excluding sternal osteomyelitis remains difficult to assess. Even in an uneventful postoperative course, the sternal bone requires several months till radiological signs of healing are seen [53]. However, "true" healing of the sternotomy may not correlate with the CT results. In the later course of disease, the sensitivity remains good and the specificity increases [52, 54]. CT is also the imaging method of choice for chronic sternal osteomyelitis (Figure 22.2). Magnetic resonance imaging plays a minor role in the early postoperative period. In chronic PSTOM, it is helpful for differentiating between bone and peristernal soft tissue infection. In our clinical practice, nuclear imaging is seldom used. Liberatore *et al.* [55] found sensitivity, specificity, and accuracy to be 100% each for deep SWI, using a technetium-99m hexamethylpropyleneamine oxime (99mTc-HMPAO)-labeled leukocyte scan in 12 patients. Although these results are promising, they should be reserved for selected patients for reasons of practicability and cost.

Management

General Concepts

The management of PSTOM should always be interdisciplinary, including specialists in the fields of cardiac surgery, infectious diseases, microbiology, plastic surgery, and intensive care. It is important to evaluate the following factors in the therapeutic strategy: (i) clinical presentation, (ii) duration of disease, and (iii) causative pathogen.

Clinical Presentation

If the clinical presentation is fulminate, treatment concepts are equivalent to those for severe sepsis with a source of infection that has to be surgically drained. Diagnostics must be performed immediately without a delay in surgical intervention. Intravenous (IV) empiric antimicrobial treatment must be administered after blood cultures are obtained. The antimicrobial compound must be effective against virulent organisms (e.g., *S. aureus*, Gram-negative bacteria, beta-hemolytic streptococci) and should be chosen on the basis of institutional policy. Conversely, if signs of sepsis syndrome are absent, a thorough work-up of the infection is possible. In these cases, antimicrobial treatment should cover low-virulence pathogens (e.g., CNS) after samples have been obtained (e.g., vancomycin).

Duration of Disease

An estimate of the disease duration indicates the extent of infection; aids in the decision about removing, replacing, or retaining sternal fixation wires; and can direct early planning for large soft tissue covering. The longer the sternal infection progresses, the more likely tissue defects are growing in the previously divided bone. It is evident that SWI must be detected early and patients rapidly treated. The earlier the management, the less complex the intervention is, and the better the outcome. As a rule, wound-healing disturbances after sternotomy—irrespective of severity—must be referred, without delay, to the responsible cardiac surgeon for evaluation.

Causative Pathogen

Some organisms are difficult to eradicate because of their ability to build significant biofilm, their slow replication, or their antimicrobial resistance patterns. Examples include small-colony variants of *S. aureus*, multidrug-resistant Gram-negative bacteria, and fungi [56, 57]. This issue gains importance in consideration that sternal vascularization may be suboptimal, and antibiotic penetration into the tissue is impaired [58]. Thus, the characteristics of a pathogen influence the treatment strategy, and in cases of a difficult-to-treat pathogen, alternative strategies should be evaluated early to prevent potential treatment failures.

Acute PSTMO

The goal is sternal preservation. Exploration of the wound, meticulous debridement of the bone and cartilage, and removal of devitalized tissue are the key features of intervention. The intervention is aggressive in the first revision and often prior to secondary closure. The degree of bone vitality is assessed by mobilizing the edges of the sternal bone and removing fibrous tissue. This measure is helpful both to prevent surgical complications at subsequent revisions and to facilitate resection of the sternal edges (if needed) prior to its closure. An unstable sternum must be fixed. Following debridement, the wound can be closed primarily or secondarily, using a VAC to bridge the wound. Several factors influence the choice of primary or secondary closure: the extent of inflammation (e.g., massive pus, cloudy fluid), amount of devitalized tissue, duration of disease, size of wound, consistency of bone (e.g., multiple fragmented pieces), and ease or difficulty of adapting wound edges (e.g., body mass index (BMI) of the patient). Closing the wound when the underlying tissue is severely inflamed is associated with high risk of failure. Although primary closure is possible in selected cases, the treatment option of choice for most patients is a VAC to bridge and close the wound secondarily. VAC therapy has been shown to positively affect the blood flow in the peristernal thoracic wall [59]. Formation of granulation tissue is enhanced, and secondary wound healing is faster than in cases with open wound treatment [60]. In a meta-analysis, VAC therapy reduced the length of hospital stay by 7.2 days (CI 95% 3.5–10.8) without significant impact on mortality [61]. On the basis of these arguments, many surgeons prefer to routinely begin VAC treatment after the primary wound debridement, observe wound healing, and await results from microbiology laboratory. Until the secondary closure, the treatment time is bridged with VAC changes and wound cleaning but often without or only subtle debridement (e.g., gentle removal of delineated necrotic tissue).

To Retain or Exchange Fixation Wires

Revision often goes in line with removal of foreign body material. The sternum has to be refixed with new cerclages, because stable bones heal better and are less susceptible to infection. The rationale for removing the fixation wires is the possibility of bacteria adhering to the foreign body material, causing a biofilm. This multicellular matrix is difficult to eradicate and contributes to infection persistence. The pathogenesis is mainly derived from studies on orthopedic implant-associated infections. In these infections, there are, however, constellations to retain foreign body material. The most important factor for this procedure is the disease duration. The prognosis is best when the duration is short (e.g., <3 weeks [62]). In our clinic, we try to retain the cerclages, if disease duration is not longer than 1 week. In cases of dehiscence between the bony edges, additional cerclages to rewire the sternum can be used (e.g., figure-of-eight and Robicsek technique). Sternal fixation wires are exchanged, if the disease duration is more than 3 weeks. We do not favor a two-stage exchange of the cerclages, because of sternal instability. Rarely, in cases of purulent osteomyelitis, a wire-free interval of 2–3 days is performed. The decision falls into a gray zone, when the incubation period is between one and 3 weeks. In these cases, several factors are evaluated for a single patient.

Early versus Delayed Secondary Wound Closure

Although VAC therapy is an important part of a successful treatment concept, as has been shown in a large population [63], complications can occur (e.g., bleeding, rarely right ventricular rupture [64]). With repeated debridement and VAC exchanges, there is a risk of sternal breakdown or superinfection. Despite optimal hygienic conditions and VAC, bacteria from the skin flora migrate into the wound. These nosocomial microorganisms may be multiresistant, due to their selection through antimicrobial treatment (e.g., Gram-negative bacteria, fungi). This selected flora becomes relevant for infection after reaching a significant load. Because the bacterial load increases with time (e.g.,

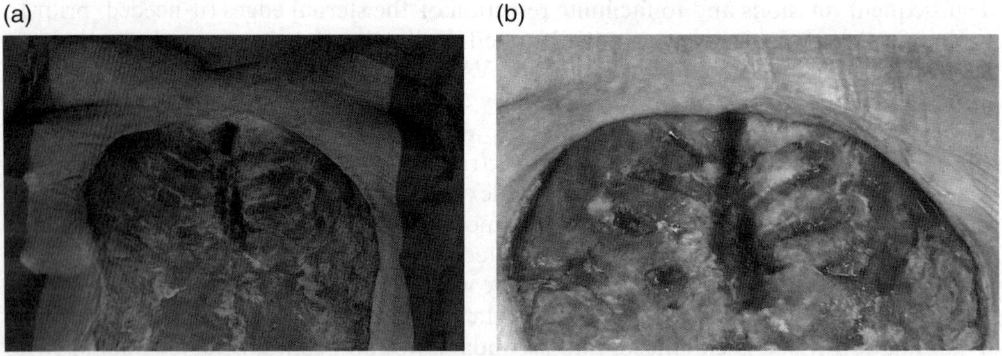

Figure 22.3. (a) An extensive mediastinal wound in a patient with chronic PSTMO and infected costal cartilage. (b) Despite meticulous debridement, the vitality of the costae is suboptimal. Complete resection cannot be performed because of thorax instability. (*See insert for color representation of the figure.*) Picture courtesy of Dr. M. Shafighi.

recurrent wound problems > 21 days in [65]), closing the wound early is of interest. On the other hand, a wound that is closed too early (e.g., prior to signs of granulation formation) will likely not heal. Several studies reported detection of wound granulation in the majority of their patients within 7–10 days [60, 66, 67]. An optimal balance between as long as necessary and as short as possible should be used in VAC therapy. This might be different for every single patient and different in various centers based on their experience and policy. The Lund group (Sweden) included a CRP cutoff level < 70 mg/l as one of the criteria to close the wound secondarily [68]. In our institution, we target closure within 6–12 days after the first surgery. The interval between VAC exchanges is 2–4 days, which enables observation of wound healing. Prior to closure, sufficient sternal vascularization and granulation formation must be present, while significant systemic clinical signs and laboratory signs of inflammation response must be absent. Hence, the extent of infection determines the final number of revisions/VAC exchanges and the time interval between primary revision and secondary wound closure. This again points toward the importance of early detection and rapid intervention of SWI. The shorter the disease duration is, the smaller the extent of infection and the earlier can the wound be closed.

Chronic PSTMO

In these infections, osteolysis has progressed over time. Removal of all foreign body materials and partial or total resection of the sternum are required. In addition, resection of infected costal cartilage or infected parts of the sternoclavicular joint may be necessary. If the infected area involves several costal parts, aggressive resection can lead to instability of the thorax (Figure 22.3). In these cases, meticulous debridement of the involved costal parts must be performed, followed by prolonged antimicrobial chemotherapy. In chronic PSTOM, the surgical intervention causes a gap, requiring flap covering. In our institution, chronic PSTOM cases receive interdisciplinary management, and the intervention is frequently performed in one stage (i.e., bone resection and flap covering during one single operation).

In patients with multiple comorbidities, it should be noted that the aggressiveness and complexity of the intervention and the surgery itself are associated with considerable morbidity and mortality. In selected patients, a prolonged antimicrobial chemotherapy might be considered.

Soft Tissue Reconstruction

Prior to evaluating "when" to close the wound, it is important early on to assess "if" the wound can be closed without flap coverage. Persistent infections, poor wound healing, and granulation formation or a large wound are indicators for a complicated course. Sternal reconstruction with a flap is frequently required in chronic infections. There are several advantages in using muscle or musculocutaneous flaps. First, the flaps bring their own well-vascularized tissues into the wound bed. Thereby, wound healing can improve. The penetration of antibiotic into the infected area is also facilitated. Second, stability of the chest wall can be achieved [69]. Finally, the dead space is obliterated, and an anatomical barrier is created, preventing the migration of the bacterial flora from one compartment to another.

The timing of the coverage depends not only on the general condition of the patient but also on the wound bed situation. It is therefore important to thoroughly assess the condition of the peristernal tissue. In patients with poor signs of wound healing and little granulation tissue, the flap may fail in healing onto the wound bed. Then flap dehiscence or partial or complete flap loss occurs. As a consequence, the infection persists or recurs. Thorough and aggressive debridement until a vital wound bed is achieved is therefore mandatory prior to flap coverage.

There are four main flap options for coverage of a sternal wound. If the wound area is relatively small, a pectoralis major muscle (PEC) may be used unilaterally or bilaterally to obliterate the dead space, followed by either direct skin closure or by a skin graft [8, 70]. In larger defects, the use of abdominal or dorsal musculocutaneous flaps is indicated. From the abdominal site, the vertical rectus abdominis myocutaneous (VRAM) flap and the omentum majus flap can be harvested and mobilized into the sternal defect [71]. From the dorsum, the use of a latissimus dorsi (LAT) flap presents a convenient solution. The omental flap is seldom used by the authors, because it requires the opening of a second body cavity, thus involving additional potential morbidity.

The choice of the flap depends on the size of defect, its anatomic location, the available blood supply, incisions from previous interventions, and patient comorbidities.

Antimicrobial Treatment Concepts

We apply for PSTMO similar treatment concept than for osteomyelitis in general. This is described in detail elsewhere (Chapter 15, Table 15.5, Chapters 17 and 21). After sampling for microbiological investigations, empiric IV antimicrobial therapy is started. The choice of compound is based on (i) clinical presentation (septic vs. nonseptic), (ii) commonly found pathogens associated with this clinical presentation (e.g., *S. aureus*, Gram-negative bacilli, or CNS), and (iii) results from local antimicrobial resistance surveillances (e.g., prevalence of MRSA). After causative pathogens have been isolated, treatment should be streamlined accordingly. In most cases, IV therapy is continued until the wound is closed secondarily (e.g., 10–14 days). Then formulation is switched to oral compounds, if the pathogen is susceptible to an agent with good bioavailability (Chapters 3, 15, and 21). Because fixation wires are considered as foreign body material, we administer rifampin in staphylococcal PSTOM. In a study including 100 patients with deep SWI due to staphylococci, the infection outcome in association with the antibiotic regimen was analyzed [72]. In the multivariate analysis, a rifampin-containing antibiotic regimen was the only factor associated with lower risk of treatment failure. The agent is always combined with another compound, and the organisms must be rifampin susceptible. Rifampin is not administered until the wound is closed and dry, and all are suction drainages are removed. The rationale behind this concept is the risk of secondary infections due to rifampin-resistant staphylococci emerging in the skin flora. The total treatment duration depends on whether infection is considered acute or chronic and how meticulous surgical treatment was. The following recommendations are considered as approximation and have to be adapted on patient's individual conditions:

- Acute PSTOM with complete resection of infected bone parts: 2–3 weeks
- Acute PSTOM with debridement without bone resection: 4–6 weeks
- Chronic PSTOM with complete resection of infected bone: 2–3 weeks
- Chronic PSTOM with incomplete bone resection: 3 months

Prior to cessation of the therapy, a clinical and laboratory (and occasionally radiological) examination is performed.

Tocco *et al.* [42] reviewed 45 patients with chronic sternal osteomyelitis who received only antimicrobial chemotherapy for 6–18 months. All patients were cured. Medical treatment only is an alternative to consider in selected polymorbid patients or patients. Irrespective of the treatment modality, a thorough follow-up (≥1 year), including clinical and radiological examination, is required to control the success of treatment.

Instructive Cases

Case 1: Acute PSTOM in a Patient with Risk Factors

A 76-year-old woman was referred to our center, because of coronary artery disease and severe aortic valve stenosis. Her medical history included insulin-dependent diabetes mellitus, obesity (BMI 34 kg/m^2), and arterial hypertension. Triple coronary artery bypass grafting and aortic valve replacement were performed. The postoperative course was uneventful. Six days later, the patient was referred to a cardiac rehabilitation center. Ten days after surgery, the patient developed fever and was immediately transferred back to our center. On clinical examination, she had tachycardia (heart rate 88/min), her blood pressure was 110/60 mmHg, and her body temperature was 38.1°C. The sternal wound was slightly edematous. Between the breasts, purulent discharge was visible (Figure 22.1). Blood cultures were obtained. Blood tests revealed the following values: WBC counts of 14.7 G/l (normal 4–10 G/l) and CRP of 107 mg/l (normal < 5 mg/l). The wound was explored and meticulously debrided and the fixation wires were tightened. After obtaining biopsy samples, empiric antimicrobial therapy was started (amoxicillin/clavulanate 2.2 g IV, 6 hourly, plus vancomycin 15 mg IV per kg bodyweight, 12 hourly) and VAC therapy initiated. *S. aureus* grew in one out of two blood cultures and in all obtained biopsy samples. Treatment was switched to flucloxacillin (2 g IV, 6 hourly). Development of granulation tissue was observed during VAC exchanges. The patient remained afebrile, and the CRP value dropped to 23 mg/l within 4 days. Eleven days after revision surgery, the wound was closed, and antimicrobial treatment switched to ciprofloxacin (750 mg

PO, 12 h hourly) and rifampin (450 mg PO, 12 hourly). After 6 weeks, follow-up examinations were unremarkable, and antimicrobials were stopped. One year later, the patient was infection-free.

Learning Points

- The medical history of this patient included diabetes mellitus and obesity, two major risk factors for SWI.
- Symptoms for PSTOM started after hospital discharge. Because of immediate referral and a timely surgical intervention, a more complex intervention and a poorer outcome could be prevented.

Case 2: Complex Management of Chronic PSTOM

A 73-year-old man was referred to our center, because of chronic PSTOM. Eighteen months previously, he received a fourfold coronary artery bypass grafting. One month after the index surgery, he contacted a physician because of sternal pain. A CT showed broken fixation wires, but no intervention was performed. Because of persistent symptoms, revision surgery was performed 5 months later. The broken wire was removed, but no specimens for microbiology were obtained. Three weeks after revision surgery and 7 months after index surgery, the wound opened spontaneously with serosanguineous discharge. The wound was again revised and primarily closed, followed by an antimicrobial treatment course. Over the next 13 months, he had several relapsing episode of sternal skin inflammation (i.e., erythematous swelling). He treated each episode with oral antibiotics but was never pain-free.

On clinical examination, his condition was stable but the sternum was not. Blood tests revealed a WBC counts of 4.7 G/l (normal 4–10 G/l) and CRP of 12 mg/l (normal < 5 mg/l). A CT showed an abscess within the sternum and presternal soft tissue inflammation (Figure 22.2). The surgical treatment included removal of all foreign body materials and resection of the upper and lower parts of the sternum. Samples for microbiology were obtained, and the large defect was covered with a musculus LAT flap. CNS grew in biopsy cultures, only susceptible to glycopeptides, co-trimoxazole, tetracycline, and rifampin but resistant to all other tested antibiotics. IV high-dose vancomycin was administered and combined with rifampin when wounds were dry. The postoperative course was uneventful. The patient was discharged, and treatment switched to co-trimoxazole (one double-strength tablet, 8 hourly) and rifampin (450 mg, 12 hourly). After 3 months, follow-up examinations were unremarkable, and antimicrobials were stopped. Two years later, the patient did not report any episodes of skin inflammation and was in good condition.

Learning Points

- Missing the diagnosis of PSTOM in the early postoperative period unavoidably results in chronic infections requiring complex surgery.
- PSTOM should not be treated with antibiotics alone. Multiple short-course treatment attempts cause resistance.
- Treatment concepts for chronic PSTOM require an interdisciplinary approach.

References

[1] Platt MA, Ziegler K. Primary sternal osteomyelitis with bacteremia and distal seeding. J Emerg Med 2012,43(2):e93–95.

[2] Vasa M, Ohikhuare C, Brickner L. Primary sternal tuberculosis osteomyelitis: A case report and discussion. Can J Infect Dis Med Microbiol 2009,20(4):e181–184.

[3] Zimmerli W, Waldvogel FA, Vaudaux P, Nydegger UE. Pathogenesis of foreign body infection: description and characteristics of an animal model. J Infect Dis 1982,146(4): 487–497.

[4] Jonkers D, Elenbaas T, Terporten P, *et al.* Prevalence of 90-days postoperative wound infections after cardiac surgery. Eur J Cardiothorac Surg 2003,23(1):97–102.

[5] Berg TC, Kjorstad KE, Akselsen PE, *et al.* National surveillance of surgical site infections after coronary artery bypass grafting in Norway: incidence and risk factors. Eur J Cardiothorac Surg 2011,40(6):1291–1297.

[6] Dhadwal K, Al-Ruzzeh S, Athanasiou T, *et al.* Comparison of clinical and economic outcomes of two antibiotic prophylaxis regimens for sternal wound infection in high-risk patients following coronary artery bypass grafting surgery: a prospective randomised double-blind controlled trial. Heart 2007,93(9):1126–1133.

[7] Friberg O, Dahlin LG, Soderquist B, *et al.* Influence of more than six sternal fixation wires on the incidence of deep sternal wound infection. Thorac Cardiovasc Surg 2006,54(7): 468–473.

[8] Strecker T, Rosch J, Horch RE, *et al.* Sternal wound infections following cardiac surgery: risk factor analysis and interdisciplinary treatment. Heart Surg Forum 2007,10(5):E366–371.

[9] Tegnell A, Isaksson B, Granfeldt H, Ohman L. Changes in the appearance and treatment of deep sternal infections. J Hosp Infect 2002,50(4):298–303.

[10] Borger MA, Rao V, Weisel RD, *et al.* Deep sternal wound infection: risk factors and outcomes. Ann Thorac Surg 1998,65(4):1050–1056.

[11] Kubota H, Miyata H, Motomura N, *et al.* Deep sternal wound infection after cardiac surgery. J Cardiothorac Surg 2013,8:132.

[12] Steingrimsson S, Gustafsson R, Gudbjartsson T, *et al.* Sternocutaneous fistulas after cardiac surgery: incidence and late outcome during a ten-year follow-up. Ann Thorac Surg 2009,88(6): 1910–1915.

[13] Pairolero PC, Arnold PG. Management of infected median sternotomy wounds. Ann Thorac Surg 1986,42(1):1–2.

[14] El Oakley RM, Wright JE. Postoperative mediastinitis: classification and management. Ann Thorac Surg 1996,61(3):1030–1036.

[15] Robicsek F. Postoperative sterno-mediastinitis. Am Surg 2000,66(2):184–192.

[16] Gummert JF, Barten MJ, Hans C, *et al.* Mediastinitis and cardiac surgery–an updated risk factor analysis in 10,373 consecutive adult patients. Thorac Cardiovasc Surg 2002,50(2): 87–91.

[17] Baskett RJ, MacDougall CE, Ross DB. Is mediastinitis a preventable complication? A 10-year review. Ann Thorac Surg 1999,67(2):462–465.

[18] Stahle E, Tammelin A, Bergstrom R, *et al.* Sternal wound complications–incidence, microbiology and risk factors. Eur J Cardiothorac Surg 1997,11(6):1146–1153.

[19] Singh K, Anderson E, Harper JG. Overview and management of sternal wound infection. Semin Plast Surg 2011,25(1):25–33.

[20] Sjogren J, Gustafsson R, Nilsson J, *et al.* Clinical outcome after poststernotomy mediastinitis: vacuum-assisted closure versus conventional treatment. Ann Thorac Surg 2005,79(6): 2049–2055.

[21] Kluytmans JA, Mouton JW, Ijzerman EP, *et al.* Nasal carriage of *Staphylococcus aureus* as a major risk factor for wound infections after cardiac surgery. J Infect Dis 1995,171(1): 216–219.

[22] Kluytmans JA, Mouton JW, VandenBergh MF, *et al.* Reduction of surgical-site infections in cardiothoracic surgery by elimination of nasal carriage of *Staphylococcus aureus*. Infect Control Hosp Epidemiol 1996,17(12):780–785.
[23] Kappeler R, Gillham M, Brown NM. Antibiotic prophylaxis for cardiac surgery. J Antimicrob Chemother 2012,67(3):521–522.
[24] Koch CG, Nowicki ER, Rajeswaran J, *et al.* When the timing is right: Antibiotic timing and infection after cardiac surgery. J Thorac Cardiovasc Surg 2012,144(4):931–937 e934.
[25] Mertz D, Johnstone J, Loeb M. Does duration of perioperative antibiotic prophylaxis matter in cardiac surgery? A systematic review and meta-analysis. Ann Surg 2011,254(1):48–54.
[26] Sakamoto H, Fukuda I, Oosaka M, Nakata H. Risk factors and treatment of deep sternal wound infection after cardiac operation. Ann Thorac Cardiovasc Surg 2003,9(4):226–232.
[27] Centofanti P, Savia F, La Torre M, *et al.* A prospective study of prevalence of 60-days post-operative wound infections after cardiac surgery. An updated risk factor analysis. J Cardiovasc Surg (Torino) 2007,48(5):641–646.
[28] Carrier M, Gregoire J, Tronc F, *et al.* Effect of internal mammary artery dissection on sternal vascularization. Ann Thorac Surg 1992,53(1):115–119.
[29] Peterson MD, Borger MA, Rao V, *et al.* Skeletonization of bilateral internal thoracic artery grafts lowers the risk of sternal infection in patients with diabetes. J Thorac Cardiovasc Surg 2003,126(5):1314–1319.
[30] Mannien J, Wille JC, Kloek JJ, van Benthem BH. Surveillance and epidemiology of surgical site infections after cardiothoracic surgery in The Netherlands, 2002–2007. J Thorac Cardiovasc Surg 2011,141(4):899–904.
[31] Bitkover CY, Gardlund B. Mediastinitis after cardiovascular operations: a case-control study of risk factors. Ann Thorac Surg 1998,65(1):36–40.
[32] Loop FD, Lytle BW, Cosgrove DM, *et al.* J. Maxwell Chamberlain memorial paper. Sternal wound complications after isolated coronary artery bypass grafting: early and late mortality, morbidity, and cost of care. Ann Thorac Surg 1990,49(2):179–186.
[33] Bray PW, Mahoney JL, Anastakis D, Yao JK. Sternotomy infections: sternal salvage and the importance of sternal stability. Can J Surg 1996,39(4):297–301.
[34] Vestergaard RF, Nielsen PH, Terp KA, *et al.* Effect of Hemostatic Material on Sternal Healing After Cardiac Surgery. Ann Thorac Surg 2014; 97(1):153–160.
[35] Fedak PW, Kieser TM, Maitland AM, *et al.* Adhesive-enhanced sternal closure to improve postoperative functional recovery: a pilot, randomized controlled trial. Ann Thorac Surg 2011,92(4):1444–1450.
[36] Bejko J, Tarzia V, De Franceschi M, *et al.* Nitinol flexigrip sternal closure system and chest wound infections: insight from a comparative analysis of complications and costs. Ann Thorac Surg 2012,94(6):1848–1853.
[37] Mavros MN, Mitsikostas PK, Alexiou VG, *et al.* Gentamicin collagen sponges for the prevention of sternal wound infection: a meta-analysis of randomized controlled trials. J Thorac Cardiovasc Surg 2012,144(5):1235–1240.
[38] Friberg O, Dahlin LG, Kallman J, *et al.* Collagen-gentamicin implant for prevention of sternal wound infection; long-term follow-up of effectiveness. Interact Cardiovasc Thorac Surg 2009,9(3):454–458.
[39] Schimmer C, Ozkur M, Sinha B, *et al.* Gentamicin-collagen sponge reduces sternal wound complications after heart surgery: a controlled, prospectively randomized, double-blind study. J Thorac Cardiovasc Surg 2012,143(1):194–200.
[40] Bennett-Guerrero E, Ferguson TB, Jr., Lin M, *et al.* Effect of an implantable gentamicin-collagen sponge on sternal wound infections following cardiac surgery: a randomized trial. JAMA 2010,304(7):755–762.
[41] Friberg O, Svedjeholm R, Kallman J, Soderquist B. Incidence, microbiological findings, and clinical presentation of sternal wound infections after cardiac surgery with and without local gentamicin prophylaxis. Eur J Clin Microbiol Infect Dis 2007,26(2):91–97.

[42] Tocco MP, Ballardini M, Masala M, Perozzi A. Post-sternotomy chronic osteomyelitis: is sternal resection always necessary? Eur J Cardiothorac Surg 2013,43(4):715–721.
[43] Trouillet JL, Vuagnat A, Combes A, *et al.* Acute poststernotomy mediastinitis managed with debridement and closed-drainage aspiration: factors associated with death in the intensive care unit. J Thorac Cardiovasc Surg 2005,129(3):518–524.
[44] Gardlund B, Bitkover CY, Vaage J. Postoperative mediastinitis in cardiac surgery—microbiology and pathogenesis. Eur J Cardiothorac Surg 2002;21(5):825–830.
[45] Gol MK, Karahan M, Ulus AT, *et al.* Bloodstream, respiratory, and deep surgical wound infections after open heart surgery. J Card Surg 1998;13(4):252–259.
[46] Farinas MC, Gald Peralta F, Bernal JM, *et al.* Suppurative mediastinitis after open-heart surgery: a case-control study covering a seven-year period in Santander, Spain. Clin Infect Dis 1995;20(2):272–279.
[47] Bor DH, Rose RM, Modlin JF, *et al.* Mediastinitis after cardiovascular surgery. Rev Infect Dis 1983;5(5):885–897.
[48] Kohman LJ, Coleman MJ, Parker FB, Jr. Bacteremia and sternal infection after coronary artery bypass grafting. Ann Thorac Surg 1990;49(3):454–457.
[49] Fowler VG, Jr., Kaye KS, Simel DL, *et al. Staphylococcus aureus* bacteremia after median sternotomy: clinical utility of blood culture results in the identification of postoperative mediastinitis. Circulation 2003;108(1):73–78.
[50] Benlolo S, Mateo J, Raskine L, *et al.* Sternal puncture allows an early diagnosis of poststernotomy mediastinitis. J Thorac Cardiovasc Surg 2003;125(3):611–617.
[51] Zappe B, Graf S, Ochsner PE, *et al. Propionibacterium* spp. in prosthetic joint infections: a diagnostic challenge. Arch Orthop Trauma Surg 2008;128(10):1039–1046.
[52] Jolles H, Henry DA, Roberson JP, *et al.*. Mediastinitis following median sternotomy: CT findings. Radiology 1996;201(2):463–466.
[53] Bitkover CY, Cederlund K, Aberg B, Vaage J. Computed tomography of the sternum and mediastinum after median sternotomy. Ann Thorac Surg 1999;68(3):858–863.
[54] Gur E, Stern D, Weiss J, *et al.* Clinical-radiological evaluation of poststernotomy wound infection. Plast Reconstr Surg 1998;101(2):348–355.
[55] Liberatore M, Fiore V, D'Agostini A, *et al.* Sternal wound infection revisited. Eur J Nucl Med 2000;27(6):660–667.
[56] Sendi P, Rohrbach M, Graber P, *et al. Staphylococcus aureus* small colony variants in prosthetic joint infection. Clin Infect Dis 2006;43(8):961–967.
[57] Phan TQ, Depner C, Theodorou P, *et al.* Failure of secondary wound closure after sternal wound infection following failed initial operative treatment: causes and treatment. Ann Plast Surg 2013;70(2):216–221.
[58] Andreas M, Zeitlinger M, Hoeferl M, *et al.* Internal mammary artery harvesting influences antibiotic penetration into presternal tissue. Ann Thorac Surg 2013;95(4):1323–1329.
[59] Petzina R, Gustafsson L, Mokhtari A, *et al.* Effect of vacuum-assisted closure on blood flow in the peristernal thoracic wall after internal mammary artery harvesting. Eur J Cardiothorac Surg 2006;30(1):85–89.
[60] Doss M, Martens S, Wood JP, *et al.*. Vacuum-assisted suction drainage versus conventional treatment in the management of poststernotomy osteomyelitis. Eur J Cardiothorac Surg 2002;22(6):934–938.
[61] Damiani G, Pinnarelli L, Sommella L, *et al.* Vacuum-assisted closure therapy for patients with infected sternal wounds: a meta-analysis of current evidence. J Plast Reconstr Aesthet Surg 2011;64(9):1119–1123.
[62] Zimmerli W, Widmer AF, Blatter M, *et al.*. Role of rifampin for treatment of orthopedic implant-related staphylococcal infections: a randomized controlled trial. Foreign-Body Infection (FBI) Study Group. JAMA 1998;279(19):1537–1541.
[63] Baillot R, Cloutier D, Montalin L, *et al.* Impact of deep sternal wound infection management with vacuum-assisted closure therapy followed by sternal osteosynthesis: a 15-year review of 23,499 sternotomies. Eur J Cardiothorac Surg 2010;37(4):880–887.

[64] Abu-Omar Y, Naik MJ, Catarino PA, Ratnatunga C. Right ventricular rupture during use of high-pressure suction drainage in the management of poststernotomy mediastinitis. Ann Thorac Surg 2003;76(3):974; author reply 974–975.

[65] Bapat V, El-Muttardi N, Young C, *et al.*. Experience with Vacuum-assisted closure of sternal wound infections following cardiac surgery and evaluation of chronic complications associated with its use. J Card Surg 2008;23(3):227–233.

[66] Cowan KN, Teague L, Sue SC, Mahoney JL. Vacuum-assisted wound closure of deep sternal infections in high-risk patients after cardiac surgery. Ann Thorac Surg 2005;80(6):2205–2212.

[67] Gustafsson R, Johnsson P, Algotsson L, *et al.*. Vacuum-assisted closure therapy guided by C-reactive protein level in patients with deep sternal wound infection. J Thorac Cardiovasc Surg 2002;123(5):895–900.

[68] Sjogren J, Malmsjo M, Gustafsson R, Ingemansson R. Poststernotomy mediastinitis: a review of conventional surgical treatments, vacuum-assisted closure therapy and presentation of the Lund University Hospital mediastinitis algorithm. Eur J Cardiothorac Surg 2006;30(6):898–905.

[69] Cabbabe EB, Cabbabe SW. Surgical management of the symptomatic unstable sternum with pectoralis major muscle flaps. Plast Reconstr Surg 2009;123(5):1495–1498.

[70] Jones G, Jurkiewicz MJ, Bostwick J, *et al.* Management of the infected median sternotomy wound with muscle flaps. The Emory 20-year experience. Ann Surg 1997;225(6):766–776.

[71] Vyas RM, Prsic A, Orgill DP. Transdiaphragmatic omental harvest: a simple, efficient method for sternal wound coverage. Plast Reconstr Surg 2013;131(3):544–552.

[72] Khanlari B, Elzi L, Estermann L, *et al.* A rifampicin-containing antibiotic treatment improves outcome of staphylococcal deep sternal wound infections. J Antimicrob Chemother 2010;65(8):1799–1806.

Index

Bone and Joint Infections: From Microbiology to Diagnostics and Treatment, First Edition.
Edited by Werner Zimmerli.

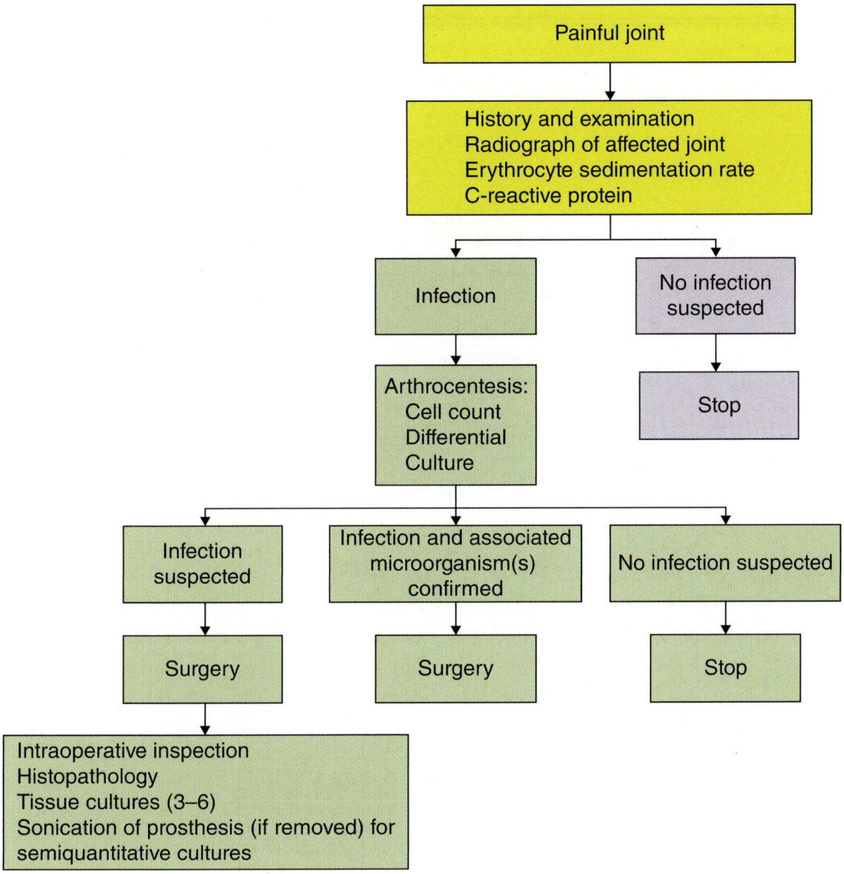

Figure 2.1. Algorithm for the diagnosis of periprosthetic joint infection.

Bone and Joint Infections: From Microbiology to Diagnostics and Treatment, First Edition.
Edited by Werner Zimmerli.

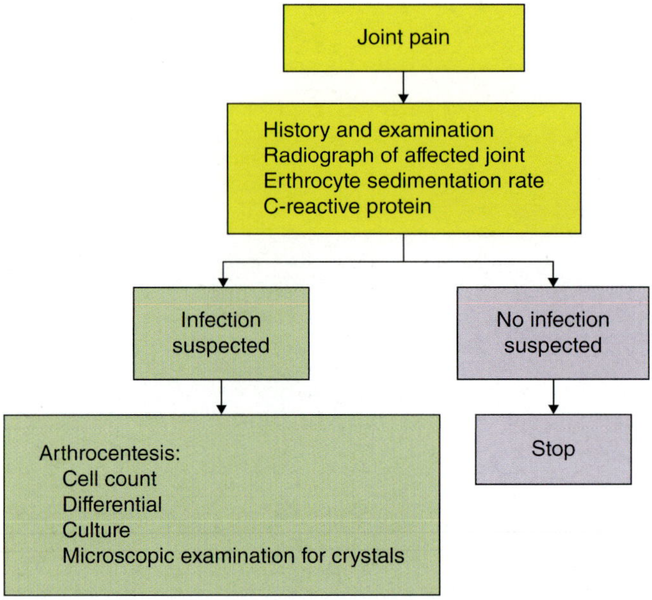

Figure 2.2. Algorithm for the diagnosis of native joint arthritis.

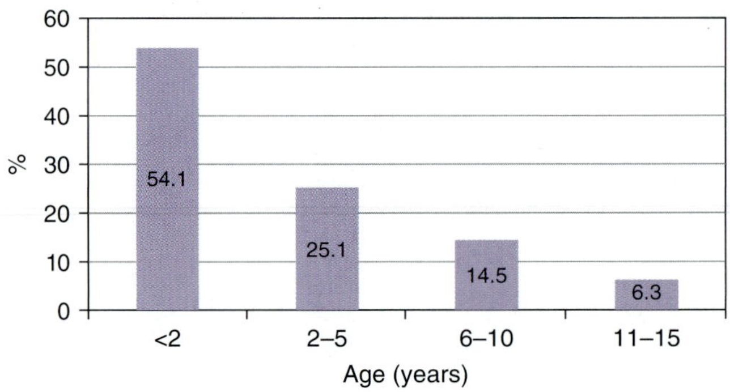

Figure 5.1. Age distribution of children with septic arthritis [5].

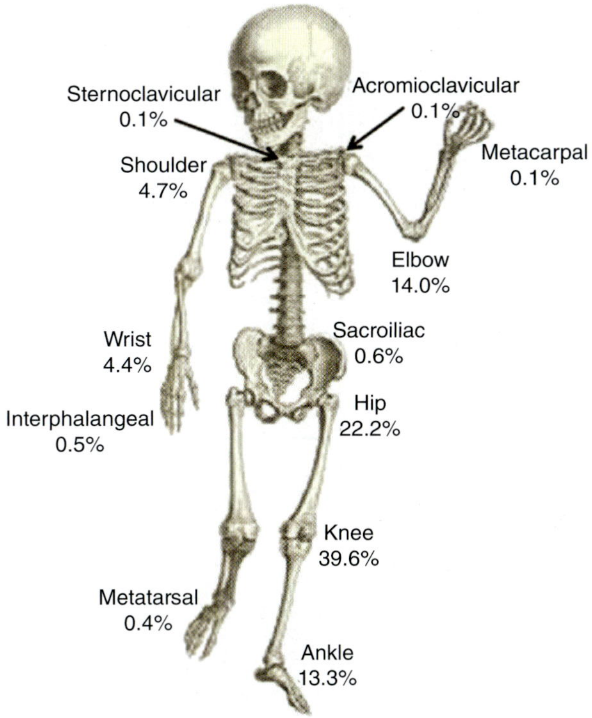

Figure 5.2. Anatomical distribution of 781 septic joints diagnosed in 725 children [5].

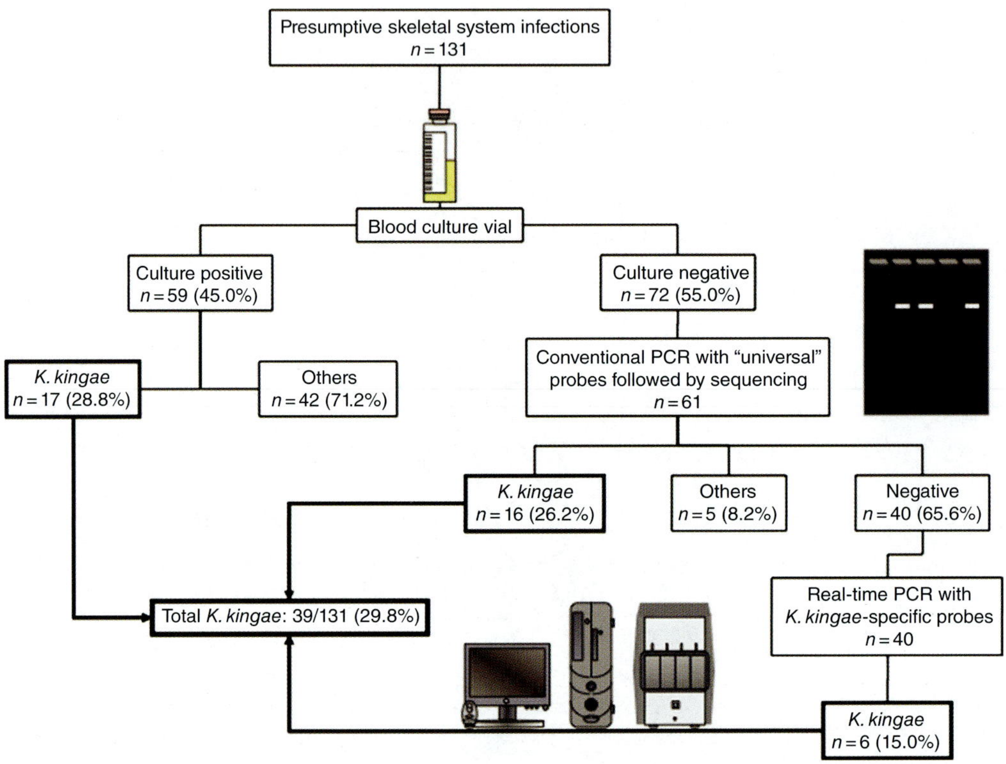

Figure 5.3. Combined use of the blood culture vial technique and nucleic acid amplification assays for diagnosing *Kingella kingae* arthritis [18].

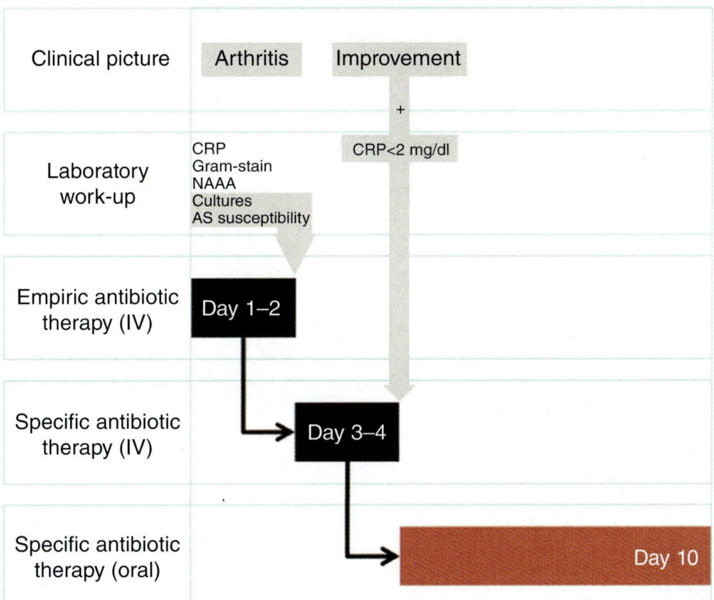

Figure 5.4. Shortened protocol for guiding antibiotic therapy in uncomplicated pediatric septic arthritis [52, 55, 65]. AB, antibiotic; CRP, C-reactive protein; NAAA, nucleic acid amplification assays.

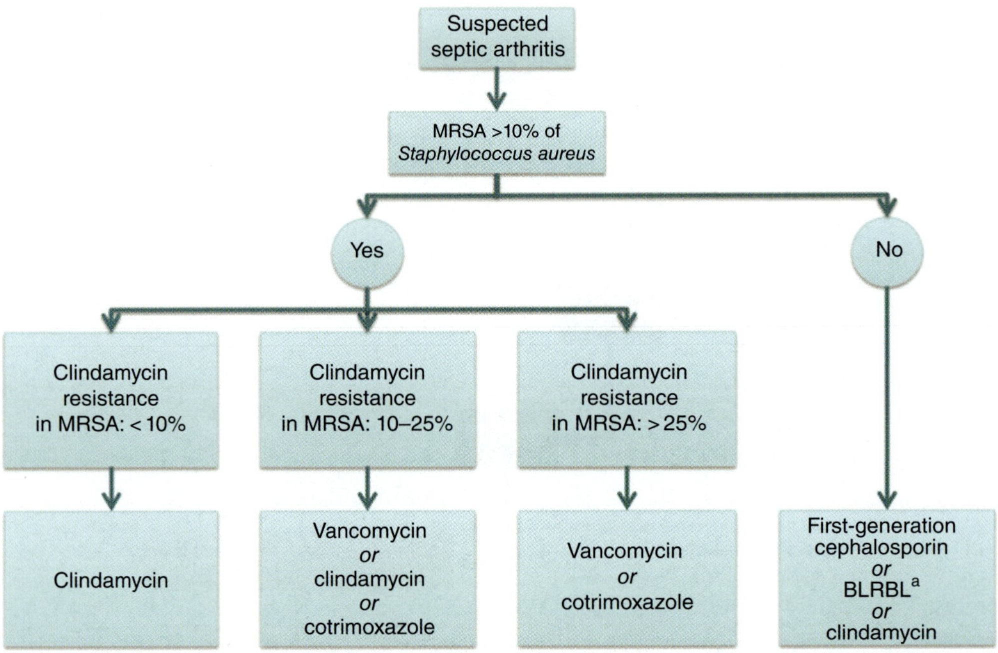

Figure 5.5. Initial administration of antibiotics based on the local prevalence of antibiotic resistance in *Staphylococcus aureus* [67]. [a] betalactamase-resistant betalactam.

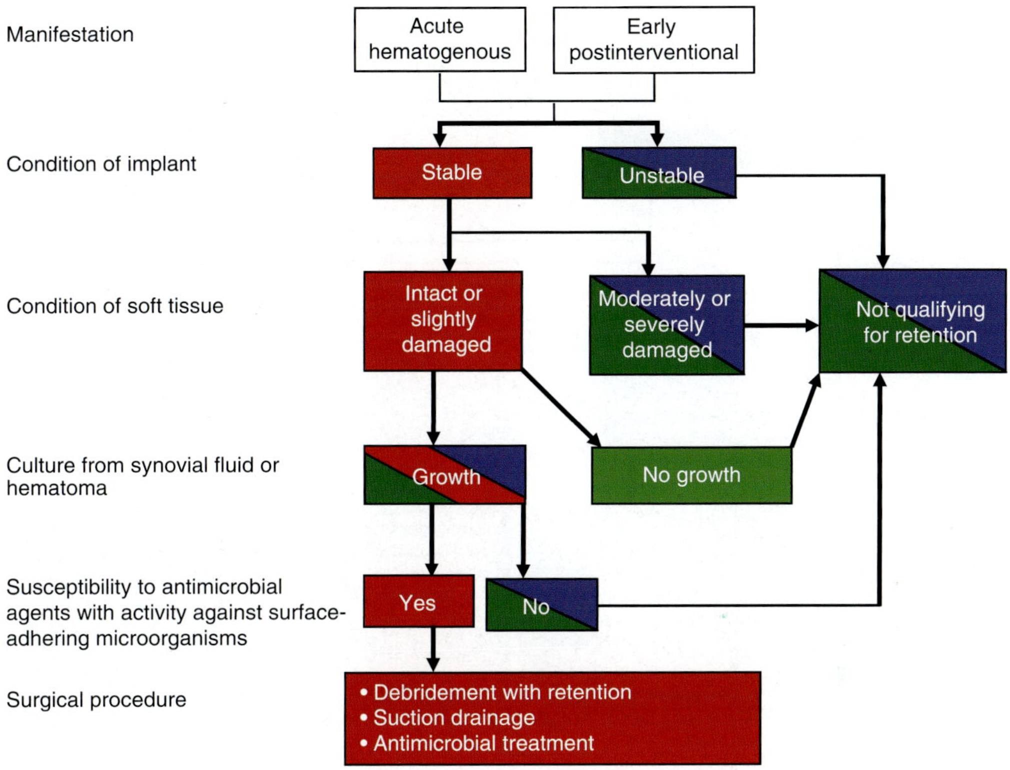

Figure 9.1. Treatment algorithm of acute hematogenous and early postinterventional PJI. Modified from Ref. 30.

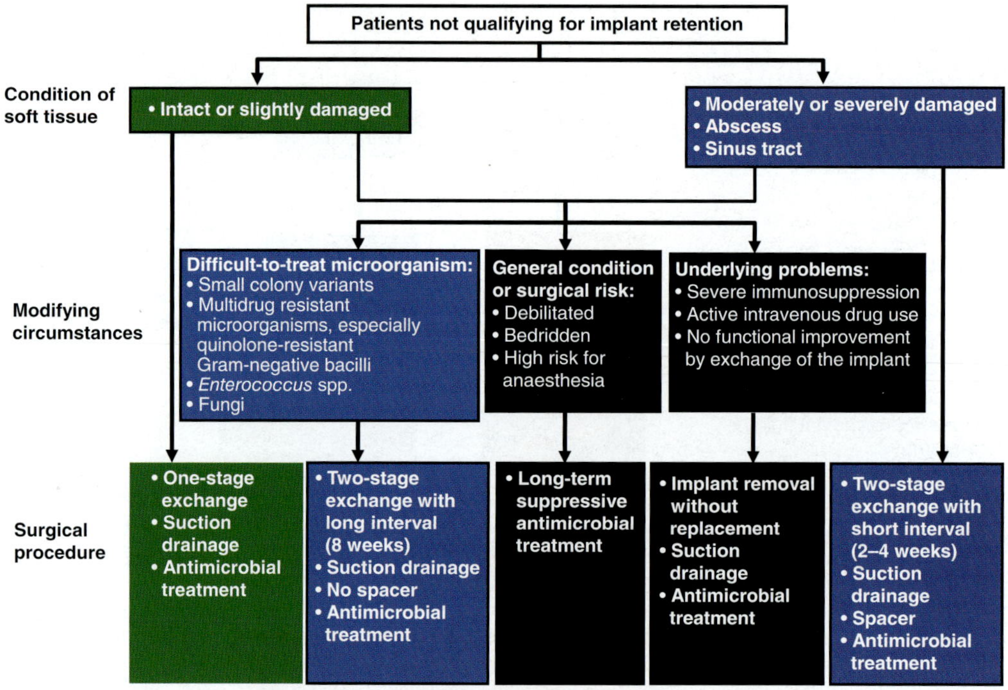

Figure 9.2. Treatment algorithm for patients with PJI not qualifying for implant retention. Modified from Ref. 30.

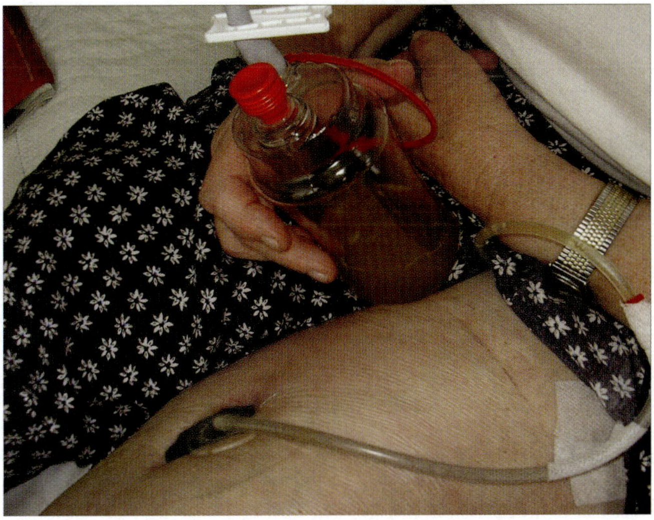

Figure 9.3. Vacuum-assisted closing (VAC) system. Clinical picture of a VAC system covering the open wound. After VAC removal the femoral component of the TKA was visible.

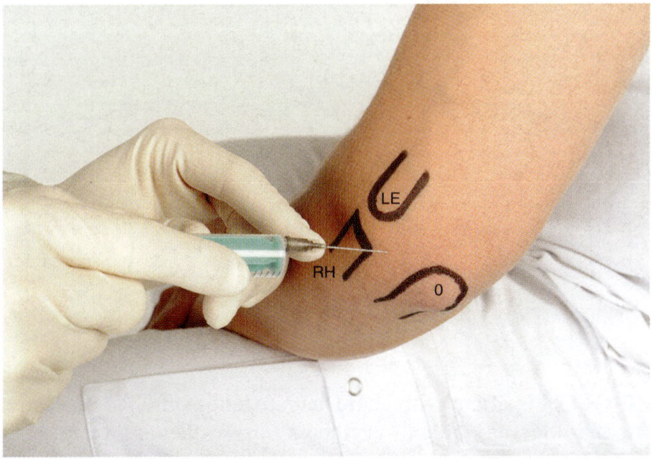

Figure 11.1. Arthrocentesis of the elbow through a dorso-radial approach. Landmarks are the olecranon (O), the lateral epicondylus (LE), and the radial head (RH). While the elbow is held in approximately 135° of flexion, a 22-gauge needle should enter in the middle of a triangle formed by the lateral epicondyle, the radial head, and the tip of the olecranon.

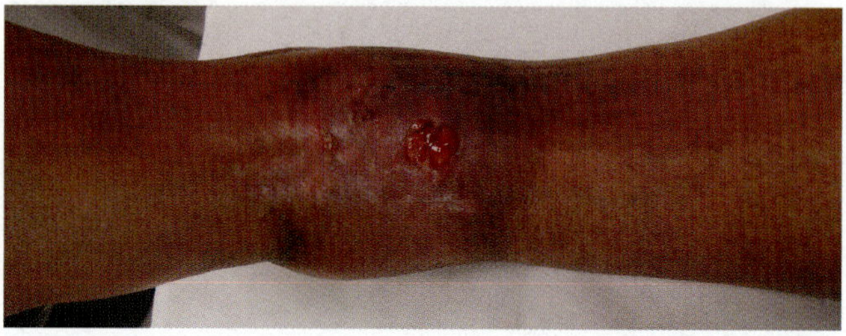

Figure 12.1. A 61-year-old patient with chronic periprosthetic ankle joint infection. Two months after implantation, he presented with a swollen, tender ankle joint and a sinus tract. The implant was removed, five samples were obtained, and a spacer was implanted. *S. aureus* grew in all biopsy samples and in sonicated fluid culture from the implant.

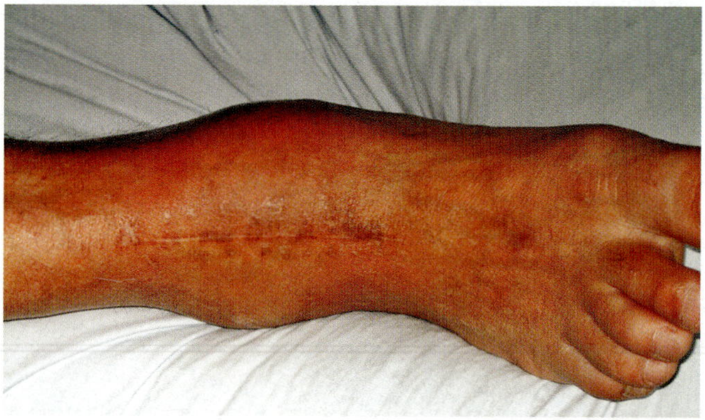

Figure 12.2. A 52-year-old patient with acute hematogenous PAJI. Fourteen months after implantation, the right ankle was suddenly erythematous, swollen, and tender.

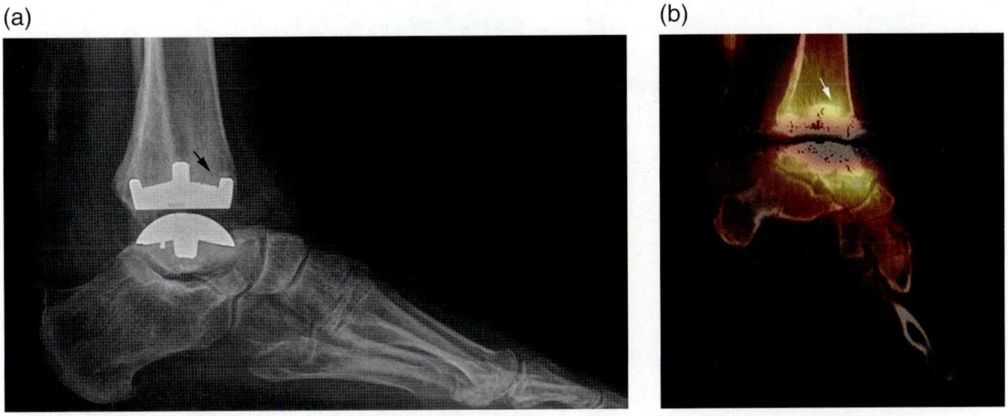

Figure 12.3. A 62-year-old woman with an exogenous chronic PAJI. (a) Lateral view of the left ankle joint, showing osteolysis around the tibial component (arrow). (b) SPECT/CT showing a significant signal uptake surrounding both implant components. The signal uptake is most around the tibial component (arrow).

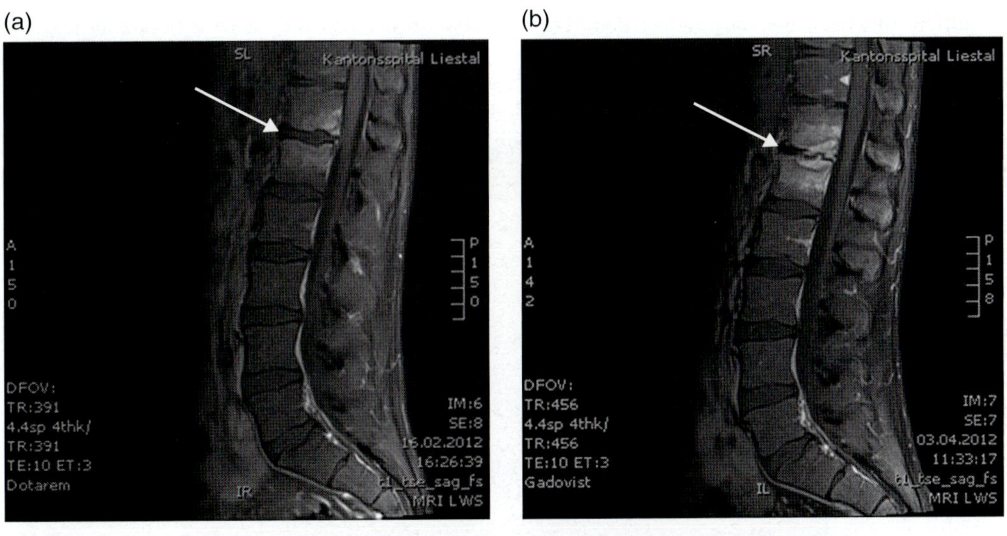

Figure 13.1. A 20-year-old man with hematogenous vertebral osteomyelitis due to *Salmonella enterica* subsp enterica Tennessee. (a) Magnetic resonance imaging (MRI) T1 with gadolinium 5 weeks after episode with diarrhea. No disk enhancement is visible (arrow). (b) MRI T1 with gadolinium 11 weeks after episode with diarrhea and after 6 weeks of adequate antimicrobial therapy. Inflammation and destruction of intervertebral disk is visible (arrow).

(a) (b)

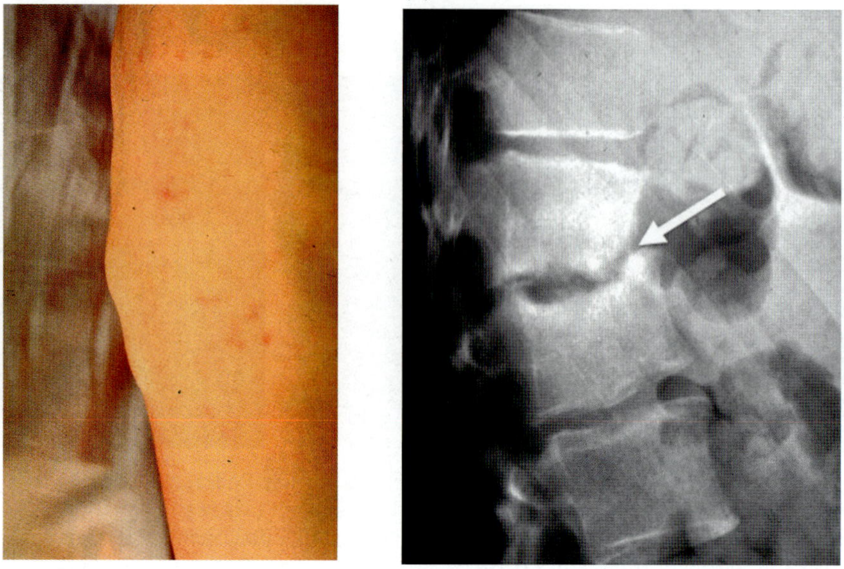

Figure 15.1. A 38-year-old IV drug addict who regularly injected heroin dissolved in lemon juice. He had a history of backache for several weeks when he presented with low-grade fever and immobilizing back pain. Biopsy of vertebra Th10 revealed *Candida albicans* in the culture. (a) gibbus due to severe ventral destruction of vertebra Th10. (b) plain radiograph of thoracic gibbus of the same patient.

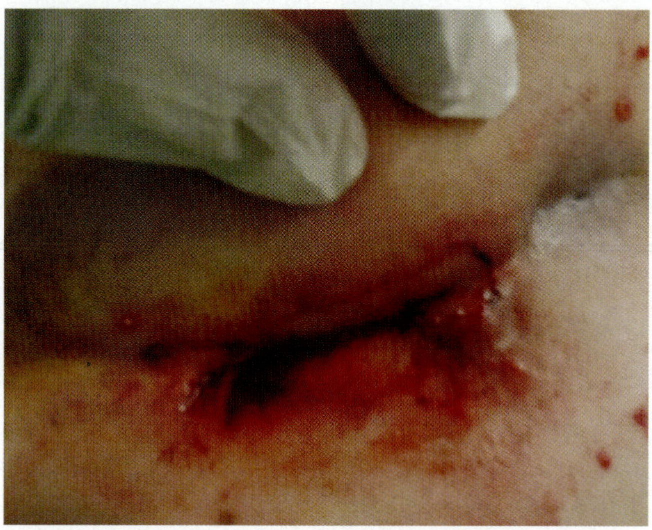

Figure 17.3. External orifice of a sinus tract on the calve/femur with maceration of the skin. Image is the property of Geneva University Hospitals and is displayed with the permission of the patient.

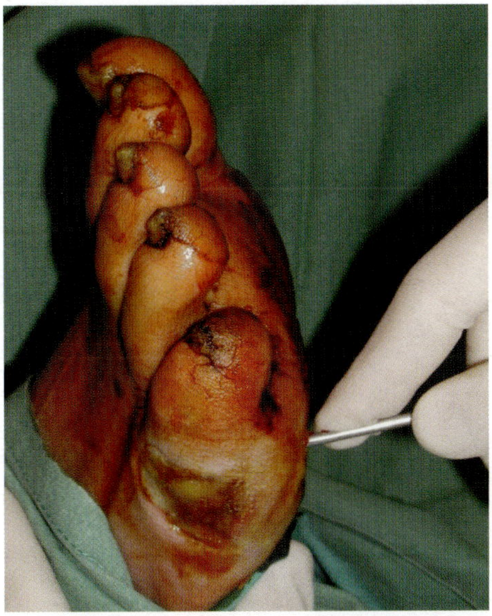

Figure 18.1. Transcutaneous bone biopsy of the metatarso-phalangeal joint of the fifth ray on the right foot in a diabetic foot patient; the trocard is introduced through a noninfected skin area opposite to the ulcer. Courtesy of Dr. Eric Beltrand, Department of Orthopedic Surgery Dron Hospital Tourcoing F-59200 France.

(a) (b)

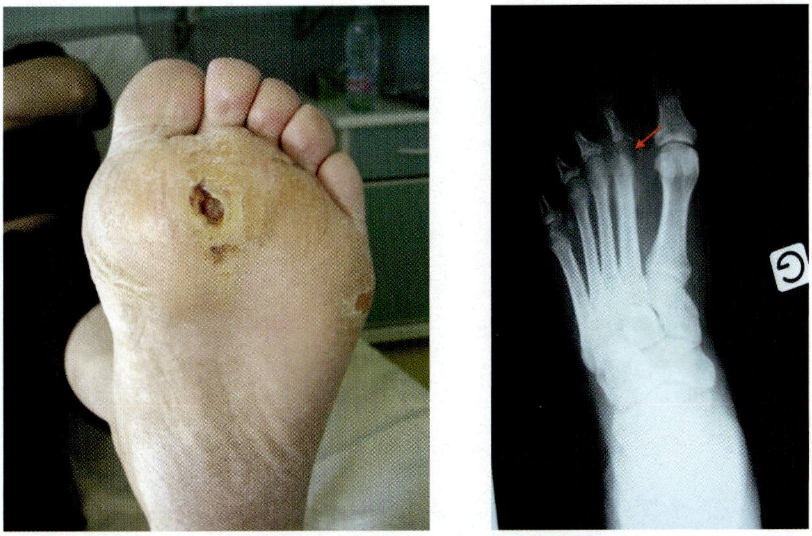

Figure 18.2. (a) Clinical picture of a diabetic patient with a chronic ulcer of the plantar surface with regard to the second metatarsal head. (b) Radiological signs of osteomyelitis on plain X-rays of the left foot in the same patient. The red arrow indicates joint destruction.

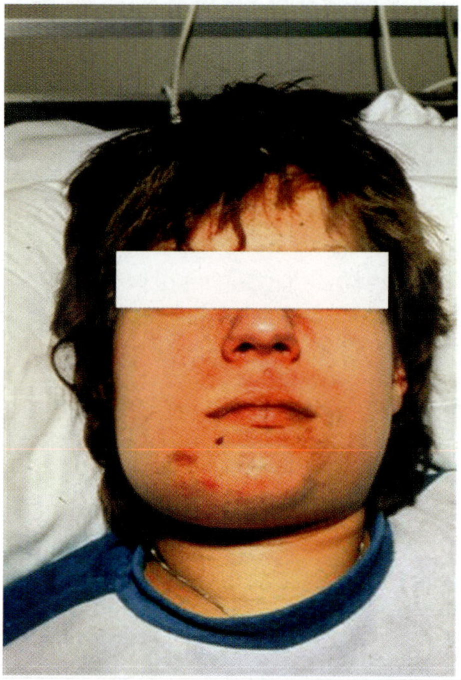

Figure 19.2. A 28-year-old man with a 6-month history of painful swelling of the right lower jaw after tooth extraction. Anaerobic cultures of tissue biopsies revealed *Actinomyces israelii* and *Fusobacterium* sp.

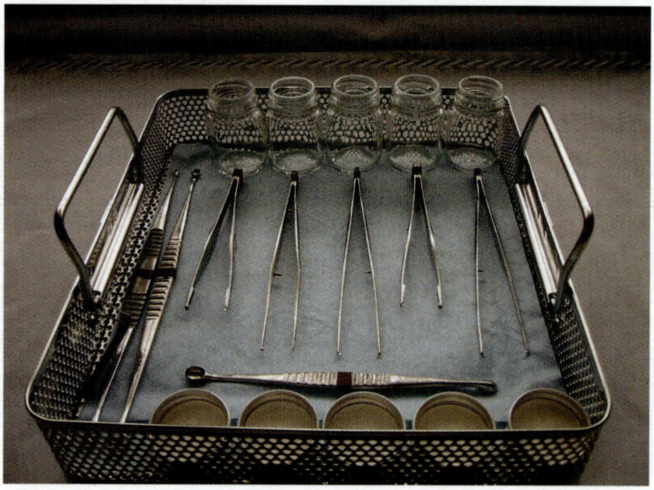

Figure 20.1. Prepared instrument set for biopsy sampling. Separate instruments (forceps, knife, curette or bone nibbler) and transport tubes are used for each biopsy sample.

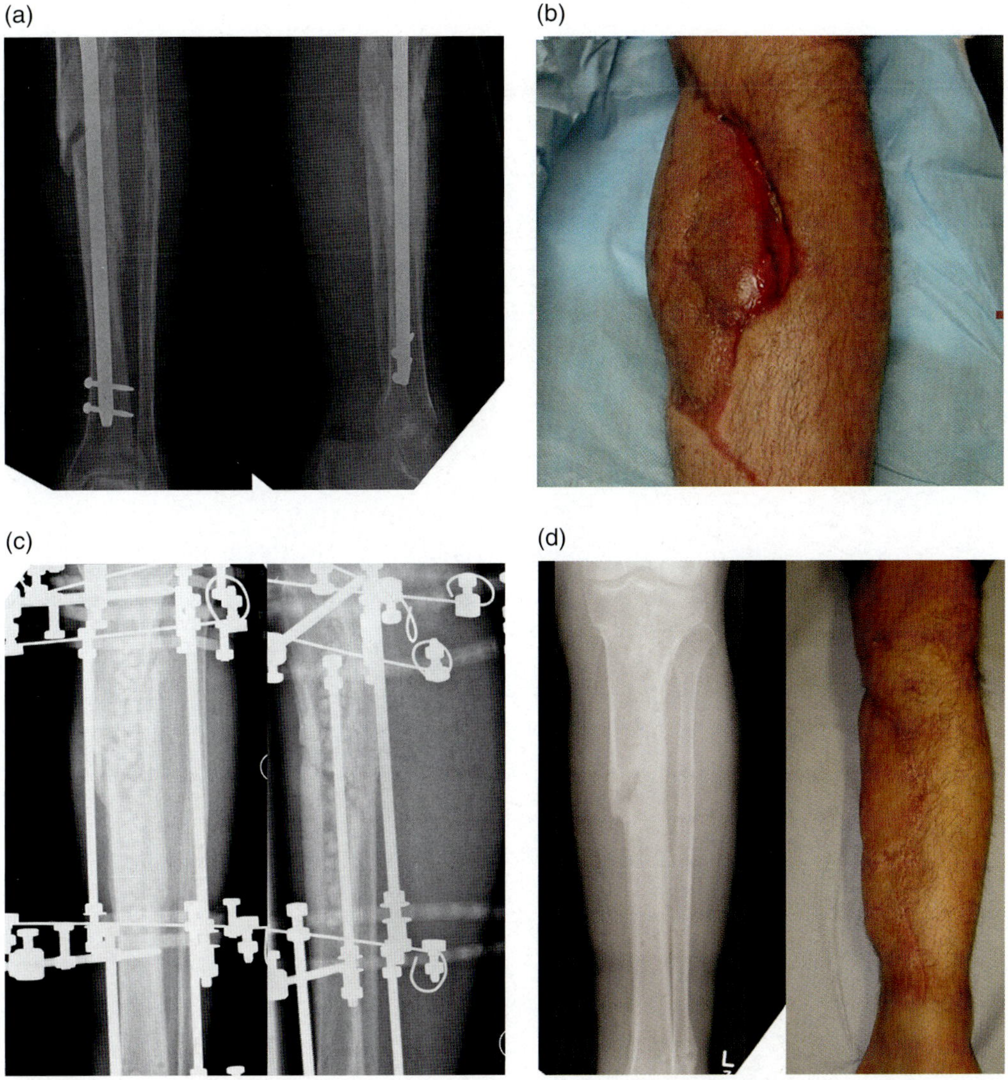

Figure 20.2. A 22-year-old man with chronic implant-associated osteomyelitis after a Gustilo–Anderson grade IIIB open tibia fracture. (a) Anteroposterior and lateral radiographs of a fragmented tibial fracture with an intramedullary nail *in situ*, taken 14 weeks after injury. (b) Wound discharge, indicating chronic infection, 7 months after primary intervention. (c) Stabilization of the fracture with an Ilizarov external fixator. (d) Final X-ray and clinical assessment at 34-month follow-up.

(a) (b)

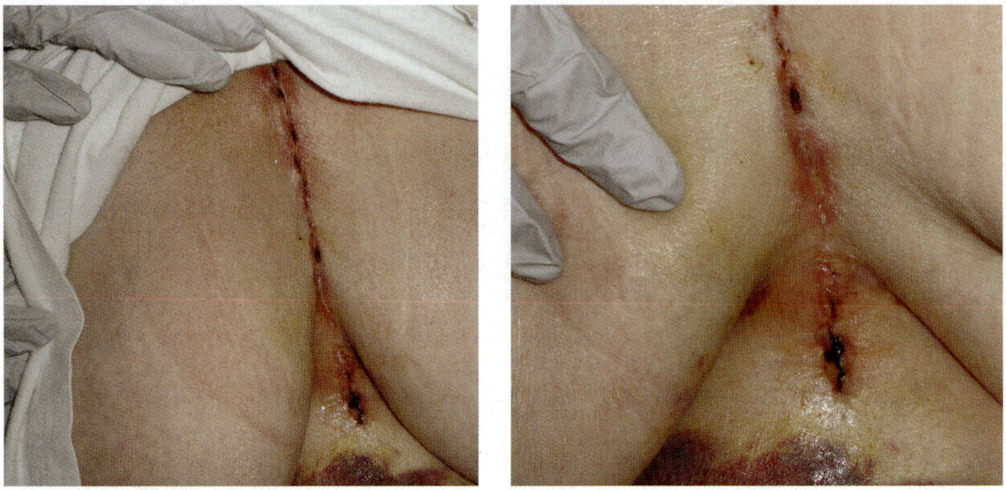

Figure 22.1. (a) Sternal wound 10 days after median sternotomy in a patient who developed fever. (b) After the breasts were moved aside, purulent discharge was visible.

(a) (b)

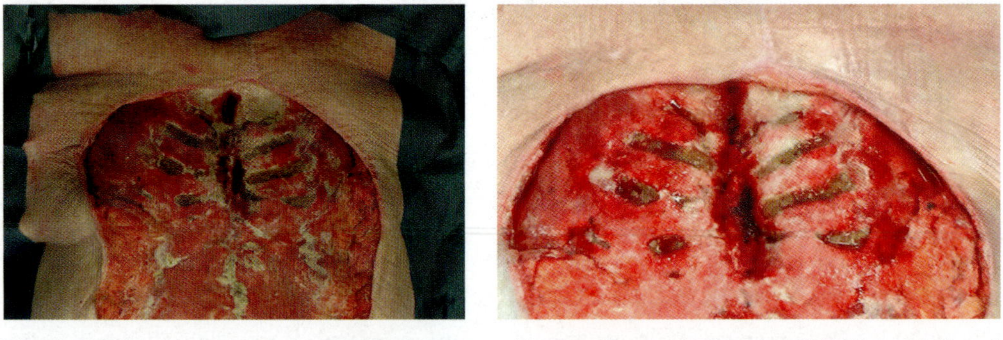

Figure 22.3. (a) An extensive mediastinal wound in a patient with chronic PSTMO and infected costal cartilage. (b) Despite meticulous debridement, the vitality of the costae is suboptimal. Complete resection cannot be performed because of thorax instability. Picture courtesy of Dr. M. Shafighi.